Automotive Technician Certification Test Preparation Manual

Don Knowles
Knowles Automotive Training
Moose Jaw, Saskatchewan
Canada

Delmar Publishers

an International Thomson Publishing company I(T)P®

Albany • Bonn • Boston • Cincinnati • Detroit • London • Madrid
Melbourne • Mexico City • New York • Pacific Grove • Paris • San Francisco
Singapore • Tokyo • Toronto • Washington

Notice to the Reader

Publisher does not warrant or guarantee any of the products described herein or perform any independent analysis in connection with any of the product information contained herein. Publisher does not assume, and expressly disclaims, any obligation to obtain and include information other than that provided to it by the manufacturer.

The reader is expressly warned to consider and adopt all safety precautions that might be indicated by the activities described herein and to avoid all potential hazards. By following the instructions contained herein, the reader willingly assumes all risks in connection with such instructions.

The publisher makes no representations or warranties of any kind, including but not limited to, the warranties of fitness for particular purpose or merchantability, nor are any such representations implied with respect to the material set forth herein, and the publisher takes no responsibility with respect to such material. The publisher shall not be liable for any special, consequential or exemplary damages resulting, in whole or in part, from the readers' use of, or reliance upon, this material.

Delmar Staff
Publisher: Robert D. Lynch
Acquisitions Editor: Vernon Anthony
Developmental Editor: Denise Denisoff
Project Editor: Thomas Smith
Production Coordinator: Karen Smith
Art and Design Coordinator: Michael Prinzo

Copyright © 1997
By Delmar Publishers
A division of International Thomson Publishing Inc.

The ITP logo is a trademark under license.

Printed in the United States of America

For more information, contact:

Delmar Publishers
3 Columbia Circle, Box 15015
Albany, New York 12212-5015

International Thomson Publishing Europe
Berkshire House 168-173
High Holborn
London WC1V7AA
England

Thomas Nelson Australia
102 Dodds Street
South Melbourne, 3205
Victoria, Australia

Nelson Canada
1120 Birchmont Road
Scarborough, Ontario
Canada M1K 5G4

International Thomson Editores
Campos Eliseos 385, Piso 7
Col Polanco
11560 Mexico D F Mexico

International Thomson Publishing GmbH
Königswinterer Strasse 418
53227 Bonn
Germany

International Thomson Publishing Asia
221 Henderson Road
#05-10 Henderson Building
Singapore 0315

International Thomson Publishing — Japan
Hirakawacho Kyowa Building, 3F
2-2-1 Hirakawacho
Chiyoda-ku, Tokyo 102
Japan

Online Services

Delmar Online
To access a wide variety of Delmar products and services on the World Wide Web, point your browser to:
    http://www.delmar.com/delmar.html
    or email: info@delmar.com

thomson.com
To access International Thomson Publishing's home site for information on more than 34 publishers and 20,000 products, point your browser to:
    http://www.thomson.com
    or email: findit@kiosk.thomson.com

A service of I(T)P®

1 2 3 4 5 6 7 8 9 10 XXX 02 01 00 99 98 97 96

Library of Congress Cataloging-in-Publication Data
Knowles, Don.
  Automotive technician certification test preparation manual /
Don Knowles
    p. cm.
  Includes index.
  ISBN: 0–8273–6934-4
  1. Automobiles—Maintenance and repair—Examination, questions, etc.
  2. Automobile mechanics—Certification—United States.
  I. Title
  TL152.K575 1997
  629.28'72'076—dc20
                   96-25838
                   CIP

Contents

Introduction 1

1 Engine Repair

2 Automatic Transmissions and Transaxles

5 Brakes

6 Electrical/Electronic Systems

7 Heating and Air Conditioning Systems

8 Engine Performance

National Institute for Automotive Service Excellence Certification

Pretest

The purpose of this pretest is to test your understanding of ASE certification and test questions. If you answer any of the pretest questions incorrectly, complete a careful study of the information in the chapter. Your understanding of ASE test categories will help you in future career planning. A knowledge of ASE test questions is essential before writing any certification tests.

1. ASE was established
   A. at the request of the car manufacturers.
   B. at the request of new car dealers.
   C. at the request of technicians.
   D. as a result of congressional hearings.

2. ASE certification is available in
   A. 8 areas of automobile certification.
   B. 6 areas of automobile certification.
   C. 2 areas of automobile specialist certification.
   D. 1 area of automobile specialist certification.

3. To maintain valid certification ASE automobile recertification tests must be completed every
   A. 2 years.
   B. 3 years.
   C. 5 years.
   D. 10 years.

4. ASE automobile specialist testing is available in
   A. Engine Machinist, Suspension and Steering, and Alternate Fuels.
   B. Advanced Engine Performance, Electrical, and Brakes.
   C. Engine Machinist, Alternate Fuels, and Advanced Engine Performance.
   D. Alternate Fuels, Engine Performance, and Advanced Engine Performance.

5. While discussing ASE test locations and schedules
   Technician A says a technician may write the tests at the most convenient location even when this location is in a different state.
   Technician B says ASE tests are held at 450 to 500 locations in the United States.
   Who is correct?
   A. A only
   B. B only
   C. Both A and B
   D. Neither A nor B

6. While discussing the writing of ASE test questions
   Technician A says ASE test questions are written and screened by a panel of experts in the automotive service industry.
   Technician B says ASE test questions are pretested by automotive instructors selected from across the United States.
   Who is correct?
   A. A only
   B. B only
   C. Both A and B
   D. Neither A nor B

7. While discussing the types of ASE test questions
   Technician A says some multiple-choice questions require the technician to select one incorrect response from four responses.
   Technician B says some multiple-choice questions contain two statements made by technician A and B, and the technician answering the question must determine if one, both, or neither of the statements are correct.
   Who is correct?
   A. A only
   B. B only
   C. Both A and B
   D. Neither A nor B

8. While discussing study procedures in preparation for ASE tests
   Technician A says you should study while watching a basketball game on television.
   Technician B says you should do all your studying in the last two days before the tests.
   Who is correct?
   A. A only
   B. B only
   C. Both A and B
   D. Neither A nor B

9. While discussing the writing of ASE tests
   Technician A says you should be careful to align the test book and answer sheet properly.
   Technician B says after you have marked the answer to a question you should be sure the question number on the answer sheet is the same as the question number in the test book.
   Who is correct?
   A. A only
   B. B only
   C. Both A and B
   D. Neither A nor B

10. The following statements about ASE test questions and results are correct EXCEPT
    A. test results in ASE files are confidential.
    B. test results are mailed to your employer.
    C. some test questions measure the technician's knowledge of basic theory.
    D. some test questions meaure the technician's knowledge of diagnostic or repair procedures.

Answers to Pretest

1. D, 2. A, 3. C, 4. C, 5. C, 6. A, 7. C, 8. D, 9. C, 10. B

Introduction

A careful study of this chapter should help you understand ASE test procedures and certification of automotive technicians. This chapter also provides helpful suggestions for ASE test preparation. Your test preparation will play a significant role in your success in passing the tests. Studying the complete book will help you prepare for all the ASE automobile certification tests.

National Institute for Automotive Service Excellence (ASE)

Introduction to ASE

During the late 1960s members of the United States Congress received a large volume of complaints about the quality of automotive repair service. As a result of these complaints, congressional hearings were held regarding automotive service. Many experts from various segments of the automotive industry testified at these hearings. From these testimonies it was revealed that the major cause of inaccurate automotive service was technician incompetence. To correct this problem, congress approved the establishment of the National Institute for Automotive Service Excellence (ASE). Originally ASE was formed by the National Auto Dealers Association and the Motor Vehicle Manufacturer's Association. Shortly thereafter, ASE became an independent nonprofit organization. Since its beginning in 1972 ASE has certified more than 540,000 automotive technicians. Of the estimated 700,000 currently employed automotive technicians over 300,000 technicians, or 43 percent, are ASE certified.

ASE Certification

ASE provides voluntary certification tests for automotive technicians in the following subject areas:

1. Engine Repair (Test A1)
2. Automatic Transmission/Transaxle (Test A2)
3. Manual Drive Train and Axles (Test A3)
4. Suspension and Steering (Test A4)
5. Brakes (Test A5)
6. Electrical/Electronic Systems (Test A6)
7. Heating and Air Conditioning (Test A7)
8. Engine Performance (Test A8)

When an automotive technician successfully passes all 8 ASE automobile certification tests, the technician is certified as a Master Technician. When a technician passes an ASE test in one of the eight areas, an Automotive Technician's shoulder patch is issued by ASE. If a technician passes all eight tests, he or she receives a Master Technician's shoulder patch (Figure 1-1, next page). Technicians must prove two years' experience in automotive service work prior to certification in any ASE test area. Successful completion of an Automotive Training program at a recognized institution may be substituted for one year of automotive service work.

Figure 1-1 ASE certification shoulder insignia worn by automobile technicians and master technicians (Courtesy of the National Institute for Automotive Service Excellence (ASE)

Recertification tests are required at 5-year intervals to maintain valid ASE certification. Since the subject areas and tasks are updated periodically, the tests may have been updated several times in a 5-year period. Therefore, the recertification exam tests the technician's understanding of current technology. Recertification tests contain approximately one-half the questions in each subject area compared to the regular certification tests.

ASE also offers specialist automobile testing with the Engine Machinist, Alternate Fuels, and Advanced Engine Performance Specialist tests.

Certification tests are provided by ASE in the Medium/Heavy Truck technology. Test categories in this subject area include: Gasoline Engines, Diesel Engines, Drive Train, Brakes, Suspension and Steering, Electrical Systems, Heating, Ventilation and Air Conditioning, and Preventative Maintenance Inspection (PMI).

ASE also provides certification tests in Body Repair and Painting/Refinishing tests. These tests include Painting and Refinishing, Nonstructural Analysis and Damage Repair, Structural Analysis and Damage Repair, and Mechanical and Electrical Components.

ASE certification also is available in Parts: Medium/Heavy Truck Parts Specialist, and Automobile Parts Specialist.

A technician may choose to complete all eight Automobile tests plus the Body Repair and Painting/refinishing tests, and Medium/Heavy Truck tests. Upon the successful completion of all these tests the technician is certified as a World Class Technician.

National Automotive Technician's Education Foundation (NATEF) Certification

ASE offers voluntary certification for automotive and autobody technician training programs. This certification of training programs is provided on the recommendations of the National Automotive Technicians Education Foundation (NATEF) (Figure 1-2). NATEF goals include the development, encouragement, and improvement of automotive technical education. This foundation evaluates automotive technical training programs in the following areas:

1. Purpose
2. Administration
3. Learning Resources
4. Finances
5. Student Services
6. Instruction
7. Equipment
8. Facilities
9. Instructional Staff
10. Cooperative Work Agreements

Figure 1-2 ASE and the National Automotive Technicians Education Foundation (NATEF) certification of training programs (Courtesy of National Institute for Automotive Service Excellence (ASE)

NATEF certification of automotive training programs involves the completion of self-evaluation forms, and a two-day review by an independent inspection team. Training programs must request certification in three of the eight automotive certification areas. The instructional material of the automotive training program must include 80 percent of the high-priority tasks in each certification area to be NATEF certified. NATEF estimates there are approximately 2,500 secondary and post secondary Automotive Technology training programs in the United States, and over 675 of these programs are NATEF certified.

ASE Test Locations and Schedules

ASE tests are given in 450 to 500 locations across the United States. Technicians may take the tests of their choice at the most convenient location, even if this location is in a different state. Tests are conducted in May and November each year. An ASE Test Registration Booklet containing a test registration form, a list of test centers, and a test schedule may be obtained by writing to the National Institute for Automotive Service Excellence, 13505 Dulles Technology Drive, Herndon, VA 22071-3415. If there is no test center near your home, arrangements for a special test center may be completed with ASE. The requirements that must be met to establish a special test center include:

1. There must be no test center within fifty miles of your home.

2. A minimum of twenty technicians must register as a group for each of the test dates requested.

3. Tests at special test centers must be given at the same time as tests at the regular test centers.

ASE Tests

ASE tests are written by a committee of experts in the automotive service industry. These committees include automotive instructors, trainers employed by car manufacturers, and test equipment manufacturers. All the test questions are carefully screened by the committee of experts. After this screening the test questions are pretested by technicians selected from across the United States. The test questions that are not confusing, or easily misinterpreted, become part of a series of questions from which the actual ASE test questions are selected.

ASE test questions are designed to measure the technician's knowledge of basic theory and diagnostic or repair procedures. Many test questions are based on a specific adjustment, repair, or diagnostic problem. Technicians must have an understanding of the adjustment, repair, or diagnostic problem to answer the question correctly.

ASE test results are confidential. Test results will not be released to anyone without written consent from the technician.

Types of ASE Test Questions

Each ASE test contains forty to eighty multiple-choice questions. In this type of question the technician must read each of the four responses carefully, and then select the response that is the correct answer. In many cases questions are answered incorrectly because the technician did not take time to thoroughly read the question and all the responses. Always take time to carefully read the complete question and responses!

One type of multiple-choice question has three wrong answers and one correct answer. Some of the responses may be almost correct. Therefore, it is important not to pick one response without reading all the responses. When you are not sure of the answer, eliminate the responses that you know are incorrect to arrive at the correct response. An example of a multiple-choice question with three wrong answers and one correct answer follows.

When a computer has a read only memory (ROM) chip the microprocessor can

   A. read information from the ROM.

   B. write information into the ROM.

   C. read information from and write information into the ROM.

   D. erase information in the ROM.

A careful reading of this question indicates that you are being asked what functions the microprocessor can perform in relation to the ROM chip. Since the information is permanently programmed into a ROM chip, the microprocessor cannot erase information in the ROM. Therefore, response D is wrong. The microprocessor can read information from the ROM, but it cannot write information into the ROM. Therefore, responses B and C are wrong and response A is correct.

Another type of multiple-choice question used on the ASE tests states that all the responses are correct EXCEPT one. Since the correct answer in this type of question is the incorrect response, you must determine which response is wrong. Always read the question and all the responses carefully. An example of a multiple choice question follows in which you must select the one incorrect response.

The following defines the purpose of engine oil EXCEPT

   A. provides a sealing action between the piston rings and cylinder walls.

   B. cleans engine components.

   C. cools engine components.

   D. increases the compression ratio.

Engine oil helps to to provide a seal between the piston rings and the cylinder walls. Engine components are cooled, cleaned, and lubricated by the engine oil. Therefore, responses A, B, and C are correct. Since compression ratio is determined by the piston stroke and combustion chamber volume, engine oil does not increase the compression ratio. We can conclude that response D is the one incorrect response that must be identified.

Another type of multiple-choice question used in ASE tests involves statements from two technicians identified as Technician A and Technician B. When answering this type of question you must determine if one, both, or neither of the statements are correct. A multiple choice question follows with statements from Technician A and Technician B.

Technician A says excessive positive camber on a front wheel may cause excessive wear on the inside edge of the tire tread.

Technician B says excessive toe-in on the front wheels may cause excessive wear on the center of the tire tread.

Who is correct?

   A. A only

   B. B only

   C. Both A and B

   D. Neither A nor B

The first step in answering this question is to determine the effect of excessive positive camber. Since excessive positive camber tilts the wheel outward at the top, this problem causes excessive tread wear on the outside edge of the tire tread. Excessive toe-in caused feathered tread wear across the entire tread, not just on the center of the tire tread. Therefore, neither of the statements by Technician A or Technician B are correct, and the right answer is D.

ASE Test Preparation

Studying

Following these suggestions will help you study for the ASE tests:

1. Leave plenty of time for studying prior to the tests depending on the number of tests you are scheduled to write. Do not attempt to rush through your study in the final days before the tests.
2. Obtain an ASE test preparation guide.
3. After studying of all the information regarding the tests you are writing, identify your weak areas and concentrate your study on these areas.
4. Set a specific study schedule that provides adequate time to complete your studies without interrupting your normal sleep and work schedule.
5. Select a quiet study location where you will not be interrupted.
6. Study in a comfortable position. Since drowsiness may reduce your concentration while lying down, avoid studying in this position.
7. A large meal may cause drowsiness. Do not attempt to study after eating a heavy meal.
8. Make notes of important facts that you learned during your studies, or highlight these facts in the ASE test preparation guide.
9. Examine all diagrams carefully in the test preparation guide. Some ASE questions relate to diagrams.
10. Be sure you really understand the theory or service procedure related to each question in the test preparation guide. The questions on the ASE test will not be exactly the same as the ones in the test preparation guide. However, the ASE questions will be related to the same theory or service procedure as the questions in the test preparation guide.
11. Take a short study break every half hour, and complete a few exercises during this break to remain alert.

Preparation Immediately Prior to the Tests

On the day and evening prior to the test your activities may affect your performance during the tests. For example, if you watch a late movie on the night before the test, you may be tired during the test, and this tiredness may reduce your ability to read and think during the test. The following suggestions regarding the day and night before the test may improve your performance during the test.

1. Be sure you know the location of the building and the room where you will write the tests. If you are writing the tests in a city that you are not familiar with, drive to the test location and find the proper visitor's parking area; then find the ASE test room location.
2. Limit your study to reviewing your notes or highlighted areas in the ASE test preparation guide.

3. Maintain your regular schedule for eating and sleeping.

4. Place everything you need for writing the tests in a briefcase or suitable bag. These items include your admission ticket, photo ID, four sharpened Number 2 pencils and a watch.

5. Do not drink alcoholic beverages.

6. Avoid stressful situations on the day before the test.

7. Maintain mental alertness in the hours before the test. For example, you will probably be more mentally alert if you participated in some moderate exercise compared to your mental alertness if you watched three hours of television.

8. Know the exact test schedule and determine the time you will spend on each test.

9. Arrive at the ASE test site fifteen to twenty minutes prior to the beginning of the tests.

Writing the Tests

Prior to the tests the monitor will provide you with some verbal instructions. Listen carefully to these instructions, and read the directions carefully in the ASE test booklet. Question the monitor about anything you do not understand before the tests begin.

Align the test book and answer sheet carefully before you begin answering any of the questions. Check this alignment periodically as you answer the questions. Do not place any marks on the answer sheet other than your answers. During the correction process the computer may mistake other marks on the answer sheet for an answer. When you need scratch paper you may write in the test booklet.

Do not rush as you write the tests. Pace yourself according to your planned schedule. If you are uncertain about the answer to a question, place a check mark beside the question in the test booklet and proceed to the next question. Answer the questions that you feel confident about the answers. When you have completed the test, go back and answer the check-marked questions. Follow these suggestions while writing the tests:

1. Read each question completely and carefully so you understand the question.

2. After you have marked the answer to a question, be sure the question number on the answer sheet matches the question number in the test booklet.

3. Do not spend too much time on any question. Maintain your planned schedule to complete the tests.

4. Answer every question. If you are uncertain about the answer to a question eliminate the wrong answers to arrive at the correct answer, as mentioned previously. Often a careful analysis of the question and possible answers will help you arrive at the correct answer. It is better to provide an answer that you are unsure of rather than leaving a question unanswered.

5. Review your answers when you have completed the test.

6. After you have answered a question do not change your answer unless you are absolutely sure it is wrong.

Glossary

National Institute for Automotive Service Excellence (ASE) An organization dedicated to the certification of automobile technicians, medium/heavy truck technicians, and body repair and painting/refinishing technicians.

ASE automobile certification tests Automotive tests administered by ASE in eight different general catagories and three specialist catagories.

Master technician A technician who has passed the ASE tests in the eight general catagories.

Recertification tests Tests administered by ASE at 5-year intervals to recertify technicians.

Specialist automobile testing ASE tests provided to certify technicians as specialists in the areas of Advanced Engine Performance, Alternate Fuels, and Engine Machinist.

National Automotive Technician's Education Foundation (NATEF) certification A division of ASE that provides voluntary certification of automotive and autobody technician training programs.

Multiple-choice question A test question with four responses; the technician must select one correct response, or in some cases one incorrect response.

Introduction

The purpose of this guide is to help you successfully complete all eight ASE Automotive Certification tests. This guide contains the same types of questions as the ASE tests. Approximately 20 percent of the questions on ASE tests are based on a diagram. Similarly, the same percentage of questions in this guide are based on a diagram. Question answers and analysis are provided at the end of each chapter and on separate answer sheets provided with the guide. A pretest at the beginning of each chapter gives the reader an indication of the amount of study he or she requires to successfully complete the ASE test in that subject area. Three sample questions from the guide follow:

1. A vehicle with a spark control system has an open wire between the knock sensor and the knock sensor module. This may result in higher than specified emissions of

   A. carbon monoxide (CO).

   B. carbon dioxide (CO_2).

   C. hydrocarbons (HC).

   D. oxides of nitrogen (NOX).

Answer D

A defective spark control system may allow excessive spark advance which causes engine detonation. Since detonation causes high cylinder temperature, this action results in NOX emissions. A, B, and C are wrong, and D is correct.

2. An engine has a hard starting problem after it is shut off for several hours. The vehicle driveability and emission levels are normal. The ignition system is always firing during cold cranking.

   Technician A says the fuel pump check valve may be leaking.

   Technician B says the ECT sensor may be defective.

   Who is correct?

   A. A only

   B. B only

   C. Both A and B

   D. Neither A nor B

Answer A

A leaking fuel pump check valve causes the fuel to drain back out of the fuel system into the tank when the engine is shut off. This action causes hard starting without any other driveability or emission problems. A is correct. A defective ECT sensor may cause hard starting, but this problem also results in other driveability and emission problems. Therefore, B is wrong.

3. All of the following problems may be caused by an engine thermostat that is stuck open EXCEPT

   A. a rich air–fuel ratio and reduced fuel economy.

   B. an inoperative EGR system.

   C. improper cooling fan operation.

   D. an improper air charge temperature sensor signal.

Answer D

This question asks for the statement that is not true. If the engine thermostat is stuck open, the air–fuel ratio is rich and fuel economy is reduced because the ECT sensor informs the PCM regarding the lower coolant temperature. This defect also affects the EGR and cooling fan operation. Therefore, A, B, and C are true, but none of these are the requested answer.

Since the air charge sensor sends a signal to the PCM in relation to air intake temperature, this signal is not affected by a thermostat that is stuck open. Therefore, D is not true, and this is the requested answer.

Engine Repair

Pretest

The purpose of this pretest is to determine the amount of review that you may require prior to writing the ASE Engine Repair Test. If you answer all the pretest questions correctly, complete the questions and study the information in this chapter to prepare for the ASE Engine Repair Test.

If two or more of your answers to the pretest questions are incorrect, complete a study of Chapters 3 through 12 in *Today's Technician Engine Repair and Rebuilding*, published by Delmar Publishers, plus a study of the questions and information in this chapter.

The pretest answers are located at the end of the pretest, and on the answer sheet supplied with this book.

1. When an engine has a sharp, metallic rapping noise only with the engine idling, the trouble could be worn
   A. main bearings.
   B. piston pins.
   C. connecting rod bearings.
   D. piston skirts.

2. On a port fuel-injected engine the cause of excessive black smoke in the exhaust could be
   A. low fuel pressure.
   B. an intake manifold vacuum leak.
   C. high oxygen sensor voltage.
   D. high fuel pressure.

3. During a cylinder power balance test on a four-cylinder port fuel injected engine one cylinder has a 75 rpm drop and the other cylinders have a 125 rpm drop.
   Technician A says the cylinder with 75 rpm drop may have a burned exhaust valve.
   Technician B says the cylinder with 75 rpm drop may have a defective fuel injector.
   Who is correct?
   A. A only
   B. B only
   C. Both A and B
   D. Neither A nor B

4. While discussing cylinder head service
   Technician A says torque-to-yield bolts may be reused if they are not damaged.
   Technician B says torque-to-yield bolts are usually tightened to a specified torque and then rotated a certain number of degrees.
   Who is correct?
   A. A only
   B. B only
   C. Both A and B
   D. Neither A nor B

5. The cause of excessive valve stem height could be
   A. an improperly installed valve guide.
   B. excessive material machined from the head surface.
   C. a worn valve guide and stem.
   D. excessive material removed from the valve seat.

6. While diagnosing timing belt condition
   Technician A says some vehicle manufacturers recommend checking timing belt wear by measuring the installed length of the belt tensioner.
   Technician B says on some engines the valves will strike the top of the pistons if the timing belt jumps several notches on the camshaft sprocket.
   Who is correct?
   A. A only
   B. B only
   C. Both A and B
   D. Neither A nor B

7. While discussing engine sealants
   Technician A says RTV sealer dries in the absence of air.
   Technician B says anaerobic sealer dries in the absence of air.
   Who is correct?
   A. A only
   B. B only
   C. Both A and B
   D. Neither A nor B

8. While measuring cylinder taper and out-of-round
   Technician A says the cylinder diameter should be measured at two vertical locations.
   Technician B says cylinder out-of-round is the difference between the cylinder diameter in the thrust and axial directions.
   Who is correct?
   A. A only
   B. B only
   C. Both A and B
   D. Neither A nor B

9. While discussing connecting rod service
   Technician A says each connecting rod should be measured for a bent condition.
   Technician B says each connecting rod should be measured for a twisted condition.
   Who is correct?
   A. A only
   B. B only
   C. Both A and B
   D. Neither A nor B

10. A cooling system is pressurized to 15 psi with a pressure tester. After 5 minutes the pressure reading on the tester is 3 psi and there are no external leaks.
    Technician A says the automatic transmission cooler may be leaking.
    Technician B says the head gasket may be leaking.
    Who is correct?
    A. A only
    B. B only
    C. Both A and B
    D. Neither A nor B

11. When diagnosing a no-start condition if a 12V test light connected from the negative primary coil terminal to ground flashes while cranking the engine the trouble could be a defective
    A. pickup coil.
    B. ignition module.
    C. ignition coil.
    D. ignition switch.

12. Technician A says an intake manifold vacuum leak may cause cylinder misfiring at idle speed.
    Technician B says an intake manifold vacuum leak may cause cylinder misfiring at 2,500 rpm.
    Who is correct?
    A. A only
    B. B only
    C. Both A and B
    D. Neither A nor B

13. An accumulation of oil in the air cleaner may be caused by
    A. a partially restricted PCV valve.
    B. a PCV valve that is stuck open.
    C. a leaking rocker arm cover gasket.
    D. a leaking PCV clean air hose.

14. All of the following statements about a battery load test are true EXCEPT the battery
    A. should be discharged at one-half the cold cranking rating during the test.
    B. temperature should be above 40°F before performing the test.
    C. specific gravity should be above 1,190 before performing the test.
    D. is satisfactory if the voltage remains above 8.8V after fifteen seconds with the battery temperature at 70°F.

Answers to Pretest

1. B, 2. D, 3. C, 4. B, 5. D, 6. C, 7. B, 8. B, 9. C, 10. C, 11. C, 12. A, 13. A, 14. D

General Engine Diagnosis

ASE Tasks, Questions, and Related Information

In this chapter each task in the Engine Repair category is followed by a question and some information related to the task. If you answer any question incorrectly, study this information very carefully until you understand the correct answer. For additional information on any task refer to *Today's Technician Engine Repair and Rebuilding*, by Delmar Publishers.

Question answers and analysis are provided at the end of this chapter and in the answer sheets provided with this book.

Task 1   Verify the driver's complaint and/or road test the vehicle, then determine the needed repairs.

1. While discussing a basic diagnostic procedure
Technician A says the most complicated diagnostic tests should be performed first.
Technician B says the customer complaint must be identified.
Who is correct?

   A. A only
   B. B only
   C. Both A and B
   D. Neither A nor B

Hint   *The technician must be familiar with a basic diagnostic procedure such as the following:*
   • *Listen carefully to the customer's complaint, and question the customer to obtain more information regarding the problem.*
   • *Identify the complaint, road test the vehicle if necessary.*
   • *Think of the possible causes of the problem.*
   • *Perform diagnostic tests to locate the exact cause of the problem. Always start with the easiest, quickest test.*
   • *Be sure the customer's complaint is eliminated; road test the vehicle if necessary.*

Task 2   Inspect the engine assembly for fuel, oil, coolant, and other leaks, then determine the needed repairs.

2. A cooling system is pressurized with a pressure tester to locate a coolant leak. After 15 minutes the tester gauge has dropped from 15 psi to 5 psi, and there are no visible signs of coolant leaks in the engine compartment.
Technician A says the engine may have a leaking head gasket.
Technician B says the heater core may be leaking.
Who is correct?

   A. A only
   B. B only
   C. Both A and B
   D. Neither A nor B

Hint   *Basic fuel, lubricating, and cooling systems and components must be understood. The location of all possible leaks in these systems must be identified. Coolant leaks may be internal or external in relation to the engine.*

Task 3   Listen to engine noises and determine the needed repairs.

3. A heavy thumping noise occurs with the engine idling, but the oil pressure is normal. This noise may be caused by

   A. worn pistons and cylinders.
   B. loose flywheel bolts.
   C. worn main bearings.
   D. loose camshaft bearings.

Hint   *You must be familiar with noises caused by defective main bearings, connecting rod bearings, pistons, piston pins, piston rings, and ring ridge in the engine block.*
   Valve train and camshaft noises must be understood. You must be able to recognize combustion chamber noise and flywheel and vibration damper noises. If you are not familiar with these engine noises, or their causes, refer to Today's Technician Engine Repair and Rebuilding, *published by Delmar Publishers.*

Task 4   **Diagnose the cause of excessive oil consumption, unusual engine exhaust color, odor, and sound and determine the needed repairs.**

4. A port fuel-injected engine has a steady "puff" noise in the exhaust with the engine idling. The cause of this problem could be
   A. a burned exhaust valve.
   B. excessive fuel pressure.
   C. a restricted fuel return line.
   D. a sticking fuel pump check valve.

Hint   *The causes of excessive oil consumption such as worn rings, scored cylinder walls, worn valve guides seals and stems, worn turbocharger seals, or oil leaks must be understood.*

Unusual exhaust colors and their causes are
- *Blue exhaust—excessive oil consumption, may be more noticeable on acceleration and deceleration.*
- *Black exhaust—rich air fuel ratio, excessive fuel consumption*
- *Gray exhaust—coolant leaking into the combustion chambers, may be more noticeable when the engine is first started.*

Normal exhaust noise contains steady pulses at the tail pipe. A puff noise in the exhaust at regular intervals usually indicates a cylinder misfire caused by a compression, ignition, or fuel system defect. Erratic exhaust pulses at the tail pipe indicate a rough idle condition caused by ignition or fuel system defects. Excessive exhaust noise indicates a leak in the exhaust system.

A high-pitched squealing noise during hard acceleration may be caused by a small leak in the exhaust system, particularly in the exhaust manifolds or exhaust pipe. An intake manifold vacuum leak causes a high-pitched whistle at idle and low speeds. This whistle gradually decreases when the engine is accelerated and the intake vacuum decreases.

Excessive sulphur smell in the exhaust indicates a rich air-fuel ratio on vehicles with catalytic converters.

Task 5   **Perform engine vacuum tests and determine needed repairs.**

5. With the engine idling, a vacuum gauge connected to the intake manifold fluctuates as illustrated in Figure 1-1. These vacuum gauge fluctuations may be caused by
   A. late ignition timing.
   B. intake manifold vacuum leaks.
   C. a restricted exhaust system.
   D. sticky valve stems and guides.

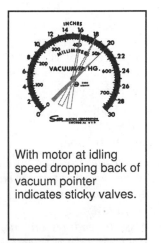

With motor at idling speed dropping back of vacuum pointer indicates sticky valves.

Figure 1-1   Vacuum gauge reading *(Courtesy of Sun Electric Corporation)*

Hint   *The vacuum gauge should be connected directly to the intake manifold to diagnose engine and related system conditions. When a vacuum gauge is connected to the intake manifold, the reading on the gauge should provide a steady reading between 17 and 22 inches of mercury (in. Hg.) with the engine idling. Abnormal vacuum gauge readings indicate these problems:*

- *A low, steady reading indicates late ignition timing.*
- *If the vacuum gauge reading is steady and much lower than normal, the intake manifold has a significant leak.*
- *When the vacuum gauge pointer fluctuates between approximately 11 and 16 in. Hg. on a carbureted engine at idle speed, the carburetor idle mixture screws require adjusting. On a fuel-injected engine the injectors require cleaning or replacing.*
- *Burned or leaking valves cause a vacuum gauge fluctuation between 12 and 18 in. Hg.*
- *Weak valve springs result in a vacuum gauge fluctuation between 10 and 25 in. Hg.*
- *A leaking head gasket may cause a vacuum gauge fluctuation between 7 and 20 in. Hg.*
- *If the valves are sticking, the vacuum gauge fluctuates between 14 and 18 in. Hg.*
- *When the engine is accelerated and held at a steady higher rpm, if the vacuum gauge pointer drops to a very low reading, the catalytic converter or other exhaust system components are restricted.*

Task 6   Perform a cylinder power balance test and determine needed repairs.

6. During a cylinder balance test on an engine with electronic fuel injection, cylinder number three provides very little rpm drop.
   Technician A says the ignition system may be misfiring on number three cylinder.
   Technician B says the engine may have an intake manifold vacuum leak.
   Who is correct?
   A. A only
   B. B only
   C. Both A and B
   D. Neither A nor B

Hint   *If the cylinder is working normally, a noticeable rpm decrease occurs when the cylinder misfires. If there is very little rpm decrease when the analyzer causes a cylinder to misfire, the cylinder is not contributing to engine power. Under this condition the engine compression, ignition system, and fuel system should be checked to locate the cause of the problem. An intake manifold vacuum leak may cause a cylinder misfire with the engine idling or operating at low speed. If this problem exists, the misfire will disappear at a higher speed when the manifold vacuum decreases. When all the cylinders provide the specified rpm drop, the cylinders are all contributing equally to the engine power.*

Task 7   Perform a cylinder compression test and determine neededrepairs.

7. During a compression test a cylinder has 40 percent of the specified compression reading. When the technician performs a wet test the compression reading on this cylinder is 75 percent of the specified reading. The cause of the low compression reading could be
   A. a burned exhaust valve.
   B. worn piston rings.
   C. a bent intake valve.
   D. a worn camshaft lobe.

Hint   *The ignition and fuel injection system must be disabled before proceeding with the compression test. During the compression test the engine is cranked through four compression strokes on each cylinder and the compression readings recorded. Lower than specified compression readings may be interpreted as follows:*

- *Low compression readings on one, or more, cylinders indicate worn rings, valves, a blown head gasket, or a cracked cylinder head. A gradual buildup on the four compression readings on each stroke indicates worn rings, whereas little buildup on the four strokes usually is the result of a burned exhaust valve.*

- *When the compression readings on all the cylinders are even, but lower than the specified compression, worn rings and cylinders are indicated.*
- *Low compression on two adjacent cylinders is caused by a leaking head gasket or cracked cylinder head.*
- *Higher than specified compression usually indicates carbon deposits in the combustion chamber.*
- *Zero compression on a cylinder usually is caused by a hole in a piston, or a severely burned exhaust valve. If the zero compression reading is caused by a hole in the piston, the engine will have excessive blowby.*

When the engine spins freely and compression in all cylinders is low, check the valve timing. If a cylinder compression reading is below specifications, a wet test may be performed to determine if the valves, or rings, are the cause of the problem. Squirt approximately 2 or 3 teaspoons of engine oil through the spark plug opening into the cylinder with the low compression reading. Crank the engine to distribute the oil around the cylinder wall and then retest the compression. If the compression reading improves considerably, the rings (or cylinders) are worn. When there is little change in the compression reading, one of the valves is leaking.

Task 8   Perform a cylinder leakage test and determine needed repairs.

8. During a leakage test cylinder number two has 50 percent leakage and air is escaping from the PCV valve opening.
   Technician A says the intake valve in number two cylinder may be leaking.
   Technician B says the rings in number two cylinder may be worn.
   Who is correct?

   A. A only
   B. B only
   C. Both A and B
   D. Neither A nor B

Hint   *During the leakage test a regulated amount of air from the shop air supply is forced into the cylinder with both exhaust and intake valves closed. The gauge on the leakage tester indicates the percentage of leakage in the cylinder. A gauge reading of 0 percent indicates there is no cylinder leakage, and if the reading is 100 percent the cylinder is not holding any air.*

If the reading on the leakage tester exceeds 20 percent check for air escaping from the tailpipe, positive crankcase valve (PCV) opening, and the top of the throttle body or carburetor. Air escaping from the tailpipe indicates an exhaust valve leak. When the air is coming out of the PCV valve opening, the piston rings are leaking. An intake valve is leaking if air is escaping from the top of the throttle body, or carburetor. Remove the radiator cap and check the coolant for bubbles, which indicate a leaking head gasket or cracked head.

Cylinder Head and Valve Train Diagnosis and Repair

ASE Tasks, Questions, and Related Information

Task 1   **Remove cylinder heads according to manufacturer's specifications and procedures; inspect cylinder heads for cracks, gasket surface areas for warpage and leakage; and check passage condition.**

9. The feeler gauge measurement in Figure 1-2 is 0.014 in. (0.35 mm). Technician A says the cylinder head should be resurfaced.
Technician B says block warpage should be measured.
Who is correct?
A. A only
B. B only
C. Both A and B
D. Neither A nor B

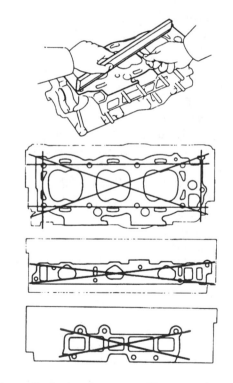

Figure 1-2   Measuring cylinder head warpage *(Courtesy of Toyota Motor Corporation)*

Hint   *An electromagnetic-type tester and iron filings may be used to check for cracks in cast iron heads. A dye penetrant may be used to locate cracks in aluminum heads.*

Use a straightedge and a feeler gauge to measure cylinder head warpage. Place the straightedge diagonally across the head to block surface at two locations. Use the same method to check warpage on the cylinder head, intake manifold, and exhaust manifold mounting surfaces. The old head gasket may be examined to determine if proper sealing existed between the head and block.

Task 2 **Install cylinder heads and gaskets and tighten retaining bolts according to manufacturer's specifications and procedures.**

10. While discussing torque-to-yield head bolts

Technician A says compared to conventional head bolts, torque-to-yield bolts provide more uniform clamping force.

Technician B says torque-to-yield bolts are tightened to a specific torque, and then rotated a certain number of degrees.

Who is correct?

A. A only

B. B only

C. Both A and B

D. Neither A nor B

Hint   *All head bolt openings in the block must be checked for the presence of oil or foreign material which may cause false torque readings or a cracked block. Use compressed air to remove any foreign material from these bolt openings.*

Many engines now have torque-to-yield head bolts that must be replaced each time they are removed. Head bolts must be tightened to the specified torque in the proper sequence. Torque-to-yield head bolts usually are tightened to a specific torque and then rotated the specified number of degrees.

Task 3 **Inspect valve springs for squareness pressure, and free height comparison; replace as necessary.**

11. While discussing the measurement in Figure 1-3

Technician A says the valve spring is being measured for installed height.

Technician B says the valve spring must be held in one position.

Who is correct?

A. A only

B. B only

C. Both A and B

D. Neither A nor B

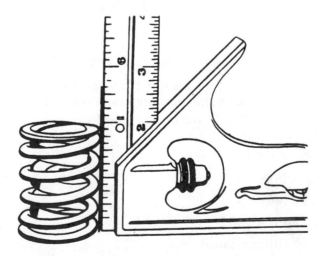

Figure 1-3   Valve spring measurement (*Courtesy of Chevrolet Motor Division, General Motors Corporation*)

12. While performing the test shown in Figure 1-4
    A. the tool table must be set 2 in. below the top of the stud.
    B. the actual torque wrench reading indicates valve spring tension.
    C. the torque wrench reading must be multiplied by two.
    D. the torque wrench must be pulled until the valve spring is at the specified height.

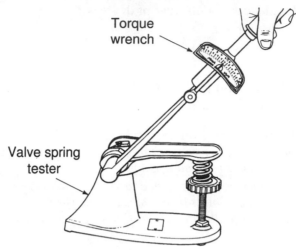

Figure 1-4   Valve spring measurement *(Courtesy of Chrysler Corporation)*

Hint   *Valve springs must be measured for free length and squareness by placing each spring against a steel square and surface plate on a level surface. Free length is the height of the valve spring measured with no tension on the spring. Squareness is the vertical straightness of the valve spring. The valve spring must be replaced if it does not have the specified free length. When the spring is rotated, a spring height variance of more that 1/16 in. indicates a bent spring that must be replaced.*

A valve spring tester is used to measure valve spring tension. Adjust the tool table until the tool table surface is aligned with the specified valve spring height marked on the threaded stud of the tester. Install the spring over the stud on the tester and lift the compressing lever to set the tone device. Install a torque wrench on the tester and pull on this wrench until a ping or click is heard. Read the torque wrench at this instant. Multiply this reading by two to obtain the spring load at test length.

Task 4   Inspect valve spring retainers, locks, and valve lock grooves.

13. Technician A says worn valve lock grooves may cause the valve locks to fly out of place with the engine running, resulting in severe engine damage.
    Technician B says worn valve lock grooves may cause a clicking noise with the engine idling.
    Who is correct?

    A. A only

    B. B only

    C. Both A and B

    D. Neither A nor B

Hint   *Valve spring retainers and locks must be checked for wear, scoring, or damage. When any of these conditions are present, replace the components. The valve lock grooves on the valve stems must be inspected for wear, particularly round shoulders. If these shoulders are uneven or rounded, replace the valve.*

Task 5 **Replace valve stem seals.**

14. While discussing valve stem seals
Technician A says worn valve stem seals may cause rapid valve stem and guide wear.
Technician B says worn valve stem seals may cause excessive oil consumption.
Who is correct?
A. A only
B. B only
C. Both A and B
D. Neither A nor B

Hint   *Prior to valve installation the valve stem and guide should be lubricated lightly with the manufacturer's specified engine oil. Install the valve spring seat and slide the seal over the top of the valve stem. Be careful not to damage the seal on the valve lock grooves. The seal must be properly positioned on the valve stem. Some seals are pushed down over the spring seat.*

Task 6 **Inspect valve guides for wear; check valve guide height and stem-to-guide clearance; determine needed repairs.**

15. While discussing valve stem-to-guide measurement
Technician A says the valve stems and guides should be measured at three vertical locations.
Technician B says the valve guide diameter should be measured with a hole, or snap, gauge.
Who is correct?
A. A only
B. B only
C. Both A and B
D. Neither A nor B

16. The measurement being performed in Figure 1-5 is
A. valve concentricity.
B. valve seat concentricity.
C. valve face to seat contact.
D. valve stem to guide clearance.

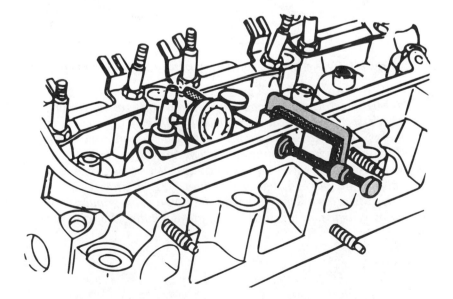

Figure 1-5 Valve measurement *(Courtesy of Chevrolet Motor Division, General Motors Corporation)*

Hint   *The valve guide should be measured near the top and bottom, and in the center with a hole gauge. Measure the valve stem diameter with a micrometer in the same three positions, and subtract the stem readings from the guide measurements to obtain the clearance. An alternate method for measuring stem-to-guide clearance is to install the valve in the guide with the valve 1/8 in. off the seat. Mount a dial indicator against the valve stem below the lock groove and move the valve stem from side-to-side while observing the clearance reading on the dial indicator.*

If the valve stem-to-guide clearance is more than specified, the valve guides may be replaced, knurled, or bored out and a thin-wall liner installed. Excessive valve stem-to-guide clearance may result in improper valve seating and lower compression. Increased oil consumption may result from excessive valve stem-to-guide clearance.

Valve guide height usually is measured from the top of the spring seat to the top of the guide.

Task 7   Inspect valves; surface or repair.

17. A valve margin of 1/64 in. may cause
    A. a clicking noise at idle speed.
    B. valve overheating and burning.
    C. improper valve seating.
    D. valve seat recession.

18. The valves in Figure 1-6 are from an engine with valve rotators. As indicated in the figure the valve stem tip wear patterns indicate
    A. valve 1 indicates a normal wear pattern.
    B. valve 1 indicates a worn out rotator.
    C. valve 2 indicates a worn valve guide.
    D. valve 3 indicates a weak valve spring.

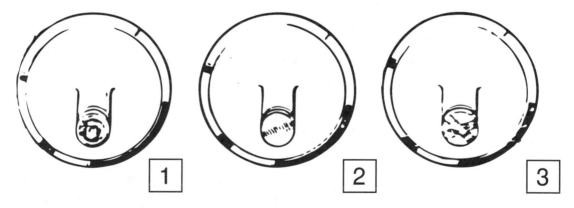

Figure 1-6 Valve stem tip condition *(Courtesy of Chevrolet Motor Division, General Motors Corporation)*

Hint   *Measure the overall length of the valves. If the length of any valve is not within manufacturer's specifications, replace the valve. Inspect the surface of the valve tip for wear and score marks. When these conditions are present, resurface the tip on a valve stem grinder, or replace the valve. Do not grind the valve stem tip until the overall valve length is less than specified.*

Measure the valve margin thickness. When this thickness is less than specified replace the valve. If the valve margin is less than 1/32 in. after resurfacing, valve overheating and reduced valve life occurs. Reface the valve face on a valve grinder. If the specified valve seat angle is 45°, many vehicle manufacturers recommend grinding the valve face to an interference angle of 44.5°.

Task 8 **Inspect valve seats; resurface or repair/replace.**

19. Valve seats are typically ground to an angle of
    A. 15° or 20°.
    B. 20° or 30°.
    C. 30° or 45°.
    D. 45° or 60°.

Hint *The proper sized pilot is placed in the valve guide prior to valve seat resurfacing. A grinding wheel is placed over this pilot, and the wheel is rotated with an electric drive tool. A grinding stone of the proper size and angle must be installed on the grinding wheel. This grinding wheel must fit on the valve seat without touching any other part of the head surface.*

Valve seat inserts may be removable or integral with the cylinder head. Removable valve seat inserts may be removed with a special puller or a pry bar. A special driver is used to install the valve seat insert, and the insert should be staked after installation.

Task 9 **Check valve face-to-seat contact and valve seat concentricity (runout); service seats and valves as necessary.**

20. While discussing a valve seat contact area that is too low on the valve face:
    Technician A says 60° and 45° grinding stone should be used to raise the valve seat contact area.
    Technician B says the low valve seat contact area on the valve face may cause a clicking noise at idle speed.
    Who is correct?
    A. A only
    B. B only
    C. Both A and B
    D. Neither A nor B

Hint *After the valve seats are resurfaced, install blue on the valve face and install the valve against the seat. Remove the valve and observe the seat pattern on the valve face. If the blue appears 360° around the valve face, the valve is concentric. When the blue does not appear 360° around the valve face, replace the valve. A valve seat concentricity tester containing a dial indicator may also be used to measure valve seat concentricity (Figure 1-7).*

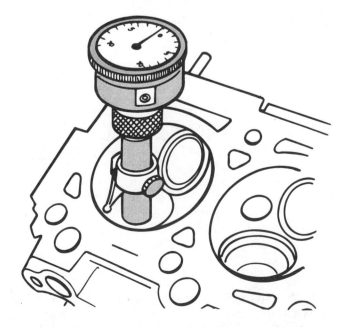

Figure 1-7 Measuring valve seat concentricity *(Courtesy of Ford Motor Company)*

The seat contact area on the valve face should be the width specified by the vehicle manufacturer. This contact area must be located in the center of the valve face (Figure 1-8).

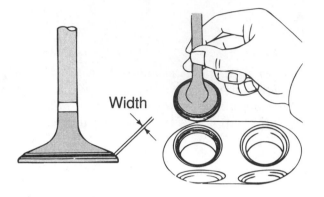

Figure 1-8   Checking valve seat contact area and width on the valve face *(Courtesy of Toyota Motor Corporation)*

When the seat contact area is too high on the valve face, use a 30° and a 45° grinding stone to lower the seat contact area (Figure 1-9). If the seat contact area is too low on the valve face, use a 60° and a 45° grinding stone to raise the seat area (Figure 1-10).

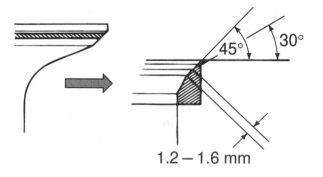

Figure 1-9   Using a 30° and a 45° grinding stone to lower the seat contact area on the valve face *(Courtesy of Toyota Motor Corporation)*

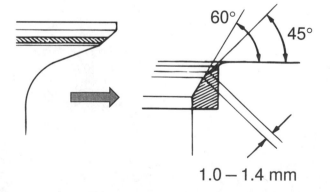

Figure 1-10   Using a 60° and a 45° grinding stone to raise the seat contact area on the valve face *(Courtesy of Toyota Motor Corporation)*

Task 10 **Check valve spring assembled height and valve stem height; service valve and spring assemblies as necessary.**

21. Measurement B in Figure 1-11 is more than specified.
Technician A says this problem may bottom the lifter plunger.
Technician B says a shim should be installed under the valve spring.
Who is correct?
   A. A only
   B. B only
   C. Both A and B
   D. Neither A nor B

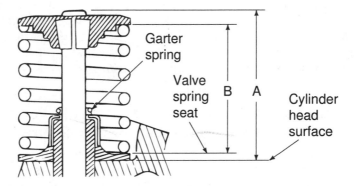

Figure 1-11 Valve spring measurement *(Courtesy of Chrysler Corporation)*

Hint *Measure the installed valve stem height from the cylinder head surface on the spring seat to the valve stem tip. If this measurement is more than specified, the valve stem is stretched, or too much material has been removed from the seat. When this measurement is still excessive with a new valve, replace the seat or cylinder head. Excessive valve stem height moves the plunger downward in the valve lifter.*

Measure the installed valve spring height from the lower edge of the top retainer to top edge of the spring seat. When this measurement is excessive, install the thickness of shim specified by the manufacturer under the valve spring seat. Excessive installed valve spring height reduces valve spring tension, which may result in valve float and cylinder misfiring at higher speeds.

Task 11 **Inspect pushrods, rocker arms, rocker arm pivots, and shafts for wear, bending, cracks, looseness, and blocked oil passages; repair or replace.**

22. While discussing valve train service
Technician A says excessive valve spring tension may cause bent pushrods.
Technician B says a improper valve timing may cause bent pushrods.
Who is correct?
   A. A only
   B. B only
   C. Both A and B
   D. Neither A nor B

Hint *Pushrods should be inspected for a bent condition and wear on the ends. Roll the pushrod on a level surface to check for a bent condition. Bent pushrods usually indicate interference in the valve train such as a sticking valve or improper valve adjustment.*

Check the rocker arm shafts for wear and scoring in the rocker arm contact area. Worn rocker arms, shafts, or pivots cause improper valve adjustment and a clicking noise in the valve train.

Task 12   Inspect and replace hydraulic or mechanical lifters.

23. Technician A says hydraulic valve lifter bottoms should be flat or concave.
    Technician B says a sticking lifter plunger may cause a burned exhaust valve.
    Who is correct?
    A. A only
    B. B only
    C. Both A and B
    D. Neither A nor B

Hint   *When the valve train is serviced the valve lifters should be removed, cleaned, and tested for leakdown. If the lifter bottoms are pitted or worn, replace the lifters. Lifter bottoms must be convex. Replace lifters with flat or concave bottoms. When the lifter bottoms are scored or concave, replace the camshaft.*

Sticking lifter plungers cause a clicking noise, especially when the engine is started. Burned valves may be caused by sticking lifter plungers. If the lifters are cleaned and reassembled they should be tested in a leakdown tester. This tester checks the time required to bottom the plunger with a specific weight applied to the plunger. When the lifters leak down too quickly, a clicking noise may be heard in the valve train with the engine idling. If the camshaft is replaced, new valve lifters should be installed.

Task 13   Adjust valves on engines with mechanical or hydraulic lifters according to manufacturer's specifications and procedures.

24. While adjusting mechanical valve lifters
    Technician A says when the valve clearance is checked on a cylinder, the piston in that cylinder should be at TDC on the exhaust stroke.
    Technician B says some mechanical valve lifters have removable shim pads available in various thicknesses to provide the proper valve clearance.
    Who is correct?
    A. A only
    B. B only
    C. Both A and B
    D. Neither A nor B

Hint   *In engines with mechanical valve lifters some rocker arms have an adjustment screw and a lock screw on the valve stem end of the rocker arm. With the piston at TDC on the compression stroke, place a feeler gauge of the specified thickness between the adjusting screw and the valve stem (Figure 1-12). If the valve adjustment is correct the feeler gauge slides between the adjusting screw and the valve stem with a light push fit. Some engines with mechanical valve lifters have removable metal pads in each lifter or spring retainer. Pads with different thicknesses are available to provide the specified valve clearance.*

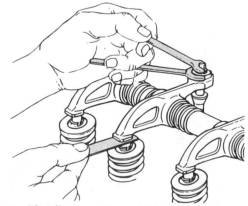

Figure 1-12   Valve adjustment with adjusting screw and locknut in the end of the rocker arm *(Courtesy of Chrysler Corporation)*

Some valve trains have hydraulic valve lifters and individual rocker arm pivots retained with self-locking nuts. These valve trains require an initial adjustment of the rocker arm nut to position the lifter plunger. With the valve closed, loosen the rocker arm nut until there is clearance between the end of the rocker arm and the valve stem. Slowly turn the rocker arm nut clockwise while rotating the pushrod. Continue rotating the rocker arm nut until the end of the rocker arm contacts the end of the valve stem, and the push rod becomes harder to turn. Continue turning the rocker arm nut clockwise the number of turns specified in the service manual. In some engines this specification is one turn plus or minus one quarter turn.

Task 14 **Inspect and replace camshaft drives (includes checking gear wear and backlash, sprocket and chain wear, overhead cam drive sprockets, drive belts, belt tension, and tensioners).**

25. When the timing gear teeth are meshed directly with the crankshaft gear teeth Technician A says the timing gear backlash may be measured with a dial indicator. Technician B says on this type of engine the timing gear backlash may be measured with a micrometer.
    Who is correct?

    A. A only
    B. B only
    C. Both A and B
    D. Neither A nor B

Hint   *When the camshaft gear teeth are directly meshed with the crankshaft gear the camshaft gear backlash may be measured with a dial indicator positioned against one of the camshaft gear teeth.*

Many timing belts have a hydraulic tensioner that is operated by engine oil pressure. Some manufacturers recommend measuring the installed length of the tensioner to determine the belt wear. If the tensioner length exceeds the manufacturer's specifications, replace the timing belt (Figure 1-13).

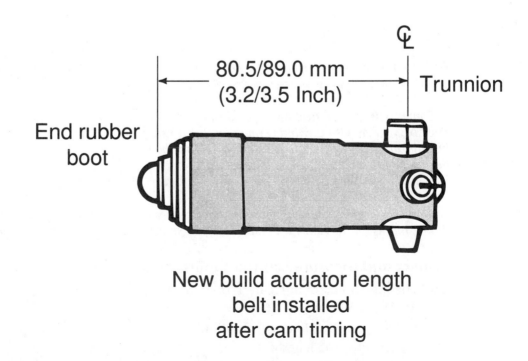

Figure 1-13 Measuring installed belt tensioner length to determine timing belt wear *(Courtesy of Olds-mobile Division, General Motors Corporation)*

Some timing belts must be installed with an identification mark facing toward the belt cover. Prior to belt installation the timing marks on the camshaft and crankshaft sprockets must be properly aligned with the marks on the engine casting (Figure 1-14).

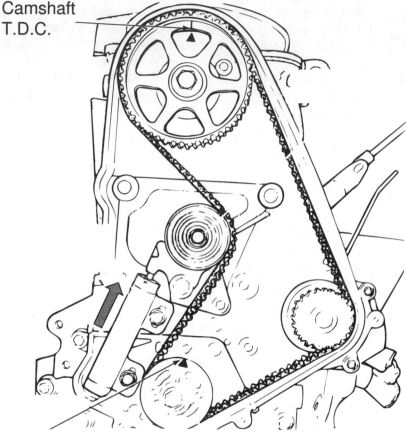

Figure 1-14   The timing marks on the camshaft and crankshaft sprockets must be properly aligned with the marks on the engine casting *(Courtesy of Chrysler Corporation)*

Some engines have steel timing chains. On these engines the crankshaft sprocket timing mark must be aligned with the timing mark on the camshaft gear prior to the camshaft gear and chain installation. Timing chain stretch and wear may be measured on these engines with a socket and flex handle installed on one of the camshaft sprocket retaining bolts. Rock the camshaft sprocket back and forth without moving the crankshaft gear, and measure the movement on one of the chain links on the camshaft sprocket.

On other engines the timing chains have copper colored links that must be aligned with the timing marks on the camshaft and crankshaft sprockets when the chain and gears are installed.

Task 15   Inspect and measure camshaft journals and lobes.

26. The measurement in Figure 1-15 is checking
    - A. the camshaft lobe lift.
    - B. the camshaft journal condition.
    - C. the pushrod length.
    - D. the valve lifter condition.

Hint

Figure 1-15   Valve train measurement *(Courtesy of Chevrolet Motor Division, General Motors Corporation)*

The outer camshaft journals should be placed in V-blocks to measure the camshaft runout (Figure 1-16). To measure the camshaft runout position a dial indicator against the other camshaft journals and rotate the camshaft.

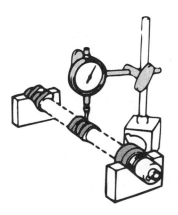

Figure 1-16   Measuring camshaft runout *(Courtesy of Toyota Motor Corporation)*

A micrometer may be placed from the highest point on the camshaft lobe to the side opposite the lobe to measure the lobe wear (Figure 1-17). If the wear on any lobe exceeds specifications, replace or regrind the camshaft. Most manufacturers do not recommend camshaft regrinding.

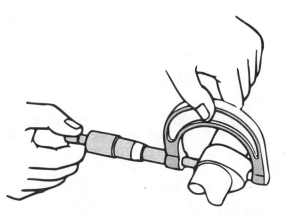

Figure 1-17   Measuring camshaft lobe height *(Courtesy of Toyota Motor Corporation)*

Use a micrometer to measure the diameter of each camshaft journal. Measure this diameter in several locations. If the diameter is less than specified, replace the camshaft.

Task 16   Inspect and measure camshaft bearing surfaces for damage, out-of-round, and alignment; repair or replace according to manufacturer's specifications.

27. When discussing camshaft bearing clearance
    Technician A says excessive camshaft bearing clearance may result in lower than specified oil pressure.
    Technician B says excessive camshaft bearing clearance may cause a clicking noise when the engine is idling.
    Who is correct?
    A. A only
    B. B only
    C. Both A and B
    D. Neither A nor B

Hint

The camshaft bearing out-of-round should be measured at several locations around the bearing with a telescoping gauge. When the out-of-round exceeds specifications bearing replacement is necessary. On overhead cam engines with removable camshaft bearing caps these caps must be torqued to specifications before measuring the bearing out-of-round.

When the lower half of the camshaft bearings are positioned on top of the cylinder head, a straightedge may be placed on these bearing surfaces with the bearing caps removed to measure bearing alignment. Measure the clearance between the straightedge and each bearing bore to determine the bore alignment. When the camshaft bearing bores are improperly aligned replace the cylinder head.

On overhead cam engines with removable camshaft bearing caps, Plastigage may be used to measure the bearing clearance (Figure 1-18).

Figure 1-18   Measuring camshaft bearing clearance with Plastigage *(Courtesy of Toyota Motor Corporation)*

Task 17   Measure camshaft timing according to manufacturer's specifications and procedures.

28. Technician A says improper valve timing may cause reduced engine power.
    Technician B says improper valve timing may cause bent valves in some engines.
    Who is correct?
    A. A only
    B. B only
    C. Both A and B
    D. Neither A nor B

Hint   *With the timing gear cover removed, the camshaft timing may be measured by checking the position of the marks on the camshaft and crankshaft sprockets. These marks must be aligned as indicated in the vehicle manufacturer's service manual (Figure 1-19).*

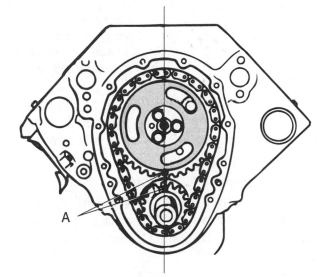

Figure 1-19   Alignment of timing marks on crankshaft and camshaft sprockets *(Courtesy of Chevrolet Motor Division, General Motors Corporation)*

Valve timing may be checked by observing the valve position in relation to the piston position. With any piston at TDC on the exhaust stroke the intake valve should be opening, and the exhaust valve should be closing. This valve position with the piston at TDC on the exhaust stroke is called valve overlap. If the valves do not open properly in relation to the crankshaft position, the valve timing is not correct. Under this condition the timing chain (or belt) cover should be removed to check the position of the camshaft sprocket in relation to the crankshaft sprocket position. When the valve timing is incorrect, the timing chain (or belt) and/or sprockets must be replaced.

Engine Block Diagnosis and Repair

ASE Tasks, Questions, and Related Information

Task 1   **Inspect and replace pans, covers, gaskets, and seals.**

29. Technician A says a springless seal may be used in the timing gear cover.
    Technician B says when a seal is installed the seal lip must face in the direction of oil flow.
    Who is correct?
    A. A only
    B. B only
    C. Both A and B
    D. Neither A nor B

Hint   *Gaskets are used to seal minor variations between two flat surfaces. Oil pan gaskets, or rocker arm cover gaskets usually are manufactured from cork, rubber, or a combination of rubber and cork or rubber and silicone. Always inspect the gasket mounting surfaces on pans and rocker arm covers for warpage. Warped mounting surfaces must be straightened.*

Seals may be classified as springless or spring loaded. Springless seals are used in front wheel hubs where they seal a heavy lubricant. The garter spring in a spring-loaded seal provides extra lip sealing force to compensate for lip wear, shaft runout, and bore eccentricity. A fluted lip seal may be used to direct oil back into a housing. A sealer is painted on the case of some seals to prevent leaks between the seal case and the bore. When a seal is installed, the garter spring must face toward the fluid flow. Split rubber seals, or rope seals, may be used in rear main bearings. Prior to seal installation the shaft and seal bore must be inspected for scratches and roughness. Seal lips should be lubricated before installation, and the proper seal driver must be used to install the seal.

Task 2

Assemble engine parts using formed-in-place (tube-applied) gaskets/sealants according to manufacturer's recommendations.

30. When installing RTV sealer
    A. the components to be sealed should be washed with an oil-base solvent.
    B. the RTV bead 1/8 in. wide should be placed in the center of the sealing surface.
    C. the RTV bead should be placed on one side of any bolt holes.
    D. the RTV bead should be allowed to dry for 10 minutes before component installation.

Hint   *Room-temperature vulcanizing (RTV) sealer may be used in place of conventional gaskets on oil pans and rocker arm covers. RTV sealer dries in the presence of air by absorbing moisture from the air. All the old material must be removed from the surface area before the RTV sealer is applied. A chlorinated solvent must be used to clean the RTV sealer mounting area. If oil-based solvents are used to clean this mounting area, an oily residue is left on the area that prevents RTV sealer adhesion. Apply a 1/8 in. diameter bead of RTV sealer to the center of one surface to be sealed. This bead must surround bolt holes in the mounting area. Do not apply excessive amounts of RTV sealer, and always assemble the components within five minutes after the RTV application or curing will occur.*

Anaerobic sealer may be used in place of a gasket between two machined surfaces. This sealer dries in the absence of air, and the same cleaning procedure must be used for RTV and anaerobic sealer.

Task 3

Inspect engine block for cracks, passage condition, core and gallery plug condition, and surface warpage; determine needed repairs.

31. Technician A says a warped cylinder head mounting surface on an engine block may cause valve seat distortion.
    Technician B says a warped cylinder head mounting surface on an engine block may cause coolant and combustion leaks.
    Who is correct?
    A. A only
    B. B only
    C. Both A and B
    D. Neither A nor B

Hint   *A cast iron block may be inspected for cracks with a electromagnetic crack detector. A dye penetrant may be used to check for cracks in an aluminum block. The block may be pressurized with a pressure tester to check for cracks. During an engine rebuild procedure the core plugs and oil gallery plugs should be replaced.*

Inspect the core plug bores before installing the new plugs. If the bore is damaged it may be repaired by boring it to the next specified oversize plug. Oversize core plugs are stamped with the letters OS. Before installing these plugs coat the sealing edge with an nonhardening water-resistant sealer. Oil gallery plugs should be coated with an oil-resistant sealer. A block may have dish-type, cup-type, or expansion-type core plugs. Install the core plugs with the proper special driving tool.

The cylinder head mounting surfaces on the block must be checked for warpage with a straightedge and a feeler gauge (Figure 1-20). If the warpage on these surfaces exceeds manufacturer's specifications the block surfaces must be resurfaced.

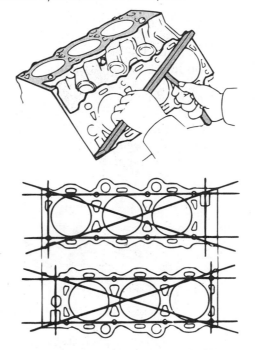

Figure 1-20   Measuring warpage on the top surfaces of the block *(Courtesy of Toyota Motor Corporation)*

Task 4   Inspect and repair damaged threads.

32. While discussing heli-coil installation
    Technician A says the first step is to use a tap, and thread the opening to match the external threads on the heli-coil.
    Technician B says the heli-coil should be installed with the proper size drill bit.
    Who is correct?
    A. A only
    B. B only
    C. Both A and B
    D. Neither A nor B

Hint   *If threads are damaged the opening may be drilled and threaded, and then a heli-coil may be installed to provide a thread the same as the original (Figure 1-21).*

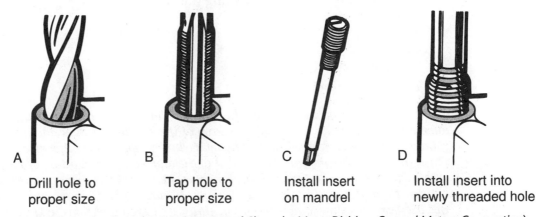

A   Drill hole to proper size

B   Tap hole to proper size

C   Install insert on mandrel

D   Install insert into newly threaded hole

Figure 1-21   Thread repair *(Courtesy of Chevrolet Motor Division, General Motors Corporation)*

Task 5   Remove cylinder wall ridges.

33. If new rings are installed without removing the the ring ridge, the following may result:
    A. the piston skirt may be damaged.
    B. the piston pin may be broken.
    C. the connecting rod bearings may be damaged.
    D. the piston ring lands may be broken.

Hint   *If the amount of cylinder wear does not require cylinder reboring, remove the ring ridge at the top of each cylinder with a ridge reamer (Figure 1-22). Always consult the engine manufacturer's service manual regarding ring ridge removal. While using the ridge reamer be careful not to mark the cylinder wall below the ring ridge. Remove the ring ridge only. Do not remove any metal from the cylinder wall below the ring ridge.*

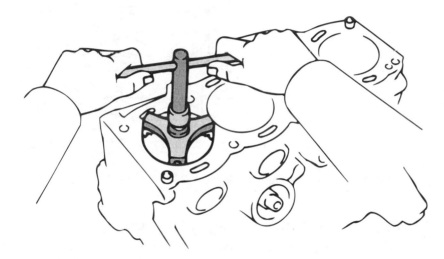

Figure 1-22   Ring ridge removal *(Courtesy of Toyota Motor Corporation)*

Task 6   Inspect cylinder walls for damage and wear; determine needed repairs.

34. While discussing cylinder measurement
    Technician A says the cylinder taper is the difference between the cylinder diameter at the top of the ring travel compared to the cylinder diameter at the center of the ring travel.
    Technician B says cylinder out-of-round is the difference between the axial cylinder bore diameter at the top of the ring travel compared to the thrust cylinder bore diameter at the bottom of the ring travel.
    Who is correct?
    A. A only
    B. B only
    C. Both A and B
    D. Neither A nor B

Hint   *Use a dial bore gauge to measure the cylinder diameter in three vertical locations. These locations are just below the ring ridge at the top of the cylinder, in the center of the ring travel, and just above the lowest part of the ring travel. Cylinder taper is the difference in the cylinder diameter at the top of the ring travel compared to the diameter at the bottom of the ring travel.*

In each of the three cylinder vertical measurement locations measure the cylinder diameter in the thrust direction and in the axial direction (Figure 1-23). Cylinder out-of-round is the difference between the cylinder diameter in the thrust and axial directions. If the cylinder out-of-round exceeds specifications, rebore the cylinder.

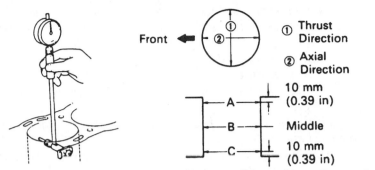

Figure 1-23 Cylinder measurement *(Courtesy of Toyota Motor Corporation)*

Task 7 Hone and clean cylinder walls.

35. While deglazing and cleaning a cylinder
    A. 120 grit stones may be used on the cylinder hone.
    B. wash the cylinder with soapy water after deglazing.
    C. wash the cylinder with an oil-based solvent after deglazing.
    D. 400 grit emery paper and hand pressure may be used.

Hint
If cylinder wear, out-of-round, and taper does not exceed specifications, the cylinders may be deglazed. Cylinders may be deglazed with 220 or 280 grit stones installed on a cylinder hone, or a hone-type brush. After deglazing, the cylinder should be cleaned with hot, soapy water and a stiff-bristle brush. Clean the residue from the cylinder with a soft lint-free cloth. Rinse the block and dry it thoroughly. Coat all machined surfaces in the block with a light coating of the manufacturer's recommended engine oil.

If one cylinder requires reboring, most manufacturers recommend reboring all the cylinders to the same size. Cylinder reboring usually is done with a honing machine or reboring bar. The proper stones must be used on the honing machine depending on the block material. Honing stones have an identification number. Stones with a higher number have a finer grit. When the honing operation is completed, the cylinders should have a 60° crosshatch pattern. After cylinder honing the same procedure for block cleaning should be followed as discussed previously in cylinder deglazing.

Task 8 Inspect and measure camshaft bearings for wear, damage, out-of-round, and alignment; determine needed repairs.

36. The tool shown in Figure 1-24 is used to
    A. remove camshaft bearings.
    B. install camshaft bearings.
    C. remove and install camshaft bearings.
    D. measure camshaft bearing alignment.

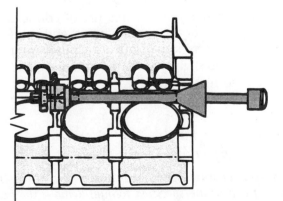

Figure 1-24 Engine block service tool *(Courtesy of Chevrolet Motor Division, General Motors Corporation)*

Hint　　　*If the camshaft bearings are mounted in the block, the oil hole in the bearing must be aligned with the oil hole in the block. Many overhead cam engines do not have removable camshaft bearings. Inspect the camshaft bearings or bearing bores for scoring, roughness, and wear. Camshaft bearings, or bearing bores, should be measured at two different locations with a telescoping gauge. Measure the camshaft journals with a micrometer, and subtract the journal diameter from the bearing diameter to obtain the clearance. If the wear exceeds specifications, replace the bearings, or cylinder head.*

When the camshaft bearing bores are in the cylinder head, the bearing caps should be removed and a straightedge positioned across all the bearing bores in the cylinder head to measure bearing alignment. Insert a feeler gauge between the straightedge and each bearing bore to measure any misalignment.

Task 9　**Inspect crankshaft for surface cracks and journal damage; check oil passage condition; measure journal wear; determine needed repairs.**

37. When measuring the crankshaft journal in Figure 1-25 the difference between measurements
    A.　A and B indicates horizontal taper.
    B.　C and D indicates vertical taper.
    C.　A and C indicates out-of-round.
    D.　A and D indicates vertical taper.

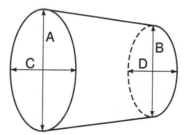

Figure 1-25　Crankshaft journal measurement *(Courtesy of Ford Motor Company)*

Hint　　　*Use a micrometer to measure each crankshaft journal for vertical taper, horizontal taper, and out-of-round. The out-of-round should be measured at two locations on each side of the journal. When the journal out-of-round or taper exceeds specifications, or journal scoring is evident, journal grinding is required.*

Task 10　**Inspect and measure main and connecting rod bearings for damage, clearance, and end/side play; determine needed repairs including the proper selection of bearings.**

38. Technician A says the curvature of connecting rod bearings is slightly larger than the curvature of the bearing bores, and this feature is called bearing spread.
    Technician B says that when a connecting rod bearing half is installed the bearing edges extend slightly from the mounting area, and this feature is called bearing crush.
    Who is correct?
    A.　A only
    B.　B only
    C.　Both A and B
    D.　Neither A nor B

Hint　　　*To measure bearing clearance install a strip of Plastigage across the journal and then tighten the bearing cap to the specified torque. Remove the bearing cap and measure the width of the Plastigage on the journal (use the scale provided on the Plastigage package) to determine the bearing clearance (Figure 1-26).*

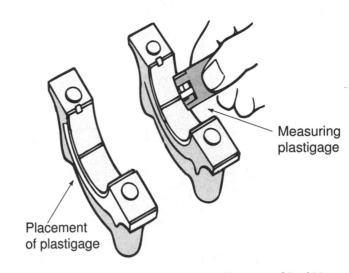

Figure 1-26   Bearing clearance measurement *(Courtesy of Ford Motor Company)*

A feeler gauge should be inserted between the side of the connecting rod and the edge of the crankshaft journal to measure the side clearance (Figure 1-27). If this clearance exceeds specifications, the sides of the connecting rod or crankshaft journal are worn. Crankshaft end play may be measured by inserting a feeler gauge between the crankshaft journal and the thrust lip on one of the main bearings. On some engines a dial indicator is used to measure crankshaft end play.

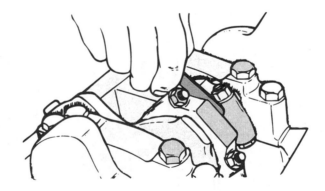

Figure 1-27   Measuring connecting rod side clearance *(Courtesy of Chevrolet Motor Division, General Motors Corporation)*

Task 11   Identify piston and bearing wear patterns that indicate connecting rod alignment and main bearing bore problems; inspect rod alignment and bearing bore condition (limited to identifying rod alignment setup).

39. A bent connecting rod may cause
    - A. uneven connecting rod bearing wear.
    - B. uneven main bearing wear.
    - C. uneven piston pin wear.
    - D. excessive cylinder wall wear.

Hint   *When the main bearing bores are misaligned, excessive wear occurs on some main bearings compared to the other main bearings. Uneven wear on the edges of the piston skirt next the pin hole may be caused by connecting rod misalignment. Connecting rod misalignment results in V-shaped connecting rod bearing wear. Measure the connecting rod bore for out-of-round and proper bore size. Assemble the connecting rod on the rod alignment measuring tool. Place a feeler gauge between the upper projection on the fixture and the machined tool surface to measure the rod*

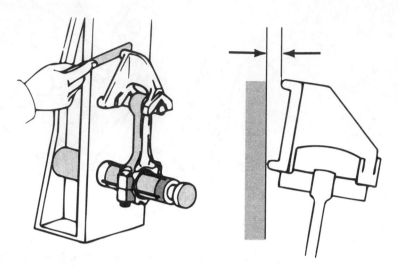

Figure 1-28   Measuring connecting rod bend *(Courtesy of Toyota Motor Corporation)*

bend (Figure 1-28). Install a feeler gauge between the side projections on the fixture and the the machined tool surface to measure rod twist (Figure 1-29). If rod bend or twist exceeds specifications, replace the connecting rod.

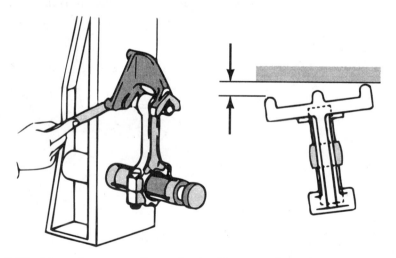

Figure 1-29   Measuring connecting rod twist (Courtesy of Toyota Motor Corporation)

Task 12   Inspect, measure, service, or replace pistons, piston pins, and pin bushings.

40. The tool in Figure 1-30 is used to
   A. widen the piston ring grooves.
   B. deepen the piston ring grooves.
   C. remove and replace piston rings.
   D. remove carbon from the ring groove.

Hint   *Piston ring grooves should be cleaned using a ring groove cleaning tool. Install a new piston ring in each ring groove, and position a feeler gauge between each ring and the ring groove to measure the ring groove clearance. Use a micrometer to measure the piston diameter at right angles to the piston pin bores and 1 in. below the bottom edge of the lowest ring groove (Figure 1-31).*

Some connecting rods have a bushing in the rod opening. The oil hole in the bushing must be aligned with the oil hole in the connecting rod. The bushing may be honed in a pin hole machine to obtain the specified pin to bushing clearance. When a bushing is positioned in the upper rod

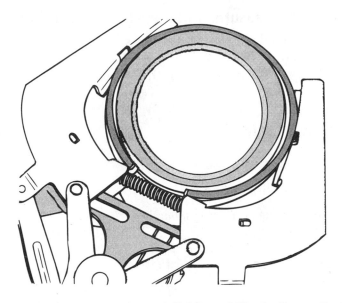

Figure 1-30 Piston service tool *(Courtesy of Chrysler Corporation)*

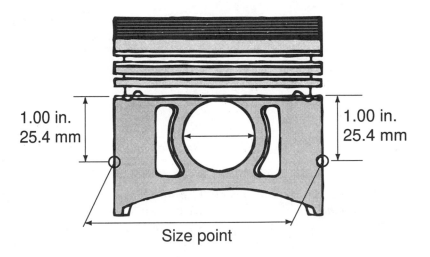

1.00 in.
25.4 mm

1.00 in.
25.4 mm

Size point

Figure 1-31 Measuring piston diameter *(Courtesy of Chevrolet Motor Division, General Motors Corporation)*

opening some manufacturers recommend heating the pistons in a piston heater prior to piston pin installation.

In some engines the piston pins are pressed into the upper rod opening. The upper end of the rod must be heated before pressing the pin into the rod and piston. The pin and connecting rod must be centered in the piston bores, and the marks on the piston and connecting rod must be properly positioned before the connecting rod and piston are assembled.

Task 13 Inspect, measure, and install or replace piston rings.

41. While discussing piston ring service

    Technician A says the ring gap should be measured with the ring positioned at the top of the ring travel in the cylinder.

    Technician B says the two compression rings are interchangeable on most pistons.

    Who is correct?

    A. A only
    B. B only
    C. Both A and B
    D. Neither A nor B

Hint   *Position the piston ring squarely in the cylinder at the bottom of the ring travel, and measure the ring gap with a feeler gauge (Figure 1-32). If the gap is less than specified, file a small amount of metal from the end of the ring. When the gap is more than specified the ring or cylinder bore is worn, or the ring is the wrong size.*

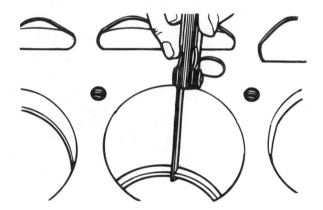

Figure 1-32 Measuring ring gap *(Courtesy of Ford Motor Company)*

Piston rings are removed and installed with a ring expander. Piston rings must be installed in the proper vertical direction on the piston (Figure 1-33). Piston ring gaps must be positioned properly around the piston to minimize blowby. The rings and piston should be coated with the specified engine oil; a ring compressor is then tightened on the piston to compress the rings. Protecting sleeves must be installed on the connecting rod bolts prior to piston installation. Be sure the piston and connecting rod marks are properly positioned and then tap the piston gently into the cylinder.

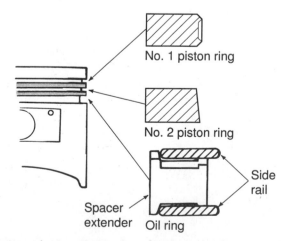

No. 1 piston ring

No. 2 piston ring

Side rail

Spacer extender   Oil ring

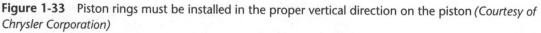

Figure 1-33 Piston rings must be installed in the proper vertical direction on the piston *(Courtesy of Chrysler Corporation)*

Task 14   Inspect, repair, or replace crankshaft vibration damper.

42. Technician A says the vibration damper counterbalances the back-and-forth twisting motion of the crankshaft each time a cylinder fires.
Technician B says if the seal contact area on the vibration damper hub is scored, the damper assembly must be replaced.
Who is correct?
A. A only
B. B only
C. Both A and B
D. Neither A nor B

Hint   *Inspect the rubber between the inner hub and outer inertia ring on the vibration damper. If this rubber is cracked, oil soaked, deteriorated, or loose, replace the damper. Inspect the seal contact area on the vibration damper hub for ridging and scoring. When these conditions are present replace the damper, or machine the damper hub and install the proper sleeve on the hub to provide a new seal contact area. A special puller and installer tool are required to remove and install the vibration damper.*

Task 15   Inspect crankshaft flange and flywheel for burrs; repair as necessary.

43. Technician A says metal burrs on the crankshaft flange may cause excessive wear on the ring gear and starter drive gear teeth.
Technician B says metal burrs on the crankshaft flange may cause improper torque converter to transmission alignment.
Who is correct?
A. A only
B. B only
C. Both A and B
D. Neither A nor B

Hint   *Inspect the crankshaft flange and the flywheel to crankshaft mating surface for metal burrs (Figure 1-34). Remove any metal burrs with fine emery paper. Be sure the threads in the crankshaft flange are in satisfactory condition. Replace the flywheel bolts and retainer if any damage is visible on these components. Install the flywheel, retainer and bolts, and tighten the bolts in sequence to the manufacturer's specified torque.*

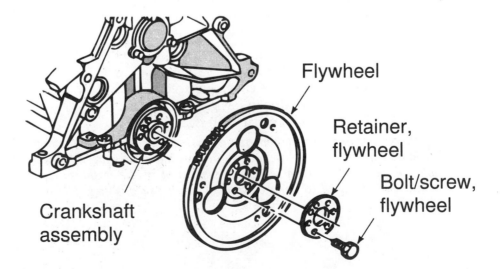

Figure 1-34   Flywheel and crankshaft mounting flange *(Courtesy of Chevrolet Motor Division, General Motors Corporation)*

Task 16 Inspect flywheel for cracks, wear (includes flywheel ring gear), and measure flywheel runout; determine needed repairs.

44. Technician A says excessive flywheel runout may cause grabbing, erratic clutch operation.

Technician B says the pressure plate should always be reinstalled in the original position on the flywheel.

Who is correct?

A. A only

B. B only

C. Both A and B

D. Neither A nor B

Hint *Inspect the flywheel for scoring and cracks in the clutch contact area. Minor score marks and ridges may be removed by resurfacing the flywheel.*

Mount a dial indicator on the engine flywheel housing, and position the dial indicator stem against the clutch contact area on the flywheel (Figure 1-35). Rotate the flywheel to measure the flywheel runout. If the flywheel runout exceeds specifications replace the flywheel.

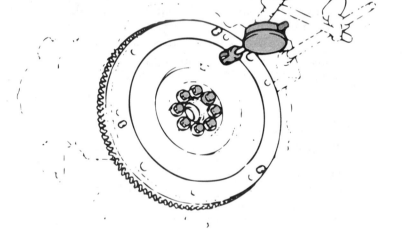

Figure 1-35 Measuring flywheel runout (*Courtesy of Toyota Motor Corporation*)

Task 17 Inspect, remove, and replace crankshaft pilot bearing/bushing.

45. A worn pilot bearing may cause a growling, rattling noise

A. while driving at a steady 20 mph.

B. in reverse with the clutch engaged.

C. while accelerating in low gear.

D. with the clutch pedal depressed.

Hint *Place your finger in the inner pilot bearing race and rotate the race. If the bearing feels rough or loose, replace the bearing. A new transmission main shaft may be positioned in the pilot bushing. If the mainshaft has excessive movement in the bushing, replace the bushing. A special puller may be used to remove the pilot bearing or bushing. The proper driver must be used to install the pilot bearing or bushing.*

Task 18   **Remove auxiliary (balance, intermediate, idler, counterbalance, or silencer) shaft(s); inspect shaft(s) and support bearings for damage and wear.**

46. Technician A says improper balance shaft timing causes severe engine vibrations. Technician B says the balance shafts are timed in relation to the camshaft. Who is correct?
    A. A only
    B. B only
    C. Both A and B
    D. Neither A nor B

Hint   *The balance shafts should be checked for runout with the same procedure used for measuring camshaft runout. The balance shaft journals should be measured for taper with the same procedure for measuring crankshaft journal taper. When the balance shafts are installed they must be properly timed to the crankshaft.*

Lubrication and Cooling System Diagnosis and Repair

ASE Tasks, Questions, and Related Information

Task 1   **Perform oil pressure tests; determine needed repairs.**

47. All of the following are causes of low engine oil pressure EXCEPT
    A. worn camshaft bearings.
    B. worn crankshaft bearings.
    C. weak oil pressure regulator spring tension.
    D. restricted pushrod oil passages.

Hint   *When testing engine oil pressure the oil sending unit usually is removed and an oil pressure gauge is connected to the sending unit opening. The engine should be at normal operating temperature, and the oil pressure test usually is performed at idle speed and a higher speed such as 2,500 rpm.*

Task 2   **Inspect, measure, repair, or replace oil pumps (includes gears, rotors, and housing), pressure relief devices, and pump drives.**

48. The following are normal oil pump component measurements EXCEPT
    A. inner rotor diameter.
    B. clearance between the rotors.
    C. inner and outer rotor thickness.
    D. outer rotor to housing clearance.

Hint   *Inspect the oil pump pressure relief valve for sticking and wear. If this valve sticks in the closed position, oil pressure is excessive. A pressure relief valve stuck in the open position results in low oil pressure.*

   Measure the thickness of the inner and outer rotors with a micrometer. When this thickness is less than specified on either rotor replace the rotors or the oil pump. The following oil pump measurements should be performed with a feeler gauge:

   - *Measure pump cover flatness with a feeler gauge positioned between a straightedge and the cover.*
   - *Measure the clearance between the outer rotor and the housing.*
   - *Measure the clearance between the inner and outer rotors with the rotors installed.*
   - *Measure the clearance between the top of the rotors and a straightedge positioned across the top of the oil pump.*

Task 3   Perform cooling system tests; determine needed repairs.

49. The tester in Figure 1-36 may be used to test the following items EXCEPT
    A. cooling system leaks.
    B. radiator cap pressure relief valve.
    C. coolant specific gravity.
    D. heater core leaks.

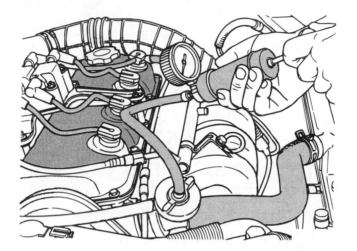

Figure 1-36   Pressure tester *(Courtesy of Chrysler Corporation)*

Hint   *A pressure tester may be connected to the radiator filler neck to check for cooling system leaks. Operate the tester pump and apply 15 psi to the cooling system. Inspect the cooling system for external leaks with the system pressurized. If the gauge pressure drops more than specified by the vehicle manufacturer, the cooling system has a leak. If there are no visible external leaks, check the front floor mat for coolant dripping out of the heater core. When there are no external leaks, check the engine for combustion chamber leaks.*

The radiator pressure cap may be tested with the pressure tester. When the tester pump is operated, the cap should hold the rated pressure. Always relieve the pressure before removing the tester.

Task 4   Inspect, replace, and adjust drive belts and pulleys according to manufacturer's specifications.

50. A loose alternator belt may cause
    A. a discharged battery.
    B. a squealing noise while decelerating.
    C. a damaged alternator bearing.
    D. engine overheating.

Hint   *Since the friction surfaces are the sides of a V-belt, the belt must be replaced if the sides are worn and the belt is contacting the bottom of the pulley.*

The belt tension may be checked with the engine shut off, and a belt tension gauge placed over the belt at the center of the belt span (Figure 1-37). A loose or worn belt may cause a squealing noise when the engine is accelerated.

The belt tension also may be checked by measuring the amount of belt deflection with the engine shut off. Use your thumb to depress the belt at the center of the belt span. If the belt tension is correct, the belt should have 1/2 in. deflection per foot of belt span.

Ribbed V-belts usually have a spring-loaded belt tensioner, with a belt wear indicator scale on the tensioner housing. If a power steering pump belt requires tightening always pry on the pump ear, not on the housing.

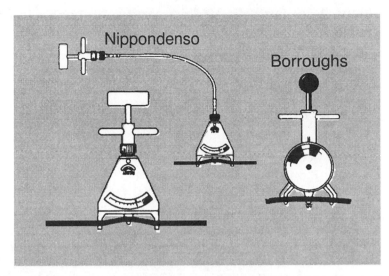

Figure 1-37   Measuring belt tension *(Courtesy of Toyota Motor Corporation)*

Task 5   Inspect and replace engine cooling and heater system hoses.

51. Technician A says a collapsed upper radiator hose may be caused by a defective pressure release valve in the radiator cap.

    Technician B says a collapsed upper radiator hose may be caused by a plugged hose between the radiator filler neck and the recovery reservoir.

    Who is correct?

    A. A only

    B. B only

    C. Both A and B

    D. Neither A nor B

Hint   *All cooling system hoses should be inspected for soft spots, swelling, hardening, chafing, leaks, and collapsing. If any of these conditions are present, hose replacement is necessary. Hose clamps should be inspected to make sure they are tight. Some radiator hoses contain a wire coil inside them to prevent hose collapse as the coolant temperature decreases. Remember to include heater hoses, and the bypass hose, in the hose inspection. Prior to hose removal the coolant must be drained from the radiator.*

Task 6   Inspect, test, and replace thermostat, bypass, and housing.

52. The thermostat is stuck open on a port fuel-injected engine. This problem may cause

    A. a rich air-fuel ratio.

    B. a lean air-fuel ratio.

    C. excessive fuel pressure.

    D. engine overheating.

Hint   *The thermostat may be submerged with a thermometer in a container filled with water. Heat the water while observing the thermostat valve, and the thermometer. The thermostat valve should begin to open when the temperature on the thermometer is equal to the rated temperature stamped on the thermostat. Replace the thermostat if it does not open at the rated temperature. Many thermostats are marked for installation in the proper direction. Inspect the bypass hose for cracks, deterioration, and restrictions; replace the hose if these conditions are present.*

Task 7   **Inspect coolant; drain, flush, and refill cooling system with recommended coolant.**

53. While discussing cooling system service:
    Technician A says if the cooling system pressure is reduced the coolant boiling point is increased.
    Technician B says when more antifreeze is added to the coolant the coolant boiling point is increased.
    Who is correct?
    A. A only
    B. B only
    C. Both A and B
    D. Neither A nor B

Hint   *If the radiator tubes, and coolant passages in the block and cylinder head, are restricted with rust and other contaminants, these components may be flushed. Cooling system flushing equipment is available for this purpose. Always operate the flushing equipment according to the equipment manufacturer's directions, and be sure that your service procedure conforms to pollution laws in your state. Engine coolant must be recycled or handled as a hazardous waste material. Coolant reconditioning machines are available to remove harmful particles and restore corrosion additives so the coolant can be returned to the cooling system.*

Task 8   **Inspect, test, and replace water pump.**

54. Technician A says a defective water pump bearing may cause a growling noise when the engine is idling.
    Technician B says the water pump bearing may be ruined by coolant leaking past the pump seal.
    Who is correct?
    A. A only
    B. B only
    C. Both A and B
    D. Neither A nor B

Hint   *With the engine shut off grasp the fan blades, or the water pump hub, and try to move the blades from side to side. This action checks for looseness in the water pump bearing. If there is any side-to-side movement in the bearing, water pump replacement is required.*

Check for coolant leaks, rust, or residue at the water pump drain hole in the bottom of the pump, and at the inlet hose connected to the pump. When coolant is dripping from the pump drain hole, replace the pump. The water pump may be tested with the pressure tester connected to the radiator filler neck.

Task 9   **Inspect, test, and replace radiator, pressure cap, and expansion tank.**

55. An excessively high coolant level in the recovery reservoir may be caused by any of these problems EXCEPT
    A. restricted radiator tubes.
    B. a thermostat that is stuck open.
    C. a loose water pump impeller.
    D. an inoperative electric-drive cooling fan.

Hint   *The radiator cap should be inspected for a damaged sealing gasket, or vacuum valve. If the pressure cap sealing gasket, or seat, are damaged, the engine will overheat, and coolant is lost to the coolant recovery system. Under this condition the coolant recovery container becomes over-filled with coolant.*

If the cap vacuum valve is sticking, a vacuum may occur in the cooling system after the engine is shut off and the coolant temperature decreases. This vacuum may cause collapsed cooling system hoses. A pressure tester may be used to test the pressure cap, and pressure test the entire cooling system.

The coolant level should be at the appropriate mark on the recovery container, depending on engine temperature.

Task 10   Clean, inspect, test, and replace fan (both electrical and mechanical), fan clutch, fan shroud, and cooling related sensors.

56. In Figure 1-38 an open ground circuit on the engine temperature switch may cause
    A. continual cooling fan motor operation.
    B. completely inoperative cooling fan motor.
    C. a burned out cooling fan motor.
    D. engine overheating.

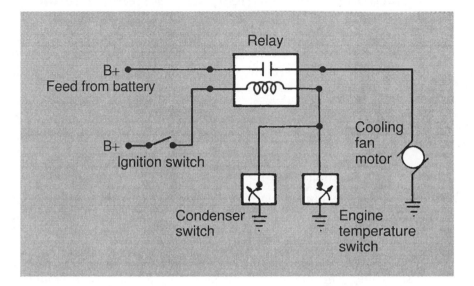

Figure 1-38   Electric-drive cooling fan circuit *(Courtesy of SAE International)*

Hint   *If the radiator shroud is loose, improperly positioned, or broken, air flow through the radiator is reduced, and engine overheating may result.*

The viscous-drive fan clutch should be visually inspected for leaks. If there are oily streaks radiating outward from the hub shaft, the fluid has leaked out of the clutch.

With the engine shut off rotate the cooling fan by hand. If the viscous clutch allows the fan blades to rotate easily both hot and cold, the clutch should be replaced. A slipping viscous clutch results in engine overheating. If there is any looseness between the viscous clutch and the shaft, replace the viscous clutch.

If the electric-drive cooling fan does not operate at the coolant temperature specified by the vehicle manufacturer, engine overheating will result, especially at idle and lower speeds when air flow through the radiator is reduced.

Task 11   Inspect, test, or repair auxiliary oil coolers.

57. An engine oil cooler helps to prevent
    A. oxidation of the engine oil.
    B. excessive oil pressure.
    C. oil pump wear.
    D. main bearing wear.

Hint   *Some engines, such as diesels and turbocharged engines, have an oil cooler. This oil cooler may by contained in one of the radiator tanks, or mounted separately near the front of the engine. External oil coolers should be inspected for leaks and restricted air flow passages.*

The maximum oil operating temperature is approximately 250°F (121°C). Hot oil combines with oxygen in the air to form carbon and a sticky varnish. This process is called oil oxidation. An engine oil cooler reduces oil temperature and helps to prevent oil oxidation especially on engines operating under heavy load.

Ignition System Diagnosis and Repair

ASE Tasks, Questions, and Related Information

Task 1 Diagnose no-starting, hard starting, and engine misfire on vehicles with electronic or point-type ignition systems; determine needed repairs.

58. While discussing a condition where a 12V test light connected from the negative primary coil terminal to ground flashes while cranking the engine, but a test spark plug does not fire when connected from the coil secondary wire to ground
Technician A says the distributor cap and rotor may be defective.
Technician B says the coil may be defective.
Who is correct?
A. A only
B. B only
C. Both A and B
D. Neither A nor B

Hint   *When diagnosing a no-start condition a 12V test light may be connected from the negative primary coil terminal to ground. If this test light flashes while cranking the engine, the primary circuit is triggered on and off by the module. Under this condition a defect in the secondary ignition circuit is likely preventing spark plug firing. If a test spark plug connected from each spark plug wire to ground does not fire when cranking the engine, check the coil, cap and rotor, and spark plug wires. When the test light does not flash while cranking the engine, the module or pickup coil are defective.*

Task 2 Inspect, test, repair, or replace ignition primary circuit wiring and components.

59. If ohmmeter 1 in Figure 1-39 shows an infinite reading, the pickup coil is
A. shorted to ground.
B. shorted.
C. open.
D. not grounded.

Hint   *The coil primary winding and the pickup coil should be tested for open circuits, short circuits, and a shorted to ground condition with an ohmmeter. The module should be tested with a module tester. The primary circuit wires may be tested with a voltmeter or an ohmmeter.*

Task 3 Inspect, test, and service distributor including drives, shaft, bushings, cam, breaker plate, and vacuum and mechanical advance/retard units.

60. The following statements about distributor advances are true EXCEPT
A. the vacuum advance controls spark advance in relation to engine load.
B. the mechanical advance controls spark advance in relation to engine rpm.
C. the mechanical advance rotates the reluctor in the opposite direction to shaft rotation.
D. the vacuum advance rotates the pickup plate in the opposite direction to shaft rotation.

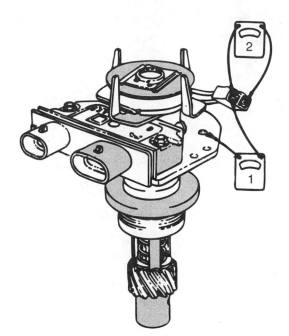

Figure 1-39   Testing distributor pickup coil *(Courtesy of Chevrolet Motor Division, General Motors Corporation)*

Hint   *The mechanical advance controls the spark advance in relation to engine rpm. In this advance pivoted weights fly outward as the engine speed increases, and these weights rotate the cam or reluctor ahead of the distributor shaft. The vacuum advance controls spark advance in relation to engine load. This advance rotates the pickup or breaker plate in the opposite direction to distributor shaft rotation to provide spark advance.*

Task 4   **Inspect, test, adjust, service, repair, or replace ignition points and condenser.**

61. While discussing ignition point adjustment
    Technician A says an increase in point gap increases the cam dwell reading.
    Technician B says point dwell must be long enough to allow the magnetic field to build up in the coil.
    Who is correct?
    A. A only
    B. B only
    C. Both A and B
    D. Neither A nor B

Hint   *The point gap may be adjusted with a feeler gauge and the point rubbing block positioned on one of the cam lobe high points. The points may also be adjusted with a dwell meter connected from the negative primary coil terminal to ground. The cam dwell reading indicated on the dwell meter is the number of degrees that the cam rotates while the points are closed on each cam lobe.*

Task 5   Inspect, test, service, repair or replace ignition system secondary circuit wiring and components.

62. A low maximum secondary coil voltage may be caused by
   A. a low primary resistance.
   B. a low primary input voltage.
   C. wide spark plug gaps.
   D. an open spark plug wire.

Hint   *An ohmmeter should be used to test the secondary coil winding for shorts, opens, and grounds. The secondary coil voltage may be tested with a test spark plug or an oscilloscope. The resistance of the spark plug wires and coil secondary coil wire may be tested with an ohmmeter.*

Task 6   Inspect, test, and replace ignition coil.

63. The ohmmeter in Figure 1-40 is connected to test
   A. the primary winding for shorts.
   B. the primary winding for grounds.
   C. the secondary winding for shorts.
   D. the secondary winding for grounds.

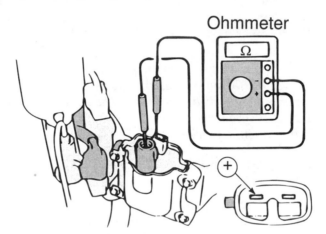

Figure 1-40   Testing ignition coil *(Courtesy of Toyota Motor Corporation)*

Hint   *The coil secondary winding should be tested for shorts, grounds, and opens with an ohmmeter. Always inspect the coil tower for cracks. Secondary coil voltage should be tested with a test spark plug or an oscilloscope.*

Task 7   Check and adjust ignition system timing and timing advance.

64. Technician A says if the basic timing is less than specified, the engine may detonate. Technician B says if the basic timing is less than specified, the engine may overheat. Who is correct?
   A. A only
   B. B only
   C. Both A and B
   D. Neither A nor B

Hint   *The timing usually is adjusted with the engine idling at the specified idle speed. On a distributor with advances the vacuum advance hose should be disconnected while checking the basic timing. With the timing light beam directed at the timing marks the distributor is rotated until the timing mark is in the specified position. After the timing is adjusted tighten the distributor clamp bolt. Most fuel injected and electronically controlled engines require a special procedure such as disconnecting an in-line timing connector while checking the basic timing.*
Many timing lights have the capability to check the spark advance. An advance control on the

light slows down the flashes of light as the knob is rotated. When the light flashes are slowed with the engine running at higher speed, the timing marks move back to the basic timing setting and the degree scale on the timing light indicates the degrees of advance.

Task 8   Inspect, test, and replace electronic ignition system pickup sensor or trigger devices.

65. The pickup coil air gap should be measured with
    A. a brass feeler gauge.
    B. a steel feeler gauge.
    C. a dial indicator.
    D. a dwell meter.

Hint   *In some electronic ignition systems the pickup gap may be adjusted with a nonmagnetic feeler gauge positioned between the reluctor high point and the pickup coil. Pickup coils containing a winding and permanent magnet may be tested for shorts, grounds, and opens with an ohmmeter. Hall Effect pickups may be tested with a voltmeter. The voltage signal from these pickups should cycle from a very low voltage to the manufacturer's specified voltage when the engine is cranked.*

Task 9   Inspect, test, and replace electronic ignition system control unit (module).

66. Technician A says some ignition modules will not operate unless they are grounded to the vehicle chassis.
    Technician B says a defective ignition module may operate satisfactorily when cold, but cause the engine to stop running when the module is hot.
    Who is correct?
    A. A only
    B. B only
    C. Both A and B
    D. Neither A nor B

Hint   *The ignition module may be tested with a module tester. Most of these testers check the ability of the module to cycle the primary circuit on and off. Some modules that are mounted separately from the distributor must be grounded on the vehicle chassis.*

Fuel and Exhaust Systems Diagnosis and Repair

ASE Tasks, Questions, and Related Information

Task 1   Inspect, test, and replace fuel pumps and pump controls; inspect, service, and replace fuel filters.

67. All of these statements about fuel pump testing on a port fuel injected engine are true EXCEPT
    A. the fuel system should be depressurized prior to the fuel pump pressure test.
    B. if the fuel pump pressure is less than specified the air-fuel ratio is too rich.
    C. the pressure gauge should be connected to the Schrader valve on the fuel rail.
    D. lower than specified fuel pressure may be caused by a restricted fuel filter.

Hint   *Mechanical or electric fuel pumps should be tested for pressure and flow or volume. The pressure gauge is connected in series in the fuel inlet line on the carburetor or throttle body assembly on throttle-body-injected engines. On most port-injected engines the pressure gauge is connected to the Schrader valve on the fuel rail.*

Task 2   Inspect, clean or replace fuel injection air induction system, intake manifold, and gaskets; check air/fuel ratio (mixture) and idle speed.

68. Technician A says an intake manifold vacuum leak may cause a cylinder misfire with the engine idling.
Technician B says an intake manifold vacuum leak may cause a cylinder misfire during hard acceleration.
Who is correct?
A. A only
B. B only
C. Both A and B
D. Neither A nor B

Hint   *The intake manifold should be inspected for cracks, and the surface of the manifold that fits against the cylinder head should be checked for warpage with a straightedge and a feeler gauge. The idle mixture should be adjusted on carbureted engines with the engine at normal operating temperature and idling at the specified speed. After the idle mixture adjustment is completed, the vehicle must meet emission standards.*

Task 3   Inspect and service or replace air filters and filter housings.

69. While cleaning a pleated paper-type air filter element the air gun should be held
A. 6 in. from the outside of the air filter element.
B. directly against the outside of the air filter element.
C. directly against the inside of the air filter element.
D. 6 in. from the inside of the air filter element.

Hint   *Air filter elements should be changed at the manufacturer's recommended service intervals. When the vehicle is operating in extremely dusty conditions, the air filter should be changed at more frequent intervals. If the air filter element contains small holes, dust enters the engine and causes very fast cylinder wall and piston wear. A restricted air filter element reduces air flow into the engine and causes a rich air/fuel ratio.*

Task 4   Inspect, service, and replace exhaust manifold, manifold heat control valves (heat risers), exhaust pipes, mufflers, converters, resonators, tail pipes, and heat shields.

70. All of these statements regarding manifold heat control valves are true EXCEPT
A. a manifold heat control valve improves fuel vaporization in the intake manifold especially when the engine is cold.
B. a manifold heat control valve stuck in the closed position causes a loss of engine power.
C. a manifold heat control valve stuck in the open position may cause an acceleration stumble.
D. a manifold heat control valve stuck in the closed position reduces intake manifold temperature.

Hint   *Many original mufflers, and catalytic converters are integral with the interconnecting pipes. When these components are replaced they must be cut from the exhaust system with a cutting tool. The inlet and outlet pipes on the replacement muffler, or converter, must have a 1.5 in. overlap on the connecting pipes. Before cutting the pipes to remove the muffler, or converter, measure the length of the new component and always cut these pipes to provide the required overlap.*

Task 5   Test the operation of positive crankcase ventilation (PCV) systems.

71. In Figure 1-41 the hose from the PCV valve to the intake manifold is restricted. This problem could result in
A. an acceleration stumble.
B. oil accumulation in the air cleaner.
C. engine surging at high speed.
D. engine detonation during acceleration.

Hint   *If the PCV valve is stuck in the open position, excessive air flow through the valve causes a lean air-fuel ratio and possible rough idle operation, or engine stalling. When the PCV valve, or hose, is restricted, excessive crankcase pressure forces blowby gases through the clean air hose and filter into the air cleaner. Worn rings, or cylinders, cause excessive blowby gases and increased crankcase pressure, which forces blowby gases through the clean air hose and filter into the air cleaner. A restricted PCV valve, or hose, may result in the accumulation of moisture and sludge in the engine, and engine oil.*

Some vehicle manufacturers recommend removing the PCV valve from the rocker arm cover, and placing your finger over the valve with the engine idling. When there is no vacuum at the PCV valve, the valve, hose, or manifold inlet are restricted. Replace the restricted component, or components.

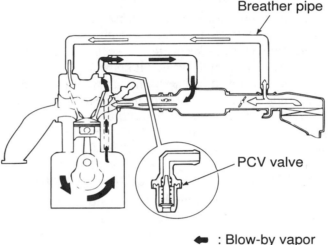

Figure 1-41   PCV system *(Courtesy of American Honda Motor Co., Inc.)*

Remove the PCV valve from the rocker arm cover and the hose. Shake the valve beside your ear, and listen for the tapered valve rattling inside the valve housing. If no rattle is heard, PCV valve replacement is required.

Task 6   **Inspect, service, and replace positive crankcase ventilation (PCV) valves/ metering devices.**

72. Technician A says the PCV valve moves toward the closed position when the throttle is opened from the idle to half throttle.
Technician B says the PCV valve is moved toward the closed position by spring tension.
Who is correct?
    A. A only
    B. B only
    C. Both A and B
    D. Neither A nor B

Hint   *Vehicle manufacturers recommend different PCV valve checking procedures. Always follow the procedure in the vehicle manufacturer's service manual. Some vehicle manufacturers recommend removing the PCV valve from the rocker arm cover and the hose. Connect a length of hose to the inlet side of the PCV valve and blow air through the valve with your mouth while holding your finger near the valve outlet. Air should pass freely through the valve. If air does not pass freely through the valve, replace the valve.*

Connect a length of hose to the outlet side of the PCV valve, and try to blow back through the valve. It should be difficult to blow air through the PCV valve in this direction. When air passes easily through the valve, replace the valve.

Task 7   **Inspect, service, and replace positive crankcase ventilation (PCV) filters (breather cap), tubes, and hoses.**

73. Technician A says the PCV system may draw unfiltered air into the engine through a leaking rocker arm cover gasket.
    Technician B says the PCV system may draw unfiltered air into the engine through a loose oil filler cap.
    Who is correct?
    A. A only
    B. B only
    C. Both A and B
    D. Neither A nor B

Hint   *Leaks at engine gaskets, such as rocker arm cover or crankcase gaskets, will result in oil leaks, and the escape of blowby gases to the atmosphere. However, the PCV system also draws unfiltered air through these leaks into the engine. This action could result in wear of engine components, especially when the vehicle is operating in dusty conditions.*

When diagnosing a PCV system, the first step is to check all the engine gaskets for signs of oil leaks. Be sure the oil filler cap fits and seals properly. Check the clean air hose, and the PCV hose for cracks, deterioration, loose connections, and restrictions.

Check the PCV clean air filter for contamination, and replace this filter if necessary. If there is evidence of oil in the air cleaner, check the PCV valve and hose for restriction.

Task 8   **Inspect turbochargers for proper operation; repair as necessary.**

74. Reduced turbocharger boost pressure may be caused by
    A. a wastegate valve stuck closed.
    B. a wastegate valve stuck open.
    C. a leaking wastegate diaphragm.
    D. a disconnected wastegate linkage.

Hint   *Excessive blue smoke in the exhaust may indicate worn turbocharger seals. The technician must remember that worn valve guide seals, or piston rings, also cause oil consumption and blue smoke in the exhaust.*

Check for intake system leaks. If there is a leak in the intake system before the compressor housing, dirt may enter the turbocharger and damage the compressor, or turbine, wheel blades. When a leak is present in the intake system between the compressor wheel housing and the cylinders, turbocharger pressure is reduced.

Turbocharger boost pressure may be tested with a pressure gauge connected to the intake manifold. The boost pressure should be tested during hard acceleration while driving the vehicle.

Battery and Starting System Diagnosis and Repair

ASE Tasks, Questions, and Related Information

Task 1   **Inspect, test, clean, fill, or replace battery.**

75. During the battery test in Figure 1-42
    A. the battery should be discharged at two-thirds of the cold cranking rating.
    B. the battery should be discharged at one-half of the amp-hour rating.
    C. the voltage should remain above 9.6V at 70°F battery temperature.
    D. the load should be applied to the battery for 20 seconds.

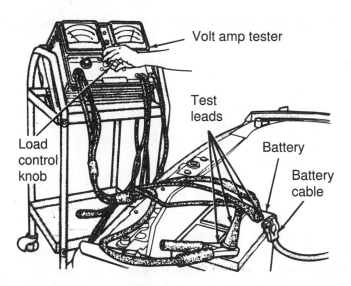

Figure 1-42 Battery test *(Courtesy of Chrysler Corporation)*

Hint   *The hydrometer tests the specific gravity of the battery electrolyte. A fully charged battery should have a specific gravity of 1.265. A voltmeter may be connected across the battery terminals to test the open circuit battery voltage. The specific gravity and the open circuit voltage indicate the battery state of charge (Figure 1-43).*

State of charge	Specific gravity*	Open-circuit voltage
100%	1.265	12.6
75%	1.225	12.4
50%	1.190	12.2
25%	1.155	12.0
Dead	1.120	11.9

Figure 1-43 Battery specific gravity and open circuit voltage *(Courtesy of Toyota Motor Corporation)*

A battery load or capacity tester is used to test the battery capacity. During this test the battery is discharged at one-half the cold cranking amperes for 15 seconds, and the voltage is recorded at the end of this time. A satisfactory battery at 70°F has 9.6V at the end of the capacity test.

Task 2   Slow and fast charge battery.

76. All of these statements about battery charging are true EXCEPT
   A. the battery is completely charged when the specific gravity reaches 1,225.
   B. explosive hydrogen gas is produced in the battery cells during the charging process.
   C. oxygen gas is produced in the battery cells during the charging process.
   D. do not charge a maintenance free battery if the hydrometer in the battery top indicates yellow.

Hint   *A battery may be slow charged at one-tenth of the amp-hour rating. For example, an 80 amp-hour battery should be charged at 8 amps. The battery should be charged until the specific gravity is above 1.250, or until there is no further increase in specific gravity, or battery voltage, in one hour.*

Maintenance-free batteries require different charging rates compared to low-maintenance batteries. Some vehicle manufacturers recommend using the reserve capacity rating of the battery to determine the amp-hours charge required by the battery. For example, a battery with a reserve capacity of 75 minutes requires 25 amps × 3 hours charge = 75 amp-hours. Continue charging the battery until the green dot appears in the hydrometer, and the battery passes a capacity test.

Task 3   Jump start a vehicle with jumper cables and a booster battery or auxiliary power supply.

77. While discussing battery boosting
Technician A says the negative booster cable should be connected to the negative terminal of the battery in the vehicle being boosted.
Technician B says the negative booster cable should be connected before the positive booster cable.
Who is correct?
A. A only
B. B only
C. Both A and B
D. Neither A nor B

Hint   *When jump starting a vehicle always turn off all electrical accessories in the vehicle being boosted and the boost vehicle. Electrical accessories left on during the boost procedure may be damaged. A booster battery must be connected with the proper polarity, and the correct procedure. Reversed battery polarity, or improper procedure, may result in severe electrical system damage on boost vehicle, and the vehicle being boosted. This procedure may also result in battery explosion, personal injury, and property damage. Do not boost a vehicle if the battery built-in hydrometer indicates clear, or light yellow. This action may result in battery explosion with resulting personal injury and property damage. Always follow the battery boost procedure in the vehicle manufacturer's service manual.*

Task 4   Inspect, clean, repair, and replace battery cables and clamps.

78. All these statements about battery cable service are true EXCEPT
A. the terminals should be removed from the battery with a puller.
B. remove the negative cable before the positive cable.
C. remove the positive cable before the negative cable.
D. after removal wash the battery with a baking soda and water solution.

Hint   *After the battery terminal nuts have been loosened, the battery terminals may be removed with a terminal puller. Do not hammer or twist battery cable ends to loosen them. This action will loosen the terminal in the battery cover, and cause electrolyte leaks. Battery terminals and cable ends may be cleaned with a battery terminal cleaner.*

Task 5   Inspect, test, and /or replace starter relays, solenoids, and circuits.

79. In Figure 1-44 the resistance is being tested in
A. the starter ground circuit.
B. the solenoid circuit.
C. the battery to starter circuit.
D. the starter relay circuit.

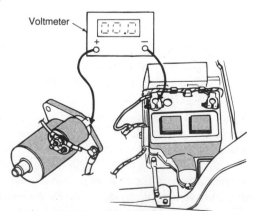

Voltmeter

Figure 1-44   Starting circuit test *(Courtesy of Chrysler Corporation)*

Hint   *Most starter relays, solenoids, and circuits may be tested by measuring the voltage drop across the various components and wires with the starting motor cranking the engine. If the voltage drop across any component, or connecting wire, exceeds manufacturer's specifications, replace the component or wire.*

Task 6   Test, remove, and/or replace starter.

80. During the test in Figure 1-45 the starter current is less than specified and the cranking speed is slower than normal. This problem indicates
    A. the field coils are shorted.
    B. the armature windings are shorted.
    C. the field coils have high resistance.
    D. the solenoid windings are open.

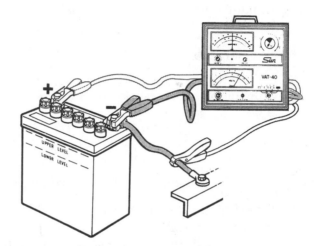

Figure 1-45   Battery-starter tester connections for starter current draw test *(Courtesy of Toyota Motor Corporation)*

Hint   *The engine must be at normal operating temperature and the battery fully charged before a starter current draw test is performed. The ignition system must be disabled prior to the starter current draw test.*

...wers and Analysis

1. B   The easiest, quickest test should be performed first, therefore A is wrong. The customer complaint must be identified during a basic diagnostic procedure. B is the correct answer.

2. C   Technicians A and B are both correct because neither a leaking head gasket nor heater core will cause leaks in the engine compartment, but they will reduce the test pressure. C is the right answer.

3. B   Worn pistons and cylinders would cause a rapping noise while accelerating, and worn main bearings cause a thumping noise when the engine is started. Therefore, A and C are wrong.

Loose camshaft bearings usually do not cause a noise unless severely worn, so D is wrong.

Loose flywheel bolts may cause a thumping noise at idle. B is the correct answer.

4. A   High fuel pressure or a restricted return fuel line causes a rich air-fuel ratio. Therefore, B and C are wrong.

A sticking fuel pump check valve may cause hard starting, so D is wrong.

A burned exhaust valve causes a "puff" noise in the exhaust. A is right.

5. D   Late ignition timing, or and intake manifold vacuum leak, causes a low steady vacuum gauge reading. Therefore, A and B are wrong.

A restricted exhaust system causes a low vacuum reading at high speeds, so C is wrong.

Sticking valves cause a vacuum gauge fluctuation between 12 in. and 18 in. Hg. at idle. D is the correct answer.

6. C   A cylinder with very little rpm drop indicates cylinder misfiring which may be caused by the ignition system or an intake manifold vacuum leak. Therefore, both A and B are right, and C is the correct answer.

7. B   If the compression increases significantly during a wet test, worn rings and cylinders are indicated. Therefore, A, C, and D are wrong; B is the right answer.

8. B   A leaking intake valve causes air to escape from the air intake. Therefore, A is wrong.

Worn rings cause air to escape from the PCV valve during the cylinder leakage test. B is the right answer.

9. C   If the cylinder head warpage is 0.014 in. the head must be resurfaced. Block warpage may have caused the head distortion. Therefore, A and B are correct, and C is the right answer.

10. C   Torque-to-yield bolts do provide more clamping force, because they are tightened to a specific torque and then rotated a certain number of degrees. A and B are both correct, and C is the right answer.

11. D   The valve spring must be rotated while measuring height and squareness. Neither A nor B are correct; D is the right answer.

12. C   The tool table must be set at the specified valve spring height, and the torque wrench pulled until the tool clicks. At this time the torque wrench reading is multiplied by 2 to indicate the spring tension. Therefore, A, B, and D are wrong, and C is right.

13. A   Worn valve spring locks will not cause a clicking noise because they are continually loaded by the valve spring tension. Therefore, B is wrong.

Worn valve lock grooves may cause the locks to fly out of place with the engine running causing severe engine damage. A is right.

14. B   Worn valve stem seals allow oil to run down the stems into the cylinder causing excessive oil consumption, but worn seals will not increase stem and guide wear. Therefore, A is wrong, and B is the right answer.

15. C   Valve stems should be measured at three locations, and valve guides are measured with a snap or hole gauge. Both A and B are correct, and C is the right answer.

16. D   Valve stem to guide clearance is measured in Figure 1-5. Therefore, A, B, and C are wrong, and D is the right answer.

17. B   A reduced valve margin causes valve overheating and burning. This problem does not cause improper valve seating, valve seat recession, or a clicking noise at idle speed. Therefore, A, C, and D are wrong, and B is the right answer.

18. A   Valve 1 indicates normal wear, valve 2 indicates partial valve rotation, and valve 3 indicates complete lack of valve rotation. Therefore, B, C, and D are wrong, and A is the right answer.

19. C   Most valve seats have an angle of 30° or 45°. Therefore, A, B, and D are wrong, and C is the right answer.

20. A   Improper valve seat contact area on the valve face does not cause a clicking noise at idle speed. B is wrong. When the valve seat contact area is too low on the valve face, 60° and 45° stones are used to move the seat contact area upward. A is the right answer.

21. B   Excessive installed valve spring height would not bottom the lifter unless the installed valve stem height is also more than specified. Therefore, A is wrong.

If the installed valve spring height is more than specified, a shim may be installed under the spring to correct this problem. B is right.

22. C   Bent pushrods may be caused by sticking valves or excessive valve spring tension. Improper valve timing may cause the valves to strike the pistons in some engines resulting in bent pushrods. Therefore, both A and B are correct, and C is the right answer.

23. B   Lifter bottoms must be convex. Therefore, A is wrong.

A sticking lifter plunger may hold the valve open resulting in a burned valve. B is the right answer.

24. B   When measuring valve clearance the piston must be TDC compression with both valves closed. Therefore, A is wrong. Some mechanical lifters have removable shim pads. B is the right answer.

25. A   When the timing gear teeth are meshed with the crankshaft gear teeth the gear backlash is measured with a dial indicator. Therefore, B is wrong, and A is the right answer.

26. A   On some engines the cam lobe lift is measured with a dial indicator positioned against the top of the pushrod. Therefore, B, C, and D are wrong, and A is right.

27. A   Excessive camshaft bearing clearance causes low oil pressure, but this problem does not cause a clicking noise with the engine idling. Therefore, B is wrong, and A is the right answer.

28. C   Improper valve timing may cause the valves to strike the pistons in some engines causing bent valves, and this problem also reduces engine power. Both A and B are correct, and C is the right answer.

29. D   The timing gear cover seal contains a garter spring, and the seal lip must face toward the oil flow. Therefore, A and B are both wrong, and D is the right answer.

30. B   RTV sealer cures in five minutes so components using this sealer must be assembled quickly. Chlorinated solvent must be used to clean RTV sealed components. The RTV bead must surround any bolt holes. Therefore, A, C, and D are wrong.

The RTV bead 1/8 in. wide should be placed in the center of the sealing surface. B is correct.

31. C   A warped head mounting surface on the block may cause valve seat distortion when the head bolts are torqued, and this problem may also cause coolant and combustion leaks. Therefore, both A and B are correct, and C is the right answer.

32. D   The first step in heli-coil installation is to drill the opening, and a special tool is used to install the heli-coil. Therefore, both A and B are wrong, and D is the right answer.

33.   D   Installing new rings without removing the ring ridge may cause broken ring lands. Therefore, A, B, and C are wrong, and D is the right answer.

34.   D   Cylinder taper is the difference between the cylinder diameter at the top of the ring travel compared to the diameter at the bottom of the ring travel. Therefore, A is wrong.

Cylinder out-of-round is the difference between the axial bore diameter and the thrust bore diameter at the same cylinder position. Therefore, B is also wrong, and D is the right answer.

35.   B   Honing stones with a lower number have a coarser grit. Cylinders may be deglazed with 220–280 grit stones, and the cylinder should be washed with soapy water after deglazing. Therefore, A, C, and D are wrong, and B is the right answer.

36.   C   The tool in Figure 1-24 is used to remove and install camshaft bearings. Therefore, A, B, and D are wrong, and C is the right answer.

37.   C   A and B measure vertical taper, C and D measure horizontal taper, A and C measure out-of-round, A and D is not a valid measurement. Therefore, A, B, and D are wrong, and C is the right answer.

38.   C   Bearing spread is the slightly larger curvature of the bearing insert compared to the bearing bore. Bearing crush is the bearing design that allows the bearing edges to be extended slightly above the mounting area when the bearing is installed. Therefore, both A and B are correct, and C is the right answer.

39.   A   A bent connecting rod causes uneven connecting rod bearing wear, but this problem does not influence main bearing, piston pin, or cylinder wall wear. Therefore, B, C, and D are wrong, and A is the right answer.

40.   C   The tool in Figure 1-30 is a ring removal and replacement tool. Therefore, A, B, and D are wrong, and C is right.

41.   D   Ring gap should be measured with the ring at the bottom or the ring travel, and compression rings must not be interchanged on most pistons. Therefore, both A and B are wrong, and D is the right answer.

42.   A   If the damper seal contact area is scored, the damper hub may be machined and a sleeve installed to provide a new seal contact area. Therefore, B is wrong.

The vibration damper counterbalances the back-and-forth twisting crankshaft motion. A is the right answer.

43.   C   Metal burrs on the crankshaft flange cause flywheel and torque converter misalignment, and this flywheel misalignment may cause excessive ring gear and starter drive gear wear. Therefore, both A and B are correct, and C is the right answer.

44.   C   Flywheel runout causes an uneven clutch plate mating surface resulting in clutch grabbing, and the pressure plate must be installed in the original position to maintain proper balance. Therefore, both A and B are correct, and C is the right answer.

45.   D   With the clutch engaged, the clutch plate is held firmly between the flywheel and pressure plate, and the crankshaft and complete clutch assembly is turning together. Under this condition the pilot bearing inner and outer races are turning at the same speed, thus preventing bearing noise under this condition. Therefore, A, B, and C are wrong.

With the clutch pedal depressed, the clutch disc and input shaft are free to move up and down in the worn pilot bearing resulting in a rattling noise. D is the right answer.

46.   A   The balance shafts are timed in relation to the crankshaft, and improper balance shaft timing results in severe engine vibrations. Therefore, B is wrong and A is the right answer.

47.   D   Worn camshaft bearings, crankshaft bearings, or a weak pressure regulator spring would cause low oil pressure. Therefore, A, B, and C are wrong.

Restricted pushrod passages will not cause low oil pressure, therefore response D is the right answer.

48. A   B, C and D are valid oil pump measurements, and so these responses are wrong.

If the clearance between the rotors is normal, there is no need to measure the inner rotor diameter. Therefore, A is not a normal measurement, and this is the correct answer.

49. C   A, B, and D are valid cooling system tests with the pressure tester. Therefore, these are not the requested answer.

The cooling system pressure tester does not measure coolant specific gravity, and C is the requested answer.

50. A   A squealing on acceleration is also a result of a loose alternator belt. Therefore, B is wrong.

A loose alternator belt would not damage the alternator bearing, and so C is wrong.

The alternator belt may not drive the water pump, so a loose alternator belt may not cause engine overheating. D is wrong.

A loose alternator belt reduces alternator output and causes a discharged battery. A is the right answer.

51. B   A defective pressure release valve in the radiator cap allows coolant to escape from the radiator to the recovery reservoir. Therefore, A is wrong.

A plugged recovery reservoir hose prevents coolant flow from the reservoir into the radiator as the engine cools down resulting in a collapsed upper hose. B is right.

52. A   If the thermostat is stuck open, the coolant temperature remains lower than normal and the coolant temperature sensor informs the computer regarding this low coolant temperature. The computer provides a richer air-fuel ratio in relation to the low coolant temperature resulting in reduced fuel economy. Therefore, B, C, and D are wrong, and A is the right answer.

53. B   When the cooling system pressure is reduced, the coolant boiling point is decreased. Therefore, A is wrong.

Adding more antifreeze to the coolant solution, increases the boiling point. B is the correct answer.

54. C   A defective water pump bearing may cause a growling noise at idle speed, and the bearing may have been contaminated by coolant leaking past the pump seal. Therefore, both A and B are correct, and C is the right answer.

55. B   Responses A, C, or D cause engine overheating and excessive coolant in the recovery reservoir. Therefore, these responses are not the requested answer.

A thermostat that is stuck open reduces coolant temperature and does not cause high coolant level in the recovery reservoir. Since B is not a cause of the problem, this is the right answer.

56. D   If the temperature switch ground circuit is open and the A/C is off, the circuit from the fan relay winding to ground cannot be completed resulting in an inoperative fan motor and engine overheating. When the A/C is on, the fan relay winding may be grounded through the condenser switch to provide cooling fan operation and some reduction in engine temperature. Therefore, A, B, and C are wrong, and D is right.

57. A   An engine oil cooler reduces oil temperature which prevents oil oxidation. Therefore, B, C, and D are wrong, and A is the right answer.

58. B   With the test spark plug connected to the coil secondary wire the cap and rotor are not subjected to secondary voltage and cannot affect this voltage. Therefore, A is wrong.

Since the 12V test light is flashing while cranking, the primary circuit is triggered on and off. If the coil does not fire the test spark plug, the coil must be defective. B is the correct answer.

59. D   With the ohmmeter leads connected from the pickup lead to ground an infinite reading indicates the pickup is not grounded. A low meter reading would indicate a grounded pickup. Therefore, A, B, and C are wrong, and D is the right answer.

60. C   The vacuum advance provides spark advance in relation to engine load, and the mechanical advance supplies spark advance in relation to engine rpm. The vacuum advance rotates the pickup plate in the opposite direction to shaft rotation. Therefore, statements A, B, and D are correct.

The mechanical advance rotates the reluctor ahead of the distributor shaft in the same direction as shaft rotation. Since statement C is not correct, this is the right answer.

61. B   An increase in point gap decreases the cam dwell, and thus A is wrong.

The dwell must be long enough to allow the magnetic field to build up in the coil. B is the right answer.

62. B   An open spark plug wire, or wide spark plug gaps increase the normal required secondary voltage, but do not affect the maximum secondary voltage. Therefore, C and D are wrong.

Low primary resistance increases primary current and magnetic strength resulting in more secondary energy, and so A is wrong.

Low primary voltage reduces primary current flow and magnetic strength resulting in low maximum secondary coil voltage. B is the right answer.

63. C   The ohmmeter in Figure 1-40 is connected to test the secondary winding for shorts or open circuits. Therefore, A, B, and D are wrong, and C is the right answer.

64. B   Late ignition timing may cause engine overheating. Excessive timing causes engine detonation. Therefore, A is wrong, and B is the right answer.

65. A   The pickup gap must be measured with a nonmagnetic feeler gauge (such as brass), so B, C, and D are wrong, and A is right.

66. C   Some ignition modules have to be grounded to the chassis, and other modules may operate when cold but fail when they are hot. Both A and B are correct, and C is the right answer.

67. B   The fuel system should be depressurized prior to a fuel pump test, the pressure gauge should be connected to the Schrader valve on the fuel rail, and lower than specified fuel pressure may be caused by a restricted fuel filter. Therefore, statements A, C, and D are correct.

Low fuel pressure cause a lean air-fuel ratio. Therefore, statement B is wrong, and this is the correct answer.

68. A   An intake manifold vacuum leak may cause a cylinder misfire at idle speed, but not during acceleration when the manifold vacuum is reduced. Therefore, B is wrong, and A is the right answer.

69. D   When blowing out an air filter element the air gun should be 6 in. from the inside of the element. Therefore, A, B, and C are wrong, and D is right.

70. D   Statements A, B, and C are correct, and so these are not the requested answer.

A manifold heat control valve stuck in the closed position increases intake manifold temperature. Therefore, D is the only incorrect response, and this is the requested answer.

71. B   When the hose between the PCV valve and the intake manifold is restricted, crankcase pressure increases causing oil accumulation in the air cleaner. Therefore, A, C, and D are wrong, and B is the right answer.

72. D   The PCV valve moves toward the open position as the throttle is opened from idle to half throttle, or half throttle to wide open throttle. Therefore, both A and B are wrong, and D is the correct answer.

73. C   The PCV valve may draw unfiltered air into the engine through a leaking rocker arm cover gasket or through a loose oil filler cap. The PCV valve is moved toward the closed position by intake manifold vacuum. Therefore, both A and B are correct, and C is the right answer.

74. B   A waste gate that is stuck closed, a leaking wastegate diaphragm, or a disconnected wastegate linkage, increases boost pressure. Therefore, A, C, and D are wrong.

Reduced turbocharger boost pressure may be caused by a wastegate that is stuck open, and so B is the correct answer.

75. C   The discharge rate is one-half the cold cranking rating for 15 seconds during a battery load test, and so A, B, and D are wrong.

During the battery load test the battery voltage should remain above 9.6V at 70°F, and so C is the right answer.

76. A   Statements B, C, and D are correct, and so these statements are not the requested answer. A battery is fully charged when the electrolyte is 1,265 specific gravity. Therefore, statement A is incorrect, and this is the requested answer.

77. D   While battery boosting the negative cable should be connected to a ground on the engine of the vehicle being boosted, and the negative cable should be connected last. Therefore, both A and B are wrong, and D is the right answer.

78. C   Statements A, B, and D are correct, and so these statements are not the requested answer.

The negative battery cable should always be disconnected first. Therefore, statement C is incorrect, and this is the requested answer.

79. A   The voltmeter leads are connected from the starting motor ground to the battery ground to test voltage drop and resistance in the starter ground. Therefore, B, C, and D are wrong, and A is the right answer.

80. C   Low current draw, low cranking speed, and high cranking voltage usually indicate excessive resistance in the starting circuit. Therefore, A, B, and D are wrong, and C is the right answer.

Glossary

Abrasion Wearing or rubbing that damages the surface area of a part.
Abrasión El desgaste o frotamiento de una parte que daña la superficie de una área.

Additive A material added to the engine oil to provide additional properties not originally found in the oil.
Aditivo Una materia añadida al aceite de motor para proporcionar unas propiedades adicionales que no se encuentran originalmente en el aceite.

Adhesion The property of oils to cling to surfaces.
Adhesión La propriedad de los aceites que permite que se adhieren a las superficies.

AERA Automotive Engine Rebuilder's Association.
AERA La Asociación de Reacondicionadores Automotivos.

Aerobic sealants Sealants that require the presence of oxygen to cure.
Sellantes aeróbicos Los sellantes que requieren la presencia del oxígeno para curarse.

Aftermarket Equipment and parts sold to consumers after the production of the vehicle.
Repuestos no originales Los accesorios y partes vendidos al consumidor después de la producción de un vehículo.

Align bore The process of boring, reaming, or honing the main bearing bores to center all bores onto a true centerline.
Calibrado en serie El proceso de taladrar, escariar o bruñir los taladros de los muñones del cigüeñal para alinearlos todos en una linea recta central.

Align honing Machining process of the main bearing journals used to remove in excess of 0.050 inch, but this will result in changing the location of the crankshaft in the block. Align honing restores original bore size by removing metal from the entire circumference of the bore.
Rectificado en serie Un proceso de rectificar en máquina los muñones del cigüeñal para quitar más de 0.050 de una pulgada, pero ésto resulta en que cambia de lugar el cigüeñal en el monoblock. El rectificado en serie restaura el tamaño original del taladro removiendo el metal de toda su circunferencia.

Aligning bars Tools used to determine the proper alignment of the crankshaft saddle bores.
Barras para alinear Las herramientas que sirven para determinar el alineamiento correcto de los taladros del asiento del cigüeñal.

Alkaline A chemical class which is the opposite of acidic, but is just as corrosive to certain materials.
Alcalino Un tipo de químico que es lo contrario del ácido, pero igualmente corrosivo para algunas materiales.

Alloys Mixtures of two or more metals. For example, brass is an alloy of copper and zinc.
Aleación La mezcla de dos o más metales. Por ejemplo, el latón es una aleación de cobre con zinc.

Aluminum hydroxide Corrosion products that are carried to the radiator and deposited when they cool off. They appear as dark grey when wet and white when dry.
Hidróxido de aluminio Los productos de corrosión que se llevan al radiador y se depositan al enfriarse. Son de color gris obscuro mojados y se blanquean al secarse.

Anaerobic sealants Sealants that will only cure in the absence of oxygen.
Sellantes anaeróbicos Los sellantes que sólo se curan en la ausencia del oxígeno.

Antifriction bearing A bearing constructed with balls or rollers between the journal and the bearing surface.
Cojinete antifricción Un cojinete construido con bolas o rodillos entre las superficies de apoyo.

Atmospheric pressure The weight of the air.
Presión atmosférica El peso del aire.

Austenitic steel A corrosion resistant steel made with carbide or carbon alloys.
Acero austenítico Un acero resistente a la corrosión hecho con el carburo o las aleaciones de carbono.

Backlash The clearance between two parts.
Culateo/juego La holgura entre dos partes.

Backpressure The pressure created within the engine cylinder as a result of a restricted exhaust system. With the creation of backpressure, vacuum cannot be produced as efficiently.
Contrapresión La presión creada dentro del cilindro del motor causada por una restricción en el sistema de escape. Al crear esta presión, el vacío no se puede producir eficazmente.

Balance Weight introduced to prevent vibration of moving parts.
Equilibrio Un peso metido para prevenir la vibración de las partes en movimiento.

Balancing The process of removing vibrations which are caused by the entire engine's reciprocating and rotational mass.
Equilibrar El proceso de quitar las vibraciones causadas por la masa recíproca y rotativa del motor.

Battery terminal test A test which checks for poor electrical connections between the battery cables and terminals. Use a voltmeter to measure voltage drop across the cables and terminals.
Prueba de los terminales de la batería Una prueba que averigua las conexiones eléctricas malas entre los cables y los terminales de la batería. Usa un voltímetro para medir la caída del voltaje entre los cables y los terminales.

Bead blasters Parts cleaners which use beads of abrasive media carried by air pressure to clean the parts. The abrasives knock off the contaminates.
Chorro de perlitas Una limpiadora de partes que usa una materia abrasiva impulsada por el aire bajo presión para limpiar las partes. Los abrasivos desprenden los contaminantes.

Bearing Soft metallic shells used to reduce friction created by rotational forces.
Cojinete Una pieza hueca de metal blanda que sirve para reducir la fricción creada por las fuerzas giratorias.

Bearing cap The removable half of the saddle that holds the bearing in place.
Tapa del cojinete La mitad removible del asiento que sostiene al cojinete en su lugar.

Bearing crush The extension of the bearing half beyond the seat which is crushed into place when the seat is tightened.
Aplastamiento del cojinete La parte de extensión del cojinete debajo del asiento que se aplasta en su lugar al apretar el asiento.

Bearing lining A layer of alloy which is adhered to the bearing back and forms the bearing surface.
Revestimiento del cojinete Una capa de aleación que se adhiere a la parte exterior de un cojinete y forma la superficie del cojinete.

Bearing spread The distance between the outside parting edges is larger than the diameter of the bore.
Envergadura del cojinete La distancia al través de los rebordes exteriores es más grande que el diámetro del taladro.

Big end The end of the connecting rod that attaches to the crankshaft.
Extremo grande Refiere a la extremidad de la biela que se conecta al cigüeñal.

Blow-by The unburned fuel and combustion products that leak past the piston rings and enter the crankcase.
Soplado El combustible no consumido y los productos de la combustión que escapen por los anillos de pistón y entran al cárter.

Blueprinting A technique of building an engine using stricter tolerances than those used by most manufacturers. This results in a smoother running, longer lasting, and higher output engine.
Especificado por el plan detallado Una técnica de construcción del motor usando las tolerancias más exactas de las que usan la mayoría de los fabricantes. Esto resulta en un motor que marcha mejor, dura más, y de mejor rendimiento.

Bob weights Devices that are attached to the throws of the crankshaft to simulate the rotating and reciprocating mass of the piston assembly.
Pesos de contra-balanzón Los dispositivos que se conectan al brazo excéntrico del cigüeñal para simular la masa giratoria y recíproca del conjunto de los pistones.

Bore The diameter of a hole.
Taladro El diámetro de un agujero.

Borescope Special tool that uses fiber optics to allow the technician to see the internal condition of the engine without having to disassemble it.
Calibrescopio Una herramienta especial que usa la tecnología fibro-óptica para permitir que el técnico vea las condiciones internas del motor sín tener que desarmarlo.

Boring The process of enlarging a hole.
Escariar El proceso de agrandar un agujero.

Boss The cast or forged part of a piston that can be machined for balance.
Resalto La parte colada o forjada de un pistón que se puede rebajar por máquina para el equilibrio.

Bottom Dead Center (BDC) Term used to indicate the piston is at the very bottom of its stroke.
Punto Muerto Inferior (PMI) El término que indica que el pistón esta en el punto más inferior de su carrera.

Brake horsepower The usable power produced by an engine.
Caballos al freno La potencia disponible producido por un motor.

Broach To finish the inside surface of a bore by forcing a multiple-edge cutting tool through it.
Fresar con barrena Acabar una superficie interior de un taladro al atravesarla con una herramienta que tiene múltiples hojas cortantes.

Burned valves Valves that have warped and melted, leaving a groove across the valve head.
Válvulas quemadas Las válvulas que se han deformado y fundido, dejándo una ranura al través de la cabeza de la válvula.

Burnish To smooth or polish with a sliding tool under pressure.

Bruñir Alisar o pulir con una herramienta deslizandola bajo presión.

Bushing A removable liner for a bearing.
Casquillo Un forro removable para un cojinete.

By-pass valve A safety feature to prevent engine failure. The valve opens when there is a pressure differential of 5 to 15 psi between the outside and inside of the filter element.
Válvula de paso Un componente de seguridad para prevenir los fallos del motor. La válvula se abre cuando hay una diferencial de presión de entre 5 a 15 libras por pulgada cuadrada entre el exterior y el interior del elemento de filtro.

Calibrate To determine the scale of an instrument giving quantitative measurements.
Calibrar Determinar la escala de un instrumento dando medidas cuantitativas.

Caliper Measuring tool capable of taking readings of inside, outside, depth, and step measurements in 0.001 inch increments.
Calibre Una herramienta de medir capaz de tomar las medidas interiores, exteriores, de profundidad y de paso en incrementos de 0.001 de una pulgada.

Cam ground pistons Pistons cast or forged into a slight oval or cam shape to allow for expansion. As the piston warms it will become round.
Pistones rectificados por leva Los pistones colados o forjados en una forma lijeramente ovulada o en forma de la leva para permitir la expansión. Al calentarse el piston se redondea.

Camshaft The shaft containing lobes to operate the engine valves.
Arbol de levas Un eje que tiene lóbulos que operan las válvulas del motor.

Camshaft degreeing The altering of the point where the camshaft activates the valves in relation to the crankshaft.
Graduación del árbol de levas Cambiar el punto en donde el árbol de levas acciona las válvulas con respeto al cigüeñal.

Capacity test A test which checks the battery's ability to perform when loaded.
Prueba de capacidad Un prueba que averigua la habilidad de la bateria a funcionar bajo carga.

Carbon A nonmetallic element that forms inside of the combustion chamber as a product of burning fuel.
Carbono Un elemento nometálico que forma en el interior de la cámara de combustión como un producto del combustible al quemarse.

Carbon monoxide An odorless, colorless, and toxic gas that is produced as a result of incomplete combustion.
Monóxido de carbono Un gas sin olor, sin color y tóxico que se produce como resultado de una combustión incompleta.

Carbonize The process of carbon formation.
Carbonizar El proceso de formar el carbono.

Carburetor A fuel delivery device that mixes fuel and air to the proper ratio to produce a combustible gas.
Carburador Un dispositivo para entregar el combustible que mezcla el combustible y el agua en proporciones correctas para producir un gas combustible.

Caustic solutions Cleaning solutions usually consisting of a mixture of water, sodium hydroxide, and sodium carbonate. This solution is extremely alkaline with a pH rating of 10 or more.
Soluciones cáusticas Las soluciones de limpieza normalmente compuestas de una mezcla del agua, el óxido de sodio, y el carbonato sódico. Esta solución es extremadamente alcalino con una clasificación PH de 10 o más.

Cc-ing A method of measuring the volume of the combustion chamber by measuring the amount of oil the chamber can hold.
Midiendo en centímetros cúbicos Un método para medir el volumen de la cámara de combustión al medir la cantidad de aceite que puede contener la cámara.

Centrifugal force A force that tends to move a body away from its center of rotation.
Fuerza centrífuga Una fuerza cuya tendencia es de mudar un cuerpo fuera de su centro de rotación.

Chamfer A bevel or taper at the edge of a bore.
Chaflán Un bisel o chaflán en el borde de un taladro.

Channeling Local leakage around a valve head caused by extreme temperatures developing at isolated locations on the valve face and head.
Acanalado Una fuga local alrededor de una cabeza de válvula causada por las temperaturas severas que ocurren en puntos aislados en la cara y la en cabeza de la válvula.

Chemical cleaning The process of using chemical action to remove soil contaminants from the engine components.
Limpieza química El proceso de usar la acción química para desprender los contaminantes sucios de los componentes del motor.

Clearance The space allowed between two parts.
Holgura El espacio permitido entre dos partes.

Clearance volume The volume of the combustion chamber when the piston is at TDC. The size of the clearance volume is a factor in determining compression ratio.
Volumen de la holgura El volumen de la cámara de combustión cuando el pistón esta en PMS. El tamaño del volumen de la holgura es un factor en determinar el índice de compresión.

Combustion The process of burning.
Combustión El proceso de quemar.

Combustion chamber The volume of the cylinder above the piston with the piston at TDC.
Cámara de combustión El volumen del cilindro arriba del piston cuando el piston esta en PMS.

Composites Made-made materials using two or more different components tightly bound together. The result is a material consisting of characteristics that neither component possesses on its own.
Compuestas Las materiales artificiales que usan dos o más componentes distinctos combinados estructuralmente. El resultado es una material que tiene las características que ninguno de los componentes posee individualmente.

Compound A mixture of two or more ingredients.
Compuesta Una mezcla de dos o más ingredientes.

Compression The reduction in volume of a gas.
Compresión La reducción del volumen de un gas.

Compression ratio A comparison between the volume above the piston at BDC and the volume above the piston at TDC.
Indice de compresión Una comparación entre el volumen arriba del piston en el PMI y el volumen arriba del piston en el PMS.

Compression rings The upper rings of the piston designed to hold the compression in the cylinder.
Anillos (aros) de compresión Los anillos superiores de un piston diseñados a mantener la compresión dentro del cilindro.

Compression testing A diagnostic test to determine the engine cylinder's ability to seal and to maintain pressure.
Prueba de compresión Una prueba diagnóstica que determina la habilidad del cilindro del motor de sellar y mantener la presión.

Concentric Two or more circles having a common center.
Concéntrico Dos o más círculos que comparten un centro común.

Conductor A material capable of supporting the flow of electricity.
Conductor Una materia que permite el flujo de la electricidad.

Conformability The ability of the bearing material to conform itself to slight irregularities of a rotating shaft.
Conformidad La habilidad de una material de un cojinete de conformarse a las pequeñas irregularidades de un eje giratorio.

Connecting rod The link between the piston and crankshaft.
Biela La conexión entre el pistón y la cigüeñal.

Contraction A reduction of mass.
Contracción Una reducción de la masa.

Convection A transfer of heat by circulating heated air.
Convección Una transferencia del calor por medio de la circulación del aire calentado.

Coolant The liquid that is circulated through the engine to absorb heat and transfer it to the atmosphere.
Fluido refrigerante El líquido circulado por el motor que absorba el calor y lo transfere a la atmósfera.

Cooling fans Fans used to force air flow through the radiator to help in the transfer of heat from the coolant to the air.
Ventiladores de enfriamiento Los ventiladores que sirven para forzar un corriente de aire por el radiador para transferir el calor del fluido refrigerante al aire.

Core plugs Metal plugs screwed or pressed into the block or cylinder head at locations where drilling was required or sand cores were removed during casting.
Tapones del núcleo Los tapones metálicos fileteados o prensados en el monoblock o en la cabeza de los cilindros ubicados en donde se había requerido taladrar o quitar los machos de arena durante la fundición.

Counterbore To enlarge a bore to a given depth.
Ensanchar Extender un taladro a una profundidad designada.

Counterweight Weight cast or forged into the crankshaft to reduce rotational vibration.
Contrapeso Un peso colado o forjado en un cigüeñal para reducir las vibraciones giratorias.

Crankcase The area of the lower engine block that contains the oil and fumes from the combustion process.
Cárter El área inferior del monoblock que contiene el aceite y los vapores del proceso de combustión.

Cranking vacuum test Used to compare results with the running vacuum test. Low vacuum during cranking can indicate external leaks such as broken or disconnected vacuum hoses that may not be indicated on running engine vacuum tests.
Prueba de arranque en vacío Se efectúa para comparar los resultados con los resultados de la prueba de marcha en vacío. Un nivel bajo de vacío en el arranque puede indicar la presencia de unas fugas externas tal como las mangueras de vacío desconectadas o quebradas que no se pueden descubrir en la prueba de marcha en vacío.

Crankpins Another term for connecting rod bearing journals.
Codo de cigüeñal Otro término para indicar los muñones de la biela.

Crankshaft A mechanical device that converts the reciprocating motions of the pistons into rotary motion.
Cigüeñal Un dispositivo mecánico que convierte los movimientos recíprocos de los pistones a un movimiento giratorio.

Crankshaft endplay The measure of how far the crankshaft can move lengthwise in the block.
Juego en el extremo del cigüeñal La medida de cuánto puede moverse el cigüeñal a lo largo del monoblock.

Crankshaft throw The measured distance between the centerline of the rod bearing journal and the centerline of the crankshaft.
Codo de cigüeñal Una distancia medida entre el centro del muñon del cojinete de biela y el centro del cigüeñal.

Crown The center area of a bearing half.
Centrado de la punta El área central de un mitad del cojinete.

Crush The press fit allowance required to maintain bearing position in the bore.
Aplastamiento El margen requirido en un ajuste prensado para mantener al cojinete en su posición en el taladro.

Current draw test A test which measures the amount of current that the starter draws when actuated. It determines the electrical and mechanical condition of the starting system.
Prueba de carga al corriente Una prueba para medir la cantidad del corriente que consuma el motor de arranque al ser puesto en acción. Determina la condición eléctrica y mecánica del sistema de arranque.

Cycle A sequence that is repeated. In the four-stroke engine, four strokes are required to complete one cycle.
Ciclo Una secuencia que se repite. En el motor de cuatro carreras, se requieren cuatro carreras para completar un ciclo.

Cycling A systematic driving method that varies the load on the engine.
Ciclar Un método sistemático de conducir que diversifica la carga en el motor.

Cylinder A hole bored into the engine block which the piston travels in, and which makes up part of the combustion chamber where the air/fuel mixture is compressed.
Cilindro Un agujero taladrado en el monoblock en el cual viaja el piston, y que forma parte de la cámara de combustión en donde se comprime la mezcla de aire/combustible.

Cylinder block The main structure of the engine. Most of the other engine components are attached to the block.
Monoblock La estructura principal del motor. La mayoría de los otros componentes del motor se conectan al monoblock.

Cylinder bore dial gauge An instrument used to measure the cylinder bore for wear, taper, and out-of-round.
Calibre carátula del taladro del cilindro Un instrumento de medida que sirve para averiguar si el taladro del cilindro esta gastado, cónico, o ovulado.

Cylinder head On most engines the cylinder head contains the valves, valve seats, valve guides, valve springs, and the upper portion of the combustion chamber.
Cabeza del cilindro En la mayoría de los motores la cabeza del cilindro contiene las válvulas con sus asientos, sus guías, sus resortes y la parte superior de la cámara de combustión.

Cylinder leakage test A test which determines the condition of piston ring, intake or exhaust valve, and head gasket. It uses a controlled amount of air pressure to determine the amount of leakage.
Prueba de fugas en el cilindro Una prueba para determinar la condición del anillo de pistón, la válvula de entrada o escape, y la junta de la cabeza. Se emplea una cantidad controlada de presión de aire para determinar la cantidad de la fuga.

Cylinder ridge An area of no wear resulting from the piston ring not travelling the full height of the cylinder.
Cilindro con reborde Una área sin desgaste que resulta cuando el anillo no suba la altura completa del cilindro.

Deceleration A reduction of speed.
Deceleración Una reducción de la velocidad.

Deck The top of the engine block where the cylinder head is attached.
Cubierta La parte superior del monoblock en donde se conecta la cabeza del cilindro.

Deck clearance A measure of the distance the top of the piston is below, or above, the deck of the engine block (when the piston is at TDC).
Holgura de la cubierta La distancia que queda la parte superior del piston abajo, o arriba, de la cubierta del monoblock (cuando el piston esta en PMS).

Deflection The bending or movement away from normal due to loading.
Desviación El abarquillamiento o movimiento fuera de lo normal debido a la carga.

Deglazing The process of roughening the cylinder wall without changing its diameter.
Lijar El proceso de poner áspero el muro del cilindro sin cambiar su diámetro.

Density The compactness or relative mass of matter in a given volume.
Densidad Lo compacto o la masa relativa de la materia en un volumen designado.

Depth micrometers An instrument designed to measure the depth of a bore.
Micrómetro de profundidad Una herramiento diseñada para medir la profundidad de un taladro.

Detonation A defect which occurs if the air/fuel mixture in the cylinder is burned too fast.
Detonación Un defecto que ocurre si la mezcla aire/combustible en el cilindro se quema demasiado rápido.

Diagnosis The use of instruments, service manual, and experience to determine the action of parts or systems to determine the cause of the failure.
Diagnosis El uso de los instrumentos, el manual de servicio, y la experiencia para determinar el acción de los partes o los sistemas y descubrir la causa de un fallo.

Dial calipers Vernier style calipers with a dial gauge installed to make reading of the vernier scale easier and faster.
Calibres de carátula Los calibres tipo vernier equipados con una carátula para facilitar y acelerar la lectura de un escala vernier.

Dial indicator A measuring instrument consisting of a dial face with a needle. The dial is usually calibrated in 0.001 inch increments. A spring loaded plunger or toggle lever transfers movement to the dial needle.
Indicador de carátula Un intrumento de medida que consiste de una carátula con un aguja. La carátula suele ser calibrada en incrementos de 0.001 de una pulgada. Un pistón tubular cargado de resorte o por una palanca acodada transfere el movimiento a la aguja de la carátula.

Die A tool used to repair or cut new external threads.
Matriz Una herramienta para reparar o cortar las roscas exteriores.

Displacement A measure of engine volume. The larger the displacement, the greater the power output.
Desplazamiento Una medida del volumen del motor. Lo más grande el desplazamiento, lo mejor la producción de potencia.

Distortion A warpage or change in form from original.
Distorción Un abarquillamiento o un cambio en la forma de la original.

Distributor   The mechanism within the ignition system that controls the primary circuit and directs the secondary voltage to the correct spark plug.
Distribuidor   El mecanismo dentro del sistema de arranque que controla al circuito primario y dirige el voltaje secundario a la bujía correcta.

Dry compression testing   A compression test performed with no additional oil added to the cylinders.
Prueba de compresión en seco   Una prueba de compresión que se efectúa sin añadir aceite adicional a los cilindros.

Dry sleeves   Replacement cylinder sleeves that do not come into contact with engine coolant. They are surrounded by the cylinder bore.
Camisas secas   Las camisas de repuesto del cilindro que no se ponen en contacto con el fluido refrigerante del motor. Se ajusten en el taladro del cilindro.

Duration   The length of time, expressed in degrees of crankshaft rotation, the valve is open.
Duración   La cantidad del tiempo, representado por los grados de rotación del cigüeñal, que esta abierta la válvula.

Dynamometer   Test equipment that places a load on the engine, causing the engine to work. It measures the amount of rotating force placed against the load.
Dinamómetro   El equipo de prueba que aplica una carga en el motor, causando que el motor trabaja. Mide la cantidad de la fuerza giratoria puesta contra la carga.

Eccentric   One circle within another circle with neither having the same center.
Excéntrico   Un círculo que queda dentro de otro círculo sin compartir el mismo centro.

Eccentricity   The physical characteristics designed into some bearings calling for an inside assembled vertical diameter that is slightly smaller than the horizontal diameter.
Excentricidad   Las características físicas diseñadas en algunos cojinetes que requieren que el diámetro interior asemblado sea un poco más pequeño que el diámetro horizontal.

Eddy   A current that runs against the main current.
Interrupción   Un corriente que fluye contra el corriente principal.

Efficiency   A ratio of the amount of energy put into an engine as compared to the amount of energy produced by the engine.
Rendimiento   Un índice de la cantidad de energía introducido en el motor comparado a la cantidad de energía que produce el motor.

Electrolysis   The result of two different metals in contact with each other. The lesser of the two metals is eaten away.
Electrólisis   El resultado de dos metales distinctos que se ponen en contacto. El menor de los dos metales se consume.

Emission system   Helps to reduce the harmful emissions resulting from the combustion process.
Sistema de emisión   Ayuda en reducir las emisiones nocivos que resultan del proceso de combustión.

Energy   The ability to do work.
Energía   La habilidad de hacer un trabajo.

Engine   The power plant that propels the vehicle.
Motor   El central de energía que propulsa el vehículo.

Engine block   The main structure of the engine that houses the pistons and crankshaft. Most other engine components attach to the engine block.

Monoblock   La estructura principal del motor que contiene los pistones y el cigüeñal. La mayoría de los otros componentes del motor se conectan al monoblock.

Engine hoist   A special lifting tool designed to remove the engine through the hood opening. Most are portable or fold-up for easy storage. The lifting of the boom is performed by a special long reach hydraulic jack.
Grúa   Una herramienta especial de izar diseñada para remover el motor por la apertura del cofre. Suelen ser portátiles o se doblan fácilmente para guardarse. El brazo de la grúa se levanta con un gato hidráulico de larga extensión.

Engine stand   A special holding fixture that attaches to the back of the engine, supporting it at a comfortable working height. In addition, most stands allow the engine to be rotated for easier disassembly and assembly.
Bancada para motor   Un accesorio de apoyo especial que se conecta a la parte trasera del motor. Permite que se soporta el motor en una altura ideal para trabajar. Además, la mayoría de las bancadas permiten la rotación del motor para facilitar el desmontaje y montaje.

Epoxies   Synthetic resins that produce the strongest adhesives in current use, as well as plastics and corrosion coatings. Epoxy adhesives are thermosetting; that is, after initial hardening, they cannot be remelted by heat. They have excellent resistance to solvents and weathering agents, and high electrical and temperature resistance.
Resinas epósicas   Las resinas sintéticas que producen los adhesivos más fuertes hasta la fecha, así como los recubrimientos de plástico y anticorrosivos. Los adhesivos epósicos son termoendurecible; o sea, despues de curarse inicialmente, no se derriten al exponerse al calor. Poseen una resistencia excelente a los disolventes y los agentes destructivos de la clima, y una resistencia muy fuerte contra la electricidad y la temperatura.

Exhaust manifold   A component which collects and then directs engine exhaust gases from the cylinders.
Múltiple de escape   Un componente que colecciona y luego dirige los gases de escape del motor desde los cilindros.

Exhaust manifold gasket   A part which seals the connection between the cylinder head and the exhaust manifold.
Junta del múltiple de escape   Una parte que sella la conexión entre la cabeza del cilindro y el múltiple de escape.

Exhaust system   A system which removes the by-products of the combustion process from the cylinders.
Sistema de escape   Un sistema que remueva los subproductos del proceso de combustión de los cilindros.

Exhaust valve   An engine part which controls the expulsion of spent gases and emissions out of the cylinder.
Válvula de escape   Una parte del motor que controla la expulsión de los gases consumidos y los emisiones del cilindro.

Expansion   An increase in size.
Dilatación   Un incremento en el tamaño.

Externally balanced engine   Engines balanced by using counterweights on the flywheel and vibration dampener in conjunction with the crankshaft counterweights.
Motor de equilibración externo   Los motores que se equilibran empleando los contrapesos en el volante y un amortiguador de vibraciones junto con los contrapesos del cigüeñal.

Face shield   A clear plastic shield that protects the entire face.
Careta   Un escudo transparente de plástico que proteja la cara entera.

False guides   Devices which are similar to inserts.
Guías falsas   Los dispositivos muy parecidos a las piezas insertas.

Fast burn combustion chamber A chamber designed to increase the speed of combustion by creating a turbulence as the air/fuel mixture enters the chamber.
Cámara de combustión rápido Una cámara diseñada para aumentar la rapidez de la combustión creando una turbulencia mientras que la mezcla de aire/combustible entra a la cámara.

Fatigue The deterioration of metal under excessive loads.
Fatiga El deterioro de un metal bajo cargas excesivas.

Feedback A system using an O_2 sensor to feedback information concerning the quality of the previous combustion process to the engine controller.
Retroalimentación Refiere a una sistema que emplea un sensor de oxígeno para mandar la información a un controlador del motor pertinente a la calidad del proceso antecedente de combustión.

Feeler gauge A measuring tool consisting of a series of metal strips cut to precise thicknesses. A feeler gauge pack will usually consists of blades between 0.002 to 0.025 inch in .001 or .002 inch increments.
Galga calibrada Una serie de hojas metálicas cortadas a los espesores precisos. Un conjunto de galgas calibradas suele tener las hojas entre el 0.002 a 0.025 de una pulgada en incrementos de .001 o .002 de una pulgada.

Ferrous metals Metals which contain iron. Cast iron and steel are examples of ferrous metals. Magnets are attracted to ferrous metals.
Metales férreos Los metales que contienen el hierro. El hierro colado y el acero son ejemplos de los metales férreos. Estos metales atraerán un imán.

Fillets Small, rounded corners machined on the edges of journals to increase strength.
Cantos redondeados Las esquinas pequeñas, redondeadas labradas por maquina en los bordes de un muñon para reinforzarla.

Fire extinguisher A portable apparatus that contains chemicals, water, foam, or special gas that can be discharged to extinguish a small fire.
Extinctor de encendios Un aparato portátil que contiene los químicos, el agua, la espuma o un gas especial que se puede discargar en un fuego pequeño para apagarlo.

Flap A strip of emery cloth, or other abrasive material, wound around a slotted mandrel. The loose end of the flap is allowed to slap against the surface being machined.
Aleta Un tiro de tela de esmeril, o otra material abrasiva, envuelto alrededor de un mandrino hendido. Se permite que la extremidad suelta de la aleta golpea la superficie que se esta trabajando por máquina.

Flex fans Fans designed to widen their pitch at slow engine speeds when more air flow is required through the radiator fins. At higher engine speeds the pitch is decreased to reduce the horsepower required to turn the fan.
Ventiladores ajustables Un diseño de los ventiladores que amplían su paso en las velocidades más bajas que requieren más flujo del aire por los devanados del radiador. En las velocidades más altas el paso se disminuye reduciendo los requerimientos del par de dar las vueltas al ventilador.

Flex plate A stamped steel coupler bolted to the rear of the crankshaft. The flex plate provides a mounting for the torque convertor.
Placa flexible Un acoplador de acero embutido empernado a la parte trasera del cigüeñal. La placa flexible provee el asiento del convertidor de par.

Floating piston pin A piston pin that is not locked in the connecting rod or the piston, allowing it to turn or oscillate in both the connecting rod and piston.

Perno flotante del piston Un perno del piston que no esta clavado en la biela o en el piston, permitiendolo girar o osilar en la biela y el piston.

Floor jack A portable hydraulic tool used to raise and lower a vehicle.
Gato Una herramienta portátil hidráulica que sirve para levantar y bajar un vehículo.

Flywheel A heavy circular component located on the rear of the crankshaft that keeps the crankshaft rotating during non-productive strokes.
Volante Un componente pesado circular ubicado en la parte trasera del cigüeñal que procure que el cigüeñal sigue girando durante las carreras no productivas.

Followers Similar to rocker arms, followers are used on many OHC engines. Followers run directly off of the camshaft.
Seguidor Parecidos a los balancines, los seguidores se emplean en muchos motores OHC. Los seguidores se accionan directamente por el árbol de levas.

Foot-pound (ft-lb) A measure of the amount of energy required to lift 1 pound 1 foot.
Pie-libra (pies-lb) Una medida de la cantidad de energía o fuerza que requiere mover una libra la distancia de un pie.

Four gas engine analyzer A device that measures the exhaust of the engine. It enables the technician to look at the effects of the combustion process by measuring hydrocarbons, carbon monoxides, carbon dioxide, and oxygen levels in the exhaust.
Analizador de cuatro gases del motor Un dispositivo que mide los vapores de escape del motor. Permite que el técnico vea los efectos del proceso de combustión midiendo los niveles de los hidrocarburos, el monóxido de carbono, el bióxido de carbono y el oxígeno en los vapores del escape.

Free-wheeling engine An engine in which valve lift and angle prevent valve-to-piston contact if the timing belt or chain breaks.
Motor de piñón libre Un motor en el cual el levantamiento y el ángulo de las válvulas previenen el contacto entre la válvula y el pistón si se quiebra la correa o la cadena de sincronización.

Fuel system A system that includes the intake system which brings air into the engine and the components that deliver the fuel to the engine.
Sistema de combustible Un sistema que incluye al sistema de admisión que introduce el aire dentro del motor y a los componentes que entregan el combustible al motor.

Gallery plugs Metal plugs used to cap the drilled oil passages in the cylinder head or engine block. They can be threaded or pressed into the hole.
Tapones de la canalización de aceite Los tapones metálicos que sirven para estancar los pasajes taladrados del aceite en la cabeza del cilindro o en el monoblock. Pueden ser fileteados o prensados en el agujero.

Galling A displacement of metal.
Raspar El desplazamiento del metal.

Gasket A rubber, felt, cork, or metallic material used to seal surfaces of stationary parts.
Empaque Una material de caucho, fieltro, corcho o metal que sirve para sellar las superficies de las partes fijas.

Glaze Polishing of the cylinder wall resulting from piston ring travel in the cylinder in conjunction with combustion heat and engine oil.
Porcelana El pulido del pared del cilindro que resulta del viaje del anillo del piston en el cilindro en combinación con el calor de combustión y el aceite del motor.

Grade marks Radial lines on the bolt head that indicate the strength of the bolt.
Marcos de grado Las lineas radiales en la cabeza del perno que indican su fuerza.

Grinding A machining process of removing metal from the crankshaft journals by use of a special machine and stones.
Rectificado a esmeril Un proceso maquinario de rebajar el metal de los muñones del cigüeñal usando una máquina especial y las piedras de afiler.

Grit Cleaning sand that is angular in shape and is used for aggressive cleaning.
Grano La arena de limpieza cuyos partículos son de forma angular y que se emplea en la limpieza agresiva.

Hand tools Tools that use only the force generated from the body to operate. They multiply the force through leverage to accomplish the work.
Herramienta de mano Las herramientas que sólo emplean la fuerza proporcionado por el cuerpo. Utilizan el acción de palanca para multiplicar la fuerza recibido e efectuar el trabajo.

Harmonic Balancer (vibration dampener) A component attached to the front of the crankshaft used to reduce the torsional or twisting vibration that occurs along the length of the crankshaft.
Equilibrador armónico (amortiguador de vibraciones) Un componente conectado a la parte delantera de un cigüeñal que sirve para reducir las vibraciones torsionales que ocurren por la longitud del cigüeñal.

Hazardous material A material that could cause injury or death to a person, or could damage or pollute land, air, or water.
Materiales peligrosas Una material que podría causar daños o la muerte a una persona o podría causar los daños o la contaminación de la tierra, el aire o el agua.

Hazardous Materials Inventory Roster A required poster that lists all hazardous materials the employee may come into contact with. It also lists the area of the work place where the material is used.
Escalafón de inventario de las materiales peligrosas Un cartel mandatorio que indica todas las materiales peligrosas con las cuales un empleado puede ponerse en contacto. Tambien indica donde en el área del espacio de trabajo se usa dicha material.

HC The abbreviation for hydrocarbons. Hydrocarbons are present in particles of unburned gasoline.
HC La abreviación de hidrocarburos. Los hidrocarburos se encuentran en los partículos no consumidos de la gasolina.

Head gasket Gasket used to prevent compression pressures, gases, and fluids from leaking. It is located on the connection between the cylinder head and engine block.
Junta de la cabeza Una junta que se emplea en prevenir que se escapen las presiones, los gases y los fluidos de compresión. Se ubica en la conexión entre la cabeza de los cilindros y el monoblock.

Heat treated Metal hardened by a process of heating it to a high temperature, then quenching it in a cool bath.
Tratado térmico (cementado) El metal endurecido por el proceso de calentándolo a una temperatura muy elevada, luego amortiguándolo en un baño frio.

Helicoil A spiral thread used to restore damaged threads to original size.
Helicoil Un hilo espiral que se emplea para restaurar las roscas dañadas a su tamaño original.

Hoist A lift used to raise the entire vehicle.
Elevador Una grúa que se emplea en levantar el vehículo completo.

Horsepower The measure of the rate of work.
Caballo de fuerzas La medida del tiempo en que se realiza el trabajo.

Hot spray tank Cleaning tank that sprays caustic solutions onto the components along with soaking them.
Tanque de rocío caliente Un tanque de limpieza que rocía las soluciones cáusticas sobre los componentes mientras que éstos remojan.

Hydraulic lifters Lifters that use oil to absorb the resultant shock of valve train operation.
Elevadores hidráulicas Los elevadores que usan el aceite para absorber los golpes causados por la operación del tren de válvulas.

Hydrocarbon A chemical composition, made up of hydrogen and carbon.
Hidrocarburo Una composición quimica, compuesta del hidrógeno y el carbono.

Hydrometer Measures the specific gravity of a liquid.
Hidrómetro Mide la gravedad específica de un líquido.

Hydrostatic lock The result of attempting to compress a liquid in the cylinder. Since liquid is not compressible, the piston is not able to travel in the cylinder.
Bloqueo hidrostático El resultado de un intento de comprimir un líquido en el cilindro. Como no se puede comprimir un líquido, el piston no puede moverse en el cilindro.

Ignition The process of igniting the air/fuel mixture in the combustion chamber.
Encendido El proceso de encender la mezcla de aire/combustible en la cámara de combustión.

Ignition system System that delivers the spark used to ignite the compressed air/fuel mixture.
Sistema de encendido Un sistema que entrega la chispa que sirve para encender la mezcla comprimida de aire/combustible.

Indexing The process of offsetting the crankshaft so the rod journals are centered in the grinding machine. The offset from the main bearing journal is the same distance as from the center of the main journal to the center of the crank throw.
Rectificación graduada El proceso de descentrar el cigüeñal para que los muñones de las bielas sean centrales en la rectificadora. La desviación del muñón del cojinete principal queda la misma distancia que la distancia del centro del muñón del cojinete principal al centro del codo del cigüeñal.

Indicated horsepower The amount of horsepower the engine can theoretically produce.
Potencia disponible La cantidad del caballo de fuerzas que puede producir un motor según la teoría.

Induction hardening A process which uses an electromagnet to heat the seat through induction. The seat is heated to a temperature of about 1700°F (930°C). The harden depth is about 0.060 inch (1.5 mm).
Endurecimiento por inducción Un proceso que usa un electroimán para calentar el asiento por medio de la inducción. El asiento se calienta a una temperatura de aproximadamente 1700°F (930°C). La profundidad del endurecimiento es aproximadamente de un 0.060 pulgada (1.5 mm).

Inertia The tendency of objects in motion to remain in motion and of objects at rest to remain at rest.
Inercia La tendencia de que los ojectos en movimiento quedan en movimiento y los objetos en reposo permanezcan en reposo.

Inlet check valve Keeps the oil filter filled at all times so when the engine is started an instantaneous supply of oil is available.
Válvula de retención Mantiene siempre lleno al filtro de aceite para que cuando se arranque el motor dispone de un suministro instantáneo de aceite.

Insert bearings An interchangeable type of bearing. The bearing is a self-contained part that is inserted into the bearing housing.
Cojinetes de inserción Un cojinete de tipo intercambiable. El cojinete es de una parte entera que se puede insertar en la cubierta del cojinete.

Insert guides Removable valve guides that are pressed into the cylinder head.
Guías para inserción Las guías desmontables de las válvulas que han sido prensadas en la cabeza de los cilindros.

Inside micrometer Precision instrument designed to measure the inside diameter of a hole.
Micrómetro interior Una herramienta de medir diseñada para medir el diámetro interior de un agujero.

Intake manifold Component that delivers the air or air/fuel mixture to each engine cylinder.
Múltiple de admisión Un componente que entrega el aire o la mezcla del aire/combustible a cada cilindro del motor.

Intake manifold gasket Gasket which fits between the manifold and cylinder head to seal the air/fuel mixture or intake air.
Junta del múltiple de admisión Una junta que queda entre el múltiple y la cabeza del cilindro para sellar la mezcla de aire/combustible o el aire de admisión.

Intake valve The control passage of the air/fuel mixture entering the cylinder.
Válvula de entrada El pasaje de control para la mezcla de aire/combustible entrando al cilindro.

Integral guides Valve guides that are manufactured and machined as part of the cylinder head.
Guías integrales Las guías de las válvulas fabricadas e incorporadas a máquina como parte de la cabeza del cilindro.

Interference angle Valve design where the seat angle is 1 degree greater than the valve face angle to provide a more positive seal.
Angulo de interferencia Un diseño de la válvula en el cual el ángulo del asiento de la válvula es un grado además del ángulo de la cara de la válvula para proveer un sello más positivo.

Interference engine An engine design in which, if the timing belt or chain breaks or is out of phase, the valves will contact the pistons. In addition, in multi-valve interference engines, valve- to-valve contact is possible if the belt or chain is out of phase.
Interferencia del motor Un diseño del motor en el cual, si se quiebra la correa o la cadena de sincronización, o si esta fuera de fase, las válvulas se pondrán en contacto con los pistones. Además, en los motores de interferencia de múltiples válvulas, es posible el contacto de válvula a válvula si la correa o la cadena esta fuera de fase.

Internal combustion engines An engine that burns its fuels within the engine.
Motores de combustión interna Un motor que quema su combustible dentro del motor.

Internally balanced engine Engines balanced by the counterweights only.
Motor de equilibración interno Los motores que se equilibran solamente por medio de los contrapesos.

Jack stands (safety stands) Support devices used to hold the vehicle off the floor after it has been raised by the floor jack.
Torres (soportes de seguridad) Los dispositivos de soporte que se emplean para mantener el vehículo levantado del piso despues de que haya sido levantado del piso por un gato hidráulico.

Jet valves Valves used by some manufacturers to direct a stream of oil to the underside of the piston head. Jet valves are common on turbocharged engines to keep the piston cool to prevent detonation and piston damage.
Válvulas de chorro Las válvulas empleados por algunos fabricantes para dirigir un chorro de aceite a la parte inferior de la cabeza del piston. Las válvulas de chorro suelen usarse en los motores turbocargados para mantener frío al piston y para prevenir la detonación y los daños al piston.

Journal An inner bearing operated by a shaft.
Muñón Un cojinete interior operado por una flecha.

Kinetic Energy Energy that is working.
Energía cinética La energia trabajando.

Knock A descriptive term used to identify various noises occurring in an engine.
Golpe Un término descriptivo que sirve para identificar los varios ruidos que ocurren en un motor.

Knurl A special bit that rolls a thread into the guide and causes the metal to rise.
Moleta Una broca especial que enreda un hilo en la guía y causa que se levanta el metal.

Knurling A machining process that decreases the size of a bore by forcing a bit that swells the metal much like a tap does when it cuts threads.
Moletear Un proceso de rebajar a máquina el tamaño de un taladro forzando una broca que hincha al metal de una manera muy parecida de un macho cortando las roscas.

Lapping The process of fitting one surface to another by rubbing them together with an abrasive material between the surfaces.
Pulido El proceso de ajustar una superficie a otra rozándolas juntas con una material adhesiva entre las superficies.

Leakdown The relative movement of the lifter's plunger in respect to the lifter body.
Tiempo de fuga El movimiento relativo del émbolo del levantaválvulas con relación al cuerpo del levantaválvulas.

Leakdown testing Testing which determines the lifter's ability to hold hydraulic pressure and maintain zero lash.
Prueba de fuga La prueba que verifica la habilidad del levantaválvulas de mantener la presión y un juego cero.

Lean-burn miss The result of incomplete combustion due to the lack of oxygen.
Marcha irregular de mezcla pobre El resultado de la combustión incompleta debido a la falta de oxígeno.

Lifters Mechanical (solid) or hydraulic connections between the camshaft and the valves. Lifters follow the contour of the camshaft lobes to lift the valve off its seat.
Levantaválvulas Las conexiones mecánicas (sólidas) o hidráulicas entre el árbol de levas y las válvulas. Los levantaválvulas siguen el contorno de los lóbulos del árbol de levas para levantar la válvula de su asiento.

Line boring A machining process of the main bearing journals that restores the original bore size by cutting metal using cutting bits.
Taladrar en serie Un proceso de rectificación a máquina de los muñones del cigüeñal que restaura el tamaño original del taladro cortando el metal con brocas para cortar.

Liners Thin tubes placed between two parts.
Manguitos Los tubos delgados puestos entre dos partes.

Lip seals Molded synthetic rubber seal with a slight raise (lip) that is the actual sealing point. The lip provides a positive seal while allowing for some lateral movement of the shaft.

Junta con reborde Las juntas de caucho sintético moldeado que tienen un reborde ligero (un labio) que es el punto efectivo del sello. El reborde provee un sello positivo mientras que permite algún movimiento lateral de la flecha.

Lobe The part of the camshaft that raises the lifter.
Lóbulo La parte del árbol de levas que alza al levantaválvulas.

Lower end Refers to the main bearing journals of the cylinder block.
Extremo inferior Refiera a los muñones del cigüeñal del monoblock.

Lubrication system System which supplies oil to high friction and wear locations.
Sistema de lubricación Un sistema que proporciona el aceite a las áreas de alta fricción y desgaste.

Machinist's rule A multiple scale ruler used to measure distances or components that do not require precise measurement.
Regla de acero Una regla con graduaciones múltiples para medir las distancias o los componentes que no requieren una medida precisa.

Main bearing A bearing used as a crankshaft support.
Cojinete del cigüeñal Un cojinete que se emplea como un soporte para el cigüeñal.

Main bearing clearance The distance between the main bearing journal and the main bearings.
Holgura del cojinete principal La distancia entre el muñón del cigüeñal y los cojinetes principales del cigüeñal.

Main bearing journal The crankshaft journal that is supported by the main bearing.
Muñón del cojinete principal El muñón del cigüeñal apoyado por el cojinete del cigüeñal.

Main bearing saddle bore The housing that is machined to receive a main bearing.
Taladro del asiento del cojinete principal Un cárter se ha labrado a máquina para aceptar un cojinete de cigüeñal.

Major thrust surface The side of the piston skirt that pushes against the cylinder wall during the power stroke.
Superficie de empuje principal El lado de la faldilla que empuja contra el muro del cilíndro durante la carrera de fuerza es la superficie de empuje principal.

Manifolds Tubular channels used to direct gases into or out of the engine.
Múltiples Los canales tubulares que sirven para dirigir los gases dentro o fuera del motor.

Material Safety Data Sheet (MSDS) A sheet which contains detailed information concerning hazardous materials. The MSDS must be maintained by the employer.
Hoja de Dato de Seguridad de los Materiales (MSDS) Una hoja que contiene la información detallada referente a los materiales peligrosos. El patrón debe conservar el MSDS.

Mechanical efficiency A comparison of the power actually delivered by the crankshaft to the power developed within the cylinders at the same rpm.
Rendimiento mecánica Una comparación entre la energía que actualmente entrega el cigüeñal y la energía desarrollado dentro de los cilindros en la misma rpm.

Microinch One millionth of an inch. The microinch is the standard measurement for surface finish in the American customary system.
Micropulgada La millonésima parte de una pulgada. Una micropulgada es una unedad común para el acabado de la superficie en el sistema de medida americana.

Micrometer One millionth of a meter. The micrometer is the standard measurement for surface finish in the metric system.
Micrómetro Una millonésima parte de un metro. La medida común para el acabado de la superficie en el sistema métrica.

Micrometer Precision measuring instrument designed to measure outside, inside, or depth measurements.
Micrómetro Un instrumento de medidas precisas diseñada para tomar medidas exteriores, interiores o de profundidad.

Micron A thousandth of a millimeter or about .0008 inch.
Micrón La milésima parte de un milímetro o aproximadamente un .0008 de una pulgada.

Minor thrust side The side opposite the side of the rod that is stressed during the power stroke.
Lado de empuje menor El lado opuesto del lado cuyo biela recibe el esfuerzo durante la carrera de potencia.

Minor thrust surface The area of the piston skirt that pushes against the cylinder wall during the compression stroke.
Superficie de empuje menor El faldón del piston que empuja contra el muro del cilindro durante la carrera de compresión.

Necking A valve stem defect where the stem narrows near the head.
Vástago ahusado Un defecto del vástago de válvula en el cual el vástago se adelgaza cerca de la cabeza del cilindro.

Net valve lift The actual amount a valve lifts off its seat. It is found by subtracting the lash specification and amount of component deflection from the gross valve lift.
Producto neto del levantamiento de la válvula La cantidad actual que se levanta una válvula de su asiento. Para determinarlo se substrae la especificación del juego de las válvulas más la cantidad de desviación del componente del total bruto de la cantidad del levantamiento de la válvula.

No-crank A condition where the ignition switch is placed in the START position, but the starter does not turn the engine. This may be accompanied with a buzzing noise that indicates the starter motor drive has engaged the ring gear, but the engine does not rotate. There may also be clicking or no sounds from the starter motor or solenoid.
No acoda Una condición en la cual el interruptor del encendido se ha puesto en la posición de START(arranque), pero el encendedor no arranque al motor. Esto puede acompañarse con un zumbido que indica que el impulsor del encendedor esta accionando a la corona, pero el motor no gira. Tambien se puede oir o sonidos de chasquido o silencio total del motor del encendedor o del solenoide.

Nucleate boiling The process of maintaining the overall temperature of a coolant to a level below its boiling point, but allowing the portions of the coolant actually contacting the surfaces (the nuclei) to boil into a gas.
Ebullición nucleido El proceso de mantener un nivel de temperatura general de un fluido refrigerante menos de la de su punto de ebullición, pero permitiendo que las porciones del fluido refrigerante que actualmente estan en contacto con las superficies (el nucleido) hiervan para formar un gas.

Occupational safety glasses Eye protection device designed with special high impact lenses and frames and side protection.
Lentes de seguridad Un dispositivo de protección para los ojos diseñado con los cristales y el armazón resistentes a los impactos fuertes, y que provee protección a los lados de los ojos.

Off-square When the seat and valve stem are not properly aligned they are off-square. This condition causes the valve stem to flex as the valve face is forced into the seat by spring and combustion pressures.

Fuera de escuadra Refiere a que el asiento y el vástago de la válvula no estén alineadas correctamente. Esta condición causa que el vástago de la válvula se dobla cuando la cara de la válvula se asienta bajo la fuerza de las presiones del resorte y la combustión.

Oil breakdown Conditon of oil which results from high temperatures for extended periods of time. The oil will combine with oxygen and can cause carbon deposits in the engine.

Deterioro del aceite Un resultado de las temperaturas elevadas por largos períodos de tiempo. El aceite combinará con el oxígeno y puede causar los depósitos del carbono en el motor.

Oil clearance The difference between the inside bearing diameter and the journal diameter.

Holgura de aceite La diferencia entre el diámtero interior de un cojinete y el diámetro del muñón.

Oil gallery The main oil supply line in the engine block.

Canalización de aceite La linea principal del suministro de aceite en el monoblock.

Oil pan gaskets Gaskets used to prevent leakage from the crankcase at the connection between the oil pan and the engine block.

Empaque de la tapa del cárter Los empaques que se emplean para prevenir las fugas del cárter en la conexión entre el colector de aceite y el monoblock.

Oil pressure test Test to determine the condition of the bearings and other internal engine components.

Prueba de presión de aceite Una prueba que se efectua para determinar la condición de los cojinetes u otros componentes interiores del motor.

Oil pump A rotor or gear type positive displacement pump used to take oil from the sump and deliver it to the oil galleries. The oil galleries direct the oil to the high wear areas of the engine.

Bomba de aceite Un rotor o una bomba de tipo desplazamiento positivo de engrenajes que sirve para tomar el aceite de un suministro y entregarlo a las canalizaciones de aceite. Las canalizaciones de aceite dirigen el aceite a las áreas del motor que imponen gastos severos.

One-piece valves A valve stem design where the head is not welded to the stem. In a two-piece valve, the head is welded to the stem.

Válvulas de una pieza Un diseño del vástago de válvula que no tiene una cabeza soldada al vástago. Una válvula de dos piezas tiene una cabeza soldada al vástago.

Open circuit voltage test Test to determine the battery's state of charge. It is used when a hydrometer is not available or cannot be used.

Prueba de voltaje de circuito abierto Una prueba para determinar el estado de carga de la bateria. Se usa cuando no es disponible o no se puede usar un hidrómetro.

Open pressure Spring tension when the valve spring is compressed and the valve is fully open.

Presión abierta La tensión de un resorte de la válvula al estar comprimido con la válvula completamente abierta.

Opposed cylinder engine Engine block design in which the cylinders are across from each other. Also referred to as horizontally opposed or "pancake" engines.

Motor de cilindros opuestos Un diseño de monoblock que coloca los cilindros en lados opuestos. Tambien conocido con el nombre motores de cilindros opuestos horizontalmente o motores achatados.

Out-of-round The condition when measurements of a diameter differ at different locations. The term out-of-round applies to inside or outside diameters.

Ovulado Una condición en la cual las medidas de un diámetro varían en lugares distinctos. El término ovulado aplica a los diámetros interiores o exteriores.

Out-of-round gauges Instruments used to measure the concentricity of connecting rod bores.

Calibradores de ovulado Los instrumentos para medir la concentricidad de los taladros de las bielas.

Outside micrometers Tool designed to measure the outside diameter or thickness of a component.

Micrómetros exteriores Una herramienta diseñada para medir los diámetros exteriores o el espesor de un componente.

Overhang The area of the face between the seat contact and the margin.

Sobresaliente El área de la cara entre el contacto del asiento y el margen.

Overhead hoist A lifting tool that uses a chain fall or electric motor to hoist the engine out of the hood opening. The hoist can be attached to a moveable A-frame or on a I-beam across the shop ceiling.

Grúa (montacarga) en alto Una herramienta de izar que utiliza una caída de cadena o un motor eléctrico para levantar el motor por la apertura del capó. El aparato de izar puede conectarse a un armazón en forma de A o a un hierro en T a través del techo del taller.

Overhead valve engine An engine with the camshaft located in the engine block and the valves in the cylinder head.

Motor con válvulas en cabeza Un motor que tiene el árbol de levas ubicado en el monoblock y las válvulas en la cabeza del cilindro.

Oversize bearings Bearings that are thicker than standard to increase the outside diameter of the bearing to fit an oversize bearing bore. The inside diameter is the same as standard bearings.

Cojinete de medidas superiores Los cojinetes que son de un espesor más grueso de lo normal para aumentar el diámetro exterior del cojinete para quedarse en un taladro que rebasa la medida. El diámetro interior es lo mismo del cojinete normal.

Parts washers Parts washers generally use a mild solvent to soak the components. Some provide for agitation and spraying of the solvent.

Lavadora de partes Las lavadoras de partes generalmente remojan los componentes en un solvente debil. Algunos proveen la agitación y rocían el solvente.

Peen To stretch or clinch over by pounding.

Martillazo Estirar o remachar por machacado.

Peening The process of removing stress in a metal by striking it.

Martillar El proceso de quitar la fatiga de un metal golpeandolo.

pH A value expressing acidity or basicity in terms of the relative amounts of hydrogen ions (H+) and hydroxide ions (OH-) present in a solution.

pH Un valor que expresa la acidez o lo básico en términos del las cantidades relativas de los iones de hidrógeno (H+) y los iones hidróxidos (OH-) presentes.

Pick-up tube A tube used by the oil pump to deliver oil from the bottom of the oil pan. The bottom of the tube has a screen to filter larger contaminates.

Tubo de captación Un tubo empleado por la bomba de aceite para entregar el aceite del fondo del colector de aceite. La parte inferior del tubo tiene un rejilla para filtrar los contaminantes más grandes.

Pin boss A bore machined into the piston that accepts the piston pin to attach the piston to the connecting rod.

Mamelón Un taladro tallado a máquina en el pistón que acomoda la espiga del pistón para conectarlo a la biela.

Pinning A cold crack repair process that avoids altering the characteristics of the metal. This process uses a series of tapered iron pins threaded into overlapping holes along the length of the crack.

Chavetear Un proceso de reparar sin calor a las grietas que no cambia las características del metal. Este proceso usa una serie de espárragos cónicos fileteados en los agujeros extendidos uno sobre otro por toda la longitud de la grieta.

Piston An engine component in the form of a hollow cylinder that is enclosed at the top and open at the bottom. Combustion forces are applied to the top of the piston to force it down. The piston, when assembled to the connecting rod, is designed to transmit the power produced in the combustion chamber to the crankshaft.

Pistón Un componente del motor que consiste de un cilindro hueco cerrado en la parte de arriba y abierto en la parte de abajo. Las fuerzas de combustión se aplican en la parte superior del piston para forzarlo hacia abajo. El piston, al conectarse a la biela, es diseñado para transmitir la fuerza producida en la cámara de combustión al cigüeñal.

Piston balance pads Some manufacturers provide balance pads just below the pin boss. The piston can be balanced by removing material from this area.

Placa (pastilla) de equilibración del pistón Algunos fabricantes proveen las placas para equilibrar ubicadas justo abajo del mamelón. Se puede equilibrar al pistón quitando la material de esta área.

Piston collapse A condition describing a collapse or reduction in diameter of the piston skirt due to heat or stress.

Caída del pistón Una condición que describe el fallo o la reducción en el diámetro de la falda del pistón debido al calor o la fatiga.

Piston dwell time The length of time in crankshaft degrees the piston remains at top dead center without moving.

Angulo de cierre del pistón La cantidad del tiempo en los grados del ángulo del cigüeñal que el piston se queda en la posición de punto muerto superior sin moverse.

Piston head (crown) The top of the piston that forms the bottom of the combustion camber.

Cabeza de piston La parte superior del pistón que forma la parte inferior de la cámara de combustión.

Piston land Area used to confine and support the piston rings in their grooves.

Meseta del pared del pistón El área de un pistón que sirve para restringir y sostener los anillos del pistón en sus muescas.

Piston offset A piston design that offsets the pin bore to provide more effective downward force onto the crankshaft by increasing the leverage applied to the crankshaft.

Desviación del pistón Un diseño del piston que desvía el taladro del eje para proveer una fuerza descendente en el cigüeñal más eficáz aumentando la acción de palanca que se aplica en el cigüeñal.

Piston pin Component which connects the piston to the connecting rod. There are three basic designs used: a piston pin anchored to the piston and floating in the connecting rod, a piston pin anchored to the connecting rod and floating in the piston, and a piston pin full floating in the piston and connecting rod.

Eje del pistón Un componente que conecta el pistón a la biela. Se usan tres diseños básicos: un eje de pistón fijo al pistón y libre en la biela, un eje de pistón fijo a la biela y libre en el piston, y un eje de pistón completamente libre en el pistón y la biela.

Piston pin knock A noise caused by a worn piston pin or bushing, worn piston pin boss, and worn bearings.

Golpe del eje del pistón Un ruido causado por un eje o manguito de pistón desgastado, un mamelón desgastado, y los cojinetes desgastados.

Piston rings Components which seal the compression and expansion gases, and prevent oil from entering the combustion chamber.

Anillos (aros) del pistón Un componente que sella los gases de compresión y expansión, y previene que entra el aceite en la cámara de combustión.

Piston skirt A component which forms a bearing area in contact with the cylinder wall and helps to prevent piston from rocking in the cylinder.

Faldilla del pistón Un componente que forma una área de apoyo en contacto con el muro del cilindro y ayuda en prevenir que el pistón oscila en el cilindro.

Piston slap A sound which results from the piston hitting the side of the cylinder wall.

Golpeteo del pistón Un ruido resultando del pistón golpeando contra el muro del cilindro.

Piston stroke The distance the piston travels from TDC to BDC.

Carrera del pistón La distancia que viaja el piston del PMS al PMI.

Plasma A material containing positive ions and unbound electrons in which the total number of positive and negative charges are almost equal. The properties of plasma are sufficiently different from those of solids, liquids, and gases for it to be considered a fourth state of matter.

Plasma Una material comprendida de los iones positivos y los electrones libres en la cual el número total de cargas positivas y negativas son casi iguales. Las propriedades del plasma son bastante diferentes de las de los sólidos, los líquidos y los gases para que se considera un cuarto estado de materia.

Plastigage A string-like plastic that is available in different diameters used to measure the clearance between two components. The diameter of the plastic gage is exact, thus any crush of the gage material will provide an accurate measurement of oil clearance.

Plastigage Un plástico en forma de hilo disponible en diámetros distinctos que se emplea en medir la holgura entre dos componentes. El diámetro del calibre plástico es preciso, asi cualquier aplastamiento del material del calibre provee una medida precisa de la holgura del aceite.

Plateau honing A cylinder honing process that uses a coarse stone to produce the finished cylinder size and a very fine stone to remove an immeasurable amount and plateau the cut.

Esmerilado de nivelación Un proceso de esmerilar el cilindro usando una piedra tosca para producir el tamaño del cilindro acabado y una piedra muy fina para quitar una cantidad inmensurable y nivelar el corte.

Plunge grinding A crankshaft grinding method that uses a dressed stone the exact shape of the new journal surface. The stone is fed straight into the journal.

Rectificación empujado Un metodo de afilar el cigüeñal usando una piedra de afilar en la forma exacta de la superficie del muñón nuevo. La piedra de afilar se empuja directamente dentro del muñón.

Pneumatic tools Tools powered by compressed air.

Herramientas neumáticas Las herramientas que derivan su poder del aire bajo presión.

Polishing The process of removing light roughness from the journals by using a fine emery cloth. Polishing can be used to remove minor scoring of journals which do not require grinding.

Bruñido El proceso de quitar la aspereza ligera de los muñones usando una tela abrasiva muy fina. El bruñido puede emplearse en quitar las rayas superficiales de los muñones que no requieren rectificación.

Poppet valve A valve design consisting of a circular head with a stem attached in the center. Poppet valves are used to control the opening or closing of a passage by linear movement.

Válvula champiñón Un diseño de una válvula que consiste de una cabeza redonda con un vástago conectado en el centro. Las válvulas champiñones controlan la apertura o cerradura de un pasaje por medio de un movimiento linear.

Positive crankcase ventilation (PCV) system An emission control system that routes blow-by gases and unburned oil/fuel vapors to the intake manifold to be added to the combustion process.

Sistema (PCV) de ventilación positiva de la caja del cigüeñal Un sistema de emisión que lleva los gases soplados y los vapores del aceite/combustible no quemados al múltiple de entrada para que se pueden añadir al proceso de combustión.

Positive displacement pumps Pumps that deliver the same amount of oil with every revolution, regardless of speed.

Bombas de desplazamiento positivo Las bombas que entregan la misma cantidad del aceite con cada revolución, sin que importa la velocidad.

Power balance test Test used to determine if all cylinders are producing the same amount of power output. In an ideal situation, all cylinders would produce the exact same amount of power.

Prueba del equilibrio de fuerza Una prueba para determinar si todos los cilindros producen la misma cantidad de potencia de salida. En una situación ideal, todos los cilindros producirían exactamente la misma cantidad de poder.

Power tools Tools that use other forces than that generated from the body. They can use compressed air, electricity, or hydraulic pressure to generate and multiply force.

Herramientas de motor Las herramientas que usan otras fuerzas que las producidas por el cuerpo. Pueden usar el aire bajo presión, la electricidad, o la presión hidráulica para engendrar y multiplicar la fuerza.

Pre-ignition Defect that is the result of spark occurring too soon.

Autoencendido Un defecto que resulta de una chispa que ocurre demasiado temprano.

Profilometer A tool capable of electrically sensing the distances between peaks to determine the finish of a cut.

Perfilómetro Una herramiento capaz de detectar electrónicamente las distancias entre dos puntos altos para determinar cuando terminar un corte.

Pushrod A connecting link between the lifter and rocker arm. Engines designed with the camshaft located in the block use pushrods to transfer motion from the lifters to the rocker arms.

Varilla de presión Una conexión entre las levantaválvulas y el balancín empuja válvulas. Los motores diseñados con el árbol de levas en el bloque usan las varillas de presión para transferir el movimiento de la levantaválvulas a los balancines.

Radiator A component consisting of a series of tubes and fins that transfer the heat from the coolant to the air.

Radiador Un componente que consiste de una serie de tubos y aletas que transferen el calor del fluido refrigerante al aire.

Ream Process of accurately finishing a hole with a rotating fluted tool.

Escariar El proceso de acabar un taladro precisamente con una herramiente acanalada giratoria.

Reciprocating An up-and-down or back-and-forth motion.

Alternativo Un movimiento oscilante de arriba a abajo o de un lado a otro.

Reed valve An one-way check valve. The reed opens to allow the air/fuel mixture to enter from one direction, while closing to prevent movement in the other direction.

Válvula de lengüeta Una válvula de una vía. La lengüeta se abre para permitir entrar la mezcla de aire/combustible de una dirección, mientras que se cierre para prevenir el movimiento de la otra dirección.

Relief valve Valve used to prevent excessive oil pressure. Since the oil pump is positive displacement, pressures could increase to a hazardous level at higher engine speeds. The relief valve opens to return oil to the sump and drop pressure in the system.

Válvula de rebose Una válvula que previene una presión excesiva de aceite. Como la bomba de aceite es de desplazamiento positivo, las presiones podrían aumentar a un nivel peligroso en las velocidades más altas. La válvula abre para regresar el aceite al resumidero y bajar la presión del sistema.

Resource Conservation and Recovery Act (RCRA) Law that makes users of hazardous materials responsible for the material from the time it becomes a waste until disposal is complete.

Acta de Conservación y Recobro de Recursos (RCRA) Un ley que hace responsable a los que usan las materiales peligrosas desde el tiempo que se convierte en un producto residual hasta que se haya completado su disposición.

Ridge reamer A cutting tool used to remove the ridge at the top of the cylinder.

Escariador de reborde Una herramienta de cortar que sirve para quitar el reborde en la parte superior del cilindro.

Ring noise A noise caused by worn rings or cylinders. Other causes include broken piston ring lands and too little tension of the ring against the cylinder wall.

Ruido de los anillos Un ruido causado por los anillos o cilindros desgastados. Otra causas incluyen las mesetas de pistones rotas o una tensión insuficiente del anillo contra el muro del cilindro.

Ring seating The process of lapping the rings against the cylinder wall accomplished by the movement of the piston in the cylinder.

Asiento del anillo El proceso de asentar los anillos a pulso con el muro del cilindro que se lleva acabo por el movimiento del piston dentro del cilindro.

Rocker arm Pivots that transfer the motion of the pushrods or followers to the valve stem.

Balancín Un punto pivote que transfiere el movimiento de las levantaválvulas o de los seguidores al vástago de la válvula.

Rocker arm ratio An mathematical comparison of rocker arm dimensions. The rocker arm ratio compares the center-to-valve-stem measurement against the center-to-pushrod measurement.

Indice del balancín Una comparación de las dimensiones del balancín. El índice del balancín compara las dimensiones del balancín del centro-al-vástago de la válvula con las dimensiones del centro-al-levantaválvulas.

Rod beaming The process of polishing the beams of the rods to prevent stress risers. This is done by blending the casting seam on the sides of the rods.

Pulido de las varillas El proceso de pulir los resaltos de las varillas para prevenir la deformación de colada. Esto se lleva acabo puliendo la mazarota en los lados de las varillas.

Rod length to stroke ratio A mathematical comparison between the length of the connecting rod and the length of the engine's stroke. It is determined by dividing the connecting rod length by the stroke.

Indice de longitud de la biela a la carrera Una comparación matemática entre la longitud de la biela y la longitud de la carrera del motor. Se determine dividiendo la longitud de la biela por la carrera.

Rotary valve A valve that rotates to cover and uncover the intake port. A rotary valve is usually designed as a flat disc that is driven from the crankshaft.

Válvula rotativa Una válvula que gira para cubrir y descubrir la puerta de admisión. Suelen ser diseñadas como un disco plano impulsado por el cigüeñal.

Saddle The portion of the crankcase bore that holds the bearing half in place.

Asiento del cojinete (silleta) La parte del taladro del cigüeñal que mantiene en su lugar a la mitad con el cojinete.

SAE Society of Automotive Engineers.

SAE Asociación de Ingenieros Automotrices.

Safety goggles Safety devices which provide eye protection from all sides. Goggles fit against the face and forehead to seal off the eyes from outside elements.

Gafas de seguridad Proporciona la protección a los ojos de todos lados siendo que quedan apretados contra la cara y el frente para formar un sello para los ojos contra los elementos exteriores.

Scale The distance of the marks from each other on a measuring tool.

Escala La distancia entre las marcas en una herramienta de medir.

Score A scratch, ridge, or groove marring a finish surface.

Raya Un rasguño, una arruga, o una muesca que echa a perder una superficie.

Scuffing Scraping and heavy wear between two surfaces.

Erosión El rozamiento y desgaste fuerte entre dos superficies.

Seal Component used to seal between a stationary part and a moving one.

Junta Un componente que se emplea para sellar entre una parte fija y una que mueva.

Sealant A special liquid material commonly used to fill irregularities between the gasket and its mating surface. Some sealants are designed to be used in place of a gasket.

Compuesto obturador Un material líquido especial que normalmente se emplea para rellenar las irregularidades entre un empaque y su superficie de contacto. Algunos compuestos son diseñados de uso sin empaque.

Seat A surface upon which another part rests.

Asiento Una superficie sobre la cual queda otra parte.

Seat pressure Term which indicates spring tension with the spring at installed height and the valve closed.

Presión del asiento Un término que indica la tensión del resorte en su altura de instalación con la vávula cerrada.

Seat runout (concentricity) A measure of how circular the valve seat is in relation to the valve guide.

Excentricidad del asiento (concentricidad) Una medida de lo circular del asiento de la válvula con relación a la guía de la válvula.

Second order vibration Vibration which occurs twice per revolution.

Vibración de segunda orden Una vibración que ocurre dos veces por revolución.

Seize When one surface moving upon another causes scratches. The metal transfer can become severe enough to cause the moving component to stop.

Rayar Cuando los movimientos de una superficie sobre otra causan las rayas. La transferencia del metal puede ser tan severa que para al componente en movimiento.

Service manual One of the most important tools for today's technician. The service manual provides information concerning engine identification, service procedures, and specifications. In addition, the service manual provides information concerning wiring harness connections and routing, component location, and fluid capacities. Service manuals may be obtained from the vehicle manufacturer or through aftermarket suppliers.

Manual de servicio Una de las herramientas más importantes del tecnico moderno. Proporciona la información sobre la identificación del motor, los procedimientos del servicio, y las especificaciones. Además, el manual de servicio provee la información sobre las conexiones y los rumbos del mazo de alambres, ubicación de los componentes y las capacidades de los fluidos. Los manuales se pueden obtener del fabricante del vehículo o por los proveedores de repuestas.

Service sleeve (speedy-sleeve) A metal sleeve that is pressed over a damaged sealing area to provide a new, smooth surface for the seal lip.

Camisa de servicio (manguito rápido) Un camisa de metal que se coloca sobre una área de sello dañada que provee una superficie nueva y lisa para el borde del sello.

Short An electrical defect that allows electrical current to by- pass its normal path.

Cortocircuito Un defecto eléctrico que permite que el corriente eléctrico sobrepasa su rumbo normal.

Shot Round beads which are used for cleaning when etching of the metal is not desired.

Granalla Las bolitas que se emplean en la limpieza cuando el grabado del metal no es deseado.

Shot-peening A tempering process that uses shot under pressure to tighten the outside surface of the metal. Shot peening is used to help strengthen the component and reduce chances of stress or surface cracks.

Chorreo con granalla Un proceso de templado que usa la granalla bajo presión para estrechar la superficie exterior del metal. El chorreo con granalla fortalece al componente y disminuye la aparencia del fatiga o de grietas en la superficie.

Silicon A non-metallic element that can be doped to provide good lubrication properties. The melting point of silicon is 2,570oF (1,410°C).

Silicio Un elemento no metálico que se puede agregar para proporcionar las propriedades buenas de la lubricación. El silicio se funde hacia 2,570°F (1,410°C).

Sizing point The location the manufacturer designates for measuring the diameter of the piston to determine clearance.

Punto de calibración El lugar indicado por el fabricante para efectuar las medidas del diámetro del piston para determinar la holgura.

Sleeving The process of boring the cylinder to accept a sleeve.

Preparar para camisa El proceso de taladrar un cilindro para aceptar una camisa.

Slipper skirt A piston skirt ground to provide additional clearance between the piston and the counterweights of the crankshaft. Without this recessed area, the piston would contact the crankshaft when shorter connecting rods are used.

Faldilla deslizante Una faldilla del piston rectificada para proveer una holgura adicional entre el piston y los contrapesos del cigüeñal. Sin esta área rebajada, el piston podría rozar contra el cigüeñal cuando se emplean las bielas más cortas.

Slow cranking A defect that occurs when the starter drive engages the ring gear, but the engine turns at too slow of a speed to start. Some manufacturers provide specifications for engine cranking speed.

Arranque lento Un defecto que ocurre cuando el acoplamiento del motor de arranque impulsa a la corona del volante, pero el motor gira con una velocidad demasiado lento para arrancar. Algunos fabricantes proveen las especificaciones para la velocidad del arranque del motor.

Small end Term that refers to the end of the connecting rod that accepts the piston pin.
Extremo pequeño Un término que refiere a la extremidad de la biela que accepta el eje del piston.

Small-hole gauge An instrument used to measure holes or bores that are smaller than a telescoping gauge can measure.
Calibre de taladros chicos Un instrumento que se emplea para medir los agujeros o taladros que son demasiado pequeños para medirse con un calibrador telescópico.

Soak tanks Cleaning tanks equipped with a large basket which holds the parts while they are submerged into a caustic solution or detergent. Some soak tanks are equipped with an agitation system.
Tanques (cubos) de remojo Los tanques equipados con un capacho para sostener las partes que se sumergen en una solución cáustica o en el detergente. Algunos tanques tienen un sistema de agitación.

Solid lifters (mechanical lifters) Components which provide a rigid connection between the camshaft and the valves.
Levantaválvulas macizas (o mecánicas) Los componentes que proveen una conexión rígida entre el árbol de levas y las válvulas.

Specific gravity A unit measurement for determining the sulfuric acid content of an electrolyte.
Gravedad específica Una unedad de medida para determinar el contenido del ácido sulfúrico en el electrolito.

Spontaneous combustion A fire that occurs spontaneously. For example, heat is slowly generated from oxidation of oil on rags; the heat continues to increase until the flash point is reached and the rags suddenly begin to burn.
Combustión espontánea Un fuego que ocurre espontáneamente. Por ejemplo, el calor se produce lentamente por la oxidación del aceite en los trapos; el calor continua a aumentarse hasta que llega a la temperatura de inflamabilidad y los trapos comienzan a quemarse.

Spread Descriptive term applied when the diameter at the outside parting edges of a bearing shell exceeds the inside diameter of the mating housing bore.
Aplastamiento Un término descriptivo que se aplica cuando el diámetro de los bordes exteriores del casquillo de un cojinete es más grande que el diámetro interiordel taladro en el superficie de contacto del cárter.

Spring free length The height the spring stands when not loaded.
Longitud libre del resorte La altura del resorte cuando no tiene carga.

Spring shims Components used to correct installed height of the valve spring. Spring shims are used to correct for machining tolerance.
Chapas de relleno Los componentes que se emplean para ajustar la altura de instalación del resorte de la válvula. El sobreespesor del maquinado se puede ajustar por medio de estas chapas.

Spring squareness Refers to how true to vertical the entire spring is.
Escuadrado del resorte Refiere a si la posición del resorte entero esta en línea recta al vertical.

Squish area The area of the combustion chamber where the piston is very close to the cylinder head. The air/fuel mixture is rapidly pushed out of this area as the piston approaches TDC, causing turbulence and forcing the mixture toward the spark plug. The squish area can also double as the quench area.

Area de compresión El área de la cámara de combustión en donde el piston esta muy cerca a la cabeza del cilindro. La mezcla del aire/combustible se expulsa rapidamente de esta área al aproximarse el piston al PMS, causando una turbulencia y empujando la mezcla hacia la bujía. El área de compresión tambien puede servir de área de extinción.

Steam cleaners A pressure washer that uses a soap solution, heated under pressure to a temperature higher than its normal boiling point. The super-heated solution boils once it leaves the nozzle as it shoots against the object being cleaned.
Limpiadoras de vapor Una limpiadora a presión que utiliza una solución de jabón, calentado bajo presión a una temperatura más elevada de su punto de ebullición. La solución sobrecalentada hierve al salir de la boquilla projectada hacia el objeto para limpiar.

Stellite A hard facing material made from a cobalt-based material with a high chromium content.
Estelita Una material de recarga compuesta de una material de base cobáltico con un contenido muy alto del cromo.

Still timing The process of adjusting base ignition timing without the engine running.
Regulación sin marcha El proceso de ajustar el avance del encendido fundamental sin que esté en marcha el motor.

Stone dressing Dressing a stone refers to using a diamond tool to clean and restore the stone's surface.
Reacondicionar la muela Para recondicionar una muela se emplea una herramienta de diamante para limpiar y rectificar la agudeza de la superficie de la muela.

Stress risers Defects in the component resulting in a weakness of the metal. The defect tends to decrease the tensile strength of the metal, and stress applied to the area of the defect tends to cause breakage.
Deformación de colada Los defectos en el componente que causan una debilidad del metal. El defecto suele disminuir la resistencia a la tracción del metal y al aplicar una carga en el área del defecto muchas veces causa la quebradura.

Stroke The distance traveled by the piston from TDC to BDC.
Carrera La distancia que viaja el piston del PMS al PMI.

Sulfation A chemical action within a battery that interferes with the ability of the cells to deliver current and accept a charge.
Sulfatación Una reacción química dentro de la batería que estorba la habilidad de las celulas de entregar el corriente y aceptar una carga.

Surface-to-volume ratio A mathematical comparison between the surface area of the combustion chamber and the volume of the combustion chamber. The greater the surface area, the more area the mixture can cling to, and mixture which clings to the metal will not burn completely because the metal cools it. Typical surface-to-volume ratio is 7.5:1.
Indice superficie-volumen Una comparación matemática entre el área de la superficie de la cámara de combustión y el volúmen de la cámara de combustión. Lo mayor el área de la superficie, lo mayor el área en donde puede pegarse la mezcla y la mezcla pegada al metal no se quemará completamente puesto que la enfría el metal. El índice típico del superficie-volumen es el 7.5:1.

Sweep grinding A crankshaft grinding method using a stone that is swept back and forth across the journal surface.
Rectificado de barrido Un método de rectificación del cigüeñal empleando una muela que mueve de un lado al otro encima de la superficie de un muñón.

Synchronization timing Timing design used on vehicles with computer controlled fuel injection systems. A pick-up in the distributor is used to synchronize the crankshaft position to the camshaft position.

Regulación (tiempo) sincronizado   Un diseño de regulación empleado en los vehículos equipados con sistemas de inyección de combustible de control computerizado. Un captador en el distribuidor sirve para sincronizar la posición del cigüeñal con la posición del árbol de levas.

Tap   A tool used to repair or cut new internal threads.

Macho   Una herramienta que sirve para reparar o cortar las roscas interiores nuevas.

Telescoping gauge   A precision tool used in conjunction with outside micrometers to measure the inside diameter of a hole. Telescoping gauges are sometimes called snap gauges.

Calibrador telescópico   Una herramienta de precisión que se emplea junta con los micrómetros exteriores para medir los diámetros interiores de un agujero. Tambien se llaman calibradores de brocha.

Tensile strength   The metal's resistance to be pulled apart.

Resistencia a la tracción   La resistencia de un metal a ser estirado.

Thermal efficiency   A measurement comparing the amount of energy present in a fuel and the actual energy output of the engine.

Rendimiento térmico   Una medida que compara la cantidad de la energía presente en un combustible y la potencia actual producido por el motor.

Thermal cleaners (pyrolytic ovens)   A parts cleaner that uses high temperatures to bake the grease and grime into ash.

Limpiadores térmicos (hornos pirolíticos)   Una limpiadora de partes que usa las temperaturas elevadas para convertir la grasa y el lodo en cenizas.

Thermodynamics   The study of the relationship between heat energy and mechanical energy.

Termodinámica   El estudio de la relación entre la energía del calor y la energía mecánica.

Thermostat   A control device that allows the engine to reach normal operating temperatures quickly and maintains the desired temperatures.

Termostato   Un dispositivo de control que permite que el motor llegue rápidamente a las temperaturas de funcionamiento normales y mantenga las temperaturas deseadas.

Thread chaser (thread restorer)   A tool designed to roll the threads back into shape, not to cut new threads.

Peine de roscar a mano   Una herramienta diseñada a repujar las roscas dañadas hacia su estado original, no a cortar las roscas nuevas.

Thread depth   The height of the thread of a bolt from its base to the top of its peak.

Profundidad de la rosca   La longitud de la rosca de un perno desde su fondo a la parte superior.

Thread insert (helicoil)   A device that allows for major thread repairs while keeping the same size fastener.

Inserto preroscado (helicoil)   Un dispositivo que permite las reparaciones completas de las roscas manteniendo los retenes del mismo tamaño.

Three-minute charge test   A reasonably accurate method for diagnosing a sulfated battery on conventional batteries.

Prueba de carga de tres minutos   Un método bastante preciso de diagnosticar una batería sulfatada con las baterías ordinarias.

Throating   Machining process using a 60 degree stone to narrow the contact surface.

Rectificación   Un proceso de rebajar a máquina empleando una muela de 60 grados para disminuir la superficie de contacto.

Throw   The distance from the center of the crankshaft main bearing to the center of the connecting rod journal.

Codo del cigüeñal   La distancia del centro del muñón principal del cigüeñal al centro del muñón de la biela.

Throw-off   The quantity of oil that escapes at the end of the bearings, and lubricates adjacent engine parts while the engine is running.

Expulsión   La cantidad del aceite que escapa de las extremidades de los cojinetes y lubrifica las partes contiguas del motor al estar en marcha el motor.

Thrust bearing   A double flanged bearing used to prevent the crankshaft from sliding back and forth.

Cojinete de empuje   Un cojinete con doble brida que sirve para prevenir que el cigüeñal desliza de un lado a otro.

Timing   The process of identifying when an event is to occur.

Sincronización   El proceso de identificar cuando debe ocurrir un evento.

Timing chain   The chain the drives the camshaft off of the crankshaft.

Cadena de sincronización   La cadena procedente del cigüeñal que acciona al árbol de levas.

Tolerance   A permissible variation between the two extremes of a specification or dimension.

Tolerancia   Una variación que se permite entre los dos extremos de una especificación o una dimensión.

Top dead center (TDC)   Term used to indicate the piston is at the very top of its stroke.

Punto muerto superior (PMS)   El término que indica que el piston esta en la cima de su carrera.

Topping   Refers to the use of the 30 degree stone to lower the contact surface on the valve face.

Despunte   Refiere al uso de una muela de 30 grados para rebajar la superficie de contacto en la cara de la válvula.

Torque   A rotating force around a pivot point. For example, the twisting force applied to a bolt or shaft is called torque.

Torsión   Una fuerza que gira alrededor de un punto de pivote. Por ejemplo, la fuerza giratoria que se aplica en un perno o en un eje se llama torsión (par).

Torque convertor   A series of components that work together to reduce slip and multiply torque. It provides a fluid coupling between the engine and the automatic transmission.

Convertidor del par   Una serie de componentes que trabajan juntos para disminuir el deslizamiento y multiplicar el par. Provee un acoplamiento flúido entre el motor y la transmisión automática.

Torque plates   Metal blocks about two inches thick that are bolted to the cylinder block at the cylinder head mating surface to prevent twisting during honing and boring operations.

Chapas de torsión   Los bloques de metal midiendo unas dos pulgadas de grueso empernados al monoblock en la superficie de contacto de la cabeza del cilindro que previenen que se retuerce durante las operaciones de esmerilado y taladreo.

Torque wrench   Wrench that measures the amount of twisting force applied to a fastener.

Llave de torsión   Una llave que sirve para medir la cantidad de la fuerza de torsión que se aplica a una fijación.

Total engine displacement   The sum of displacements for all cylinders in an engine.

Cilindrada total del motor   La suma de todos los desplazamientos de los cilindros de un motor.

Transverse mounted engine Engine placement where the block faces from side to side instead of front to back within the vehicle.
Motor de montaje transversal La ubicación del motor en la cual la longitud del monoblock queda de un lado a otro dentro del vehículo en vez de estar colocado de frente a atrás.

Undersize bearing Bearing with the same outside diameter as standard bearings but constructed of thicker bearing material in order to fit an undersize crankshaft journal.
Cojinete de dimensión inferior Un cojinete que tiene el diámetro exterior del mismo tamaño que los cojinetes normales pero que se ha fabricado de una material más gruesa para que puede quedar en un muñón de cigüeñal más pequeño.

Vacuum A pressure lower than atmospheric pressure. Vacuum in the engine is created when the volume of the cylinder above the piston is increased. This results in the atmospheric pressure pushing the air/fuel mixture into the area of lowered pressure above the piston.
Vacío Una presión más baja que la presión atmosférica. El vacío en un motor se crea al aumentar el volumen del cilindro arriba del pistón. Esto resulta en que la presión atmosférica empuja la mezcla aire/combustible dentro del área arriba del piston.

Vacuum testing Testing that determines the engine's ability to provide sufficient pressure differentials to allow the induction of the air/fuel mixture into the cylinder. Results of vacuum testing indicate internal engine condition, fuel delivery abilities, ignition system condition, and valve timing.
Prueba del vacío Una prueba que determina la habilidad del motor en proveer las diferenciales de presión adecuadas que permiten la inducción de la mezcla aire/combustible al cilindro. Los resultados indicarán la condición interna del motor, las habilidades de entregar el combustible, la condición del sistema de encendido y la sincronización de las válvulas.

Valley pans A pan located under the intake manifold in the valley of a "V" type engine used to prevent the formation of deposits on the under side of the intake manifold.
Cárter en V Un cárter ubicado en la parte inferior del múltiple de admisión en la parte más baja de un motor tipo "V" que previene la formación de los depósitos en la parte inferior del múltiple de admisión.

Valve Device that controls the flow of gases into and out of the engine cylinder.
Válvula Un dispositivo que controla el flujo de los gases entrando y saliendo del cilindro del motor.

Valve cover gaskets Components that seal the connection between the valve cover and cylinder head. The valve cover gasket is not subject to pressures, but must be able to seal hot, thinning oil.
Juntas de la tapa de válvula Los componentes que sellan la conexión entre la tapa de las válvulas y la cabeza del cilindro. La junta de la tapa de válvula no se sujeta a la presión, pero si tiene que sellar el aceite caliente muy fluido.

Valve cupping A deformation of the valve head caused by heat and combustion pressures.
Embutición de la válvula Una deformación de la cabeza de la válvula causada por el calor y las presiones de la combustión.

Valve float A condition that allows the valve to remain open longer than it is intended. Valve float is the effect of inertia on the valve.
Flotación de la válvula Una condición que permite que queda abierta la válvula por más tiempo de lo designado. Es el efecto que tiene la inercia en la válvula.

Valve guide A part of the cylinder head that supports and guides the valve stem.
Guía de válvula Una parte de la cabeza del cilindro que apoya y guía al vástago de la válvula.

Valve guide bore gauge Instrument that provides a quick measurement of the valve guide. It can also be used to measure taper and out-of-round.
Verificador del calibrador para la guía de válvula Un instrumento que provee una medida rápida de la guía de la válvula. Tambien puede servir para medir lo cónico y lo ovulado.

Valve guide clearance Measurement of the difference between the valve stem diameter and the guide bore diameter.
Holgura de la guía de válvula La medida de la diferencia entre el diámetro del vástago de la válvula y el diámtero del taladro de la guía.

Valve overhang The area of the face between the seat contact and the margin.
Desborde de la válvula El área de la cara entre el contacto del asiento y el margen.

Valve overlap The length of time, measured in degrees of crankshaft revolution, which the intake and exhaust valves of the same combustion chamber are open simultaneously.
Períodos de abertura de las válvulas El período del tiempo, que se mide en los grados de las revoluciones del cigüeñal, en el cual las válvulas de admisión y de escape de la misma cámara de combustión estan abiertas simultáneamente.

Valve seat Machined surface of the cylinder head that provides the mating surface for the valve face. The valve seat can be either machined into the cylinder head or a separate component that is pressed into the cylinder head.
Asiento de la válvula La superficie acabada a máquina de la cabeza del cilindro que provee una superficie de contacto para la cara de la válvula. El asiento puede ser labrado a máquina en la cabeza del cilindro o puede ser un componente aparte para prensar en la cabeza de la válvula.

Valve seat grinding The process of restoring the surface of the valve seat by removing small amounts of material through the use a special grinding stones.
Rectificado de los asientos de las válvulas El proceso de restaurar la superficie al quitar las cantidades pequeñas de material por medio de las muelas especiales.

Valve seat recession The loss of metal from the valve seat which causes the seat to recede into the cylinder head.
Encastre del asiento de la válvula La pérdida del metal del asiento de la válvula, lo que causa que el asiento retroceda dentro de la cabeza del cilindro.

Valve seat runout gauge Instrument that provides a quick measurement of the valve seat concentricity.
Calibrador de la excentricidad del asiento de la válvula Un instrumento que provee una medida rápida de la concentricidad del asiento de la válvula.

Valve spring A coil of specially constructed metal used to force the valve closed, providing a positive seal between the valve face and seat.
Resorte de válvula Un rollo de metal de construcción especial que provee la fuerza para mantener cerrada la válvula, así proveyendo un sello positivo entre la cara de la válvula y el asiento.

Valve spring tension tester Device used to measure the open and closed valve spring pressures.
Probador de tensión del resorte de válvula Un dispositivo que se emplea en medir las presiones de los resortes de las válvulas en la posición abierta o cerrada.

Valve train The series of components that work together to open and close the valves.
Tren de válvulas Un serie de componentes que trabajan juntos para abrir y cerrar las válvula.

Vehicle Information Number (VIN) An alpha-numeric code consisting of seventeen characters used to properly identify the vehicle and its major components.

Número de Identificación del Vehículo (VIN) Un código alfanumérico que consiste de diez y siete caracteres que sirve para identificar correctamente al vehículo y sus componentes principales.

Vehicle lift points The areas that the manufacturer recommends for safe vehicle lifting. Lift points are structurally strong enough to sustain the stress of lifting. They are usually illustrated in the service manual or by the jack manufacturer. If in doubt, ask your instructor.

Puntos para izar el vehículo Las áreas recomendadas por el fabricante en donde se puede levantar al vehículo con seguridad. Son áreas que tienen la fuerza estructural para sostener la carga de izar. Los puntos para izar suelen ser ilustrados en el manual de servicio o por el fabricante del gato. Si hay dudas, consulte su instructor.

Vernier calipers Calipers that use a vernier scale which allows for measurement to a precision of 10 to 25 times as fine as the base scale.

Pie de rey Los calibres que emplean una escala vernier permitiendo una medida precisa de 10 a 25 veces más finas que la escala fundamental.

Vibration dampener See harmonic balancer.

Amortiguador de vibraciones Vea equilibrador armónico.

Viscosity The measure of oil thickness.

Viscosidad La medida de lo espeso de un aceite.

Volumetric efficiency A measurement of the amount of air/fuel mixture that actually enters the combustion chamber compared to the amount that could be drawn in.

Rendimiento volumétrico Una medida de la cantidad de la mezcla del aire/combustible que actualmente entra en la cámara de combustión comparada con la cantidad que podría entrar.

Warning label A label that is supplied by the manufacturer of the hazardous material. Each warning label lists the chemical name, applicable hazard warnings, hazardous ingredients, and manufacturer's name and address.

Marbete de aviso Un marbete proveido por el fabricante de las materiales peligrosas. Cada marbete especifica el nombre químico, los avisos pertinentes del peligro, y el nombre y la dirección del fabricante.

Wear-in The required time needed for the ring to conform to the shape of the cylinder bore.

Tiempo de estreno El tiempo que se requiere para que el anillo se conforme a la forma del taladro del cilindro.

Wet compression testing Compression test done after adding a small amount of oil to the cylinder. Wet compression testing is performed if the cylinder fails the dry compression test.

Prueba de compresión húmedo La prueba de compresión que se lleva acabo añadiendo una pequeña cantidad del aceite al cilindro. Se efectúa si el cilindro reprueba la prueba de compresión en seco.

Wet sleeves Replacement cylinder sleeves that are surrounded by engine coolant.

Camisas húmedas Las camisas de repuesta del cilindro que se rodean por el fluido refrigerante del motor.

2 Automatic Transmissions and Transaxles

Pretest

The purpose of this pretest is to determine the amount of review that you may require prior to writing the ASE Automatic Transmission and Transaxle Test. If you answer all the pretest questions correctly, complete the questions and study the information in this chapter to prepare for the ASE Automatic Transmission and Transaxle Test.

If two or more of your answers to the pretest questions are incorrect, complete a study of Chapters 3 through 8 in *Today's Technician Automatic Transmissions and Transaxles* published by Delmar Publishers, plus a study of the questions and information in this chapter.

The pretest answers are located at the end of the pretest. These answers also are in the answer sheets supplied with this book.

1. An automatic transaxle has a loss of automatic transmission fluid (ATF) and there are no visible external leaks.
   Technician A says the vacuum modulator diaphragm may be leaking.
   Technician B says the transaxle cooler may be leaking.
   Who is correct?
   A. A only
   B. B only
   C. Both A and B
   D. Neither A nor B

2. The ATF in a transmission is milky in color. The cause of this problem could be
   A. burned clutch discs.
   B. burned bands.
   C. a worn case.
   D. a leaking transmission cooler.

3. A non-computer-controlled transaxle has higher-than-specified fluid pressure in all gear selector positions.
   Technician A says the pump may be defective.
   Technician B says the pressure regulator valve may be stuck.
   Who is correct?
   A. A only
   B. B only
   C. Both A and B
   D. Neither A nor B

4. A computer-controlled transaxle provides torque converter clutch (TCC) lockup at 25 mph. The specified lockup vehicle speed is 46 mph.
   Technician A says the TCC lockup solenoid may be defective.
   Technician B says the vehicle speed sensor (VSS) may be defective.
   Who is correct?

A. A only
B. B only
C. Both A and B
D. Neither A nor B

5. If an electronic defect occurs in a four-speed computer-controlled transaxle, the transaxle will
    A. operate only in reverse.
    B. provide 1 - 2 and 2 -3 upshifts.
    C. provide only one forward gear.
    D. provide only first gear.

6. When a vacuum gauge is connected into the transmission modulator vacuum hose with a T-fitting, the vacuum is a steady 12 in. Hg. with the engine idling.
    Technician A says the ignition timing may be later than specifed.
    Technician B says the engine may have a sticking valve.
    Who is correct?
    A. A only
    B. B only
    C. Both A and B
    D. Neither A nor B

7. During an air pressure test on the forward clutch, the clutch application is not heard, but there is no air escaping. The cause of the problem could be
    A. damaged forward clutch piston seals.
    B. a cracked forward clutch drum.
    C. a plugged forward clutch fluid passage.
    D. damaged forward clutch hub steel rings.

8. All the upshifts occur at a higher speed than specified in a non-computer-controlled transaxle.
    Technician A says the cause of this problem may be high governor pressure.
    Technician B says the cause of this problem may be low throttle pressure.
    Who is correct?
    A. A only
    B. B only
    C. Both A and B
    D. Neither A nor B

9. A vehicle has a computer-controlled transaxle, cruise control, and an electronic instrument panel. The cruise control module is separate from the powertrain control module. The torque converter clutch, cruise control, and speedometer are inoperative. The cause of this problem could be a defective
    A. vehicle speed sensor.
    B. powertrain control module.
    C. engine coolant temperature sensor.
    D. crankshaft sensor.

10. A four-speed automatic transmission slips in second and fourth gear.
    Technician A says the 2 - 4 band adjustment may be too loose.
    Technician B says the 2 - 4 servo piston seal may be leaking.
    Who is correct?
    A. A only
    B. B only
    C. Both A and B
    D. Neither A nor B

11. A computer-controlled transmission suddenly shifts into neutral while driving with the gear selector in the overdrive position. The cause of this problem could be a defective
    A. manual valve position sensor (MLPS).
    B. shift solenoid.
    C. transmission oil temperature (TOT) sensor.
    D. output speed sensor (OSS).

12. During a transmission stall test the engine rpm is higher than specified. All of these defects could be the cause of the problem EXCEPT
    A. a seized one-way stator clutch in the torque converter.
    B. slipping clutch discs.
    C. slipping band.
    D. lower than specified fluid pressure.

13. The input shaft end play in an automatic transaxle is less than specified. The cause of this problem could be
    A. a worn pump.
    B. a worn transaxle case.
    C. improper selective washer thickness.
    D. improper forward clutch clearance.

14. A fuel-injected engine with a four-speed automatic transaxle has a loss of power at high speed. All of the following defects could be the cause of the problem EXCEPT
    A. seized one-way stator clutch in the torque converter.
    B. low fuel system pressure.
    C. slipping one-way stator clutch in the torque converter.
    D. restricted exhaust system.

Answers to Pretest

1. C, 2. D, 3. B, 4. B, 5. C, 6. A, 7. C, 8. C, 9. A, 10. A, 11. A, 12. A, 13. C, 14. C

General Transmission/Transaxle Diagnosis

ASE Tasks, Questions, and Related Information

In this chapter each task in the Automatic Transmission and Transaxle category is followed by a question and some information related to the task. If you answer any question incorrectly, study this information very carefully until you understand the correct answer. For additional information on any task refer to Today's *Technician Automatic Transmissions and Transaxles,* by Delmar Publishers.

Question answers and analysis are provided at the end of this chapter and in the answer sheets provided with this book.

Task 1 **Listen to driver's complaint and road test vehicle; determine needed repairs.**

1. The customer complains about fluid usage in an automatic transmission, and there are no visible signs of fluid leaks.
   Technician A says the transmission vent may be plugged or restricted.

Technician B says the vacuum modulator diaphragm may be leaking.

Who is correct?

A. A only

B. B only

C. Both A and B

D. Neither A nor B

Hint   *(Refer to the general diagnostic procedure provided in Chapter 1). Prior to the road test check the transmission fluid level and condition. Some transmission problems are caused by improper fluid level or contaminated fluid. Since transmission problems may be caused by improper engine performance, correct any engine performance problems prior to the road test. The engine and transmission should be at normal operating temperature prior to the road test. During the road test operate the transmission in all gear selector positions, and operate the engine under various operating conditions. Operate the vehicle under the conditions where the driver complaint occurred. Record all abnormal transmission operation, noises, vibrations.*

Task 2   Diagnose noise and vibration problems; determine needed repairs.

2. An automatic transmission has a whining noise that occurs in all gears while driving the vehicle. This noise is also present with the engine running and the vehicle stopped.

   Technician A says the rear planetary gear set may be defective.

   Technician B says the oil pump may be defective.

   Who is correct?

   A. A only

   B. B only

   C. Both A and B

   D. Neither A nor B

3. A rear-wheel drive vehicle has a vibration that increases in relation to vehicle speed. This vibration also is present when the engine is accelerated with the vehicle stopped and the gear selector in neutral or park. The cause of this problem could be

   A. improper torque converter balance.

   B. improper drive shaft balance.

   C. improper drive shaft angles.

   D. worn engine mounts.

Hint   *Refer to Table 4-2, page 97, in* Today's Technician Automatic Transmissions and Transaxles, *by Delmar Pubishers, for complete diagnosis of automatic transmission noises.*

   When diagnosing noise and vibration problems pay careful attention to the exact conditions when the noise or vibration occurs. For example, if a vibration occurs while driving the vehicle and with the engine running and the vehicle not moving, the cause of the vibration is likely in the engine or torque converter. When the vibration changes with a change in engine speed, the problem may be in the torque converter. If the vibration changes with a change in vehicle speed, the problem probably is in the driveline or transmission output shaft.

Task 3   Diagnose unusual fluid usage, level, and condition problems; determine needed repairs.

4. Fluid sometimes escapes from the dipstick tube on an automatic transaxle.

   Technician A says the transaxle fluid may be contaminated.

   Technician B says the transaxle cooler may be defective.

   Who is correct?

   A. A only

   B. B only

   C. Both A and B

   D. Neither A nor B

5. Lubricant is leaking from the torque converter access cover. When this cover is removed, the shell of the converter is wet with fluid, but the front of the converter is dry. The cause of the problem could be
   A. a leaking transmission oil pump seal.
   B. a leaking rear main bearing.
   C. a loose rear main bearing.
   D. a leaking converter drain plug.

6. The fluid in an automatic transaxle is a dark brown color and smells burned.
   Technician A says this problem may be caused by a worn front planetary sun gear.
   Technician B says this problem may be caused by worn friction-type clutch plates.
   Who is correct?
   A. A only
   B. B only
   C. Both A and B
   D. Neither A nor B

Hint   *Refer to Table 3-1, page 60, in* Today's Technician Automatic Transmissions and Transaxles, *by Delmar Publishers, for diagnosis of transmission and transaxle oil leaks.*
   Refer to Table 3-2, page 60, in Today's Technician Automatic Transmissions and Transaxles, *by Delmar Publishers, for diagnosis of transmission and transaxle overheating.*
   Normal automatic transmission fluid is pink or red. If the fluid is dark brown or blackish and has a burned odor, the fluid has been overheated, possibly from burned clutches or bands. Milky-colored fluid likely is caused by coolant contamination from a leaking transmission cooler. Silvery metal particles in the fluid indicate damaged metal transmission components. If the dipstick feels sticky, and is difficult to wipe clean, the fluid contains varnish. This varnish formation indicates transmission fluid and filter changes have been neglected.

Task 4   Perform pressure tests; determine needed repairs.

7. An automatic transaxle has low pressure in third gear only.
   Technician A says the transaxle may have an internal leak.
   Technician B says the TV cable may be misadjusted.
   Who is correct?
   A. A only
   B. B only
   C. Both A and B
   D. Neither A nor B

8. All the transmission pressures are normal at idle speed, but low at wide-open throttle.
   Technician A says the TV cable may need adjusting.
   Technician B says the vacuum modulator may be defective.
   Who is correct?
   A. A only
   B. B only
   C. Both A and B
   D. Neither A nor B

9. During a transmission pressure test the pressure gradually decreases at higher engine speeds. The cause of this problem could be
   A. a worn oil pump.
   B. a restricted oil filter.
   C. a stuck pressure regulator.
   D. a plugged modulator hose.

Hint   *Transmission or transaxle pressure tests may be performed to diagnose internal problems such as rough or improperly timed shifts. These two problems may be caused by excessive line pressure which may be tested with a pressure test. To perform the pressure test the technician requires two* *pressure gauges, a tachometer, and specifications for the transmission being tested. If the transmission has a vacuum modulator, then a vacuum gauge and a hand vacuum pump also are required. Various plugs in the transaxle case must be removed to install the pressure gauges (Figure 2-1).*

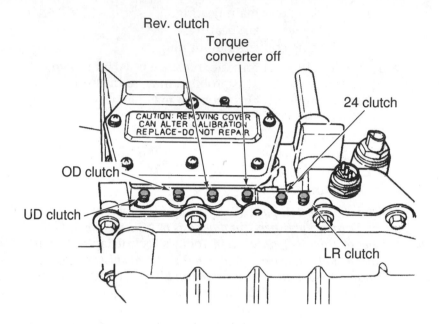

Figure 2-1   Pressure test plug locations *(Courtesy of Chrysler Corporation)*

If the transaxle pressures are low at slow idle the problem may be in the pump, filter, fluid level, or pressure regulator. An internal leak may also be the cause of this problem.

When the pressure is low only in a specific gear, the problem is likely an internal leak. A partially plugged transaxle oil filter may cause a gradual reduction in pressure at higher engine speeds.

If all the pressures are low at wide open throttle (WOT), pull on the TV cable or disconnect the vacuum modulator hose and test the pressures. When the pressures increase, check the TV cable adjustment, cable condition, vacuum modulator and hose.

Task 5   Perform stall tests; determine needed repairs.

10. During a stall test the engine rpm is less than specified.
    Technician A says the turbine in the torque converter may be defective.
    Technician B says some of the transmission clutches may be slipping.
    Who is correct?
    A. A only
    B. B only
    C. Both A and B
    D. Neither A nor B

11. During a stall test the stall speed is above specifications.
    Technician A says the exhaust system may be restricted.
    Technician B says the engine may have low compression.
    Who is correct?
    A. A only
    B. B only
    C. Both A and B
    D. Neither A nor B

Hint   *During a stall test a tachometer is connected to the ignition system, and this meter must be located where it can be seen by the driver. Apply the parking brake and place blocks in front of the vehicle's tires. Press and hold the brake pedal and place the gear selector in drive. Press the accelerator pedal to the WOT position, and note the rpm on the tachometer at which the engine stalled.*

When the stall speed is below specifications, the torque converter stator clutch may be slipping, or the exhaust system may be restricted. An engine that is improperly tuned or has low compression also reduces stall speed.

If the stall speed is above specifications the clutches and bands in the transmission may be slipping.

Task 6   Perform lock-up converter tests; determine needed repairs.

12. A torque converter clutch does not lock up at any vehicle speed or engine temperature. The cause of the problem could be
    A. a defective TCC solenoid.
    B. a defective stator clutch.
    C. a leaking second speed servo piston seal.
    D. a sticking pressure regulator valve.

13. The engine shudders immediately after TCC lockup.
    Technician A says the engine may have an ignition defect.
    Technician B says the fuel injection system may have a lean condition.
    Who is correct?
    A. A only
    B. B only
    C. Both A and B
    D. Neither A nor B

14. The TCC system in Figure 2-2 locks up at 28 mph rather than the specified speed of 42 mph. The cause of this problem could be a defective
    A. MAP sensor.
    B. P/N switch.
    C. ECT sensor.
    D. VSS sensor.

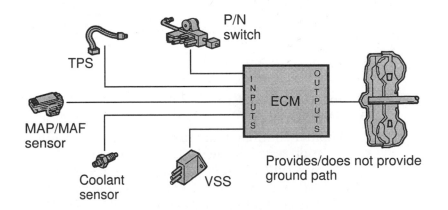

Figure 2-2 Typical PCM inputs for torque converter clutch control *(Courtesy of Hydra-Matic Division, General Motors Corporation)*

15. Engine stalling sometimes occurs when the brakes are applied and the engine is decelerated from 55 mph.
    Technician A says the TPS sensor may be defective.
    Technician B says the brake switch may be defective.

Who is correct?

A. A only

B. B only

C. Both A and B

D. Neither A nor B

Hint   *Nearly all modern lockup torque converters are controlled by the powertrain control module (PCM). While the converter is unlocked the fluid is directed through the input shaft and out in front of the converter lockup plate. This action keeps the lockup plate away from the front of the converter. When the converter is locked, the fluid is directed into the hub area of the converter, and over the top of the turbine. Fluid is then directed against the back of the lockup plate. This fluid movement forces the lockup plate against the front of the converter so the friction material on this plate contacts the front of the converter.*

The PCM operates a torque converter clutch (TCC) solenoid in the transmission. This solenoid supplies, or does not supply, fluid pressure to a valve in the valve body. This fluid pressure determines the valve position which in turn supplies fluid to the appropriate converter location to lock or unlock the converter.

The PCM uses input information from the engine coolant temperature (ECT) sensor, throttle position sensor (TPS), vehicle speed sensor (VSS), manifold absolute pressure (MAP) sensor, park neutral switch, and brake pedal switch to control the TCC. The PCM does not lock the torque converter if the engine coolant is below a specific temperature, the throttle is wide open, or the vehicle speed is below a certain value. When the brake pedal is depressed, the brake pedal switch signals the PCM to release the TCC. In many transaxles TCC lockup is provided in third or fourth gear.

Task 7   **Diagnose electronic, mechanical, and vacuum control systems; determine needed repairs.**

16. A computer-controlled transaxle remains in second gear at all forward vehicle speeds. The cause of the problem may be

    A. a worn oil pump.

    B. a restricted filter.

    C. a defective shift solenoid.

    D. an improper linkage adjustment.

17. While diagnosing a computer-controlled transaxle a diagnostic trouble code representing the TCC solenoid is obtained on a scan tester.

    Technician A says one of the TCC solenoid wires may be grounded inside the transaxle.

    Technician B says there may be an open wire in the transaxle electrical connector.

    Who is correct?

    A. A only

    B. B only

    C. Both A and B

    D. Neither A nor B

18. A computer-controlled transaxle does not shift into fourth gear. When using a transmission tester all the shifts occur normally. The cause of this problem could be

    A. an open shift solenoid winding inside the transaxle.

    B. an open wire from a shift solenoid to the transaxle connector.

    C. a restricted fluid passage through a shift solenoid.

    D. an open shift solenoid wire from the transaxle electrical connector to the PCM.

19. The vacuum gauge is connected to the modulator system on a non-computer-controlled transmission with a T-fitting (Figure 2-3). With the engine idling, the vacuum is 18 in. Hg. With the engine running at 2,000 rpm with a steady throttle, and the vehicle speed at 55 mph, the vacuum is 2 in. Hg. The shifts occur at a higher speed than specified, and the stall test rpm is lower than specified.

    Technician A says the ignition timing may be later than specified.

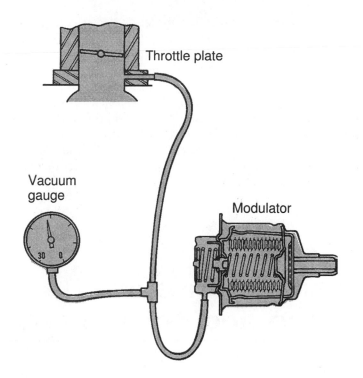

Figure 2-3   Testing vacuum modulator vacuum system   *(Courtesy of Hydra-Matic Division, General Motors Corporation)*

Technician B says the exhaust system may be restricted.
Who is correct?
A. A only
B. B only
C. Both A and B
D. Neither A nor B

20. During an air pressure test on the reverse clutch, the clutch application is not heard, and a hissing noise is evident.
Technician A says the transmission case may be cracked.
Technician B says the reverse clutch drum may be cracked.
Who is right?
A. A only
B. B only
C. Both A and B
D. Neither A nor B

Hint   *Transaxle electronic systems may be tested with a scan tester, multimeter, or a transmission tester. The scan tester is connected to the data link connector (DLC) under the instrument panel, and data regarding transaxle inputs and outputs are displayed on the scan tester. Diagnostic trouble codes (DTCs) related to transaxle defects also are displayed on the scan tester.*

If a DTC indicates a defect in a specific area (such as the TCC solenoid), an ohmmeter may be used to test the solenoid winding, and the connecting wires from the solenoid to the PCM to locate the exact cause of the problem. Many computer-controlled transmissions and transaxles have a default mode in the computer that causes the transmission to remain in one forward gear if an electrical defect occurs that could result in a hazardous condition. In this mode many transmissions and transaxles remain in second gear.

Some vehicle manufacturers recommend the use of a transmission tester. This tester is connected in series with the transaxle wiring harness connector at the transaxle. The tester replaces the PCM, and the technician can operate the tester to perform various transaxle shifts and functions. If the transaxle defect is present when using the tester, the problem is inside the transaxle. When the problem disappears while using the transaxle tester, the input sensors, wiring harness, or PCM are the cause of the problem.

When pressure tests, and a road test indicate the problem is in one of the transaxle apply components, air pressure tests may be performed to test these components. With the valve body removed, apply clean moisture-free air at 40 psi to the appropriate passage in the transaxle housing. These passages are identified in the service manual. When air pressure is applied to a component such as the forward clutch, the technician should be able to hear the clutch piston application. If excessive air can be heard escaping, there is an internal leak, which is probably located at the forward clutch piston seal.

Transmission/Transaxle Maintenance and Adjustment

ASE Tasks, Questions, and Related Information

Task 1   **Inspect, adjust, and replace manual valve shift linkage.**

21. Technician A says an improper shift linkage adjustment may cause premature transmission clutch failure.
    Technician B says an improper shift linkage adjustment may cause higher than normal fluid pressure.
    Who is correct?
    A.   A only
    B.   B only
    C.   Both A and B
    D.   Neither A nor B

Hint   *Most automatic transmissions or transaxles have a cable or rod-type shift linkage, connected from the gear selector lever to the transmission lever. If the shift linkage adjustment is not correct, the manual valve is improperly positioned (Figure 2-4). Improper manual valve position may result in low fluid pressure, improper clutch application, and excessive clutch wear. To check the shift linkage adjustment, remove the shift linkage or cable from the transmission lever. Place the gear selector in the specified position, which often is the park position. Move the transmission lever to the park position and install and tighten the linkage on the transmission lever. Move the gear selector through all the gear positions; be sure there is a detent in each position.*

Task 2   **Inspect, adjust, and replace cables or linkages for throttle valve (TV), kickdown, and accelerator pedal.**

22. When the throttle valve cable is improperly adjusted so throttle pressure is higher than normal, the transmission shifts occur
    A.   at a lower vehicle speed than specified.
    B.   at the specified vehicle speed.
    C.   at the same vehicle speed.
    D.   at a higher vehicle speed than specified.

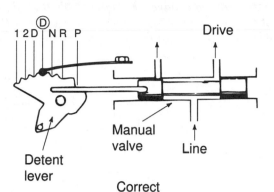

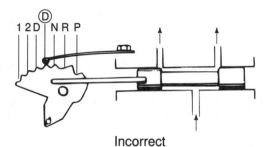

Figure 2-4 Wrong manual valve position caused by improper shift linkage adjustment *(Courtesy of Hydra-Matic Division of General Motors Corporation)*

Hint

The TV cable connects the accelerator pedal to the throttle valve in the transmission valve body. In some transmissions the TV cable movement controls the throttle valve and the downshift valve. In some applications the vacuum modulator controls the throttle valve and the TV cable controls the downshift valve.

The throttle valve position controls throttle pressure. Upshifts occur when governor pressure overcomes throttle pressure on a specific shift valve. Downshifts occur when throttle pressure overcomes governor pressure and moves a certain shift valve.

An improperly adjusted TV cable may cause lower than normal throttle pressure and early upshifts. This misadjustment may also result in high throttle pressure and delayed, harsh upshifts.

A typical TV cable adjustment involves releasing the cable lock tab, and pulling the cable fully in the readjust direction (Figure 2-5). Hold the throttle lever fully clockwise against its stop and press the lock tab into the locked position.

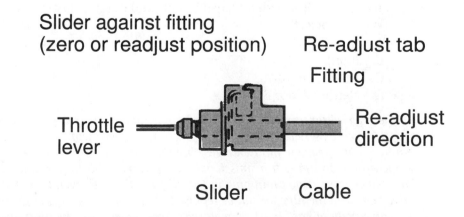

Figure 2-5 Throttle valve cable adjustment *(Courtesy of Chevrolet Motor Division of General Motors Corporation)*

Some transmissions have a kickdown switch in place of a downshift linkage and valve. When the throttle is wide open, the throttle linkage closes the kickdown switch and operates a downshift solenoid in the transmission to provide the necessary downshift (Figure 2-6).

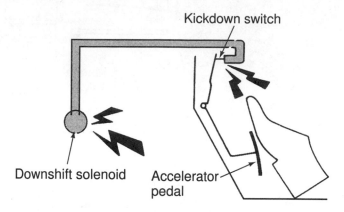

Figure 2-6 Kickdown switch *(Courtesy of Nissan Motor Co., Ltd.)*

In computer-controlled transmissions and transaxles the computer operates shift solenoids to control all upshifts and downshifts in response to sensor input signals, and a throttle valve and downshift valve are not required.

Task 3   Adjust bands.

23. An improper band adjustment may cause
    A. shifts at a lower vehicle speed than specified.
    B. transmission slipping in some gears.
    C. shifts at a higher vehicle speed than specified.
    D. transmission slipping in all gears.

Hint   *Many band adjustments may be performed externally, whereas in other transmissions the oil pan must be removed to complete this adjustment. Improper band adjustment may cause harsh shifting, slipping, band burning, and premature failure.*

To complete the band adjustment loosen the adjuster locknut, and tighten the adjuster to the specified torque to seat the band. Loosen the band adjuster the specified number of turns, and hold the adjuster in this position while tightening the locknut to the specified torque.

Task 4   Replace fluid and filter(s).

24. Technician A says that transmission filters may be cleaned and reused.
    Technician B says that transmission fluid oxidizes faster at lower temperatures.
    Who is correct?
    A. A only
    B. B only
    C. Both A and B
    D. Neither A nor B

Hint   *The interval for transmission and transaxle fluid changes depends on the type of transmission and the type of service. Severe service such as trailer towing, or commercial vehicle operation, requires more frequent fluid changes, because this type of service involves higher transmission temperatures. In many transmissions and transaxles the oil pan must be removed to drain the fluid, since there is no drain plug in the oil pan. The filter usually is bolted to the bottom of the valve body and the filter should be replaced not cleaned. A gasket or O-ring between the filter and the valve body or case must be replaced. Fiber material in the oil pan indicates worn bands or clutches. Steel particles in the oil pan indicates damaged components such as gear sets or the oil pump. Aluminum particles in the oil pan indicates a damaged case.*

Task 5   **Inspect, adjust, and replace electronic sensors, wires, and connectors.**

25. A permanent magnet generator-type speed sensor generates
    A. a DC voltage.
    B. an on/off signal.
    C. an AC voltage.
    D. a square wave signal.

Hint   *Many sensors (such as vehicle speed sensors and transmission input speed or output speed sensors) are permanent magnet generators. With the vehicle supported properly on a lift, operate the vehicle with the transmission in drive. Connect an AC voltmeter to the sensor terminals. The AC voltage generated by the sensor should increase gradually and smoothly in relation to vehicle speed. Replace the sensor if the proper signal is not obtained. Some vehicle speed sensors contain eight permanent magnets rotating past a reed switch. This type of sensor should provide eight pulses per revolution on an ohmmeter connected to the sensor terminals.*

In-Vehicle Transmission/Transaxle Repair

ASE Tasks, Questions, and Related Information

Task 1   **Inspect, adjust, and replace vacuum modulator, valve, lines, and hoses.**

26. A vehicle is operating where the temperature is 0°F, and the transmission has a vacuum modulator. The transmission experiences repeated clutch piston seal failures, and the driver complains about harsh, late shifting.
    Technician A says moisture may be freezing in the vacuum modulator diaphragm chamber.
    Technician B says the manual valve shift linkage may require adjusting.
    Who is correct?
    A. A only
    B. B only
    C. Both A and B
    D. Neither A nor B

Hint   *The vacuum hose and vacuum modulator may be tested by connecting a vacuum gauge to the modulator hose with a T-fitting. If the vacuum is less than specified, check the engine vacuum, connecting hose, and modulator. When a hand-operated vacuum pump is connected to the modulator, it should hold 18 in. Hg. of vacuum. Some vacuum modulators may be adjusted with a hand-operated vacuum pump and the proper length of gauge pins.*

Task 2   **Inspect, adjust, repair, and replace governor cover, seals, sleeve, valve, weights, springs, retainers, and gear.**

27. The shifts in an automatic transaxle occur at a higher speed than specified. All of these items could be the cause of the problem EXCEPT
    A. a sticking governor valve.
    B. excessive governor spring tension.
    C. worn governor weights and pins.
    D. weak governor spring tension.

Hint   *When a road test and transmission pressure tests indicate a governor problem, the governor should be removed, cleaned, and inspected. Improper shift points is one of the most common complaints related to governor operation. The transmission must be disassembled to access some governors. Other transmissions have a governor that is removable from the outside of the transmission (Figure 2-7).*

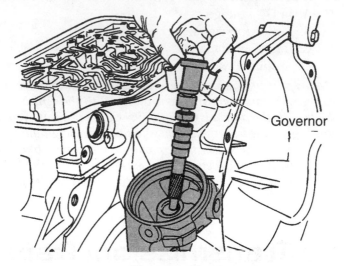

Figure 2-7   Externally removeable governor *(Courtesy of Ford Motor Company)*

Many governors contain a primary and secondary valve. The primary governor valve must be removed first, followed by the secondary valve retaining pin, spring, and secondary valve. All worn components must be replaced. After cleaning and lubricating with ATF, these valves must move freely in their bores. Some governors are mounted on the transmission output shaft; a drive ball in this shaft forces the governor to rotate with the shaft (Figure 2-8).

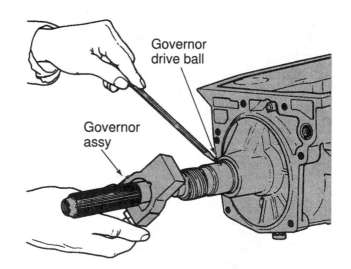

Figure 2-8   Governor drive ball in the output shaft *(Courtesy of Ford Motor Company)*

Task 3   Inspect and replace external seals and gaskets.

28. A transaxle experiences repeated pump seal failure.
    Technician A says the pump body bushing may be worn.
    Technician B says the governor pressure is higher than specified.
    Who is correct?
    A. A only
    B. B only
    C. Both A and B
    D. Neither A nor B

Hint   *Seal mounting bores must be inspected for scratches, metal burrs, and cracks. All seals must be removed with the proper puller, and installed with the appropriate seal driver. Since a plugged vent may cause seal leakage, always inspect the vent for restrictions.*

Task 4   Inspect, repair, and replace extension housing.

29. An extension housing bushing and the bushing contact area on the drive shaft slip yoke are severely pitted and scored, and the housing seal is leaking. There are no other transmission problems or complaints.

    Technician A says there may be excessive resistance in the battery ground between the chassis and the engine.

    Technician B says the transmission fluid may be contaminated because fluid and filter changes have not been performed.

    Who is correct?
    A. A only
    B. B only
    C. Both A and B
    D. Neither A nor B

Hint   *Inspect the extension housing for cracks, and metal burrs in the seal bore and transmission case mating surfaces. Replace the extension housing bushing if it is worn. Inspect the speedometer drive assembly for worn components.*

Task 5   Inspect, test, flush, and replace cooler, lines, and fittings.

30. With the engine idling, the ATF flow through a transmission cooler should be
    A. one quart in 60 seconds.
    B. one pint in 60 seconds.
    C. one quart in 40 seconds.
    D. one quart in 20 seconds.

Hint   *Transmissions may have a cooler in one of the radiator tanks, an external cooler, or both. If the internal cooler develops a leak, the ATF becomes contaminated with coolant, and the coolant is contaminated with ATF. When ATF is contaminated with coolant, the ATF has a milky color, and the coolant may be visible as bubbles on the transmission dipstick. If ATF enters the coolant, the ATF may be seen floating on top of the coolant when the radiator cap is removed. A leaking internal transmission cooler causes a loss of ATF without any visible signs of external leaks.*

A transmission cooler may be leak tested by plugging one cooler line and supplying up to 75 psi air pressure to the other line. Bubbles and/or air escaping indicate the leak's location.

Restricted transmission coolers may cause overheating of the transmission resulting in severe damage to clutches, bands, and seals. To test the transmission cooler for restriction, disconnect the outlet line and connect a hose from the outlet into an empty container. Start the engine and allow it to run for 20 seconds. One quart of ATF should flow through the cooler into the container. If the cooler flow is less than one quart, the cooler or inlet line is restricted. Transmission coolers may be flushed with compressed air or an approved cleaning solution and compressed air.

Compressed air should be used to remove debris from external cooler air passages. All cooler lines should be inspected for leaks and restrictions.

Task 6   Inspect and replace speedometer drive gear, driven gear, and retainers.

31. An erratic speedometer could be caused by all of these defects EXCEPT
    A. missing drive gear retaining clip.
    B. dry speedometer cable.
    C. worn speedometer gears.
    D. worn driven gear retaining bushing.

Hint   *The speedometer drive and driven gears should be inspected for worn or damaged teeth. In some transmissions the speedometer drive gear is machined into the output shaft, whereas in other transmissions this drive gear is splined onto the output shaft or held in place with a clip (Figure 2-9). If the speedometer drive gear is a machined part of the output shaft, the complete shaft must be replaced if this gear is damaged. Inspect all gear retaining clips for looseness or damage. The speedometer driven gear and bushing assembly is retained in the transaxle housing with a bolt (Figure 2-10). If the speedometer driven gear bushing assembly is loose, it should be replaced.*

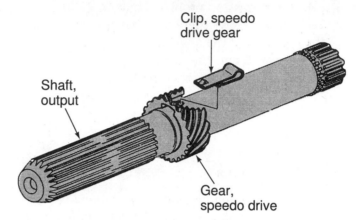

Figure 2-9   Speedometer drive gear held on the output shaft with a clip *(Courtesy of Hydra-Matic Division of General Motors Corporation)*

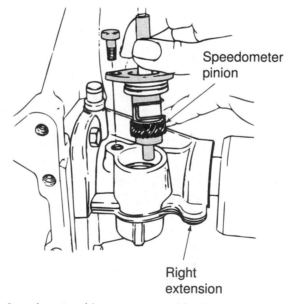

Figure 2-10   Speedometer driven gear assembly *(Courtesy of Chrysler Corporation)*

Task 7   Inspect valve body mating surfaces, bores, valves, spring, sleeves, retainers, brackets, check balls, screens, spacers, and gaskets; replace as necessary.

32. The clearance between the valves and matching valve body bores should not exceed
    A. 0.001 in.
    B. 0.003 in.
    C. 0.005 in.
    D. 0.008 in.

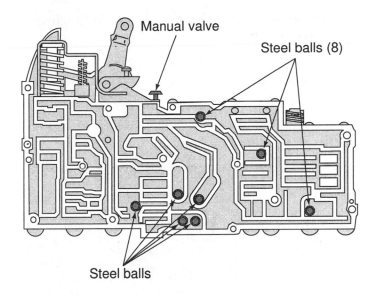

Figure 2-11   Valve body and steel balls *(Courtesy of Chrysler Corporation)*

Hint   *When pressure tests indicate there may be a problem in the valve body, this component should be removed, cleaned, and inspected. The valve body may be removed with the transmission in the vehicle, or after transmission removal for an overhaul. When the valve body is removed, be careful not to lose the steel balls (Figure 2-11). The appropriate service manual usually provides a diagram showing the proper ball positions.*

When the valves, springs, pins, and balls are removed from the valve body note the position of each component, and lay these components in the proper order on clean, lint-free shop towels.

Clean the valve body in an approved cleaning solution. Check all valves and bores for scratches and metal burrs. Minor scratches or metal burrs may be removed from valves with crocus cloth. If a valve does not move freely in its bore after cleaning, replace the valve. Check for wear on all valves and bores. The clearance between each valve and the matching bore should not exceed 0.001 in. If a bore is worn, replace the valve body.

Always install new valve body gaskets, and place the gasket, or gaskets, on the valve body to be sure they fit properly and do not cover any openings. Valve body and transmission case mating surfaces should be checked for warpage with a straightedge.

Task 8   Check/adjust valve body bolt torque.

33. After a non-computer-controlled transaxle and valve body overhaul, the transaxle does not complete a 1-2 upshift, and shifts from first gear to third gear. All other shifts are normal, and this problem was not present before the overhaul.
    Technician A says the valve body torque may be excessive.
    Technician B says the governor pressure may be too low.
    Who is correct?
    A. A only
    B. B only
    C. Both A and B
    D. Neither A nor B

Hint   *Be sure each valve body bolt is installed in the proper position. These bolts usually are different lengths. Installing these bolts in the wrong holes may cause improper torque or transmission case damage. Be sure the gaskets and spacer are properly positioned prior to valve body installation. Clean petroleum jelly may be used to hold the gaskets in place during valve body installation. Valve body bolts must be tightened to the specified torque in the proper sequence. If this bolt torque is excessive, valve body warping and valve sticking may occur.*

Task 9   **Inspect servo bore, piston, seals, pin, spring, and retainers; repair/replace as necessary.**

34. The procedure shown in Figure 2-12 is to determine the proper:
   A. servo piston thickness.
   B. band anchor length.
   C. servo piston selective pin.
   D. servo piston adjustment.

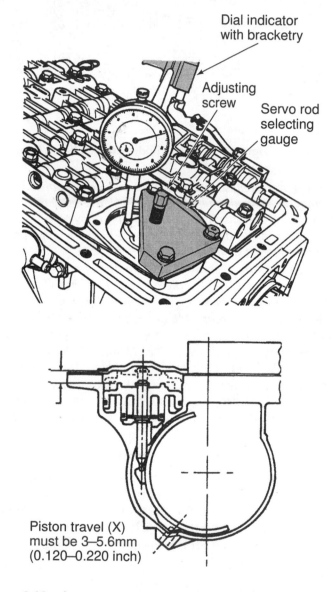

Figure 2-12   Servo measurement *(Courtesy of Ford Motor Company)*

Hint   *A servo is a hydraulically operated piston and pin that is used to apply a band. Servo pistons and bores must be cleaned and inspected for cracks, scratches, metal burrs, and wear. Servo piston-to-bore clearance usually is 0.003 to 0.005 in. Servo piston seals may be Teflon® or cast iron with hooked ends. Some servo pistons have molded rubber seals. All piston seals and bores must be lubricated with the proper ATF prior to installation. Transmissions without band adjustment screws have a selective servo apply pins, and the technician must determine the proper pin length to provide proper band application.*

Task 10   Inspect accumulator bore, piston, seals, spring, and retainer; repair/replace as necessary.

35. An automatic transaxle has a complaint of harsh 3-4 upshifts. All the other shifts are normal.
Technician A says the fourth accumulator piston may be stuck.
Technician B says the pressure regulator valve is sticking.
Who is correct?
A. A only
B. B only
C. Both A and B
D. Neither A nor B

Hint   *Accumulators are spring loaded pistons that provide a hydraulic cushion to reduce shift harshness. Accumulator piston, seal, and spring service is similar to servo service. Some accumulator pistons have O-ring seals which must be replaced during an overhaul (Figure 2-13).*

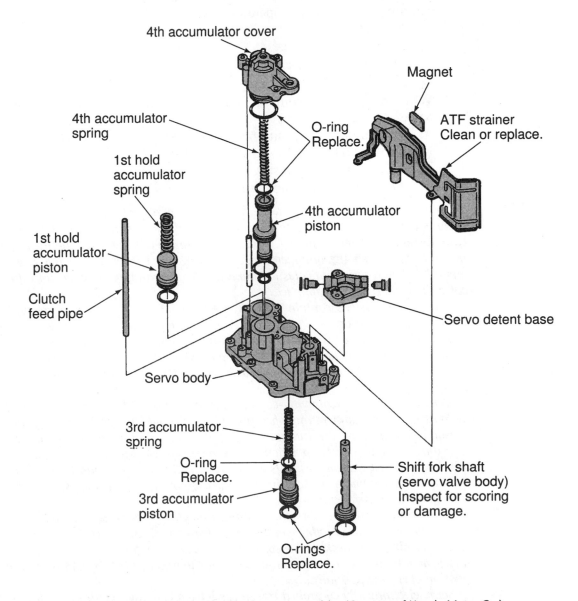

Figure 2-13   Accumulator and servo assembly *(Courtesy of Honda Motor Co.)*

Task 11 Inspect and replace parking pawl, shaft, spring, and retainer.

36. All of these statements about a parking pawl are true EXCEPT
    A. the parking pawl locks the input shaft.
    B. the parking pawl is mechanically operated.
    C. the parking pawl's projection engages in a notched drum.
    D. the parking pawl is retained on a pivot pin.

Hint   *The parking pawl opening and pivot pin should be inspected for wear. Check the pawl spring for proper tension. The projection on the pawl should be checked for wear. Inspect the linkage from the pawl to the shift lever for proper operation and wear.*

Task 12 Inspect, test, adjust, or replace electrical/electronic components including computers, solenoids, sensors, relays, switches, and harnesses.

37. A computer-controlled transaxle has a relay that supplies voltage to the solenoids and switches in the transaxle when the ignition switch is turned on. The computer senses a defect in the input speed sensor, and does not close the relay. Under this condition the transaxle operates in
    A. first gear and reverse.
    B. second gear and reverse.
    C. first and second gear.
    D. second and third gear.

38. When diagnosing a computer-controlled transmission or transaxle the first sensor to be verified should be
    A. the manual lever position sensor (MLPS).
    B. the transaxle input speed sensor (TISS).
    C. the transaxle output speed sensor (TOSS).
    D. the transaxle oil temperature sensor (TOT).

Hint   *Solenoids, such as the TCC or shift solenoids, may be tested with an ohmmeter. When the ohmmeter leads are connected to the solenoid terminals, a reading below the specified value indicates a shorted solenoid winding. If an infinite ohmmeter reading is obtained, the solenoid winding is open. When the ohmmeter leads are connected from one of the solenoid terminals to ground, a low reading indicates a grounded solenoid winding, whereas an infinite reading indicates the winding is not grounded.*

Some transmissions contain normally open, or normally closed, oil pressure switches. These switches usually inform the computer regarding the present operating gear in the transmission. These switches may be closed with air pressure for test purposes. When air pressure is applied to a normally open switch, an ohmmeter connected to the switch terminals should indicate a very low reading. With no pressure applied to the switch, an infinite ohmmeter reading should be obtained.

When diagnosing a computer-controlled transmission or transaxle one of the first tests should be to verify the adjustment and operation of the manual lever position sensor (MLPS). The MLPS sensor is operated by the manual valve shift linkage, and this sensor informs the computer regarding the gear selector position. The computer then provides the selected gear. As the gear selector is moved, various resistors inside the MLPS sensor are connected through the sensor terminals to the computer. A defective MLPS sensor, or an improper sensor adjustment, may cause improper transmission shifting.

Many transmission sensors are permanent magnet generators that produce an AC voltage signal proportional to rotational speed in the transmission. An ohmmeter may be connected to these sensor terminals to check the sensor winding for open, circuits, grounds, and shorts. An AC voltmeter may be connected to the sensor terminals to measure the AC voltage signal with the transmission operating normally.

Some electronically controlled transaxles have a computer-controlled relay that supplies voltage to the transmission solenoids and switches when the ignition switch is turned on. If the computer senses a serious electronic defect it does not energize the relay, and voltage supply to the solenoids and switches is shut off. Under this condition the transaxle operates in only second gear while moving forward. Since reverse is a function of manual valve position, this gear is still available. The relay may be tested by energizing the relay winding with a 12V battery and connecting an ohmmeter to the relay contact terminals. The ohmmeter should indicate very low resistance across the relay contacts when the winding is energized.

A scan tool may by connected to the data link connector (DLC) under the dash to obtain data and diagnostic trouble codes (DTCs) from the transmission computer.

Task 13   Inspect, replace, and align power train mounts.

39. A front-wheel-drive vehicle experiences intermittent shifting. Sometimes the transaxle shifts normally, and occasionally it misses a shift.
    Technician A says the manual valve shift linkage may need adjusting.
    Technician B says the engine or transaxle mounts may be broken.
    Who is correct?
    A. A only
    B. B only
    C. Both A and B
    D. Neither A nor B

Hint   *Worn or broken engine or transmission mounts may cause excessive engine and transaxle or transmission movement during acceleration. This excessive movement changes the effective length of the shift and throttle cables or linkages, which may result in erratic transaxle shifting. On a front-wheel-drive (FWD) vehicle worn or improperly positioned engine cradle mounts may cause improper transaxle shifting, drive axle vibration, and incorrect front suspension angles. Some engine cradles have an alignment hole to check cradle alignment with the chassis (Figure 2-14).*

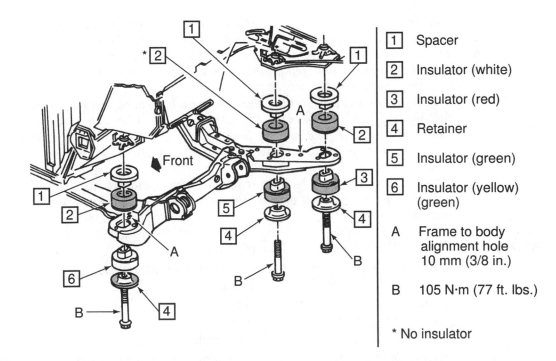

Figure 2-14   Engine cradle, mounts, and alignment hole from a front wheel drive (FWD) vehicle *(Courtesy of Oldsmobile Division, General Motors Corporation)*

Worn or broken engine or transmission mounts on a rear-wheel-drive (RWD) vehicle may cause improper shifting and incorrect drive shaft angles, resulting in vibration.

To check the mounts lift up on the engine, transmission, or transaxle and check for excessive movement in the mounts. The torque reaction of the mounts also should be checked. Set the parking brake and apply the foot brake firmly. Start the engine and place the gear selector in drive. Accelerate the engine to 2,000 rpm and observe the engine movement. Excessive engine and transmission, or transaxle movement indicates weak or broken mounts.

Off-Vehicle Transmission/Transaxle Repair, Removal, Disassembly, and Assembly

ASE Tasks, Questions, and Related Information

Task 1   **Remove and replace transmission/transaxle.**

40. All these statements about transmission/transaxle removal and replacement are true EXCEPT

   A. the negative battery cable should be disconnected prior to transmission removal.

   B. the drive shaft should be marked in relation to the differential flange.

   C. the front drive axles should be marked in relation to the front hubs.

   D. the engine support fixture should be installed before loosening the transaxle to engine bolts.

Hint   *Always disconnect the battery ground cable prior to transaxle or transmission removal. The transaxle fluid should be drained prior to removal. Raise the vehicle on a lift, and place a large drain pan under the oil pan. If the transaxle oil pan does not have a drain plug, remove the bolts on one side of the oil pan and loosen the bolts on the other side to allow the pan to drop partially downward to drain the fluid.*

On a front-wheel-drive vehicle the drive axles have to be removed prior to transaxle removal. Most manufacturers recommend loosening the outer drive axle retaining nuts with the drive wheel contacting the floor and the brakes applied. On rear-wheel-drive vehicles the drive shaft must be removed prior to transmission removal. Always mark the drive shaft in relation to the differential flange.

Always install an engine support fixture before loosening the transaxle to engine retaining bolts. Support the transaxle or transmission securely on a transmission jack during removal and installation.

Task 2   **Disassemble, clean, and inspect.**

41. If the input shaft end play in Figure 2-15 is more than specified

   A. the input shaft must be replaced.

   B. the pump must be replaced.

   C. the transmission case is worn.

   D. a thicker selective washer is required.

Hint   *Before the transmission or transaxle is disassembled, end play should be measured on various components as recommended in the vehicle manufacturer's service manual. Components usually are removed in this order; external components such as the modulator and bell housing, valve body, servos, accumulators, pump, input shaft and front clutch and gear set, rear clutch and gear set. Visually inspect all components during disassembly. Clean all components with an approved cleaning solution, and inspect all parts for wear and damage. Always inspect the oil pan for band and clutch material, and aluminum, steel, or brass cuttings.*

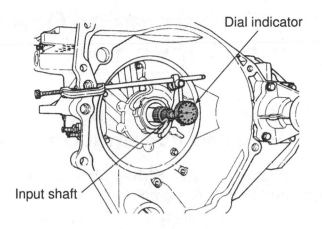

Figure 2-15   Input shaft end play measurement *(Courtesy of Chrysler Corporation)*

Task 3   Assemble after servicing.

42. While performing the measurement in Figure 2-16 the dial indicator reading is more than specified. To correct this problem install

    A. a thicker selective thrust washer.

    B. new friction and steel clutch plates.

    C. a thicker reaction plate.

    D. a new clutch drum.

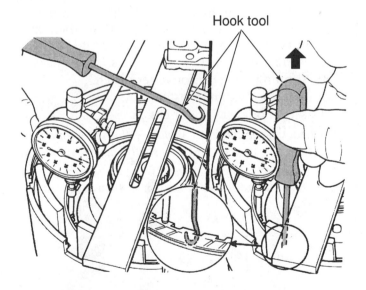

Figure 2-16   Clutch plate clearance measurement *(Courtesy of Chrysler Corporation)*

Hint   *During transmission or transaxle assembly, replace all seals, gaskets, and O-rings. Lubricate all seal lips, O-rings, and clutch plates with the specified type of ATF. All seals must be installed in the proper direction. Be sure all components are installed in their original location. Thrust washers may be held in place with clean petroleum jelly. Perform all end play measurements and band adjustments recommended in the vehicle manufacturer's service manual. A general assembly procedure is: rear gear set clutches and bands, front gear set clutches and bands, pump, valve body, servos, accumulators, and external components.*

Off-Vehicle Transmission/Transaxle Repair, Oil Pump and Converter

ASE Tasks, Questions, and Related Information

Task 1   **Inspect converter flex (drive) plate, converter attaching bolts, converter pilot, and converter pump drive surfaces.**

43. An engine has a clunking noise that usually occurs during deceleration and sometimes with the engine idling. The cause of this noise could be loose
   A. main bearings.
   B. converter to flex plate bolts.
   C. connecting rod bearings.
   D. piston pins.

Hint   *The converter flex plate should be inspected for cracks and warping. Inspect the ring gear for broken, damaged teeth. Check the torque on the flex plate bolts, and inspect the flex plate bolt holes for wear. Check the converter attaching bolts for wear and proper torque. The converter hub must be smooth in the pump seal contact area. Minor scratches in this area may be removed with fine crocus cloth. If the converter hub is scored or scratched in the seal contact area, replace the converter. Inspect the converter drive lugs for wear.*

Task 2   **Inspect, flush, measure end play, and test torque converter.**

44. A turbocharged engine has normal power during acceleration at lower speeds, but a loss of power at higher speeds. The compression, ignition, and fuel system are proven to be in satisfactory condition, there are no unusual noises, and all shifts occur normally. All of these defects could be the cause of the problem EXCEPT
   A. damaged turbocharger bearings.
   B. slipping one-way converter clutch.
   C. restricted exhaust system.
   D. seized one-way stator clutch.

Hint   *After removal the converter should be flushed with a converter flusher to remove any contaminated fluid. The converter end play should be measured with a dial indicator (Figure 2-17).*

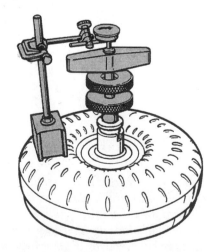

Figure 2-17   Converter end play measurement *(Courtesy of Buick Motor Division of General Motors Corporation)*

When your finger is inserted into the inner stator splined race, the stator should rotate freely in a clockwise direction, but lock up in a counterclockwise direction. Some manufacturers recommend the use of a special tool to apply a specific amount of torque to the stator in the locking direction (Figure 2-18).

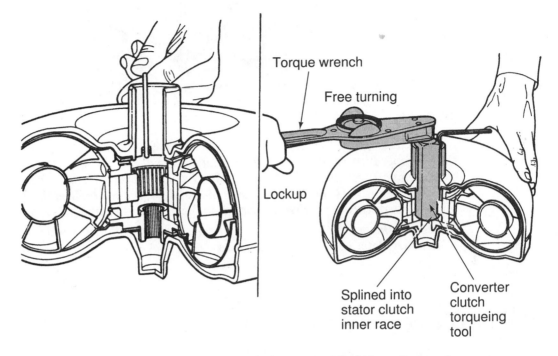

Figure 2-18   Stator check *(Courtesy of Ford Motor Company)*

Stator to turbine interference in the converter may be checked by inserting the pump and drive assembly into the converter hub. Hold the converter and pump and rotate the turbine with the input shaft. The turbine should rotate freely without any noise.

Stator to impeller interference may be checked by placing the oil pump on a bench, and inserting the converter over the stator support splines. Turn the converter until the converter hub is engaged in the pump drive. Rotate the converter counterclockwise while holding the pump. The converter should rotate freely without any noise.

Task 3   Inspect, measure, and replace oil pump housing and parts.

45. A transmission experiences a pump seal failure, and during this failure the transmission was severely overheated. The pump and seal were replaced during a transmission overhaul, and all the converter tests were satisfactory. The transmission now has a high-pitched whining noise that increases in relation to engine speed. This noise is present with the engine running in park or neutral, and while driving the car. The engine has normal power, and there are no other transmission complaints.

    Technician A says the converter hub may be misaligned.

    Technician B says the converter one-way clutch may be slipping.

    Who is correct?

    A. A only
    B. B only
    C. Both A and B
    D. Neither A nor B

Hint   *Before removing the oil pump gears, the gears should be marked in relation to each other so they may be assembled in the original position. Inspect the gears and pump surfaces for wear, scoring, and damage. Inspect the pump bushing for wear. Check the stator shaft for looseness in the pump cover.*

Figure 2-19   Clearance measurement oil pump gears to pump cover

Use a feeler gauge to measure the clearance between the outer gear and the housing. Measure the clearance between the outer gear teeth and the crescent. Use a feeler gauge and a straightedge to measure between the top of the gears and the pump cover (Figure 2-19). On a vane-type pump measure the thickness of the rotor, vanes, and slide with a micrometer. If any of these clearances and measurements are not within specifications, replace the pump.

Replace the pump O-ring, gasket, seal rings, seal, and thrust washer. Clean and inspect the pressure regulator valve, guide, and spring.

Off-Vehicle Transmission/Transaxle Repair, Gear Train, Shafts, Bushings, and Case

ASE Tasks, Questions, and Related Information

Task 1   Check end play and/or preload; determine needed service.

46. A transmission experiences repeated pump seal failure, and there are no other transmission complaints.

   Technician A says the pressure regulator valve may be sticking.

   Technician B says the drainback hole behind the seal may be plugged.

   Who is correct?

   A. A only

   B. B only

   C. Both A and B

   D. Neither A nor B

Hint   *Prior to transaxle disassembly measure the input shaft end play with a dial indicator (Figure 2-20). With the dial indicator stem positioned against the input shaft, pull this shaft in and out to measure the end play. If this measurement is not within specifications, install a number four thrust plate with the proper thickness to provide the specified end play during transaxle reassembly. The number four thrust plate is located behind the overdrive clutch hub and in front of the front sun gear (Figure 2-21). End play measurements vary depending on the transmission or transaxle. Always perform the end play measurements outlined in the vehicle manufacturer's service manual.*

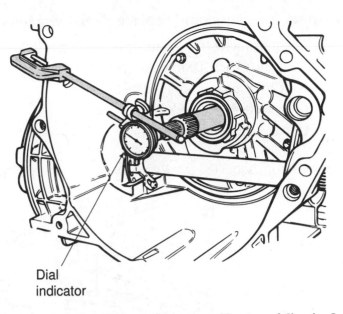

Dial
indicator

Figure 2-20   Input shaft end play measurement *(Courtesy of Chrysler Corporation)*

Front sun gear
assembly

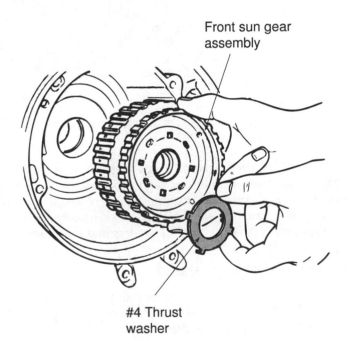

#4 Thrust
washer

Figure 2-21   Number 4 thrust washer *(Courtesy of Chrysler Corporation)*

Task 2   Inspect, measure, and replace thrust washers and bearings.

47. The purpose of the measurement in Figure 2-22 is to determine the proper
    A. clutch pack retaining ring thickness.
    B. clutch pack reaction plate thickness.
    C. selective washer thickness.
    D. steel clutch plate thickness.

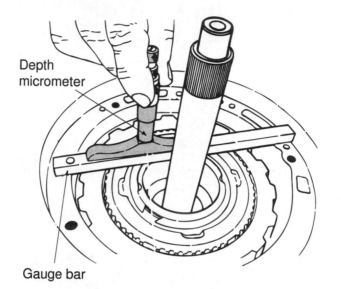

Depth micrometer

Gauge bar

Figure 2-22   Measurement at reverse clutch drum face *(Courtesy of Ford Motor Company)*

Hint   *All thrust washers should be inspected for wear, scoring, flaking, and damaged tabs. Some thrust washers are available in various thicknesses to provide the proper end play between components. When thrust washers are worn, component clearance is reduced which may cause clunking and a rubbing noise. Thrust washer thickness may be measured with a micrometer. All bearings should be inspected for looseness and roughness.*

Task 3   Inspect and replace shafts.

48. The vehicle owner complains about harsh torque converter clutch operation in the four-speed transmission partially shown in Figure 2-23. Converter clutch lockup does occur at the specified speed.

    Technician A says the vehicle speed sensor may be defective.

    Technician B say the number nine check ball in the turbine shaft is sticking.

    Who is correct?
    A. A only
    B. B only
    C. Both A and B
    D. Neither A nor B

Hint   *The bushing contact area on all shafts must be inspected for scoring, and wear. During a transmission overhaul replace all shaft O-rings, and Teflon® or cast iron sealing rings. Most Teflon® sealing rings must be cut from the shaft. Some shafts contain fluid passages and a check ball. Blow out these passages with compressed air, and be sure the check ball moves freely and seats properly.*

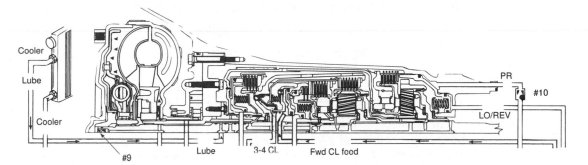

Figure 2-23 Four-speed automatic transmission with torque converter clutch lockup *(Courtesy of Chevrolet Motor Division, General Motors Corporation)*

Task 4 Inspect oil delivery seal rings, ring grooves, and sealing surface areas.

49. The tool in Figure 2-24 is used to
    A. size Teflon® oil seal rings.
    B. align the oil pump halves.
    C. check for gear interference.
    D. install Teflon® oil seal rings.

Figure 2-24 Oil pump service tool *(Courtesy of Chevrolet Motor Division, General Motors Corporation)*

Hint   *All sealing rings should be inspected for wear, scoring, scratches, and nicks. Check all ring grooves for wear. Inspect the sealing surfaces contacted by the sealing rings for wear, scoring, and ridges.*

Task 5 Inspect and replace bushings.

50. All of these statements about bushing inspection and measurement are true EXCEPT
    A. a scored shaft in the bushing contact area may indicate a lack of lubrication.
    B. shaft to bushing clearance may be measured with a wire-type feeler gauge.
    C. shaft to bushing clearance may be measured with a vernier caliper and a micrometer.
    D. normal bushing clearance is 0.015 to 0.025 in.

Hint   *All bushings should be inspected for wear, scoring, and burning. Some shafts have internal bushings on which other shafts are supported. These internal bushings must always be checked for wear. Since the shaft contains much harder material compared to a bushing, a scored shaft usually indicates lack of lubrication from a plugged passage, or a bushing in which the oil feed hole is not aligned with the oil hole in the case.*

A wire-type feeler gauge may be inserted between the bushing and the shaft to measure bushing wear. If this type of measurement is not possible, the inside diameter of the bushing, and the outside diameter of the shaft may be measured with a vernier caliper and micrometer to determine the bushing clearance. Normal shaft to bushing clearance is 0.0005 to 0.015 in. One of the most critical bushing fits is the converter hub to pump bushing clearance which is 0.0005 in.

Task 6 Inspect and measure planetary gear assembly; replace parts as necessary.

51. To provide reverse gear with a planetary gear set
   A. the carrier is held and the annulus gear drives the sun gear.
   B. the sun gear is held and the annulus gear drives the carrier.
   C. the carrier is held and the sun gear drives the annulus gear.
   D. the annulus gear is held and the sun gear drives the carrier.

Hint   *All planetary gear teeth should be checked for wear, chips, and damage. Inspect all splines in the gear sets and drums for wear and damage. Measure the end play between all planetary pinion gears and the planetary carrier. Be sure the planetary pinions rotate freely and smoothly. Inspect all thrust washers and thrust bearings for wear.*

Task 7   Inspect, repair, and replace case(s) bores, passages, bushings, vents, and mating surfaces.

52. Technician A says case porosity between two fluid passages could result in pressure bleed off and clutch burn out.

   Technician B says case porosity between two fluid passages could result in the transmission being in two gears at once resulting in jam up.

   Who is correct?
   A. A only
   B. B only
   C. Both A and B
   D. Neither A nor B

Hint   *The transmission or transaxle case must be thoroughly cleaned with an approved cleaning solution. Blow out all passages with compressed air. Be sure to remove and clean all filter screens in case passages. Inspect and blow out the vent. Blow out all threaded openings with compressed air, and inspect the thread condition in all threaded openings.*

Inspect the case for porosity and cracks. If porosity is suspected in an oil passage, plug one end of the passage and supply air pressure to the other end. When the passage is porous, air can be heard escaping, or ATF bubbles when placed on the porous area.

Inspect all bores such as servo, seal, and bearing bores for scratches, scoring, and ridges. Minor scratches may be removed with crocus cloth. Deep scratches, scoring, and ridges require case replacement. Inspect all clutch plate splines in the case for wear and damage. Use a straightedge to check the valve body mating surface for warpage.

Task 8   Inspect, repair or replace transaxle drive chains, sprockets, gears, bearings, and bushings.

53. The tool in Figure 2-25 is used to
   A. remove and install the drive chain and sprockets.
   B. measure drive chain and sprocket wear.
   C. measure wear in sprocket support bearing wear.
   D. install sprocket chain tightener.

Hint   *Inspect the drive chain and sprockets for wear and damage. The sprocket support bushings or bearings should be checked for wear, scoring, and flaking. Chain condition may be measured by deflecting the chain in each direction with a screwdriver. Measure the amount of chain deflection, and compare this measurement to specifications. If the chain deflection is excessive, chain and/or sprocket replacement is necessary.*

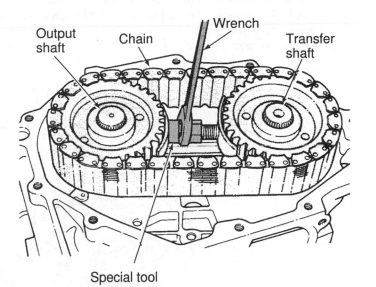

Figure 2-25   Drive chain service tool *(Courtesy of Chrysler Corporation)*

Task 9   **Inspect, measure, repair, adjust or replace transaxle final drive components.**

54. The turning torque in Figure 2-26 is less than specified. To correct this problem
   A. thicker spacers are required behind both differential side gears.
   B. a thicker spacer is required behind one side gear bearing cup.
   C. thicker spacers are required behind both side bearing cups.
   D. thicker spacers are required behind both side bearings.

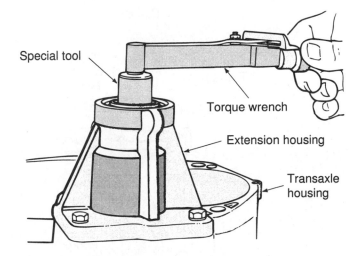

Figure 2-26   Differential turning torque measurement *(Courtesy of Chrysler Corporation)*

Hint   *Inspect the differential ring gear and pinion gear for chipped and damaged teeth. The differential side bearings and bearing cups should be inspected or scoring, pitting, and damage. Check the side gears, pinion gears, and spacers for chipped and worn teeth, and scoring or wear on the spacer surfaces. Inspect the side gear and pinion gear contact areas in the case for scoring or wear.*

Typical differential adjustments are side gear end play, differential end play, and turning torque. End play must be measured on each side gear with a special tool and a dial indicator placed against the side gear. Move the side gear up and down to measure the end play (Figure 2-27). Side gear spacers are available in various thickness to obtain the specified end play.

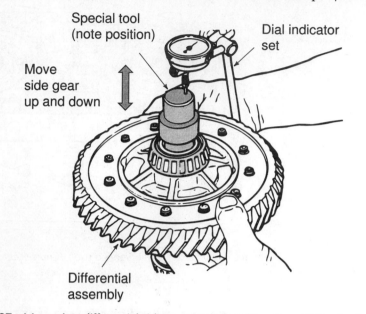

Figure 2-27   Measuring differential side gear end play *(Courtesy of Chrysler Corporation)*

The differential end play is measured with the differential assembled in the case, and a gauging shim installed behind the bearing cup in the differential bearing retainer. Pry the differential upward with a dial indicator and special tool installed against the upper side of the differential (Figure 2-28). The specified shim thickness is equal to the end play plus a specific thickness for side bearing preload. After the differential is assembled with the proper shim thickness for differential end play and side bearing preload, the differential turning torque must be measured.

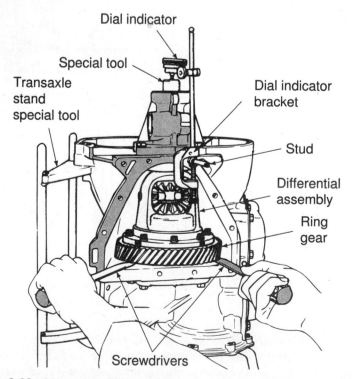

Figure 2-28   Measuring differential end play *(Courtesy of Chrysler Corporation)*

Friction and Reaction Units

ASE Tasks, Questions, and Related Information

Task 1 **Inspect clutch assembly; replace parts as necessary.**

55. The tool in Figure 2-29 is used to install clutch
    A. pistons and oil seals.
    B. hub Teflon® rings.
    C. hub cast iron rings.
    D. drum snap rings.

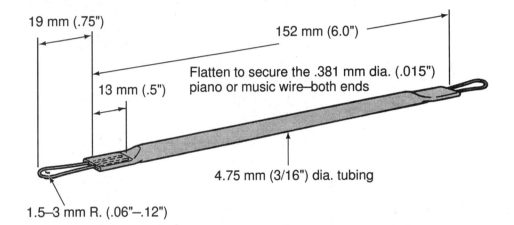

19 mm (.75")

152 mm (6.0")

13 mm (.5")

Flatten to secure the .381 mm dia. (.015")
piano or music wire—both ends

4.75 mm (3/16") dia. tubing

1.5–3 mm R. (.06"–.12")

Figure 2-29 Clutch pack service tool *(Courtesy of Oldsmobile Division, General Motors Corporation)*

Hint   *A spring compressor must be installed on some clutch packs prior to removing the retainer. The snap ring may be removed from other clutch packs without using a spring compressor. Always refer to the vehicle manufacturer's service manual. Some clutch packs are retained in the transaxle or transmission case with a snap ring.*

Inspect all the friction clutch plates for wear, flaking, damage, overheating, glazing, and warping. Inspect the steel plates for wear, scoring, damage, and overheating. Use a straightedge to measure steel plate warpage. Check all clutch plate splines for damage. Inspect wave, or reaction, plates for damage. Be sure the check balls in the drum and piston are free and seat properly. Measure the clutch piston spring height.

Do not interchange clutch pack components. Prior to reassembly soak all clutch discs for thirty minutes in the specified ATF. Replace all clutch piston and other seals. Be sure these seals fit properly; coat all seals and seal contact areas with the specified ATF. Seal lips must face in the direction that fluid pressure is against the seal.

Task 2 **Measure and adjust clutch pack clearance.**

56. All of the following problems could result in clutch disc burning EXCEPT
    A. sticking clutch drum check ball.
    B. reduced clutch pack clearance.
    C. a damaged clutch piston seal.
    D. higher than specified line pressure.

Hint   *After the clutch pack is assembled, the clutch pack clearance must be measured with a feeler gauge or dial indicator. The feeler gauge usually is inserted between the backing plate and the upper clutch disc. On some clutch packs backing plates of various thickness are available to.*

adjust the clearance, whereas in other clutch packs this clearance is adjusted with snap rings of different thicknesses.

Task 3   Air test the operation of clutch and servo assemblies.

57. The tool in Figure 2-30 is used for
    A. oil passage location.
    B. check ball location.
    C. clutch pack air pressure testing.
    D. valve body positioning.

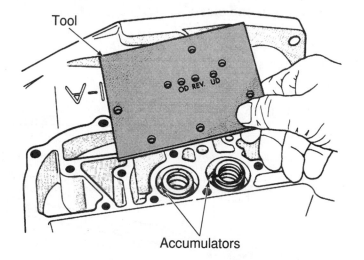

Figure 2-30   Transaxle service tool   *(Courtesy of Chrysler Corporation)*

Hint   *Some vehicle manufacturers recommend assembling the pump and front clutch assembly. Apply 35 psi air pressure to the clutch apply port, and listen to the clutch apply and release. The clutch should apply with a noticeable thud, and release quickly. If air can be heard escaping at the clutch piston, the piston seals are leaking. Some air may escape at the metal or Teflon® rings in the clutch drum hub.*

Other vehicle manufacturers recommend assembling the clutch packs, gear sets, and pump, and then applying air pressure to the appropriate apply passages in the transmission case. Servos may be air pressure tested in the same way as clutch packs.

Task 4   Inspect one-way clutch assemblies; replace parts as necessary.

58. A four-speed automatic transmission has a buzzing noise in second, third, and fourth gear. This noise does not occur in first gear.
    Technician A says the lo/roller clutch may be severely brinnelled.
    Technician B says the forward clutch discs may be badly worn.
    Who is correct?
    A. A only
    B. B only
    C. Both A and B
    D. Neither A nor B

Hint   *After disassembly the rollers in a roller clutch should be inspected for smooth finishes with no evidence of flatness. If any rollers are scored, pitted, or have flat surfaces, replace the roller clutch. When the roller surfaces indicate brinnelling, the roller clutch has been subjected to severe loads from hard driving or transmission abuse. Inspect the roller clutch races and cams for wear, scoring, pitting, and damage.*

In a sprag clutch inspect the sprag faces and races for wear, scoring, and torn face surfaces. When a roller clutch or sprag is worn or damaged, always inspect the fluid lubrication hole to the unit to ensure there is adequate lubrication.

After assembly be sure the sprag or roller clutch allows the two components to freewheel in one direction and hold in the opposite direction (Figure 2-31). Be sure the holding and freewheeling action is in the proper direction.

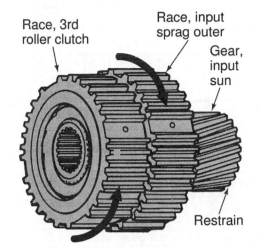

Must freewheel in direction of arrows and hold in opposite direction.

Race, 3rd roller clutch

Race, input sprag outer

Gear, input sun

Restrain

Figure 2-31 Checking sprag or roller clutch operation *(Courtesy of Hydra-Matic Division of General Motors Corporation)*

Task 5   Inspect and replace bands and drums.

59. A transaxle experiences premature band failure, and the band is worn severely on the outer edges, but not in the center.

    Technician A says the band strut may be worn unevenly resulting in band misalignment.

    Technician B says the band contact area on the clutch drum is dished.

    Who is correct?

    A. A only

    B. B only

    C. Both A and B

    D. Neither A nor B

60. A four-speed automatic transmission slips in first gear or manual first gear, and there is no engine braking when coasting in first gear (Figure 2-32, next page). The cause of these problems could be

    A. a defective O/D one-way clutch.

    B. a defective forward clutch.

    C. a defective intermediate band.

    D. a defective rear one-way clutch.

Hint   *Bands should be inspected for chips, cracks, burning, glazing, and uneven wear patterns. Be sure the band material is not worn excessively. Inspect all the band anchors, struts, levers, and pivots for wear. The clutch drum surface on which the band makes contact should be inspected for burning scoring, glazing, and distortion. Use a straightedge to check the drum for flatness in the band contact area. If the clutch drum is dished in the band contact area, band distortion and uneven, rapid band wear occurs.*

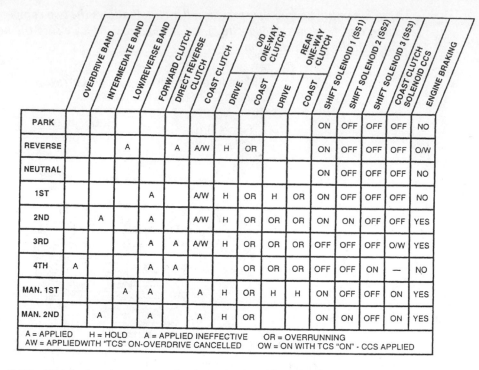

	Overdrive Band	Intermediate Band	Low/Reverse Band	Forward Clutch	Direct Reverse Clutch	Coast Clutch	O/D One-Way Clutch Drive	O/D One-Way Clutch Coast	Rear One-Way Clutch Drive	Rear One-Way Clutch Coast	Shift Solenoid 1 (SS1)	Shift Solenoid 2 (SS2)	Shift Solenoid 3 (SS3)	Coast Clutch Solenoid CCS	Engine Braking
PARK											ON	OFF	OFF	OFF	NO
REVERSE			A		A	A/W	H	OR			ON	OFF	OFF	OFF	O/W
NEUTRAL											ON	OFF	OFF	OFF	NO
1ST				A		A/W	H	OR	H	OR	ON	OFF	OFF	OFF	NO
2ND		A		A		A/W	H	OR	OR	OR	ON	ON	OFF	OFF	YES
3RD				A	A	A/W	H	OR	OR	OR	OFF	OFF	OFF	O/W	YES
4TH	A			A	A			OR	OR	OR	OFF	OFF	ON	—	NO
MAN. 1ST			A	A		A	H	OR	H	H	ON	OFF	OFF	ON	YES
MAN. 2ND		A		A		A	H	OR			ON	ON	OFF	ON	YES

A = APPLIED   H = HOLD   A = APPLIED INEFFECTIVE   OR = OVERRUNNING
AW = APPLIEDWITH "TCS" ON-OVERDRIVE CANCELLED   OW = ON WITH TCS "ON" - CCS APPLIED

Figure 2-32   Clutch, band, one-way clutch, and solenoid application *(Courtesy of Ford Motor Company)*

Answers and Analysis

1. B   A plugged or restricted transmission vent may cause pressure buildup in the transmission and loss of fluid from some of the seals, but these leaks would be visible. Therefore, A is wrong.

A leaking modulator diaphragm would allow the intake manifold vacuum to pull transmission fluid through the diaphragm into the intake manifold; this leak is not externally visible. B is right.

2. B   The rear planetary gear set is not turning with the engine running and the vehicle stopped, so a defective rear planetary gear set would not cause this noise. Therefore, A is wrong.

Since the oil pump is turning continually with the engine running, this component may be the cause of the whining noise. B is right.

3. A   Improper drive shaft balance, drive shaft angles, or worn engine mounts would not cause a vibration with the engine running and the vehicle stopped. Therefore, B, C, and D are wrong.

Improper torque converter balance would result in a vibration with the engine running and the vehicle stopped, or while driving the vehicle. A is the right answer.

4. C   Contaminated transaxle fluid may cause excessive fluid foaming, and fluid may be forced from the dipstick tube, so A is right.

A defective transaxle cooler may cause transaxle overheating which results in fluid being forced from the dipstick tube. A leaking transaxle cooler may contaminate the transaxle fluid with coolant resulting in fluid foaming and fluid escaping from the dipstick tube. Therefore, both A and B are right, and the correct answer is C.

5. A A leaking rear main bearing seal or loose rear main bearing may cause a leak from the converter access cover but the front of the converter would be wet with this problem. Therefore, B and C are wrong.

If the converter has a drain plug it usually is located in the front of the converter, so a leak at this location would also result in fluid on the front of the converter.

A leak at the transmission oil pump would cause the shell of the converter to be wet, so A is right.

6. B Dark brown ATF usually is caused by overheating resulting from burned clutches or bands. B is correct. A worn front planetary sun gear would not cause this problem. A is wrong.

7. A An improperly adjusted TV cable may cause low pressure at WOT, but this problem would not result in low pressure in 3rd gear only. Therefore, B is wrong.

A leak at an internal location such as a clutch piston may cause low pressure in one gear only. A is correct.

8. C An improperly adjusted TV cable, or a defective vacuum modulator may cause normal pressure at idle speed, but low pressure at WOT. Therefore, both A and B are right, and C is the correct answer.

9. B A defective oil pump would result in low pressures at idle speed, and so A is wrong. If the pressure regulator is stuck, the pressures will not increase during wide throttle openings. Therefore, C is wrong.

A plugged modulator hose may cause high pressures at low speed, and thus D is wrong.

A restricted oil filter may cause the normal transmission pressure to gradually decrease at higher engine speeds.

10. D A defective turbine in the converter, or slipping clutches in the transmission, would cause higher than specified stall speed. Therefore, both A and B are wrong, and the right answer is D.

A slipping converter stator clutch reduces torque multiplication in the converter, and causes the fluid in the converter to be directed against the rotation of the impeller pump and engine. This reduces stall speed.

11. D Low engine compression or a restricted exhaust system would result in a lower than specified stall speed. Higher than specified stall speed usually is caused by slipping clutches and bands. Therefore, both A and B are wrong, and D is the right answer.

12. A A defective stator clutch may cause a loss of engine power, and lower stall speed, but this problem does not affect TCC lockup. Therefore, B is wrong.

A leaking second speed servo piston seal could result in improper intermediate band application, and slipping in second gear. Since TCC lockup only occurs in third and fourth gear this problem should not affect TCC lockup. Therefore, C is wrong.

A sticking pressure regulator may reduce pressure at wide throttle opening, but it should not affect TCC lockup, so D is wrong.

A defective TCC solenoid could prevent TCC lockup. A is right.

13. C An ignition defect that causes misfiring or a fuel system defect (such as partially restricted injectors that cause a lean condition), may result in shudder after TCC lockup. Both A and B are right, and C is the correct answer.

14. D A defective MAP or ECT sensor, or P/N switch would not cause the TCC to lock up at the wrong speed. Therefore, A, B, and C are wrong.

Since the PCM uses the VSS signal to lock up the converter at the proper speed, a defective VSS could result in TCC lockup at the wrong speed. D is the right answer.

15. B   If the engine is stalling when decelerating from 55 mph, the TCC may not be releasing properly. Since the TPS sensor does not affect TCC release, a defect in this component would not cause the problem. Therefore, A is wrong.

The PCM releases the TCC when it receives a signal from the brake switch. Therefore, a defective brake switch may cause improper TCC release and engine stalling. B is the right answer.

16. C   If a computer-controlled transaxle remains in second gear at all speeds, the computer has entered by default mode because of an electrical defect. Since a worn oil pump, restricted filter, or improper linkage adjustment are mechanical problems, these would not be the cause of the problem. Therefore, A, B, and D are wrong.

An electrical defective such as a shift solenoid may cause the computer to enter the default mode resulting in one forward speed. C is right.

17. C   A grounded TCC solenoid wire inside the transaxle, or an open TCC solenoid wire in the transaxle electrical connector may cause a DTC related to the TCC solenoid. Both A and B are right, and C is the correct answer.

18. D   If all the shifts occur normally with the transmission tester connected to the transmission, the cause of the problem is external to the transmission. Since responses A, B, and C represent internal problems these answers are wrong.

An open shift solenoid wire from the transmission electrical connector to the PCM could result in normal operation with the transmission tester connected. D is right.

19. B   With late ignition timing the vacuum would be less than 18 in. Hg. at idle. Therefore, A is wrong.

Restricted exhaust could result in normal vacuum at idle, very low vacuum at part throttle, low stall speed, and delayed upshifts. B is right.

20. C   During an air pressure test a cracked transmission case surrounding the reverse clutch fluid passage, or a cracked reverse clutch drum may result in loss of clutch application and a hissing noise from the escaping air. Both A and B are right and C is the correct answer.

21. A   A misadjusted manual valve shift linkage may cause low fluid pressure. Therefore, B is wrong.

This low fluid pressure may cause clutch slipping and premature failure. A is right.

22. D   If the throttle valve cable is misadjusted so throttle pressure is higher than specified, the transmission shifts occur at a higher than specified vehicle speed, because the vehicle speed must be higher for governor pressure to overcome throttle pressure and cause the upshifts to occur. Therefore, A, B, and C are wrong. D is the right answer.

23. B   Since shift timing in relation to vehicle speed is a function of throttle pressure and governor pressure, band adjustment does not affect shift timing. Therefore, A and C are wrong.

Since the band, or bands, are not applied in all gears, improper band adjustment does not cause slipping in all gears. Therefore, D is wrong, and B is the right answer.

24. D   Transmission filters should be replaced not cleaned. Therefore, A is wrong.

Transmission fluid oxidizes faster at higher temperatures, and so B is also wrong; the correct answer is D.

25. C   A permanent magnet generator-type speed sensor generates an AC voltage in relation to vehicle speed. Therefore, A, B, and D are wrong, and C is the right answer.

26. A   An improper manual valve shift linkage adjustment may cause reduced pressure, transmission slipping, and premature clutch failure. Therefore, B is wrong.

If moisture is freezing in the modulator diaphragm, the vacuum is not applied to the diaphragm. Under this condition the pressure is high and repeated clutch piston failure plus harsh, late shifting may occur. A is the right answer.

27. D   This question asks us to select the component that is not the cause of the problem. A sticking governor valve, excessive governor spring tension, or worn weights and pins could result in reduced governor valve movement in relation to speed. This action causes the shifts to occur at a higher speed. Therefore, A, B, or C could be the cause of the problem, but they are not the requested answer.

A weak governor spring tension would cause excessive governor valve movement in relation to vehicle speed resulting in earlier shifts. D would not cause the problem, and this is the right answer.

28. A   High governor pressure affects shift timing and quality, but this problem does not cause high pressure in the pump. Pump pressure is determined by the pressure regulator valve. Therefore, B is wrong.

A worn pump body bushing allows excessive torque converter hub movement in this bushing, and this movement damages the pump seal. A is right.

29. A   If the transmission fluid is contaminated from the lack of fluid and filter changes, the extension housing bushing and drive shaft yoke may be scored but not pitted. If the fluid is contaminated, other problems would occur such as sticking valves in the valve body and improper shifting. Therefore, B is wrong.

High resistance in a battery ground between the chassis and the engine may cause the current to flow from the chassis through some of the powertrain components to the battery ground cable attached to the engine. This may cause arcing and pitting of powertrain components such as the extension housing bushing and drive shaft yoke. A is the right answer.

30. D   With the engine idling the ATF flow through a transmission cooler should be one quart in 20 seconds. Therefore, A, B, and C are wrong, and D is right.

31. A   This question asks for the response that is not the cause of the problem. A dry cable, worn gears, or worn driven gear retaining bushing could result in erratic speedometer operation. Therefore, B, C, and D are not the requested answer.

A missing drive gear retaining clip would cause the output shaft to rotate inside this gear without turning the gear. This problem would cause an inoperative speedometer. A is the right answer.

32. A   The clearance between the valves and valve body bores should not exceed 0.001 in. Therefore, B, C, and D are wrong, and A is right. ·

33. A   Low governor pressure would affect the other shifts, not just the 1-2 upshift. Therefore, B is wrong.

Excessive valve body torque may have warped the valve body causing the 1-2 shift valve to stick, and preventing the 1-2 upshift. A is the right answer.

34. C   The procedure shown in Figure 2-12 is to determine the proper servo piston selective pin. Therefore, A, B, and D are wrong, and C is the correct answer.

35. A   A sticking pressure regulator valve could result in high transaxle pressures and harsh shifting in all gears, not just in fourth gear. Therefore, B is wrong.

A sticking fourth gear accumulator piston may cause harsh shifting only in fourth gear.

36. A   This question asks for the response that is not correct. The parking pawl is mechanically operated by the shift lever, and the pawl projection engages in a notched drum. Therefore, statements B and C are right, but they are not the requested answer. The pawl is retained on a pivot pin, so D is not the requested answer.

When the pawl is engaged it locks the output shaft. Therefore, statement A is not correct, and this is the right answer.

37. B   When the computer senses an electronic defect and fails to close the transaxle relay, the transaxle operates in second gear and reverse. Therefore, A, C, and D are wrong and B is the right answer.

38. A   When diagnosing a computer-controlled transmission or transaxle for improper shifting, the first sensor to verify is the manual lever position sensor (MLPS), because this sensor informs the computer regarding the gear selector position. The computer supplies the proper gear in relation to the input. Therefore, B, C, and D are wrong, and A is right.

39. B   An improperly adjusted manual valve shift linkage may result in transaxle slipping, but the condition should be constant. Therefore, A is wrong.

Broken engine or transaxle mounts cause excessive engine and transaxle movement, which changes the effective length of the throttle valve cable and results in erratic shifting. B is the right answer.

40. C   This question asks for the statement that is not correct. Prior to transmission/transaxle removal the negative battery terminal should be removed, the drive shaft should be marked in relation to the differential flange, and the engine support fixture must be installed. Therefore, A, B, and D are correct statements, but they are not the requested answer.

The front drive axles do not require marking in relation to the front hubs. Therefore, statement C is incorrect, and this is the requested answer.

41. D   If the input shaft end play is excessive, a thicker selective washer is required on the inner end of the input shaft. Therefore, A, B, and C are wrong, and D is the right answer.

42. C   If the clutch plate clearance is more than specified, a thicker reaction plate must be installed in the clutch pack. Therefore, A, B, and D are wrong, and C is the right answer.

43. B   Loose main bearings cause a thumping noise when the engine is first started, so A is wrong.

Loose connecting rod bearings cause a knocking noise that is most noticeable during acceleration, and worn piston pins cause a constant metallic knocking noise at idle. Therefore, C and D are wrong.

Loose flex plate to converter bolts may cause a clunking noise during deceleration or possibly at idle. Therefore, B is the right answer.

44. B   This question asks for the response that is not the cause of the problem.

Damaged turbocharger bearings reduce turbocharger rpm and engine power, especially at high speed. A restricted exhaust system also reduces engine power at high speed. A seized stator one-way clutch causes the stator to oppose the fluid movement in the converter at higher speed resulting in a loss of engine power. Therefore, A, C and D are right, but they are not the requested answer.

A slipping one-way clutch causes a loss of power at lower speeds, but this defect would not result in a loss of power at high speed. Therefore, statement B is not a cause of the problem, and this is the right answer.

45. A   If the converter one-way clutch was slipping, the engine would have a power loss at lower speeds, and the question informs us that all converter tests were satisfactory. This would include a one-way clutch test. Therefore, B is wrong.

When the transmission was overheated, the converter hub was misaligned and the hub was rubbing on the pump bushing. With the transmission removed and the converter bolted to the flex plate, a converter hub misalignment measurement with a dial indicator proved the hub was misaligned. A is the right answer.

46. B   A sticking pressure regulator valve may cause high pressure and repeated pump seal failure, but this problem would also result in other complaints such as hard shifting. Therefore, A is wrong.

A plugged drainback hole behind the pump seal may cause excessive pressure on the seal and repeated seal failure. A is the right answer.

47. C   The purpose of the measurement in Figure 2-22 is to determine the proper selective washer thickness. Therefore, A, B, and D are wrong, and C is right.

48. B   A defective vehicle speed sensor could result in torque converter clutch (TCC) lockup at the wrong speed. Since the question indicates lockup speed is normal, A is wrong.

The number nine check ball in the turbine shaft seats against an orificed seat when the TCC is applied. This check ball action controls the exhausting of fluid from the front of the lockup plate which controls TCC application. A stuck check ball may allow faster exhausting of the fluid in front of the lockup plate, and harsh TCC application. B is the right answer.

49. A   The tool in Figure 2-24 is used to size Teflon® sealing rings. Therefore, B, C, and D are wrong, and A is the right answer.

50. D   This question asks for the statement that is not correct. A scored shaft in the bushing contact area may indicate a lack of lubrication, and bushing to shaft clearance may be measured with a wire-type feeler gauge or a vernier caliper and micrometer. Therefore, A, B, and C are correct, but they are not the requested answer.

Normal bushing clearance is 0.0005 to 0.015 in., so D is wrong and this is the requested answer.

51. C   If the carrier is held and the annulus gear drives the sun gear, a reverse overdrive is provided which is not practical, so A is wrong.

When the sun gear is held and the annulus gear drives the carrier, a forward overdrive is provided. Therefore, B is wrong.

When the annulus gear is held and the sun gear drives the carrier a forward overdrive is provided, so D is wrong.

If the carrier is held and the sun gear drives the annulus gear, a reverse reduction is provided. Therefore C is the right answer.

52. C   Transmission or transaxle case porosity between two fluid passages may cause pressure bleed off and clutch burn out, so A is right. This problem also may cause the transmission to be in two gears at once causing jam up. Therefore, both A and B are right, and C is the correct answer.

53. A   The tool shown in Figure 2-25 is used to remove and install the drive chain and sprockets. Therefore, B, C, and D are wrong, and A is the right answer.

54. B   To increase the turning torque a thicker shim is required behind one side bearing cup. Therefore, A, C, and D are wrong, and B is the right answer.

55. A   The tool in Figure 2-28 is used to install clutch pistons and seals. Therefore, B, C, and D are wrong, and A is the right answer.

56. D   This question asks for the response that is not correct. A sticking drum check ball, reduced clutch pack clearance, or a damaged clutch piston seal may result in clutch disc burning. Therefore, statements A, B, and C are correct, but they are not the requested answer.

High line pressure may cause harsh shifting and clutch piston seal damage, but it does not result in clutch disc burning. Statement D is incorrect, so this is the requested answer.

57. C   The tool in Figure 2-30 is used for air pressure testing the clutch packs. Therefore, A, B, and D are wrong, and C is the right answer.

58. A   Since the forward clutch is applied in all forward gears, worn forward clutch discs could result in slipping in all forward gears. However, this problem should not result in a buzzing noise. Therefore, B is wrong.

Since the lo/roller clutch is freewheeling in second, third, and fourth gear, and holding in first gear, severe brinnelling of this roller clutch may cause a buzzing noise in second, third, and fourth gear. A is the right answer.

59. B   Uneven wear on a band strut may cause band misalignment, and the resulting wear pattern on the band may be curved across the band rather than circular with the band. However, this problem would not cause severe wear on the outer band edges, so A is wrong.

A dished band contact area on the drum would result in excessive wear on the outer band edges. B is the right answer.

60. D   A defective O/D one-way clutch causes slipping in reverse, first, second, and third gears, so A is wrong.

A defective forward clutch causes slipping in all forward gears, thus B is wrong.

If the intermediate band is defective, slipping occurs in second gear, so C is wrong.

A defective rear one-way clutch causes slipping in first gear or manual first gear with no engine braking while coasting in first gear. D is correct.

Glossary

Abrasion   Wearing or rubbing away of a part.
Abrasión   El desgaste o consumo por rozamiento de una parte.

Acceleration   An increase in velocity or speed.
Aceleración   Un incremento en la velocidad.

Accumulator   A device used in automatic transmissions to cushion the shock of shifting between gears, providing a smoother feel inside the vehicle.
Acumulador   Un dispositivo que se usa en las transmisiones automáticas para suavizar el choque de cambios entre las velocidades, así proporcionando una sensación más uniforme en el interior del vehículo.

Adhesives   Chemicals used to hold gaskets in place during the assembly of an engine. They also aid the gasket in maintaining a tight seal by filling in the small irregularities on the surfaces and by preventing the gasket from shifting due to engine vibration.
Adhesivo   Los productos químicos que se usan para sujetar a los empaques en una posición correcta mientras que se efectua la asamblea de un motor. También ayuden para que los empaques mantengan un sello impermeable, rellenando a las irregularidades pequeñas en las superficies y previniendo que se mueva el empaque debido a las vibraciones del motor.

Alignment   An adjustment to a line or to bring into a line.
Alineación   Un ajuste que se efectúa en una linea o alinear.

Antifriction bearing   A bearing designed to reduce friction. This type of bearing normally uses ball or roller inserts to reduce the friction.
Cojinetes de antifricción   Un cojinete diseñado con el fin de disminuir la fricción. Este tipo de cojinete suele incorporar una pieza inserta esférica o de rodillos para disminuir la fricción.

Antiseize   Thread compound designed to keep threaded connections from damage due to rust or corrosion.
Antiagarrotamiento   Un compuesto para filetes diseñado para proteger a las conecciones fileteados de los daños de la oxidación o la corrosión.

Apply devices   Devices that hold or drive members of a planetary gear set. They may be hydraulically or mechanically applied.
Dispositivos de aplicación   Los dispositivos que sujeten o manejan los miembros de un engranaje planetario. Se pueden aplicar mecánicamente o hidráulicamente.

Arbor press   A small, hand-operated shop press used when only a light force is required against a bearing, shaft, or other part.
Prensa para calar   Una prensa de mano pequeña del taller que se puede usar en casos que requieren una fuerza ligera contra un cojinete, una flecha u otra parte.

Asbestos   A material that was commonly used as a gasket material in places where temperatures are extreme. This material is being used less frequently today because of health hazards that are inherent to the material.
Amianto   Una materia que se usaba frecuentemente como materia de empaques en sitios en los cuales las temperaturas son extremas. Esta materia se usa menos actualmente debido a los peligros al salud que se atribuyan a esta materia.

ATF   Automatic Transmission Fluid.
ATF   Fluido de Transmisión Automática

Automatic transmission   A transmission in which gear or ratio changes are self-activated, eliminating the necessity of hand-shifting gears.

Transmisión automática   Una transmisión en la cual un cambio deengranajes o los cambios en relación son por mando automático, así eliminando la necesidad de cambios de velocidades manual.

Axial   Parallel to a shaft or bearing bore.
Axial   Paralelo a una flecha o al taladro del cojinete.

Axis   The centerline of a rotating part, a symmetrical part, or a circular bore.
Eje   La linea de quilla de una parte giratoria, una parte simétrica, o un taladro circular.

Axle   The shaft or shafts of a machine upon which the wheels are mounted.
Semieje   El eje o los ejes de una máquina sobre los cuales se montan las ruedas.

Axle ratio   The ratio between the rotational speed (rpm) of the driveshaft and that of the driven wheel; gear reduction through the differential, determined by dividing the number of teeth on the ring gear by the number of teeth on the drive pinion.
Relación del eje   La relación entre la velocidad giratorio (rpm) del árbol propulsor y la de la rueda arrastrada; reducción de los engranajes por medio del diferencial, que se determina por dividir el número de dientes de la corona por el número de los dientes en el piñón de ataque.

Axle shaft   A shaft on which the road wheels are mounted.
Flecha del semieje   Una flecha en la cual se monta las ruedas.

Axle shaft end thrust   A force exerted on the end of an axle shaft that is most pronounced when the vehicle turns corners and curves.
Golpe en la flecha del semieje   Una fuerza que se aplica en el extremo de la flecha del semieje que se pronuncia más cuando un vehículo da la vuelta.

Backlash   The amount of clearance or play between two meshed gears.
Juego   La cantidad de holgura o juego entre dos engranajes endentados.

Balance   Having equal weight distribution. The term is usually used to describe the weight distribution around the circumference and between the front and back sides of a wheel.
Equilibrio   Lo que tiene una distribución igual de peso. El término suele usarse para describir la distribución del peso alrededor de la circunferencia y entre los lados delanteros y traseros de una rueda.

Balance valve   A regulating valve that controls a pressure of just the right value to balance other forces acting on the valve.
Válvula niveladora   Una válvula de reglaje que controla a la presión del valor correcto para mantener el equilibrio contra las otras fuerzas que afectan a la válvula.

Ball bearing   An antifriction bearing consisting of a hardened inner and outer race with hardened steel balls which roll between the two races, and supports the load of the shaft.
Rodamiento de bolas   Un cojinete de antifricción que consiste de una pista endurecida interior e exterior y contiene bolas de acero endurecidos que ruedan entre las dos pistas, y sostiene la carga de la flecha.

Ball joint   A suspension component that attaches the control arm to the steering knuckle and serves as the lower pivot point for the steering knuckle. The ball joint gets its name from its ball-and-socket design. It allows both up-and-down motion as well as rotation. In a MacPherson strut FWD suspension system, the two lower ball joints are nonload carrying.

Articulación esférica Un componente de la suspensión que une el brazo de mando a la articulación de la dirección y sirve como un punto pivote inferior de la articulación de la dirección. La articulación esférica derive su nombre de su diseño de bola y casquillo. Permite no sólo el movimiento de arriba y abajo sino también el de rotación. En un sistema de suspensión tipo FWD con poste de MacPherson, las articulaciones esféricas inferiores no soportan el peso.

Band A steel band with an inner lining of friction material. Device used to hold a clutch drum at certain times during transmission operation.

Banda Una banda de acero que tiene un forro interior de una materia de fricción. Un dispositivo que retiene al tambor del embrague en algunos momentos durante la operación de la transmisión.

Bearing The supporting part that reduces friction between a stationary and rotating part or between two moving parts.

Cojinete La parte portadora que reduce la fricción entre una parte fija y una parte giratoria o entre dos partes que muevan.

Bearing cage A spacer that keeps the balls or rollers in a bearing in proper position between the inner and outer races.

Jaula del cojinete Un espaciador que mantiene a las bolas o a los rodillos del cojinete en la posición correcta entre las pistas interiores e exteriores.

Bearing caps In the differential, caps held in place by bolts or nuts which, in turn, hold bearings in place.

Tapones del cojinete En un diferencial, las tapas que se sujeten en su lugar por pernos o tuercas, los cuales en su turno, retienen y posicionan a los cojinetes.

Bearing cone The inner race, rollers, and cage assembly of a tapered roller bearing. Cones and cups must always be replaced in matched sets.

Cono del cojinete La asamblea de la pista interior, los rodillos, y el jaula de un cojinete de rodillos cónico. Se debe siempre reemplazar a ambos partes de un par de conos del cojinete y los anillos exteriores a la vez.

Bearing cup The outer race of a tapered roller bearing or ball bearing.

Anillo exterior La pista exterior de un cojinete cónico de rodillas o de bolas.

Bearing race The surface upon which the rollers or balls of a bearing rotate. The outer race is the same thing as the cup, and the inner race is the one closest to the axle shaft.

Pista del cojinete La superficie sobre la cual rueden los rodillos o las bolas de un cojinete. La pista exterior es lo mismo que un anillo exterior, y la pista interior es la más cercana a la flecha del eje.

Belleville spring A tempered spring steel cone-shaped plate used to aid the mechanical force in a pressure plate assembly.

Resorte de tensión Belleville Un plato de resorte del acero revenido en forma cónica que aumenta a la fuerza mecánica de una asamblea del plato opresor.

Bellhousing A housing that fits over the clutch components and connects the engine and the transmission.

Concha del embrague Un cárter que encaja a los componentes del embrague y conecta al motor con la transmisión.

Bias voltage Voltage applied across a diode.

Tensión de polarización El voltaje aplicado através de un diodo.

Bolt torque The turning effort required to offset resistance as the bolt is being tightened.

Torsión del perno El esfuerzo de torsión que se requiere para compensar la resistencia del perno mientras que esté siendo apretado.

Brake horsepower (bhp) Power delivered by the engine and available for driving the vehicle; bhp = torque x rpm/5252.

Caballo indicado al freno (bhp) Potencia que provee el motor y que es disponible para el uso del vehículo; bhp = de par mortor x rpm/5252.

Brinnelling Rough lines worn across a bearing race or shaft due to impact loading, vibration, or inadequate lubrication.

Efecto brinel Lineas ásperas que aparecen en las pistas de un cojinete o en las flechas debido al choque de carga, la vibración, o falta de lubricación.

Bronze An alloy of copper and tin.

Bronce Una aleación de cobre y hojalata.

Burnish To smooth or polish by the use of a sliding tool under pressure.

Bruñir Pulir o suavizar por medio de una herramienta deslizando bajo presión.

Burr A feather edge of metal left on a part being cut with a file or other cutting tool.

Rebaba Una lima espada de metal que permanece en una parte que ha sido cortado con una lima u otro herramienta de cortar.

Bus A common connector used as an information source for the vehicle's various control units.

Bus Un conector común que se usa como un fuente de información para los varios aparatos de control del vehículo.

Bushing A cylindrical lining used as a bearing assembly made of steel, brass, bronze, nylon, or plastic.

Buje Un forro cilíndrico que se usa como una asamblea de cojinete que puede ser hecho del acero, del latón, del bronze, del nylon, o del plástico.

C-clip A C-shaped clip used to retain the drive axles in some rear axle assemblies.

Grapa de C Una grapa en forma de C que retiene a las flechas motrices en algunas asambleas de ejes traseras.

Cage A spacer used to keep the balls or rollers in proper relation to one another. In a constant-velocity joint, the cage is an open metal framework that surrounds the balls to hold them in position.

Jaula Una espaciador que mantiene una relación correcta entre los rodillos o las bolas. En una junta de velocidad constante, la jaula es un armazón abierto de metal que rodea a las bolas para mantenerlas en posición.

Cap An object that fits over an opening to stop flow.

Tapón Un objecto que tapa a una apertura para detener el flujo.

Carbon monoxide Part of the exhaust gas from an engine; an odorless, colorless, and deadly gas.

Óxido de carbono Una parte de los vapores de escape de un motor; es un gas sin olor, sin color y puede causar la muerte.

Cardan Universal Joint A nonconstant velocity universal joint consisting of two yokes with their forked ends joined by a cross. The driven yoke changes speed twice in 360 degrees of rotation.

Junta Universal Cardan Una junta universal de velocidad no constante que consiste de dos yugos cuyos extremidades ahorquilladas se unen en cruz. El yugo de arrastre cambia su velocidad dos veces en 360 grados de rotación.

Case-harden   To harden the surface of steel. The carburizing method used on low-carbon steel or other alloys to make the case or outer layer of the metal harder than its core.

Cementar   Endurecer la superficie del acero. El método de carburación que se emplea en el acero de bajo carbono o en otros aleaciones para que el cárter o capa exterior queda más dura que lo que esta al interior.

Case porosity   Leaks caused by tiny holes which are formed by trapped air bubbles during the casting process.

Porosidad del cárter   Las fugas que se causan por los hoyitos pequeños formados por burbújas de aire entrapados durante el proceso del moldeo.

Castellate   Formed to resemble a castle battlement, as in a castellated nut.

Acanalado   De una forma que parece a las almenas de un castillo (véa la palabra en inglés), tal como una tuerca con entallas.

Castellated nut   A nut with six raised portions or notches through which a cotter pin can be inserted to secure the nut.

Tuerca con entallas   Una tuerca que tiene seis porciones elevadas o muescas por los cuales se puede insertar un pasador de chaveta para retener a la tuerca.

Centrifugal clutch   A clutch that uses centrifugal force to apply a higher force against the friction disc as the clutch spins faster.

Embrague centrífugo   Un embrague que emplea a la fuerza centrífuga para aplicar una fuerza mayor contra el disco de fricción mientras que el embrague gira más rapidamente.

Centrifugal force   The force acting on a rotating body which tends to move it outward and away from the center of rotation. The force increases as rotational speed increases.

Fuerza centrífuga   La fuerza que afecta a un cuerpo en rotación moviendolo hacia afuera y alejándolo del centro de rotación. La fuerza aumenta al aumentar la velocidad de rotación.

Chamfer   A bevel or taper at the edge of a hole or a gear tooth.

Chaflán   Un bisél o cono en el borde de un hoyo o un diente del engranaje.

Chamfer face   A beveled surface on a shaft or part that allows for easier assembly. The ends of FWD drive shafts are often chamfered to make installation of the CV joints easier.

Cara achaflanada   Una superficie biselada en una flecha o una parte que facilita la asamblea. Los extremos de los árboles de mando de FWD suelen ser achaflandos para facilitar la instalación de las juntas CV.

Chase   To straighten up or repair damaged threads.

Embutir   Enderezar o reparar a los filetes dañados.

Chasing   To clean threads with a tap.

Embutido   Limpiar a los filetes con un macho.

Chassis   The vehicle frame, suspension, and running gear. On FWD cars, it includes the control arms, struts, springs, trailing arms, sway bars, shocks, steering knuckles, and frame. The drive shafts, constant-velocity joints, and transaxle are not part of the chassis or suspension.

Chasis   El armazón de un vehículo, la suspensión, y el engranaje de marcha. En los coches de FWD, incluye los brazos de mando, los postes, los resortes (chapas), los brazos traseros, las estabilizadoras, las articulaciones de la dirección y el armazón. Los árboles de mando, las juntas de velocidad constante, y la flecha impulsora no son partes del chasis ni de la suspensión.

Circlip   A split steel snap ring that fits into a groove to hold various parts in place. Circlips are often used on the ends of FWD drive shafts to retain the constant-velocity joints.

Grapa circular   Un seguro partido circular de acero que se coloca en una ranura para posicionar a varias partes. Las grapas circulares se suelen usar en las extremidades de los árboles de mando en FWD para retener las juntas de velocidad constante.

Clearance   The space allowed between two parts, such as between a journal and a bearing.

Holgura   El espacio permitido entre dos partes, tal como entre un muñon y un cojinete.

Clutch   A device for connecting and disconnecting the engine from the transmission or for a similar purpose in other units.

Embrague   Un dispositivo para conectar y desconectar el motor de la transmisión o para tal propósito en otros conjuntos.

Clutch packs   A series of clutch discs and plates installed alternately in a housing to act as a driving or driven unit.

Conjuntos de embrague   Una seria de discos y platos de embrague que se han instalado alternativamente en un cárter para funcionar como una unedad de propulsión o arrastre.

Clutch slippage   Engine speed increases but increased torque is not transferred through to the driving wheels because of clutch slippage.

Resbalado del embrague   La velocidad del motor aumenta pero la torsión aumentada del motor no se transfere a las ruedas de marcha por el resbalado del embrague.

Coefficient of friction   The ratio of the force resisting motion between two surfaces in contact to the force holding the two surfaces in contact.

Coeficiente de la fricción   La relación entre la fuerza que resiste al movimiento entre dos superficies que tocan y la fuerza que mantiene en contacto a éstas dos superficies.

Coil preload springs   Coil springs are made of tempered steel rods formed into a spiral that resist compression; located in the pressure plate assembly.

Muelles de embrague   Los muelles espirales son fabricadas de varillas de acero revenido y resisten la compresión; se ubican en el conjunto del plato opresor.

Coil spring   A heavy wire-like steel coil used to support the vehicle weight while allowing for suspension motions. On FWD cars, the front coil springs are mounted around the MacPherson struts. On the rear suspension, they may be mounted to the rear axle, to trailing arms, or around rear struts.

Muelles de embrague   Un resorte espiral hecho de acero en forma de alambre grueso que soporte el peso del vehículo mientras que permite a los movimientos de la suspensión. En los coches de FWD, los muelles de embrague delanteros se montan alrededor de los postes Macpherson. En la suspensión trasera, pueden montarse en el eje trasero, en los brasos traseros, o alrededor de los postes traseros.

Compound   A mixture of two or more ingredients.

Compuesto   Una combinación de dos ingredientes o más.

Concentric   Two or more circles having a common center.

Concéntrico   Dos círculos o más que comparten un centro común.

Constant-velocity joint   A flexible coupling between two shafts that permits each shaft to maintain the same driving or driven speed regardless of operating angle, allowing for a smooth transfer of power. The constant-velocity joint (also called CV joint) consists of an inner and outer housing with balls in between, or a tripod and yoke assembly.

Junta de velocidad constante   Un acoplador flexible entre dos flechas que permite que cada flecha mantenga la velocidad de propulsión o arrastre sin importar el ángulo de operación, efectuando una transferencia lisa del poder. La junta de velociadad constante (también llamado junta CV) consiste de un cárter interior e exterior entre los cuales se encuentran bolas, o de un conjunto de trípode y yugo.

Contraction   A reduction in mass or dimension; the opposite of expansion.

Contracción Una reducción en la masa o en la dimensión; el opuesto de expansión.

Control arm   A suspension component that links the vehicle frame to the steering knuckle or axle housing and acts as a hinge to allow up-and-down wheel motions. The front control arms are attached to the frame with bushings and bolts and are connected to the steering knuckles with ball joints. The rear control arms attach to the frame with bushings and bolts and are welded or bolted to the rear axle or wheel hubs.

Brazo de mando   Un componente de la suspención que une el armazón del vehículo al articulación de dirección o al cárter del eje y que se porta como una bisagra para permitir a los movimientos verticales de las ruedas. Los brazos de mando delanteros se conectan al armazón por medio de pernos y bujes y se conectan al articulación de dirección por medio de los articulaciones esféricos. Los brazos de mando traseros se conectan al armazón por medio de pernos y bujes y son soldados o empernados al eje trasero o a los cubos de la rueda.

Corrode   To eat away gradually as if by gnawing, especially by chemical action.

Corroer   Roído poco a poco, primariamente por acción químico.

Corrosion   Chemical action, usually by an acid, that eats away (decomposes) a metal.

Corrosión   Un acción químico, por lo regular un ácido, que corroe (descompone) un metal.

Cotter pin   A type of fastener, made from soft steel in the form of a split pin, that can be inserted in a drilled hole. The split ends are spread to lock the pin in position.

Pasador de chaveta   Un tipo de fijación, hecho de acero blando en forma de una chaveta que se puede insertar en un hueco tallado. Las extremidades partidas se despliegen para asegurar la posición de la chaveta.

Counterclockwise rotation   Rotating in the opposite direction of the hands on a clock.

Rotación en sentido inverso   Girando en el sentido opuesto de las agujas de un reloj.

Coupling   A connecting means for transferring movement from one part to another; may be mechanical, hydraulic, or electrical.

Acoplador   Un método de conección que transfere el movimiento de una parte a otra; puede ser mecánico, hidráulico, o eléctrico.

Coupling phase   Point in torque converter operation where the turbine speed is 90% of impeller speed and there is no longer any torque multiplication.

Fase del acoplador   El punto de la operación del convertidor de la torsión en el cual la velocidad de la turbina es el 90% de la velocidad del impulsor y no queda ningún multiplicación de la torsión.

Cover plate   A stamped steel cover bolted over the service access to the manual transmission.

Cubrejuntas   Un cubierto de acero estampado que se emperna en la apertura de servicio de la transmisión manual.

Crocus cloth   A very fine polishing paper. It is designed to remove very little metal; therefore it is safe to use on critical surfaces.

Tela de óxido férrico   Un papel muy fino para pulir. Fue diseñado para raspar muy poco del metal; por lo tanto, se suele emplear en las superficies críticas.

Deflection   Bending or movement away from normal due to loading.

Desviación   Curvación o movimiento fuera de lo normal debido a la carga.

Degree   A unit of measurement equal to 1/360th of a circle.

Grado   Una uneda de medida que iguala al 1/360 parte de un círculo.

Density   Compactness; relative mass of matter in a given volume.

Densidad   La firmeza; una cantidad relativa de la materia que ocupa a un volumen dado.

Detent   A small depression in a shaft, rail, or rod into which a pawl or ball drops when the shaft, rail, or rod is moved. This provides a locking effect.

Detención   Un pequeño hueco en una flecha, una barra o una varilla en el cual cae una bola o un linguete al moverse la flecha, la barra o la varilla. Esto provee un efecto de enclavamiento.

Detent mechanism   A shifting control designed to hold the manual transmission in the gear range selected.

Aparato de detención   Un control de desplazamiento diseñado a sujetar a la transmisión manual en la velocidad selecionada.

Diagnosis   A systematic study of a machine or machine parts to determine the cause of improper performance or failure.

Diagnóstico   Un estudio sistemático de una máquina o las partes de una máquina con el fín de determinar la causa de una falla o de un operación irregular.

Dial indicator   A measuring instrument with the readings indicated on a dial rather than on a thimble as on a micrometer.

Indicador de carátula   Un instrumento de medida cuyo indicador es en forma de muestra en contraste al casquillo de un micrómetro.

Differential   A mechanism between drive axles that permits one wheel to run at a different speed than the other while turning.

Diferencial   Un mecanismo entre dos semiejes que permite que una rueda gira a una velocidad distincta que la otra en una curva.

Differential action   An operational situation where one driving wheel rotates at a slower speed than the opposite driving wheel.

Acción del diferencial   Una situación durante la operación en la cual una rueda propulsora gira con una velocidad más lenta que la rueda propulsora opuesta.

Differential case   The metal unit that encases the differential side gears and pinion gears, and to which the ring gear is attached.

Caja de satélites   La unedad metálica que encaja a los engranajes planetarios (laterales) y a los satélites del diferencial, y a la cual se conecta la corona.

Differential drive gear   A large circular helical gear that is driven by the transaxle pinion gear and shaft and drives the differential assembly.

Corona   Un engranaje helicoidal grande circular que es arrastrado por el piñon de la flecha de transmisión y la flecha y propela al conjunto del diferencial.

Differential housing   Cast iron assembly that houses the differential unit and the drive axles. Also called the rear axle housing.

Cárter del diferencial   Una asamblea de acero vaciado que encaja a la unedad del diferencial y los semiejes. También se llama el cárter del eje trasero.

Differential pinion gears Small beveled gears located on the differential pinion shaft.

Satélites Engranajes pequeños biselados que se ubican en la flecha del piñon del diferencial.

Differential pinion shaft A short shaft locked to the differential case. This shaft supports the differential pinion gears.

Flecha del piñon del diferencial Una flecha corta clavada en la caja de satélites. Esta flecha sostiene a los satélites.

Differential ring gear A large circular hypoid-type gear enmeshed with the hypoid drive pinion gear.

Corona Un engranaje helicoidal grande circular endentado con el piñon de ataque hipoide.

Differential side gears The gears inside the differential case that are internally splined to the axle shafts, and which are driven by the differential pinion gears.

Planetarios (laterales) Los engranajes adentro de la caja de satélites que son acanalados a los semiejes desde el interior, y que se arrastran por los satélites.

Dipstick A metal rod used to measure the fluid in an engine or transmission.

Varilla de medida Una varilla de metal que se usa para medir el nivel de flúido en un motor o en una transmisión.

Direct drive One turn of the input driving member compared to one complete turn of the driven member, such as when there is direct engagement between the engine and driveshaft where the engine crankshaft and the driveshaft turn at the same rpm.

Mando directo Una vuelta del miembro de ataque o propulsión que se compara a una vuelta completa del miembro de arrastre, tal como cuando hay un enganchamiento directo entre el motor y el árbol de transmisión cuando el cigueñal y el árbol de transmisión giran al mismo rpm.

Disengage When the operator moves the clutch pedal toward the floor to disconnect the driven clutch disc from the driving flywheel and pressure plate assembly.

Desembragar Cuando el operador mueva el pedal de embrague hacia el piso para desconectar el disco de embrague del volante impulsor y del conjunto del plato opresor.

Distortion A warpage or change in form from the original shape.

Distorción Abarquillamiento o un cambio en la forma original.

Dowel A metal pin attached to one object which, when inserted into a hole in another object, ensures proper alignment.

Espiga Una clavija de metal que se fija a un objeto, que al insertarla en el hoyo de otro objeto, asegura una alineación correcta.

Dowel pin A pin inserted in matching holes in two parts to maintain those parts in fixed relation one to another.

Clavija de espiga Una clavija que se inserte en los hoyos alineados en dos partes para mantener ésos dos partes en una relación fijada el uno al otro.

Downshift To shift a transmission into a lower gear.

Cambio descendente Cambiar la velocidad de una transmisión a una velocidad más baja.

Driveline torque Relates to rear-wheel driveline and is the transfer of torque between the transmission and the driving axle assembly.

Potencia de la flecha motríz Se relaciona a la flecha motríz de las ruedas traseras y transfere la potencia de la torsión entre la transmisión y el conjunto del eje trasero.

Driven gear The gear meshed directly with the driving gear to provide torque multiplication, reduction, or a change of direction.

Engranaje de arrastre El engranaje endentado directamente al engranaje de ataque para proporcionar la multiplicación, la reducción, o los cambios de dirección de la potencia.

Drive pinion gear One of the two main driving gears located within the transaxle or rear driving axle housing. Together the two gears multiply engine torque.

Engranaje de piñon de ataque Uno de dos engranajes de ataque principales que se ubican adentro de la flecha de transmisión o en el cárter del eje de propulsión. Los dos engranajes trabajan juntos para multiplicar la potencia.

Drive shaft An assembly of one or two universal joints connected to a shaft or tube; used to transmit power from the transmission to the differential. Also called the propeller shaft.

Árbol de mando Una asamblea de uno o dos uniones universales que se conectan a un árbol o un tubo; se usa para transferir la potencia desde la transmisión al diferencial. También se le refiere como el árbol de propulsión.

Drop forging A piece of steel shaped between dies while hot.

Estampado Un pedazo de acero que se forma entre bloques mientras que esté caliente.

Dry friction The friction between two dry solids.

Fricción seca Fricción entre dos sólidos secos.

Dynamic In motion.

Dinámico En movimiento.

Dynamic balance The balance of an object when it is in motion; for example, the dynamic balance of a rotating drive shaft.

Balance dinámico El balance de un objeto mientras que esté en movimiento: por ejemplo el balance dinámico de un árbol de mando giratorio.

Eccentric One circle within another circle wherein both circles do not have the same center or a circle mounted off center. On FWD cars, front-end camber adjustments are accomplished by turning an eccentric cam bolt that mounts the strut to the steering knuckle.

Excéntrico Se dice de dos círculos, el uno dentro del otro, que no comparten el mismo centro o de un círculo ubicado descentrado. En los coches FWD, los ajustes de la inclinación se efectuan por medio de un perno excéntrico que fija el poste sobre el articulación de dirección

Efficiency The ratio between the power of an effect and the power expended to produce the effect; the ratio between an actual result and the theoretically possible result.

Eficiencia La relación entre la potencia de un efecto y la potencia que se gasta para producir el efecto; la relación entre un resultado actual y el resultado que es una posibilidad teórica.

Elastomer Any rubber-like plastic or synthetic material used to make bellows, bushings, and seals.

Elastómero Cualquiera materia plástic parecida al hule o una materia sintética que se utiliza para fabricar a los fuelles, los bujes y las juntas.

End clearance Distance between a set of gears and their cover, commonly measured on oil pumps.

Holgura del extremo La distancia entre un conjunto de engranajes y su placa de recubrimiento, suele medirse en las bombas de aceite.

Endplay The amount of axial or end-to-end movement in a shaft due to clearance in the bearings.

Juego de las extremidades La cantidad del movimiento axial o del movimiento de extremidad a extremidad en una flecha debido a la holgura que se deja en los cojinetes.

Engage When the vehicle operator moves the clutch pedal up from the floor, this engages the driving flywheel and pressure plate to rotate and drive the driven disc.

Accionar Cuando el operador del vehículo deja subir el pedal del embrague del piso, ésto acciona la volante de ataque y el plato opresor para impulsar al disco de arrastre.

Engagement chatter A shaking, shuddering action that takes place as the driven disc makes contact with the driving members. Chatter is caused by a rapid grip and slip action.

Chasquido de enganchamiento Un movimiento de sacudo o temblor que resulta cuando el disco de ataque viene en contacto con los miembros de propulsión. El chasquido se causa por una acción rápida de agarrar y deslizar.

Engine torque A turning or twisting action developed by the engine, measured in foot-pounds or kilogram meters.

Torsión del motor Una acción de girar o torcer que crea el motor, ésta se mide en libras-pie o kilos-metros.

Essential tool kit A set of special tools designed for a particular model of car or truck.

Estuche de herramientas principales Un conjunto de herramientas especiales diseñadas para un modelo específico de coche o camión.

Etching A discoloration or removal of some material caused by corrosion or some other chemical reaction.

Grabado por ácido Una descoloración o remueva de una materia que se efectua por medio de la corrosión u otra reacción química.

Extension housing An aluminum or iron casting of various lengths that encloses the transmission output shaft and supporting bearings.

Cubierta de extensión Una pieza moldeada de aluminio o acero que puede ser de varias longitudes que encierre a la flecha de salida de la transmisión y a los cojinetes de soporte.

External gear A gear with teeth across the outside surface.

Engranaje exterior Un engranaje cuyos dientes estan en la superficie exterior.

Externally tabbed clutch plates Clutch plates that are designed with tabs around the outside periphery to fit into grooves in a housing or drum.

Placas de embrague de orejas externas Las placas de embrague que se diseñan de un modo para que las orejas periféricas de la superficie se acomoden en una ranura alrededor de un cárter o un tambor.

Extreme-pressure lubricant A special lubricant for use in hypoid-gear differentials; needed because of the heavy wiping loads imposed on the gear teeth.

Lubricante de presión extrema Un lubricante especial que se usa en las diferenciales de tipo engranaje hipóide; se requiere por la carga de transmisión de materia pesada que se imponen en los dientes del engranaje.

Face The front surface of an object.

Cara La superficie delantera de un objeto.

Fatigue The buildup of natural stress forces in a metal part that eventually causes it to break. Stress results from bending and loading the material.

Fatiga El incremento de tensiones y esfuerzos normales en una parte de metal que eventualmente causen una quebradura. Los esfuerzos resultan de la carga impuesta y el doblamiento de la materia.

Feeler gauge A metal strip or blade finished accurately with regard to thickness used for measuring the clearance between two parts; such gauges ordinarily come in a set of different blades graduated in thickness by increments of 0.001 inch.

Calibrador de laminillas Una lámina o hoja de metal que ha sido acabado precisamente con respecto a su espesor que se usa para medir la holgura entre dos partes; estas galgas típicamente vienen en un conjunto de varias espesores graduados desde el 0.001 de una pulgada.

Final drive ratio The ratio between the drive pinion and ring gear.

Relación del mando final La relación entre el piñon de ataque y la corona.

Fit The contact between two machined surfaces.

Ajuste El contacto entre dos superficies maquinadas.

Fixed-type constant-velocity joint A joint that cannot telescope or plunge to compensate for suspension travel. Fixed joints are always found on the outer ends of the drive shafts of FWD cars. A fixed joint may be of either Rzeppa or tripod type.

Junta tipo fijo de velocidad constante Una junta que no tiene la capacidad de los movimientos telescópicos o repentinos que sirven para compensar en los viajes de suspensión. Las juntas fijas siempre se ubican en las extremidades exteriores de los árboles de mando en los coches de FWD. Una junta tipo fijo puede ser de un tipo Rzeppa o de trípode.

Flange A projecting rim or collar on an object for keeping it in place.

Reborde Una orilla o un collar sobresaliente de un objeto cuyo función es de mantenerlo en lugar.

Flexplate A lightweight flywheel used only on engines equipped with an automatic transmission. The flexplate is equipped with a starter ring gear around its outside diameter and also serves as the attachment point for the torque converter.

Placa articulada Un volante ligera que se usa solamente en los motores que se equipan con una transmisión automática. El diámetro exterior de la placa articulada viene equipado con un anillo de engranajes para arrancar y también sirve como punto de conección del convertidor de la torsión.

Fluid coupling A device in the power train consisting of two rotating members; transmits power from the engine, through a fluid, to the transmission.

Acoplamiento de fluido Un dispositivo en el tren de potencia que consiste de dos miembros rotativos; transmite la potencia del motor, for medio de un fluido, a la transmisión.

Fluid drive A drive in which there is no mechanical connection between the input and output shafts, and power is transmitted by moving oil.

Dirección fluido Una dirección en la cual no hay conecciones mecánicas entre las flechas de entrada o salida, y la potencia se transmite por medio del aceite en movimiento.

Flywheel A heavy metal wheel that is attached to the crankshaft and rotates with it; helps smooth out the power surges from the engine power strokes; also serves as part of the clutch and engine-cranking system.

Volante Una rueda pesada de metal que se fija al cigueñal y gira con ésta; nivela a los sacudos que provienen de la carrera de fuerza del motor; también sirve como parte del embrague y del sistema de arranque.

Flywheel ring gear A gear, fitted around the flywheel, that is engaged by teeth on the starting-motor drive to crank the engine.

Engranaje anular del volante Un engranaje, colocado alrededor del volante que se acciona por los dientes en el propulsor del motor de arranque y arranca al motor.

Foot-Pound (ft. lb.)   A measure of the amount of energy or work required to lift 1 pound a distance of 1 foot.

Pie libra   Una medida de la cantidad de energía o fuerza que requiere mover una libra a una distancia de un pie.

Force   Any push or pull exerted on an object; measured in pounds and ounces, or in newtons (N) in the metric system.

Fuerza   Cualquier acción empujado o jalado que se efectua en un objeto; se mide en pies y onzas, o en newtones (N) en el sistema métrico.

Four-wheel drive   On a vehicle, driving axles at both front and rear, so that all four wheels can be driven.

Tracción a cuatro ruedas   En un vehículo, se trata de los ejes de dirección fronteras y traseras, para que cada uno de las ruedas puede impulsar.

Frame   The main understructure of the vehicle to which everything else is attached. Most FWD cars have only a subframe for the front suspension and drive train. The body serves as the frame for the rear suspension.

Armazón   La estructura principal del vehículo al cual todo se conecta. La mayoría de los coches FWD sólo tiene un bastidor auxiliar para la suspensión delantera y el tren de propulsión. El carrocería del coche sirve de chassis par la suspensión trasera.

Free-wheel   To turn freely and not transmit power.

Volante libre   Da vueltas libremente sin transferir la potencia.

Free-wheeling clutch   A mechanical device that will engage the driving member to impart motion to a driven member in one direction but not the other. Also known as an "overrunning clutch".

Embrague de volante libre   Un dispositivo mecánico que acciona el miembro de tracción y da movimiento al miembro de tracción en una dirección pero no en la otra. También se conoce bajo el nombre de un "embrague de sobremarcha".

Friction   The resistance to motion between two bodies in contact with each other.

Fricción   La resistencia al movimiento entre dos cuerpos que estan en contacto.

Friction bearing   A bearing in which there is sliding contact between the moving surfaces. Sleeve bearings, such as those used in connecting rods, are friction bearings.

Rodamientos de fricción   Un cojinete en el cual hay un contacto deslizante entre las superficies en movimiento. Los rodamientos de manguitos, como los que se usan en las bielas, son rodamientos de fricción.

Friction disc   In the clutch, a flat disc, faced on both sides with friction material and splined to the clutch shaft. It is positioned between the clutch pressure plate and the engine flywheel. Also called the clutch disc or driven disc.

Disco de fricción   En el embrague, un disco plano al cual se ha cubierto ambos lados con una materia de fricción y que ha sido estriado a la flecha del embrague. Se posiciona entre el plato opresor del embrague y el volante del motor. También se llama el disco del embrague o el disco de arrastre.

Friction facings   A hard-molded or woven asbestos or paper material that is riveted or bonded to the clutch driven disc.

Superficie de fricción   Un recubrimiento remachado o aglomerado al disco de arrastre del embrague que puede ser hecho del amianto moldeado o tejido o de una materia de papel.

Front pump   Pump located at the front of the transmission. It is driven by the engine through two dogs on the torque converter housing. It supplies fluid whenever the engine is running.

Bomba delantera   Una bomba ubicado en la parte delantera de la transmisión. Se arrastre por el motor al través de dos álabes en el cárter del convertidor de la torsión. Provee el fluido mientras que funciona el motor.

Front-wheel drive (FWD)   The vehicle has all drive train components located at the front.

Tracción de las ruedas delanteras (FWD)   El vehículo tiene todos los componentes del tren de propulsión en la parte delantera.

FWD   Abbreviation for front-wheeldrive.

FWD   Abreviación de tracción de las ruedas delanteras.

Galling   Wear caused by metal-to-metal contact in the absence of adequate lubrication. Metal is transferred from one surface to the other, leaving behind a pitted or scaled appearance.

Desgaste por fricción   El desgaste causado por el contacto de metal a metal en la ausencia de lubricación adecuada. El metal se transfere de una superficie a la otra, causando una aparencia agujerado o con depósitos.

Gasket   A layer of material, usually made of cork, paper, plastic, composition, or metal, or a combination of these, placed between two parts to make a tight seal.

Empaque   Una capa de una materia, normalmente hecho del corcho, del papel, del plástico, de la materia compuesta o del metal, o de cualquier combinación de éstos, que se coloca entre dos partes para formar un sello impermeable.

Gasket cement   A liquid adhesive material, or sealer, used to install gaskets.

Mastique para empaques   Una substancia líquida adhesiva, o una substancia impermeable, que se usa para instalar a los empaques.

Gear   A wheel with external or internal teeth that serves to transmit or change motion.

Engranaje   Una rueda que tiene dientes interiores o exteriores que sirve para transferir o cambiar el movimiento.

Gear lubricant   A type of grease or oil blended especially to lubricate gears.

Lubricante para engranaje   Un tipo de grasa o aceite que ha sido mezclado específicamente para la lubricación de los engranajes.

Gear ratio   The number of revolutions of a driving gear required to turn a driven gear through one complete revolution. For a pair of gears, the ratio is found by dividing the number of teeth on the driven gear by the number of teeth on the driving gear.

Relación de los engranajes   El número de las revoluciones requeridas del engranaje de propulsión para dar una vuelta completa al engranaje   arrastrado. En una pareja de engranajes, la relación se calcula al dividir el número de los dientes en el engranaje de arrastre por el número de los dientes en el engranaje de propulsión.

Gear reduction   When a small gear drives a large gear, there is an output speed reduction and a torque increase which results in a gear reduction.

Velocidad descendente   Cuando un engranaje pequeño impulsa a un engranaje grande, hay una reducción en la velocidad de salida y un incremento en la torsión que resulta en una cambio descendente de los velocidades.

Gearshift   A linkage-type mechanism by which the gears in an automobile transmission are engaged and disengaged.

Varillaje de cambios   Un mecanismo tipo eslabón que acciona y desembraga a los engranajes de la transmisión.

Gear whine   A high-pitched sound developed by some types of meshing gears.
Ruido del engranaje   Un sonido agudo que proviene de algunos tipos de engranajes endentados.

Governor pressure   The transmission's hydraulic pressure which is directly related to output shaft speed. It is used to control shift points.
Regulador de presión   La presión hidráulica de una transmisión se relaciona directamente a la velocidad de la flecha de salida. Se usa para controlar los puntos de cambios de velocidad.

Governor valve   A device used to sense vehicle speed. The governor valve is attached to the output shaft.
Válvula reguladora   Un dispositivo que se usa para determinar la velocidad de un vehículo. La válvula reguladora se monta en la flecha de salida.

Graphite   Very fine carbon dust with a slippery texture used as a lubricant.
Grafito   Un polvo de carbón muy fino con una calidad grasosa que se usa para lubricar.

Grind   To finish or polish a surface by means of an abrasive wheel.
Amolar   Acabar o pulir a una superficie por medio de una muela para pulverizar.

Heat treatment   Heating, followed by fast cooling, to harden metal.
Tratamiento térmico   Calentamiento, seguido por un enfriamiento rápido, para endurecer a un metal.

Horsepower   A measure of mechanical power, or the rate at which work is done. One horsepower equals 33,000 ft.-lb. (foot-pounds) of work per minute. It is the power necessary to raise 33,000 pounds a distance of 1 foot in 1 minute.
Caballo de fuerza   Una medida de fuerza mecánica, o el régimen en el cual se efectua el trabajo. Un caballo de fuerza iguala a 33,000 lb.p. (libras pie) de trabajo por minuto. Es la fuerza requerida para transportar a 33,000 libras una distancia de 1 pie en 1 minuto.

Hub   The center part of a wheel, to which the wheel is attached.
Cubo   La parte central de una rueda, a la cual se monta la rueda.

Hydraulic press   A piece of shop equipment that develops a heavy force by use of a hydraulic piston-and-jack assembly.
Prensa hidráulica   Una herramienta del taller que provee una fuerza grande por medio de una asamblea de gato con un pistón hidráulico.

Hydraulic pressure   Pressure exerted through the medium of a liquid.
Presión hidráulica   La presión esforzada por medio de un líquido.

ID   Inside Diameter.
DI   Diámetro Interior.

Idle   Engine speed when the accelerator pedal is fully released and there is no load on the engine.
Marcha lenta   La velocidad del motor cuando el pedal accelerador esta completamente desembragada y no hay carga en el motor.

Impedance   The operating resistance of an electrical device.
Impedancia   La resistencia operativa de un dispositivo eléctrico.

Impeller   The pump or driving member in a torque converter.
Impulsor   La bomba o el miembro impulsor en un convertidor de torsión.

Increments   Series of regular additions from small to large.
Incrementos   Una serie de incrementos regulares que va de pequeño a grande.

Index   To orient two parts by marking them. During reassembly the parts are arranged so the index marks are next to each other. Used to preserve the orientation between balanced parts.
Índice   Orientar a dos partes marcándolas. Al montarlas, las partes se colocan para que las marcas de índice estén alinieadas. Se usan los índices para preservar la orientación de las partes balanceadas.

Input shaft   The shaft carrying the driving gear by which the power is applied, as to the transmission.
Flecha de entrada   La flecha que porta el engranaje propulsor por el cual se aplica la potencia, como a la transmisión.

Inspection cover   A removable cover that permits entrance for inspection and service work.
Cubierta de inspección   Una cubierta desmontable que permite a la entrada para inspeccionar y mantenimiento.

Integral   Built into, as part of the whole.
Integral   Incorporado, una parte de la totalidad.

Internal gear   A gear with teeth pointing inward, toward the hollow center of the gear.
Engranaje internal   Un engranaje cuyos dientes apuntan hacia el interior, al hueco central del engranaje.

Jam nut   A second nut tightened against a primary nut to prevent it from working loose. Used on inner and outer tie-rod adjustment nuts and on many pinion-bearing adjustment nuts.
Contra tuerca   Una tuerca secundaria que se aprieta contra una tuerca primaria para prevenir que ésta se afloja. Se emplean en las tuercas de ajustes interiores e exteriores para las barras de acoplamiento y también en muchas de las tuercas de ajuste de portapiñones.

Journal   A bearing with a hole in it for a shaft.
Manga de flecha   Un cojinete que tiene un hoyo para una flecha.

Key   A small block inserted between the shaft and hub to prevent circumferential movement.
Chaveta   Un tope pequeño que se meta entre la flecha y el cubo para prevenir un movimiento circunferencial.

Keyway   A groove or slot cut to permit the insertion of a key.
Ranura de chaveta   Un corte de ranura o mortaja que permite insertar una chaveta.

Knock   A heavy metallic sound usually caused by a loose or worn bearing.
Golpe   Un sonido metálico fuerte que suele ser causado por un cojinete suelto o gastado.

Knurl   To indent or roughen a finished surface.
Moletear   Indentar o desbastar a una superficie acabada.

Lapping   The process of fitting one surface to another by rubbing them together with an abrasive material between the two surfaces.
Pulido   El proceso de ajustar a una superficie con otra por frotarlas juntas con una materia abrasiva entre las dos superficies.

Lash   The amount of free motion in a gear train, between gears, or in a mechanical assembly, such as the lash in a valve train.
Juego   La cantidad del movimiento libre en un tren de engranajes, entre los engranajes o en una asamblea mecánica, tal como el juego en un tren de vávulas.

Linkage   Any series of rods, yokes, levers, and so on, used to transmit motion from one unit to another.
Biela   Cualquiera serie de barras, yugos, palancas, y todo lo demás, que se usa para transferir los movimientos de una unedad a otra.

Locknut   A second nut turned down on a holding nut to prevent loosening.
Contra tuerca   Una tuerca segundaria apretada contra una tuerca de sostén para prevenir que ésta se afloja.

Lock pin   Used in some ball sockets (inner tie-rod end) to keep the connecting nuts from working loose. Also used on some lower ball joints to hold the tapered stud in the steering knuckle.
Clavija de cerrojo   Se usan en algunas rótulas (las extremidades interiores de la barra de acoplamiento) para prevenir que se aflojan las tuercas de conexión. También se emplean en algunas juntas esféricas inferiores para retener al perno cónico en la articulación de dirección.

Lockplates   Metal tabs bent around nuts or bolt heads.
Placa de cerrojo   Chavetas de metal que se doblan alrededor de las tuercas o las cabezas de los pernos.

Lockwasher   A type of washer which, when placed under the head of a bolt or nut, prevents the bolt or nut from working loose.
Arandela de freno   Un tipo de arandela que, al colocarse bajo la cabeza de un perno, previene que el perno o la tuerca se aflojan.

Low speed   The gearing that produces the highest torque and lowest speed of the wheels.
Velocidad baja   La velocidad que produce la torsión más alta y la velocidad más baja a las ruedas.

Lubricant   Any material, usually a petroleum product such as grease or oil, that is placed between two moving parts to reduce friction.
Lubricante   Cualquier substancia, normalmente un producto de petróleo como la grasa o el aciete, que se coloca entre dos partes en movimiento para reducir la fricción.

Mainline pressure   The hydraulic pressure that operates apply devices and is the source of all other pressures in an automatic transmission. It is developed by pump pressure and regulated by the pressure regulator.
Línea de presión   La presión hidráulica que opera a los dispositivos de applicación y es el orígen de todas las presiones en la transmisión automática. Proviene de la bomba de presión y es regulada por el regulador de presión.

Main oil pressure regulator valve   Regulates the line pressure in a transmission.
Válvula reguladora de la línea de presión   Regula la presión en la linea de una transmisión.

Manual control valve   A valve used to manually select the operating mode of the transmission. It is moved by the gearshift linkage.
Válvula de control manual   Una válvula que se usa para escojer a una velocidad de la transmisión por mano. Se mueva por la biela de velocidades.

Meshing   The mating, or engaging, of the teeth of two gears.
Engrane   Embragar o endentar a los dientes de dos engranajes.

Meter   1/10,000,000 of the distance from the North Pole to the Equator, or 39.37 inches.
Metro   Un 1/10,000,000 de la distancia del polo del norte al ecuador.

Micrometer   A precision measuring device used to measure small bores, diameters, and thicknesses. Also called a mike.
Micrómetro   Un dispositivo de medida presisa que se emplea a medir a los taladros pequeños y a los espesores. También se llama un mike (mayk).

Misalignment   When bearings are not on the same centerline.
Desalineamineto   Cuando los cojinetes no comparten la misma linea central.

Modulator   A vacuum-diaphragm device connected to a source of engine vacuum. It provides an engine load signal to the transmission.
Modulador   Un dispositivo de diafragma de vacío que se conecta a un orígen de vacío en el motor. Provee un señal de carga del motor a la transmisión.

Mounts   Made of rubber to insulate vibrations and noise while they support a power train part, such as engine or transmission mounts.
Monturas   Hecho de hule para insular a las vibraciones y a los ruidos mientras que sujetan una parte del tren de propulsión, tal como las monturas del motor o las monturas de la transmisión.

Multiple disc   A clutch with a number of driving and driven discs as compared to a single plate clutch.
Discos múltiples   Un embrague que tiene varios discos de propulsión o de arraste al contraste con un embrague de un sólo plato.

Needle bearing   An antifriction bearing using a great number of long, small-diameter rollers. Also known as a quill bearing.
Rodamiento de agujas   Un rodamiento (cojinete) antifricativo que emplea un gran cantidad de rodillos largos y de diámetro muy pequeños.

Needle deflection   Distance of travel from zero of the needle on a dial gauge.
Desviación de la aguja   La distancia del cero que viaja una aguja de un indicador.

Neoprene   A synthetic rubber that is not affected by the various chemicals that are harmful to natural rubber.
Neoprene   Un hule sintético que no se afecta por los varios productos químicos que pueden dañar al hule natural.

Neutral   In a transmission, the setting in which all gears are disengaged and the output shaft is disconnected from the drive wheels.
Neutral   En una transmisión, la velocidad en la cual todos los engranajes estan desembragados y el árbol de salida esta desconectada de las ruedas de propulsión.

Neutral-start switch   A switch wired into the ignition switch to prevent engine cranking unless the transmission shift lever is in neutral or the clutch pedal is depressed.
Interruptor de arranque en neutral   Un interruptor eléctrico instalado en el interruptor de encendido que previene el arranque del motor al menos de que la palanca de cambio de velocidad esté en una posición neutral o que se pisa en el embrague.

Newton-meter (Nm)   Metric measurement of torque or twisting force.
Metro newton (Nm)   Una medida métrica de la fuerza de torsión.

Nominal shim   A shim with a designated thickness.
Laminilla fina   Una cuña de un espesor especificado.

Nonhardening   A gasket sealer that never hardens.
Sinfragua   Un cemento de empaque que no endurece.

Nut   A removable fastener used with a bolt to lock pieces together; made by threading a hole through the center of a piece of metal that has been shaped to a standard size.
Tuerca   Un retén removable que se usa con un perno o tuerca para unir a dos piezas; se fabrica al filetear un hoyo taladrado en un pedazo de metal que se ha formado a un tamaño especificado.

OD   Outside diameter.
DE   Diámetro exterior.

Oil seal   A seal placed around a rotating shaft or other moving part to prevent leakage of oil.
Empaque de aciete   Un empaque que se coloca alrededor de una flecha giratoria para prevenir el goteo de aceite.

One-way clutch *See* Sprag clutch.
Embrague de una via *Vea* Sprag clutch.

O-ring A type of sealing ring, usually made of rubber or a rubber-like material. In use, the O-ring is compressed into a groove to provide the sealing action.
Anillo en O Un tipo de sello anular, suele ser hecho de hule o de una materia parecida al hule. Al usarse, el anillo en O se comprime en una ranura para proveer un sello.

Oscillate To swing back and forth like a pendulum.
Oscilar Moverse alternativamente en dos sentidos contrarios como un péndulo.

Outer bearing race The outer part of a bearing assembly on which the balls or rollers rotate.
Pista exterior de un cojinete La parte exterior de una asamblea de cojinetes en la cual ruedan las bolas o los rodillos.

Out-of-round Wear of a round hole or shaft which, when viewed from an end, will appear egg-shaped.
Defecto de circularidad Desgaste de un taladro o de una flecha circular, que al verse de una extremidad, tendrá una forma asimétrica, como la de un huevo.

Output shaft The shaft or gear that delivers the power from a device, such as a transmission.
Flecha de salida La flecha o la velocidad que transmite la potencia de un dispositivo, tal como una transmisión.

Overall ratio The product of the transmission gear ratio multiplied by the final drive or rear axle ratio.
Relación global El producto de multiplicar la relación de los engranajes de la transmisión por la relación del impulso final o por la relación del eje trasero.

Overdrive Any arrangement of gearing that produces more revolutions of the driven shaft than of the driving shaft.
Sobremultiplicación Un arreglo de los engranajes que produce más revoluciones de la flecha de arrastre que los de la flecha de propulsión.

Overdrive ratio Identified by the decimal point indicating less than one driving input revolution compared to one output revolution of a shaft.
Relación del sobremultiplicación Se identifica por el punto decimal que indica menos de una revolución del motor comparado a una revolución de una flecha de salida.

Overrun coupling A free-wheeling device to permit rotation in one direction but not in the other.
Acoplamiento de sobremarcha Un dispositivo de marcha de rueda libre que permite las giraciones en una dirección, pero no en la otra dirección.

Overrunning clutch A device consisting of a shaft or housing linked together by rollers or sprags operating between movable and fixed races. As the shaft rotates, the rollers or sprags jam between the movable and fixed races. This jamming action locks together the shaft and housing. If the fixed race should be driven at a speed greater than the movable race, the rollers or sprags will disconnect the shaft.
Embrague de sobremarcha Un dispositivo que consiste de una flecha o un cárter eslabonados por medio de rodillos o palancas de detención que operan entre pistas fijas y movibles. Al girar la flecha, los rodillos o palancas de detención se aprietan entre las pistas fijas y movibles. Este acción de apretarse enclava el cárter con la flecha. Si la pista fija se arrastra en una velocidad más alta que la pista movible, los rodillos o palancas de detención desconectarán a la flecha.

Oxidation Burning or combustion; the combining of a material with oxygen. Rusting is slow oxidation, and combustion is rapid oxidation.
Oxidación Quemando o la combustión; la combinación de una materia con el oxígeno. El orín es una oxidación lenta, la combustión es la oxidación rápida.

Pascal's Law The law of fluid motion.
Ley de pascal La ley del movimiento del fluido.

Parallel The quality of two items being the same distance from each other at all points; usually applied to lines and, in automotive work, to machined surfaces.
Paralelo La calidad de dos artículos que mantienen la misma distancia el uno del otro en cada punto; suele aplicarse a las líneas y, en el trabajo automotívo, a las superficies acabadas a máquina.

Pawl A lever that pivots on a shaft. When lifted, it swings freely and when lowered, it locates in a detent or notch to hold a mechanism stationary.
Trinquete Una palanca que gira en una flecha. Levantado, mueve sín restricción, bajado, se coloca en una endentación o una muesca para mantener sín movimiento a un mecanismo.

Peen To stretch or clinch over by pounding with the rounded end of a hammer.
Martillazo Estirar o remachar con la extremidad redondeado de un martillo de bola.

Pitch The number of threads per inch on any threaded part.
Paso El número de filetes por pulgada de cualquier parte fileteada.

Pivot A pin or shaft upon which another part rests or turns.
Pivote Una chaveta o una flecha que soporta a otra parte o sirve como un punto para girar.

Planetary gear set A system of gearing that is modeled after the solar system. A pinion is surrounded by an internal ring gear and planet gears are in mesh between the ring gear and pinion around which all revolve.
Conjunto de engranajes planetarios Un sistema de engranaje cuyo patrón es el sistema solar. Un engranaje propulsor (la corona interior) rodea al piñon de ataque y los engranajes satélites y planetas se endentan entre la corona y el piñon alrededor del cual todo gira.

Planet carrier The carrier or bracket in a planetary gear system that contains the shafts upon which the pinions or planet gears turn.
Perno de arrastre planetario El soporte o la abrazadera que contiene las flechas en las cuales giran los engranajes planetarios o los piñones.

Planet gears The gears in a planetary gear set that connect the sun gear to the ring gear.
Engranages plantearios Los engranajes en un conjunto de engranajes planetario que connectan al engranaje propulsor interior (el engranaje sol) con la corona.

Planet pinions In a planetary gear system, the gears that mesh with, and revolve about, the sun gear; they also mesh with the ring gear.
Piñones planetarios En un sistema de engranajes planetarios, los engranajes que se endentan con, y giran alrededor, el engranaje propulsor (sol); también se endentan con la corona.

Plug Anything that will fit into an opening to stop fluid or airflow.
Tapón Cualquier cosa que se ajuste en una apertura para prevenir el goteo o el escape de un corriente del aire.

Pneumatic tools Power tools that rely on compressed air for power.
Herramientas neumáticas Las herramientas de motor cuyo energía proviene del aire comprimido.

Porosity   A statement of how porous or permeable to liquids a material is.

Porosidad   Una expresión de lo poroso o permeable a los líquidos es una materia.

Power train   The mechanisms that carry the power from the engine crankshaft to the drive wheels; these include the clutch, transmission, drive line, differential, and axles.

Tren impulsor   Los mecanismos que transferen la potencia desde el cigueñal del motor a las ruedas de propulsión; éstos incluyen el embrague, la transmisión, la flecha motríz, el diferencial y los semiejes.

Preload   A load applied to a part during assembly so as to maintain critical tolerances when the operating load is applied later.

Carga previa   Una carga aplicada a una parte durante la asamblea para asegurar sus tolerancias críticas antes de que se le aplica la carga de la operación.

Press fit   Forcing a part into an opening that is slightly smaller than the part itself to make a solid fit.

Ajustamiento a presión   Forzar a una parte en una apertura que es de un tamaño más pequeño de la parte para asegurar un ajustamiento sólido.

Pressure   Force per unit area, or force divided by area. Usually measured in pounds per square inch (psi) or in kilopascals (kPa) in the metric system.

Presión   La fuerza por unedad de una area, o la fuerza divida por la area. Suele medirse en libras por pulgada cuadrada (lb/pulg2) o en kilopascales (kPa) en el sistema métrico.

Pressure plate   That part of the clutch which exerts force against the friction disc; it is mounted on and rotates with the flywheel.

Plato opresor   Una parte del embraque que aplica la fuerza en el disco de fricción; se monta sobre el volante, y gira con éste.

Propeller shaft   *See* Driveshaft.

Flecha de Propulsion   *Vea* Flecha motríz.

Prussian blue   A blue pigment; in solution, useful in determining the area of contact between two surfaces.

Azul de Prusia   Un pigmento azul; en forma líquida, ayuda en determinar la area de contacto entre dos superficies.

PSI   Abbreviation for pounds per square inch, a measurement of pressure.

Lb/pulg2   Una abreviación de libras por pulgada cuadrada, una medida de la presión.

Puller   Generally, a shop tool used to separate two closely fitted parts without damage. Often contains a screw, or several screws, which can be turned to apply a gradual force.

Extractor   Generalmente, una herramienta del taller que sirve para separar a dos partes apretadas sin incurrir daños. Suele tener una tuerca o varias tuercas, que se pueden girar para aplicar la fuerza gradualmente.

Pulsation   To move or beat with rhythmic impulses.

Pulsación   Moverse o batir con impulsos rítmicos.

Race   A channel in the inner or outer ring of an antifriction bearing in which the balls or rollers roll.

Pista   Un canal en el anillo interior o exterior de un cojinete antifricción en el cual ruedan las bolas o los rodillos.

Radial   The direction moving straight out from the center of a circle. Perpendicular to the shaft or bearing bore.

Radial   La dirección al moverse directamente del centro de un círculo. Perpendicular a la flecha o al taladro del cojinete.

Radial clearance   Clearance within the bearing and between balls and races perpendicular to the shaft. Also called radial displacement

Holgura radial   La holgura en un cojinete entre las bolas y las pistas que son perpendiculares a la flecha. También se llama un desplazamiento radial.

Radial load   A force perpendicular to the axis of rotation.

Carga radial   Una fuerza perpendicular al centro de rotación.

Ratio   The relation or proportion that one number bears to another.

Relación   La correlación o proporción de un número con respeto a otro.

Reamer   A round metal-cutting tool with a series of sharp cutting edges; enlarges a hole when turned inside it.

Escariador   Una herramienta redonda para cortar a los metales que tiene una seria de rebordes mordaces agudos; al girarse en un agujero lo agranda.

Rear-wheel drive   A term associated with a vehicle where the engine is mounted at the front and the driving axle and driving wheels are at the rear of the vehicle.

Tracción trasera   Un término que se asocia con un vehículo en el cual el motor se ubica en la parte delantera y el eje propulsor y las ruedas propulsores se encuentran en la parte trasera del vehículo.

Relief valve   A valve used to protect against excessive pressure in the case of a malfunctioning pressure regulator.

Válvula de seguridad   Una válvula que se usa para guardar contra una presión excesiva en caso de que malfulciona el regulador de presión.

Retaining ring   A removable fastener used as a shoulder to retain and position a round bearing in a hole.

Anillo de retén   Un seguro removible que sirve de collarín para sujetar y posicionar a un cojinete en un agujero.

Rivet   A headed pin used for uniting two or more pieces by passing the shank through a hole in each piece and securing it by forming a head on the opposite end.

Remache   Una clavija con cabeza que sirve para unir a dos piezas o más al pasar el vástago por un hoyo en cada pieza y asegurarlo por formar una cabeza en el extremo opuesto.

Roller bearing   An inner and outer race upon which hardened steel rollers operate.

Cojinete de rodillos   Una pista interior y exterior en la cual operan los rodillos hecho de acero endurecido.

Rollers   Round steel bearings that can be used as the locking element in an overrunning clutch or as the rolling element in an antifriction bearing.

Rodillos   Articulaciones redondos de acero que pueden servir como un elemento de enclavamiento en un embrague de sobremarcha o como el elemento que rueda en un cojinete antifricción.

Rotary flow   A fluid force generated in the torque converter that is related to vortex flow. The vortex flow leaving the impeller is not only flowing out of the impeller at high speed but is also rotating faster than the turbine. The rotating fluid striking the slower turning turbine exerts a force against the turbine which is defined as rotary flow.

Flujo rotativo   Una fuerza fluida producida en el convertidor de torsión que se relaciona al flujo torbellino. El flujo torbellino saliendo del rotor no sólo viaja en una alta velocidad sino también gira más rápidamente que el turbino. El fluido rotativo chocando contra el turbino que gira más lentamente, impone una fuerza contra el turbino que se define como flujo rotativo.

RPM   Abbreviation for revolutions per minute, a measure of rotational speed.

RPM   Abreviación de revoluciones por minuto, una medida de la velocidad rotativa.

RTV sealer   Room-temperature vulcanizing gasket material, which cures at room temperature; a plastic paste squeezed from a tube to form a gasket of any shape.

Sellador RTV   Una materia vulcanizante de empaque que cura en temperaturas del ambiente; una pasta plástica exprimida de un tubo para formar un empaque de cualquiera forma.

Runout   Deviation of the specified normal travel of an object. The amount of deviation or wobble a shaft or wheel has as it rotates. Runout is measured with a dial indicator.

Corrimiento   Una desviación de la carrera normal e especificada de un objeto. La cantidad de desviación o vacilación de una flecha o una rueda mientras que gira. El corrimiento se mide con un indicador de carátula.

RWD   Abbreviation for rear-wheel drive.

RWD   Abreviación de tracción trasera.

SAE   Society of Automotive Engineers.

SAE   La Sociedad de Ingenieros Automotrices.

Score   A scratch, ridge, or groove marring a finished surface.

Entalladura   Una raya, una arruga o una ranura que desfigure a una superficie acabada.

Scuffing   A type of wear in which there is a transfer of material between parts moving against each other; shows up as pits or grooves in the mating surfaces.

Erosión   Un tipo de desgaste en el cual hay una tranferencia de una materia entre las partes que estan en contacto mientras que muevan; se manifesta como hoyitos o muescas en las superficies apareadas.

Seal   A material, shaped around a shaft, used to close off the operating compartment of the shaft, preventing oil leakage.

Sello   Una materia, formado alrededor de una flecha, que sella el compartimiento operativo de la flecha, preveniendo el goteo de aceite.

Sealer   A thick, tacky compound, usually spread with a brush, which may be used as a gasket or sealant to seal small openings or surface irregularities.

Sellador   Un compuesto pegajoso y espeso, comúnmente aplicado con una brocha, que puede usarse como un empaque o un obturador para sellar a las aperturas pequeñas o a las irregularidades de la superficie.

Seat   A surface, usually machined, upon which another part rests or seats; for example, the surface upon which a valve face rests.

Asiento   Una superficie, comúnmente maquinada, sobre la cual yace o se asienta otra parte; por ejemplo, la superficie sobre la cual yace la cara de la válvula.

Servo   A device that converts hydraulic pressure into mechanical movement, often multiplying it. Used to apply the bands of a transmission.

Servo   Un dispositivo que convierte la presión hidráulica al movimiento mecánico, frecuentemente multiplicándola. Se usa en la aplicación de las bandas de una transmisión.

Shift lever   The lever used to change gears in a transmission. Also the lever on the starting motor which moves the drive pinion into or out of mesh with the flywheel teeth.

Palanca del cambiador   La palanca que sirve para cambiar a las velocidades de una transmisión. También es la palanca del motor de arranque que mueva al piñon de ataque para engranarse o desegranarse con los dientes del volante.

Shift valve   A valve that controls the shifting of the gears in an automatic transmission.

Válvula de cambios   Una válvula que controla a los cambios de las velocidades en una transmisión automática.

Shim   Thin sheets used as spacers between two parts, such as the two halves of a journal bearing.

Laminilla de relleno   Hojas delgadas que sirven de espaciadores entre dos partes, tal como las dos partes de un muñón.

Shim stock   Sheets of metal of accurately known thickness which can be cut into strips and used to measure or correct clearances.

Materia de laminillas   Las hojas de metal cuyo espesor se conoce precisamente que pueden cortarse en tiras y usarse para medir o correjir a las holguras.

Side clearance   The clearance between the sides of moving parts when the sides do not serve as load-carrying surfaces.

Holgura lateral   La holgura entre los lados de las partes en movimiento mientras que los lados no funcionan como las superficies de carga.

Sliding-fit   Where sufficient clearance has been allowed between the shaft and journal to allow free-running without overheating.

Ajuste corredera   Donde se ha dejado una holgura suficiente entre la flecha y el muñon para permitir una marcha libre sin sobrecalentamiento.

Snap ring   Split spring-type ring located in an internal or external groove to retain a part.

Anillo de seguridad   Un anillo partido tipo resorte que se coloca en una muesca interior o exterior para retener a una parte.

Spalling   A condition where the material of a bearing surface breaks away from the base metal.

Escamación   Una condición en la cual una materia de la superficie de un rodamiento se separa del metal base.

Spindle   The shaft on which the wheels and wheel bearings mount.

Husillo   La flecha en la cual se montan las ruedas y el conjunto del cojinete de las ruedas.

Spline   Slot or groove cut in a shaft or bore; a splined shaft onto which a hub, wheel, gear, and so on, with matching splines in its bore is assembled so that the two must turn together.

Acanaladura (espárrago)   Una muesca o ranura cortada en una flecha o en un taladro; una flecha acanalada en la cual se asamblea un cubo, una rueda, un engranaje, y todo lo demás que tiene un acanaladura pareja en el taladro de manera de que las dos deben girar juntos.

Split lip seal   Typically, a rope seal sometimes used to denote any two-part oil seal.

Sello hendido   Típicamente, un sello de cuerda que se usa a veces para demarcar cualquier sello de aceite de dos partes

Split pin   A round split spring steel tubular pin used for locking purposes; for example, locking a gear to a shaft.

Chaveta hendida   Una chaveta partida redonda y tubular hecho de acero para resorte que sirve para el enclavamiento; por ejemplo, para enclavar un engranaje a una flecha.

Spool valve   A cylindrically shaped valve with two or more valleys between the lands. Spool valves are used to direct fluid flow.

Válvula de carrete   Una válvula de forma cilíndrica que tiene dos acanaladuras de cañon o más entre las partes planas. Las válvulas de carrete sirven para dirigir el flujo del fluido.

Sprag clutch   A member of the overrunning clutch family using a sprag to jam between the inner and outer races used for holding or driving action.

Embrague de puntal   Un miembro de la familia de embragues de sobremarcha que usa a una palanca de detención trabada entre las pistas interiores e exteriores para realizar una acción de asir o marchar.

Spring A device that changes shape when it is stretched or compressed, but returns to its original shape when the force is removed; the component of the automotive suspension system that absorbs road shocks by flexing and twisting.

Resorte Un dispositivo que cambia de forma al ser estirado o comprimido, pero que recupera su forma original al levantarse la fuerza; es un componente del sistema de suspensión automotívo que absorba los choques del camino al doblarse y torcerse.

Spring retainer A steel plate designed to hold a coil or several coil springs in place.

Retén de resorte Una chapa de acero diseñado a sostener en su posición a un resorte helicoidal o más.

Squeak A high-pitched noise of short duration.

Chillido Un ruido agudo de poca duración.

Squeal A continuous high-pitched noise.

Alarido Un ruido agudo continuo.

Stall A condition where the engine is operating and the transmission is in gear, but the drive wheels are not turning because the turbine of the torque converter is not moving.

Paro Una condición en la cual opera el motor y la transmisión esta embragada pero las ruedas de impulso no giran porque no mueva el turbino del convertidor de la torsión.

Stall test A test of the one-way clutch in a torque converter.

Prueba de paro Una prueba del embrague de una vía en un convertidor de la torsión.

Static A form of electricity caused by friction.

Estático Una forma de la electridad causada por la fricción.

Stress The force to which a material, mechanism, or component is subjected.

Esfuerzo La fuerza a la cual se somete a una materia, un mecanísmo o un componente.

Sun gear The central gear in a planetary gear system around which the rest of the gears rotate. The innermost gear of the planetary gear set.

Engranaje principal (sol) El engranaje central en un sistema de engranajes planetarios alrededor del cual giran los otros engranajes. El engranaje más interno del conjunto de los engranajes planetarios.

Tap To cut threads in a hole with a tapered, fluted, threaded tool.

Roscar con macho Cortar las roscas en un agujero con una herramienta cónica, acanalada y fileteada.

Teardown A term often used to describe the process of disassembling a transmission.

Desmontaje Un término común que describe el proceso de desarmar a una transmisión.

Temper To change the physical characteristics of a metal by applying heat.

Templar Cambiar las características físicas de un metal mediante una aplicación del calor.

Tension Effort that elongates or "stretches" a material.

Tensión Un esfuerzo que alarga o "estira" a una materia.

Thickness gauge Strips of metal made to an exact thickness, used to measure clearances between parts.

Calibre de espesores Las tiras del metal que se han fabricado a un espesor exacto, sirven para medir las holguras entre las partes.

Thread chaser A device, similar to a die, that is used to clean threads.

Peine de roscar Un dispositivo, parecido a una terraja, que sirve para limpiar a las roscas.

Threaded insert A threaded coil that is used to restore the original thread size to a hole with damaged threads.

Pieza inserta roscada Una bobina roscada que sirve para restaurar a su tamaño original una rosca dañada.

Thrust bearing A bearing designed to resist or contain side or end motion as well as reduce friction.

Cojinete de empuje Un cojinete diseñado a detener o reprimir a los movimientos laterales o de las extremidades y también reducir la fricción.

Thrust load A load that pushes or reacts through the bearing in a direction parallel to the shaft.

Carga de empuje Una carga que empuja o reacciona por el cojinete en una dirección paralelo a la flecha.

Thrust washer A washer designed to take up end thrust and prevent excessive endplay.

Arandela de empuje Una arandela diseñada para rellenar a la holgura de la extremidad y prevenir demasiado juego en la extremidad.

Tolerance A permissible variation between the two extremes of a specification or dimension.

Tolerancia Una variación permisible entre dos extremos de una especificación o de un dimensión.

Torque A twisting motion, usually measured in ft.-lb. (Nm).

Torsión Un movimiento giratorio, suele medirse en pies/libra (Nm).

Torque capacity The ability of a converter clutch to hold torque.

Capacidad de la torsión La abilidad de un convertidor de embraque a sostener a la torsión.

Torque converter A turbine device utilizing a rotary pump, one or more reactors (stators), and a driven circular turbine or vane, whereby power is transmitted from a driving to a driven member by hydraulic action. It provides varying drive ratios; with a speed reduction, it increases torque.

Convertidor de la torsión Un dispositivo de turbino que utilisa a una bomba rotativa, a un reactor o más, y un molinete o turbino circular impulsado, por cual se transmite la energía de un miembro de impulso a otro arrastrado mediante la acción hidráulica. Provee varias relaciones de impulso; al descender la velocidad, aumenta la torsión.

Torque curve A line plotted on a chart to illustrate the torque personality of an engine. When the engine operates on its torque curve, it is producing the most torque for the quantity of fuel being burned.

Curva de la torsión Una linea delineada en una carta para ilustrar las características de la torsión del motor. Al operar un motor en su curva de la torsión, produce la torsión óptima para la cantidad del combustible que se consuma.

Torque multiplication The result of meshing a small driving gear and a large driven gear to reduce speed and increase output torque.

Multiplicación de la torsión El resultado de engranar a un engranaje pequeño de ataque con un engranaje más grande arrastrado para reducir la velocidad y incrementar la torsión de salida.

Torque steer An action felt in the steering wheel as the result of increased torque.

Dirección la torsión Una acción que se nota en el volante de dirección como resultado de un aumento de la torsión.

Traction The gripping action between the tire tread and the road's surface.

Tracción La acción de agarrar entre la cara de la rueda y la superficie del camino.

Transaxle   Type of construction in which the transmission and differential are combined in one unit.
Flecha de transmisión   Un tipo de construcción en el cual la transmisión y el diferencial se combinan en una unedad.

Transaxle assembly   A compact housing most often used in front-wheel-drive vehicles that houses the manual transmission, final drive gears, and differential assembly.
Asamblea de la flecha de transmisión   Un cárter compacto que se usa normalmente en los vehículos de tracción delantera que contiene la transmisión manual, los engranajes de propulsión, y la asamblea del diferencial.

Transfer case   An auxiliary transmission mounted behind the main transmission. Used to divide engine power and transfer it to both front and rear differentials, either full-time or part-time.
Cárter de la transferencia   Una transmisión auxiliar montada detrás de la transmisión principal. Sirve para dividir la potencia del motor y transferirla a ambos diferenciales delanteras y traseras todo el tiempo o la mitad del tiempo.

Transmission   The device in the power train that provides different gear ratios between the engine and drive wheels as well as reverse.
Transmisión   El dispositivo en el trén de potencia que provee las relaciones diferentes de engranaje entre el motor y las ruedas de impulso y también la marcha de reversa.

Transverse   Power train layout in a front-wheel-drive automobile extending from side to side.
Transversal   Una esquema del tren de potencia en un automóvil de tracción delantera que se extiende de un lado a otro.

U-joint   A four-point cross connected to two U-shaped yokes that serves as a flexible coupling between shafts.
Junta de U   Una cruceta de cuatro puntos que se conecta a dos yugos en forma de U que sirven de acoplamientos flexibles entre las flechas.

Universal joint   A mechanical device that transmits rotary motion from one shaft to another shaft at varying angles.
Junta Universal   Un dispositivo mecánico que transmite el movimiento giratorio desde una flecha a otra flecha en varios ángulos.

Upshift   To shift a transmission into a higher gear.
Cambio ascendente   Cambiar a la velocidad de una transmisión a una más alta.

Valve body   Main hydraulic control assembly of a transmission containing the components necessary to control the distribution of pressurized transmission fluid throughout the transmission.

Cuerpo de la válvula   Asamblea principal del control hidráulico de una transmisión que contiene los componentes necessarios para controlar a la distribución del fluido de la transmisión bajo presión por toda la transmisión.

Vehicle identification number (VIN)   The number assigned to each vehicle by its manufacturer, primarily for registration and identification purposes.
Número de identificacíon del vehículo   El número asignado a cada vehículo por su fabricante, primariamente con el propósito de la registración y la identificación.

Vibration   A quivering, trembling motion felt in the vehicle at different speed ranges.
Vibración   Un movimiento de estremecer o temblar que se siente en el vehículo en varios intervalos de velocidad.

Viscosity   The resistance to flow exhibited by a liquid. A thick oil has greater viscosity than a thin oil.
Viscosidad   La resistencia al flujo que manifiesta un líquido. Un aceite espeso tiene una viscosidad mayor que un aceite ligero.

Vortex   Path of fluid flow in a torque converter. The vortex may be high, low, or zero, depending on the relative speed between the pump and turbine.
Vórtice   La vía del flujo de los fluidos en un convertido de torsión. El vórtice puede ser alto, bajo, o cero, depende de la velocidad relativa entre la bomba y la turbina.

Vortex flow   Recirculating flow between the converter impeller and turbine that causes torque multiplication.
Flujo del vórtice   El fluyo recirculante entre el impulsor del convertidor y la turbina que causa la multiplicación de la torsión.

Wet-disc clutch   A clutch in which the friction disc (or discs) is operated in a bath of oil.
Embrague de disco flotante   Un embrague en el cual el disco (o los discos) de fricción opera en un baño de aceite.

Wheel   A disc or spokes with a hub at the center which revolves around an axle, and a rim around the outside for mounting the tire on.
Rueda   Un disco o rayo que tiene en su centro un cubo que gira alrededor de un eje, y tiene un rim alrededor de su exterior en la cual se monta el neumático.

Yoke   In a universal joint, the drivable torque-and-motion input and output member, attached to a shaft or tube.
Yugo   En una junta universal, el miembro de la entrada y la salida que transfere a la torsión y al movimiento, que se conecta a una flecha o a un tubo.

3 Manual Drive Train and Axles

Pretest

The purpose of this pretest is to determine the amount of review that you may require prior to writing the ASE Manual Drive Train and Axles Test. If you answer all the pretest questions correctly, complete the questions and study the information in this chapter to prepare for the ASE Manual Drive Train and Axles Test.

If two or more of your answers to the pretest questions are incorrect, complete a study of Chapters 3 through 8 in *Today's Technician Manual Transmissions and Transaxles*, published by Delmar Publishers, plus a study of the questions and information in this chapter.

The pretest answers are located at the end of the pretest, and these answers also are in the answer sheets supplied with this book.

1. A vehicle has a final drive ratio of 3.42:1, and a first gear ratio of 3.65:1. The overall gear ratio in first gear is
   A. 7.07:1
   B. 10.56:1
   C. 12.48:1
   D. 13.76:1

2. The rear axle ratio is changed from 3.08:1 to 3.73:1.
   Technician A says the engine rpm will be higher at the same vehicle speed in any gear.
   Technician B says the engine will have less torque at the same vehicle speed in any gear.
   Who is correct?
   A. A only
   B. B only
   C. Both A and B
   D. Neither A nor B

3. A typical overdrive transmission gear ratio would be
   A. .46:1
   B. .70:1
   C. .92:1
   D. .96:1

4. The differential case and ring gear are rotating at 100 rpm, and one rear drive wheel is not turning. The opposite rotating rear drive wheel is turning at
   A. 100 rpm.
   B. 200 rpm.
   C. 300 rpm.
   D. 400 rpm.

5. Hard transaxle shifting may be caused by
   A. a bent shift rail.
   B. rough input shaft bearing.
   C. excessive output shaft end play.
   D. worn clutch release bearing.

6. A hydraulic clutch does not disengage properly resulting in hard shifting.
   Technician A says there may be air in the hydraulic clutch system.
   Technician B says there may be too much clutch pedal free play.
   Who is correct?
   A. A only
   B. B only
   C. Both A and B
   D. Neither A nor B

7. A clutch grabs and chatters while engaging.
   Technician A says the torsion springs in the clutch plate may be weak or broken.
   Technician B says the clutch release bearing may be rough.
   Who is correct?
   A. A only
   B. B only
   C. Both A and B
   D. Neither A nor B

8. When a flywheel requires resurfacing, the amount of metal machined from the flywheel surface should be
   A. .005 to .010 in.
   B. .015 to .025 in.
   C. .010 to .040 in.
   D. .010 to .090 in.

9. A bell housing bore has excessive runout. Technician A says the bell housing bore may be machined to correct the problem.
   Technician B says the offset dowels in the back of the engine block may be rotated to correct this excessive runout.
   Who is correct?
   A. A only
   B. B only
   C. Both A and B
   D. Neither A nor B

10. All of the following statements about synchronizer assemblies are true EXCEPT
    A. synchronizer hubs and sleeves should be marked before disassembly so they can be reassembled in the same position.
    B. synchronizer hubs are reversible.
    C. worn blocking rings may cause hard shifting.
    D. threads in the cone area of the blocking ring must be sharp.

11. A vehicle has a clicking noise in the differential while driving straight ahead. The cause of the problem could be damaged
    A. side gear teeth.
    B. pinion gear teeth.
    C. pinion shaft and case.
    D. ring gear and pinion teeth.

12. Technician A says in some transaxles the differe~~~
    adjusted with adjuster nuts behind the side bearin~
    Technician B says in some transaxles the different~
    adjusted by changing the shim thickness behind on~
    Who is correct?
    A.   A only
    B.   B only
    C.   Both A and B
    D.   Neither A nor B

13. The differential turning torque in a transaxle is less than s~
    Technician A says a thinner thrust washer may be installed ~~~d both differential
    side gears.
    Technician B says a thicker thrust washer may be installed behind one of the differ-
    ential side gears.
    Who is correct?
    A.   A only
    B.   B only
    C.   Both A and B
    D.   Neither A nor B

14. The transfer case on a four-wheel-drive (4WD) vehicle contains a planetary gear set:
    Technician A says in 4WD low the annulus gear is locked to the case.
    Technician B says in 4WD the input shaft is driving the sun gear and the planetary
    carrier is the output to the front and rear drive shafts.
    Who is correct?
    A.   A only
    B.   B only
    C.   Both A and B
    D.   Neither A nor B

Answers to Pretest

1. C, 2. A, 3. B, 4. B, 5. A, 6. C, 7. A, 8. C, 9. B, 10. B, 11. D, 12. D, 13. D, 14. C

Diagnosis and Repair

ASE Tasks, Questions, and Related Information

In this chapter each task in the Manual Transmission and Transaxle category is provided followed by a question and some information related to the task. If you answer any question incorrectly, study this information very carefully until you understand the correct answer. For additional information on any task refer to *Today's Technician Manual Transmissions and Transaxles,* by Delmar Publishers.

Question answers and analysis are provided at the end of this chapter and in the answer sheets provided with this book.

Task 1   **Diagnose clutch noise, binding, slippage, pulsation, and chatter problems; determine needed repairs.**

1. Clutch chatter may be caused by
   A. a worn, rough clutch release bearing.
   B. a worn, rough pilot bearing.
   C. excessive input shaft end play.
   D. weak clutch plate torsional springs.

Hint   *The main components in the clutch include the clutch disc, pressure plate, release bearing and hub, and release fork. Refer to Table 3-1, page 39 in* Today's Technician Manual Transmissions and Transaxles Shop Manual *for a complete listing and diagnosis of clutch problems.*

Task 2   **Inspect, adjust, and replace clutch pedal linkage, cables and automatic adjuster mechanisms, brackets, bushing, pivots, and springs.**

2. Lack of clutch pedal free play may cause
   A. hard shifting.
   B. improper clutch release.
   C. transaxle gear damage.
   D. clutch slipping.

3. A clutch with a self-adjusting cable has
   A. one inch of clutch pedal free play.
   B. two inches of clutch pedal free play.
   C. a constant running release bearing.
   D. an overcenter assist spring.

Hint   *Some vehicles have linkages connected from the clutch pedal to the clutch release fork (Figure 3-1). In other clutch systems the clutch pedal is connected to the release fork by a cable. Many clutch linkages or cables have an adjustment to set the clutch pedal free play. Clutch pedal free play is the amount of pedal movement before the release bearing contacts the pressure plate release fingers or diaphragm. Many late-model vehicles have a self-adjusting clutch cable. In these clutches the cable is wrapped around and attached to a toothed wheel, and a ratcheting spring-loaded pawl is engaged with the toothed wheel (Figure 3-2). Each time the clutch pedal is released, the pawl removes any slack from the cable by engaging the next tooth on the wheel. A self-adjusting clutch has no built-in free play, and a constant running release bearing.*

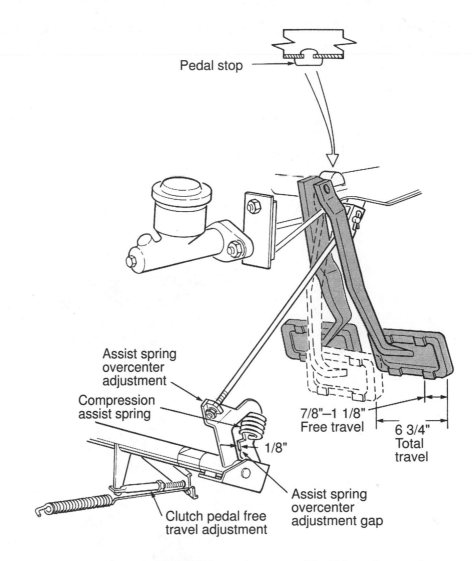

Figure 3-1 Typical clutch linkage (*Courtesy of Ford Motor Company*)

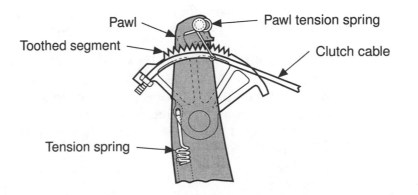

Figure 3-2 Self-adjusting clutch cable adjusting mechanism (*Courtesy of Ford Motor Company*)

Task 3   Inspect, adjust, repair, and replace hydraulic clutch slave and master cylinders, lines, and hoses.

4. A hydraulic clutch system in Figure 3-3 the clutch fails to disengage properly when the clutch pedal is fully depressed. The cause of this problem could be
   A. less than specified clutch pedal free play.
   B. air in the clutch hydraulic system.
   C. worn clutch facings.
   D. a scored pressure plate.

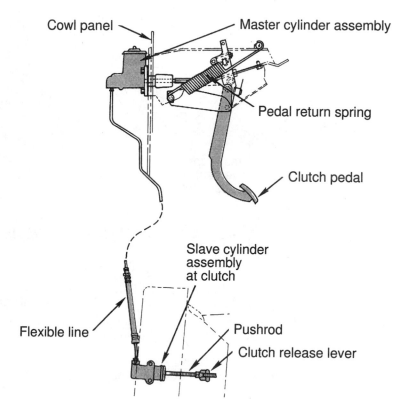

Cowl panel — Master cylinder assembly

Pedal return spring

Clutch pedal

Slave cylinder assembly at clutch

Flexible line

Pushrod

Clutch release lever

Figure 3-3 Typical clutch hydraulic system

Hint   *In a hydraulic clutch system the clutch pedal is connected to the clutch master cylinder pushrod, and a hydraulic line is connected from the clutch master cylinder to the slave cylinder.*

Task 4   Inspect, adjust, and replace release (throwout) bearing, lever, and pivot.

5. While discussing a clutch with an adjustable linkage
   Technician A says the clutch pedal freeplay adjustment sets the distance between the release bearing and the pressure plate fingers.
   Technician B says a worn release bearing is noisy with the clutch pedal released.
   Who is correct?
   A. A only
   B. B only
   C. Both A and B
   D. Neither A nor B

Hint   *The clutch release bearing is connected to the release fork, and the center opening in this bearing is mounted on a machined hub that is bolted to the front of the transmission (Figure 3-4).*

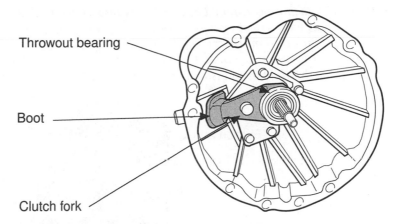

Throwout bearing

Boot

Clutch fork

Figure 3-4   Clutch release bearing and fork   *(Courtesy of Nissan Motors Co., Ltd.)*

When the clutch pedal is depressed the release bearing pushes on the pressure plate release fingers or thrust pad. This action moves the pressure plate fingers or levers against the pressure plate spring pressure. Movement of the pressure plate fingers forces the pressure plate away from the clutch disc to release the clutch, thus interrupting power flow from the engine to the transmission.

Task 5   Inspect and replace clutch pressure plate assembly.

6. Clutch chatter may be caused by
   A. excessive crankshaft endplay.
   B. loose engine main bearings.
   C. a badly scored pressure plate.
   D. improper pressure plate to flywheel position.

Hint   *A straightedge and a feeler gauge should be used to measure pressure plate warpage (Figure 3-5). If the flywheel does not have locating dowels, always punch mark the pressure plate and flywheel so the pressure plate is reinstalled in the original position.*

Figure 3-5   Measuring pressure plate warpage

Task 6   Inspect and replace clutch disc assembly.

7. The distance from the rivet heads in the clutch facing surface should be at least
   A. 0.005 in.
   B. 0.008 in.
   C. 0.012 in.
   D. 0.025 in.

Hint   *Clutch facing wear should be checked by measuring the distance from the facing surface to the rivet heads with a depth gauge or vernier caliper (Figure 3-6). When the clutch disc and pressure plate are installed on the flywheel, the clutch disc must be installed in the proper direction. An alignment tool must be installed through the clutch disc hub into the pilot bearing to align the disc properly while the pressure plate retaining bolts are tightened.*

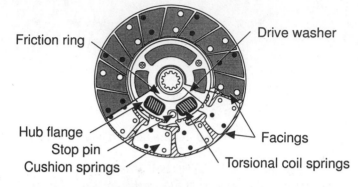

Friction ring   Drive washer

Hub flange
Stop pin
Cushion springs
Facings
Torsional coil springs

Figure 3-6 Clutch disc components

Task 7   Inspect and replace pilot bearing.

8. A worn pilot bearing may cause a rattling, growling noise while
   A. the engine is idling and the clutch pedal is fully depressed.
   B. decelerating in high gear with the clutch pedal released.
   C. accelerating in low gear with the clutch pedal released.
   D. the engine is idling in neutral with the clutch pedal released.

Hint   *The pilot bearing or bushing is mounted in the center of the flywheel. This bearing supports the front of the transmission input shaft, and the clutch disc is splined to this shaft. The pilot bearing may be a ball-type or roller-type bearing, or a bushing. The proper puller must be used to remove the pilot bearing (Figure 3-7). The pilot bearing must be installed with the proper driving tool.*

The pilot bearing is removed from the crankshaft using this slide hammer puller with the engine in the vehicle. The jaw fits inside the bearing bore for removal without damage to the crankshaft or bearing.

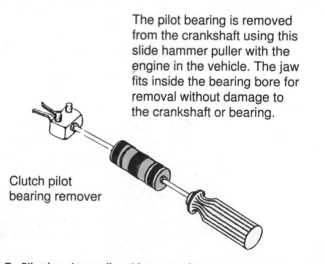

Clutch pilot
bearing remover

Figure 3-7 Pilot bearing puller (*Courtesy of Kent-Moore Division, SPX Corp.*)

Task 8   Inspect, repair, or replace flywheel and ring gear.

9. Technician A says if too much material is removed during flywheel resurfacing, the torsion springs on the clutch disc may contact the flywheel bolts resulting in noise while engaging and disengaging.
   Technician B says if excessive material is removed when the flywheel is resurfaced, the slave cylinder may not have enough travel to release the clutch properly.
   Who is correct?

      A. A only
      B. B only
      C. Both A and B
      D. Neither A nor B

Hint   *If the flywheel surface is scored, burned, or worn it should be resurfaced or replaced. During the resurfacing operation 0.010 to 0.040 in. of metal may be removed from the flywheel surface. Do not remove excessive material from the flywheel surface.*

Worn or chipped flywheel ring gear teeth may cause improper flywheel balance and engine vibrations. Improper starting motor engagement also results from worn or chipped flywheel ring gear teeth. When the flywheel is removed from the crankshaft, punch or index marks must be placed on the flywheel and crankshaft so the flywheel is reinstalled in the original position to maintain proper flywheel balance.

Task 9   Inspect engine block, clutch (bell) housing, and transmission case mating surfaces; determine needed repairs.

  10. Excessive misalignment between the bell housing and the engine block may cause
      A. reduced clutch pedal free play.
      B. a growling noise with the clutch pedal depressed.
      C. a vibration at higher speeds.
      D. clutch grabbing and chatter.

Hint   *The mating surfaces of the engine block and bell housing should be inspected for metal burrs and accumulation of foreign material. Clean these mating surfaces with an approved cleaning solution. Remove any metal burrs with a fine-toothed file. Follow the same procedure to clean and inspect the mating surfaces of the bell housing and transmission including the bell housing bore.*

Task 10   Measure flywheel-to-block runout and crankshaft end play; determine needed repairs.

  11. With the dial indicator positioned as shown in Figure 3-8 the measurement being performed is
      A. crankshaft endplay.
      B. crankshaft warpage.
      C. rear main bearing wear.
      D. rear engine block alignment.

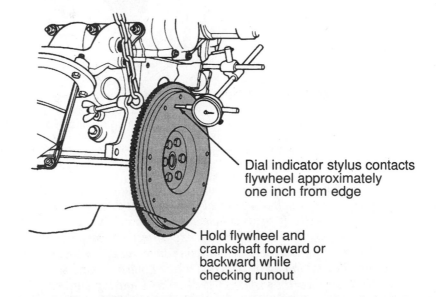

Dial indicator stylus contacts flywheel approximately one inch from edge

Hold flywheel and crankshaft forward or backward while checking runout

Figure 3-8 Dial indicator positioned against flywheel surface (*Courtesy of Ford Motor Company*)

Hint   *A dial indicator may be positioned against the clutch facing contact area on the flywheel while rotating the flywheel to measure flywheel runout. Hold the crankshaft forward or backward while measuring flywheel runout. If this runout is excessive, resurface or replace the flywheel.*

With the dial indicator positioned against the flywheel surface the crankshaft endplay may be measured by forcing the flywheel and crankshaft forward and rearward while observing the dial indicator reading.

Task 11   **Measure clutch (bell) housing bore-to-crankshaft runout and face squareness; determine needed repairs.**

12. Technician A says shims may be installed between the bell housing and engine block mating surfaces to correct bell housing face runout.
    Technician B says excessive bell housing face runout may be caused by overheated clutch disc facings.
    Who is correct?
    A. A only
    B. B only
    C. Both A and B
    D. Neither A nor B

13. To correct excessive runout on the dial indicator in Figure 3-9
    A. replace the engine mounts.
    B. adjust eccentric bell housing dowels.
    C. install bell housing shims.
    D. replace the clutch disc.

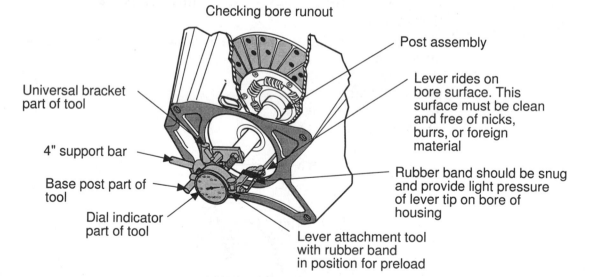

Figure 3-9   Measuring bell housing bore *(Courtesy of Ford Motor Company)*

Hint   *When bell housing face squareness is measured a dial indicator is attached to a steel post mounted in the clutch disc splined opening, and the dial indicator stem contacts the bell housing face (Figure 3-10). If the bell housing face runout is excessive, shims may be installed between the bell housing and the engine block to correct this misalignment.*

When performing the bell housing bore runout measurement, a special tool is connected between the dial indicator stem and the bell housing bore. Some engines have offset dowel pins in the block surface that mates with the bell housing surface. These offset dowel pins may be rotated to correct the bell housing bore runout. If the offset dowel adjustment is not enough to correct the bell housing bore runout, replace the bell housing.

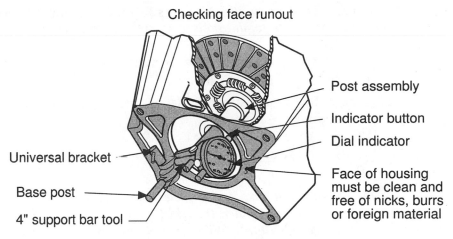

Figure 3-10 Measuring bell housing face *(Courtesy of Ford Motor Company)*

Task 12   **Inspect, replace, and align power train components.**

14. Technician A says sagged transmission mounts may cause improper drive shaft angles on a rear-wheel-drive car.
Technician B says improper drive shaft angles may cause a vibration that is constant when the vehicle is accelerated and decelerated.
Who is correct?
A. A only
B. B only
C. Both A and B
D. Neither A nor B

Hint   *Engine and transmission mounts should be inspected for broken, sagged, oil-soaked, or deteriorated conditions. Any of these mount conditions may cause a grabbing, binding clutch. On a rear-wheel-drive car defective engine or transmission mounts may cause improper drive shaft angles which result in a vibration that changes in intensity when the vehicle is accelerated and decelerated.*

Transmission Diagnosis and Repair

ASE Tasks, Questions, and Related Information

Task 1   **Diagnose transmission noise, hard shifting, jumping out of gear, and fluid leakage problems; determine needed repairs.**

15. While discussing a manual transmission that jumps out of second gear
Technician A says there is excessive end play between the second speed gear and its matching synchronizer.
Technician B says the detent springs on the shift rail may be weak.
Who is correct?
A. A only
B. B only
C. Both A and B
D. Neither A nor B

Hint   *Refer to Table 4-1 page 125 in Today's Technician Manual Transmissions and Transaxles, by Delmar Publishers, for diagnosis of manual transmission and transaxle problems.*

Task 2   **Inspect, adjust, and replace transmission shift linkages, brackets, bushings, cables, pivots, and levers.**

16. The shift lever adjustment usually is performed with the transmission in
   A. neutral.
   B. first gear.
   C. second gear.
   D. reverse gear.

Hint   *Most external shift linkages and cables require adjusting, and a similar adjustment procedure is used on some vehicles. Raise the vehicle on a lift and place the shift lever in neutral to begin the shift linkage adjustment. With a lever-type shift linkage install a 1/4 in. rod in the adjustment hole in the shifter assembly (Figure 3-11). Adjust the shift linkages by loosening the rod-retaining locknuts and moving the levers until the 1/4 in. rod fully enters the alignment holes. Tighten the locknuts and check the shift operation in all gears.*

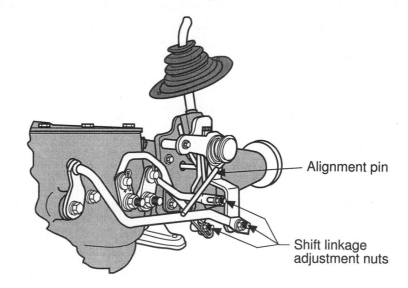

Alignment pin

Shift linkage
adjustment nuts

Figure 3-11 Typical adjustment of floor-mounted shift levers *(Courtesy of Ford Motor Company)*

Rod-type shift linkages are adjusted with basically the same procedure as lever-type linkages. When the alignment pin is in place, adjust the shift rod so the pin slides freely in and out of the alignment hole.

Task 3   **Inspect and replace transmission gaskets, seals, and sealants; inspect sealing surfaces.**

17. Repeated extension housing seal failure may be caused by
   A. a scored drive shaft yoke.
   B. excessive output shaft endplay.
   C. excessive input shaft endplay.
   D. a worn output shaft bearing.

Hint   *Fluid leaks in a manual transmission may occur at the extension housing seal, shift cover gasket, vent, backup light switch, drain plug, speedometer drive, or front bearing retainer gasket. If the extension housing seal is leaking it may be replaced with the transmission in the vehicle. When this seal is replaced always check the fit of the front drive shaft yoke in the extension housing bushing and inspect this yoke for scoring in the seal contact area. A worn extension housing bushing or a scored front drive shaft yoke will cause repeated extension housing seal failure.*

Task 4   Remove and replace transmission.

18. During manual transmission removal and replacement
    A. the drive shaft may be installed in any position on the differential pinion gear flange.
    B. the transmission weight may be supported by the input shaft in the clutch disc hub.
    C. the engine support fixture should be installed after the transmission to engine bolts are loosened.
    D. the clutch disc must be aligned with an aligning tool prior to transmission installation.

Hint   *Prior to transmission or transaxle removal the battery ground cable, shift linkages or cables, and speedometer cable must be removed. On a rear-wheel-drive car the drive shaft must be removed. Mark the rear drive shaft flange in relation to the yoke before drive shaft removal to ensure drive shaft installation in the original position. Drain the transmission lubricant. Before the transmission retaining bolts are loosened an engine support fixture must be installed to support the weight of the engine. Use a transmission jack to support the weight of the transmission during the removal process.*

Install a clutch disc alignment tool through the clutch disc hub into the pilot bearing to be sure the clutch disc is properly aligned before attempting to install the transmission (Figure 3-12).

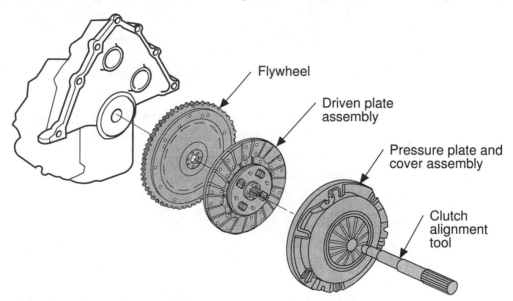

Figure 3-12   Clutch disc alignment tool   *(Courtesy of Chevrolet Motor Division, General Motors Corporation)*

Task 5   Disassemble and clean transmission components.

19. Technician A says if the synchronizer sleeve does not slide smoothly over the blocker ring and gear teeth, hard shifting will occur.
    Technician B says the synchronizer hub and sleeve must be marked in relation to each other prior to disassembly.
    Who is correct?
    A. A only
    B. B only
    C. Both A and B
    D. Neither A nor B

Hint   *Some transmissions have a seal and shim between the bell housing and the transmission case. This shim helps to control endplay. Remove the shift-lever cover, shift forks, and related mechanism. Remove the rear extension housing and front bearing retainer. Some front bearing retainers have a seal and shim behind the retainer. Remove the input gear and shaft, and output shaft assembly. Remove the counter shaft and counter gear assembly followed by the reverse idle shaft and reverse idler gear.*

Before a synchronizer assembly is removed from a shaft, mark the alignment of the collar and hub so these components may be reassembled in the original position (Figure 3-13).

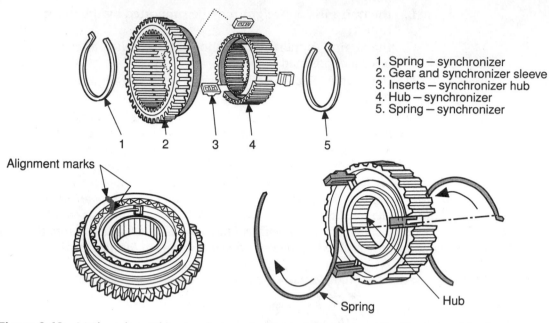

1. Spring — synchronizer
2. Gear and synchronizer sleeve
3. Inserts — synchronizer hub
4. Hub — synchronizer
5. Spring — synchronizer

Figure 3-13   Mark each synchronizer hub and collar before disassembly *(Courtesy of Ford Motor Company)*

Task 6   Inspect, repair, and /or replace transmission shift cover, forks, grommets, levers, shafts, sleeves, detent mechanisms, interlocks, and springs.

20. A bent shift rail may cause
    A. the transmission to jump out of gear.
    B. hard shifting in some gears.
    C. gear clash during some shifts.
    D. gear noise in some gears.

Hint   *Shift rails should be inspected to be sure they are not bent, broken, or worn. Inspect all the holes and notches in the shift rails for wear or damage (Figure 3-14). When each shift rail is installed in the appropriate bore in the case of shift cover, check for excessive movement between the rail and the bore. Inspect all detent springs to be sure they are not worn, bent, or weak.*

Check the interlock plates for flatness with a straightedge. If the interlock plates are worn on the surface that contacts the shift rails, replace the plates.

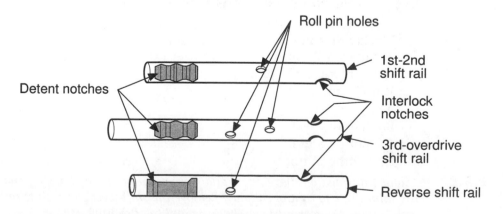

Figure 3-14   Shift fork inspection *(Courtesy of Ford Motor Company)*

Task 7　Inspect and replace input (clutch) shaft and bearings.

21. A manual transmission has a growling noise with the engine idling and the clutch released with the transmission in neutral. The noise disappears when the clutch pedal is depressed. The cause of this noise could be a worn
    A. pilot bearing in the crankshaft.
    B. input shaft pilot bearing contact area.
    C. input shaft needle bearings.
    D. main shaft ball bearing.

Hint　　*The tip of the input shaft that rides in the pilot bearing must be smooth. Worn input shaft splines may cause the clutch disc to stick on these splines resulting in improper clutch release. Be sure the input shaft teeth that mesh with the synchronizer collar are not worn or chipped. Remove the roller bearings from the inner end of the input shaft. These bearings and the bearing contact area in the input shaft must be inspected for wear, pitting, and roughness. Check the input shaft bearing for smooth rotation and looseness.*

Task 8　Inspect and replace main shaft, gears, thrust washers, bearings, and retainers.

22. The transmission in Figure 3-15 has a growling noise in first gear only. The cause of this problem could be
    A. a worn first gear blocking ring.
    B. a worn first gear synchronizer sleeve.
    C. a chipped and worn first speed gear teeth.
    D. a worn, rough main shaft bearing.

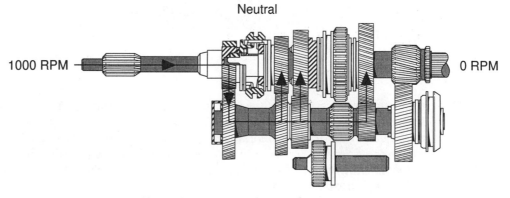

Figure 3-15　5-speed manual transmission

Hint　　*Inspect all the gear teeth on the main shaft gears for chips, pitting, cracks, and wear. A normal gear tooth wear pattern appears as a polished finish with little wear on the gear face. Inspect the bearing surfaces in the gear bores for roughness, pitting and scoring.*

　　After the gears are removed from the main shaft, inspect all the bearing surfaces on this shaft for roughness, pitting, and scoring (Figure 3-16). Inspect the gear journal areas on the main shaft for the same conditions.

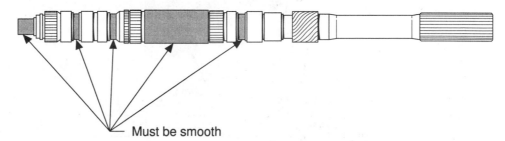

Figure 3-16　Bearing contact surfaces on the main shaft must be smooth *(Courtesy of Chrysler Corporation)*

Task 9 Inspect and replace synchronizer hub, sleeve, keys (inserts), springs, and blocking (synchronizing) rings.

23. While inspecting synchronizer assemblies
    A. the dog teeth tips on the blocking rings should be flat with smooth surfaces.
    B. the threads in the cone area of the blocking rings should be sharp and not dulled.
    C. the clearance is not important between the blocking ring and the matching gear's dog teeth.
    D. the sleeve should fit snugly on the hub and offer a certain amount of resistance to movement.

24. The clearance on the fourth speed gear shown in Figure 3-17 is less than specified.
    Technician A says this may result in noise while driving in fourth gear.
    Technician B says this problem may cause hard shifting into fourth gear.
    Who is correct?
    A. A only
    B. B only
    C. Both A and B
    D. Neither A nor B

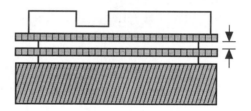

Figure 3-17 Use a feeler gauge to measure the distance between the blocking ring and the dog teeth on the matching gear

Hint *Be sure the sleeve moves freely on the hub. Inspect the insert springs to be sure they are not bent, distorted, or broken (Figure 3-18). Inspect the blocking rings for cracks, breaks, and flatness. The dog teeth on the blocking rings must be pointed with smooth surfaces. Threads in the cone area of the blocking rings must be sharp and not dulled. With the blocking ring positioned on its matching gear use a feeler gauge to measure the distance between the blocking ring and the matching gear's dog teeth.*

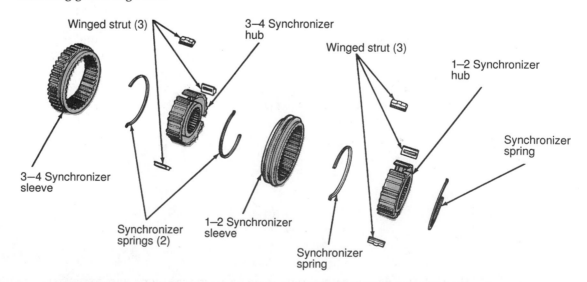

Figure 3-18 Synchronizer assembly components (Courtesy of Chrysler Corporation)

Task 10   Inspect and replace counter (cluster gear, shaft, bearings, thrust washers, and retainers).

25. The counter gear shaft and needle bearings are pitted and scored.

Technician A says the transmission may have a growling noise with the engine idling with the transmission in neutral and the clutch pedal released.

Technician B says the transmission may have a growling noise while driving in any gear.

Who is correct?

    A. A only
    B. B only
    C. Both A and B
    D. Neither A nor B

Hint   *Inspect all the gear teeth on the counter gear for chips and cracks. The wear pattern on all the gear teeth should indicate a polished finish with very little wear. Inspect the bearing surfaces in the counter gear bore for pitting, scoring, or overheating. Inspect the counter gear shaft for wear, pitting, and roughness in the bearing contact areas. All counter gear needle or ball bearings should be inspected for roughness and looseness. Replace the bearings if these conditions are present. Inspect all counter gear thrust washers and retainers for wear.*

Task 11   Inspect and replace reverse idler gear, shaft, bearings, thrust washers, and retainers.

26. In the transmission shown in Figure 3-19 the reverse idler gear is rotating in
    A. first gear.
    B. fifth gear.
    C. third and fourth gear.
    D. reverse gear.

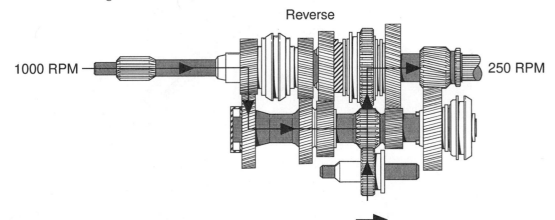

Figure 3-19 5-speed manual transmission

Hint   *The reverse idler gear teeth should be inspected for chips, pits, and cracks. Check the gear bore for roughness and scoring. The needle bearings and shaft must be inspected for roughness, scoring, and pitting.*

Task 12   Measure gear end play.

27. Excessive input shaft endplay in the transmission in Figure 3-19 may cause the transmission to jump out of
    A. first gear.
    B. second gear.
    C. fourth gear.
    D. all of the above.

Hint   *Some vehicle manufacturers provide service procedures and specifications for measuring the endplay on various transmission gears. In many endplay adjustments a dial indicator is mounted on the transmission case, and the dial indicator stem is positioned on the gear. Push the gear back and forth and observe the endplay reading on the dial indicator. Excessive counter gear endplay usually is caused by worn thrust washers.*

Task 13   Inspect, repair, and replace extension housing and transmission case mating surfaces, bores, bushings, and vents.

28. Premature wear on the extension housing bushing may be caused by
    A. worn speedometer drive and driven gears.
    B. metal burrs on the rear transmission mating surface.
    C. a plugged transmission vent opening.
    D. excessive transmission mainshaft endplay.

Hint   *Inspect the extension housing for cracks, and check the mating surfaces of this housing and the transmission case metal burrs and gouges. Remove any metal burrs or gouges with a fine-tooth file. Replace the extension housing seal and check the bushing in this housing for wear or damage.*

Task 14   Inspect and replace speedometer drive gear, driven gear, and retainers.

29. Erratic speedometer operation with the speedometer drive in Figure 3-20 may be caused by
    A. a worn adapter bushing and distance sensor.
    B. a worn extension housing bushing.
    C. excessive transmission mainshaft endplay.
    D. a distance sensor electrical defect.

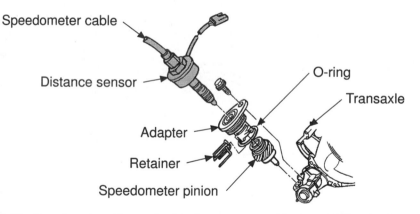

Figure 3-20 Speedometer drive and related components *(Courtesy of Chrysler Corporation)*

Hint   *The speedometer driven gear should be inspected for worn teeth and looseness. Be sure the adaptor is not loose on the cable extension. A worn adapter may cause erratic speedometer operation. During a transmission overhaul, replace the speedometer drive O-ring. Inspect the speedometer drive gear for worn or damaged teeth and looseness on the main shaft. Be sure the speedometer drive gear is securely retained on the main shaft.*

Task 15   Inspect lubrication devices.

30. While discussing manual transmission and transaxle lubricants
    Technician A says compared to a 90W gear oil ATF reduces friction and improves horsepower and fuel economy.
    Technician B says thicker gear oils have lower classification numbers.
    Who is correct?

A. A only
B. B only
C. Both A and B
D. Neither A nor B

Hint   *Some vehicle manufacturers recommend a mineral oil with an extreme pressure (EP) additive for manual transmissions. The most common gear oil classifications are SAE 75W, 75W-80, 80W-90, 85W-90, 90, or 140. Thicker gear oils have higher classification numbers (Figure 3-21). Some manual transmissions require engine oil or automatic transmission fluid (ATF). Many manual transaxles require 30W engine oil, 90W gear oil or ATF.*

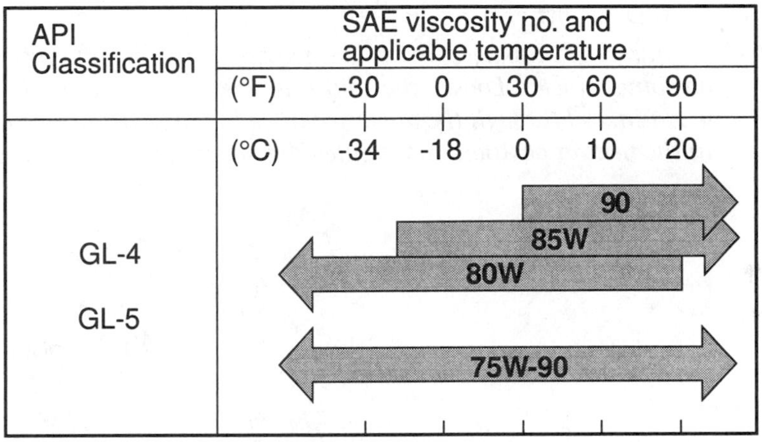

API Classification	SAE viscosity no. and applicable temperature				
	(°F) -30	0	30	60	90
	(°C) -34	-18	0	10	20
GL-4 GL-5			90 85W 80W		
			75W-90		

Figure 3-21 Transmission and transaxle lubricant ratings *(Courtesy of Subaru of America)*

Transaxle Diagnosis and Repair

ASE Tasks, Questions, and Related Information

Task 1   **Diagnose transaxle noise, hard shifting, jumping out of gear, and fluid leakage problems; determine needed repairs.**

Refer to Table 4-1 page 125 in *Today's Technician Manual Transmissions and Transaxles,* by Delmar Publishers, for diagnosis of manual transmission and transaxle problems.

31. A fully synchronized four-speed manual transaxle experiences gear clash in all forward gears and reverse. The cause of this problem could be

A. a worn fourth speed synchronizer.
B. the clutch disc sticking on the input shaft.
C. excessive main shaft endplay.
D. a worn 3-4 shifter fork.

Task 2 Inspect adjust, and replace transaxle shift linkages, brackets, bushings, cables, pivots, and levers.

32. Technician A says an improper shift linkage adjustment may cause hard transaxle shifting.
    Technician B says an improper shift linkage adjustment may result in the transaxle sticking in gear.
    Who is correct?
    A. A only
    B. B only
    C. Both A and B
    D. Neither A nor B

Hint   *Many shift linkages and cables require adjusting, and a similar adjustment procedure is used on some vehicles. Loosen the adjusting screw on the transaxle crossover cable (Figure 3-22). Slide a 1/4 in. rod through the alignment hole in the transmission lever and case to hold the transaxle in the neutral position, and tighten the adjusting screw to the specified torque (Figure 3-23).*

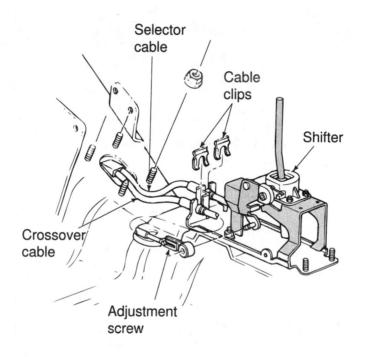

Figure 3-22 Loosening the adjustment screw on the crossover cable *(Courtesy of Chrysler Corporation)*

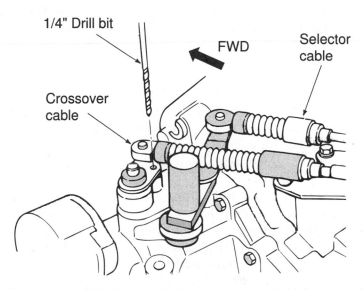

Figure 3-23 Installing 1/4 in. drill bit in transaxle lever and case to hold the transaxle in neutral *(Cour-tesy of Chrysler Corporation)*

Task 3   **Inspect and replace transaxle gaskets, seals, and sealants; inspect sealing surfaces.**

33. While discussing repeated transaxle drive axle seal leakage and replacement
    Technician A says this problem may be caused by a plugged transaxle vent.
    Technician B says this problem may be caused by a worn outer drive axle joint.
    Who is correct?
    A. A only
    B. B only
    C. Both A and B
    D. Neither A nor B

Hint   *Transaxle leaks may occur at the drive axle seals, vent, sealing surfaces between the case sec-tions, or drain plug. During a transaxle overhaul all seals and gaskets should be replaced. Seal bores should be inspected for metal burrs and scratches. Inspect all seal lip contact areas for roughness or scoring. Coat the outside diameter of each seal case with an appropriate sealer prior to installation. Always use the proper driver to install each seal. Coat the seal lips and lip contact area with the manufacturer's specified transaxle lubricant.*

Task 4   **Remove and replace transaxle.**

34. Technician A says a misaligned engine and transaxle cradle may cause drive axle vibrations.
    Technician B say a misaligned engine and transaxle cradle may cause improper front suspension angles.
    Who is correct?
    A. A only
    B. B only
    C. Both A and B
    D. Neither A nor B

Hint   *Prior to transaxle removal the battery ground cable, shift linkages or cables, speedometer cable, vehicle speed sensor, and all electrical connections must be disconnected. Drain the tran-saxle lubricant. On front-wheel-drive vehicles the front drive axles must be removed from the transaxle prior to transaxle removal. Before the transaxle retaining bolts are loosened, an engine support fixture must be installed to support the weight of the engine. On some front-wheel-drive* *vehicles the engine cradle, or part of the cradle, must be removed prior to transaxle removal. Use a transmission jack to support the weight of the transaxle during the removal process.*

Install a clutch disc alignment tool through the clutch disc hub into the pilot bearing to be sure the clutch disc is properly aligned before attempting to install the transaxle.

Task 5   Disassemble and clean transaxle.

35. Refer to Figures 3-24 – 3-29. While disassembling the transaxle
   A. the input and output shafts must be removed before the differential.
   B. the input and output shafts are retained with snap rings under the transaxle cover.
   C. the transaxle should be supported on the bench for input and output shaft removal.
   D. the input and output shafts may be pulled by hand from the transaxle bearings.

Hint   *Remove the retaining bolts between the two transaxle halves, and separate these halves (Figure 3-24). Be careful not to mark the transaxle case. Remove the differential assembly from the case. Remove the reverse idler shaft bolt, reverse idler gear shaft, and reverse fork bracket (Figure 3-25). Remove the selector shaft spacer and selector shaft. Remove the transaxle cover and use a pair of snap ring pliers to remove the snap rings in the input and output shafts (Figure 3-26). Support the transaxle on the proper bench fixture and shims (Figure 3-27). Use the proper driving fixture to press the input and output shaft assemblies out of the case (Figure 3-28). Remove the reverse brake shim, friction cone, blocking ring, and needle bearing (Figure 3-29). Wash all the transaxle components in an approved cleaning solution and blow dry these components with compressed air.*

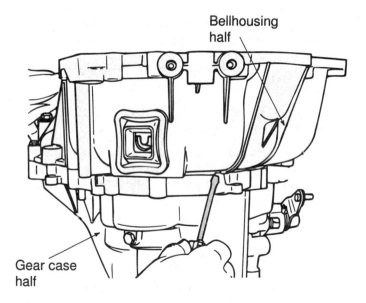

Bellhousing half

Gear case half

Figure 3-24 Separating the transaxle halves *(Courtesy of Chrysler Corporation)*

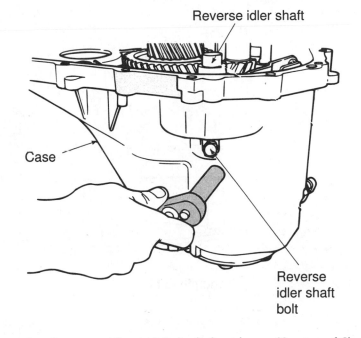

Figure 3-25 Removing the reverse idler shaft bolt, shaft and gear *(Courtesy of Chrysler Corporation)*

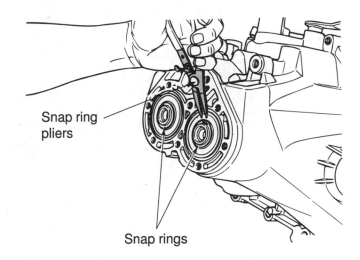
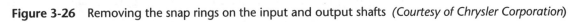

Figure 3-26 Removing the snap rings on the input and output shafts *(Courtesy of Chrysler Corporation)*

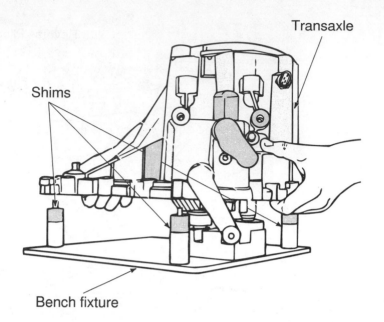

Figure 3-27 Supporting the transaxle case on the proper fixture and shims *(Courtesy of Chrysler Corporation)*

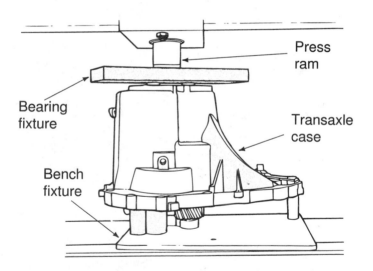

Figure 3-28 Using the proper fixture to drive the input and output shafts from the case *(Courtesy of Chrysler Corporation)*

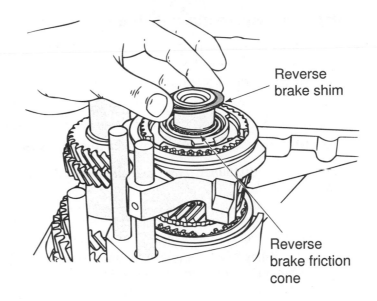

Figure 3-29 Removing the reverse brake shim and friction cone *(Courtesy of Chrysler Corporation)*

Task 6   **Inspect, repair, and /or replace transaxle shift cover, forks levers, grommets, shafts, sleeves, detent mechanisms, interlocks, and springs.**

36. While discussing a four-speed manual transaxle that jumps out of third gear: Technician A says the shift rail detent spring tension on the 3-4 shift rail may be weak. Technician B says there may be excessive wear on the fourth speed gear dog teeth. Who is correct?

   A. A only
   B. B only
   C. Both A and B
   D. Neither A nor B

Hint   *Remove the shift blocker assembly and shift forks (Figure 3-30). The same basic inspection may be performed on transmission and transaxle shifting mechanisms. Inspection of transmission shifting mechanisms was explained previously in this chapter.*

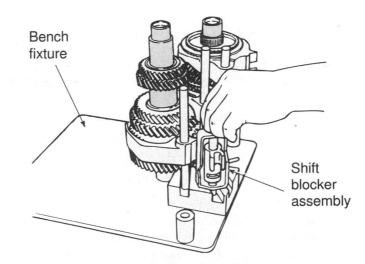

Figure 3-30 Removing shift blocker assembly and shift forks *(Courtesy of Chrysler Corporation)*

Task 7　Inspect and replace input shaft and bearings.

37. A five-speed manual transaxle has a growling and rattling noise in third gear only. The cause of this noise could be

   A. worn, chipped teeth on the third speed gear on the input shaft.

   B. worn dog teeth on the third speed gear on the input shaft.

   C. worn dog teeth on the third speed synchronizer blocking ring.

   D. worn threads in the cone area of the third speed blocking ring.

Hint　　*Worn input shaft splines may cause the clutch disc to stick on these splines resulting in improper clutch release. The gears on the input shaft should be inspected for cracks, and pitted, worn, or broken teeth (Figure 3-31). The bore in each gear and the matching surface on the input shaft should be inspected for roughness, pits, and scoring. Inspect the needle bearings mounted between the gears and the input shaft for roughness and looseness.*

Use a feeler gauge to measure the clearance between the dog teeth on the third, fourth, and fifth speed gear and the matching blocking ring.

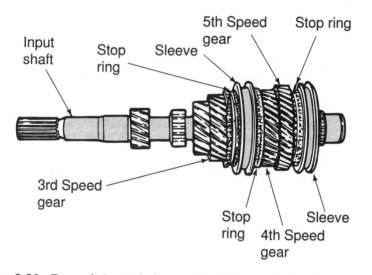

Figure 3-31　Transaxle input shaft assembly　*(Courtesy of Chrysler Corporation)*

Task 8　Inspect and replace output shaft, gears, thrust washers, bearings, and retainers.

38. The second speed gear dog teeth and blocking ring teeth are badly worn. This problem may cause

   A. a growling noise while driving in second gear.

   B. a vibration while accelerating in second gear.

   C. hard shifting in second and third gear.

   D. the transaxle to jump out of second gear

Hint　　*The inspection performed on the input shaft and gears should be repeated on the output shaft and gears. In some transaxles the output shaft assembly is serviced as a complete unit. If any gear or component on the output shaft is worn, the complete assembly must be replaced. Use a feeler gauge to measure the clearance between the dog teeth on the first and second speed gear and the matching blocking ring (Figure 3-32). Inspect the output shaft bearings in the transaxle case for roughness or looseness.*

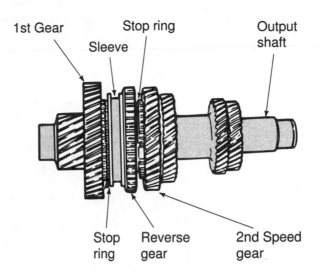

Figure 3-32 Transaxle output shaft assembly *(Courtesy of Chrysler Corporation)*

Task 9   **Inspect and replace synchronizer hub, sleeve, keys (inserts), springs, and blocking (synchronizer) rings.**

39. Technician A says synchronizer hubs are reversible on the shaft on which they are mounted.

    Technician B says synchronizer sleeves are reversible on their matching hub.

    Who is correct?

    A. A only

    B. B only

    C. Both A and B

    D. Neither A nor B

Hint   *Synchronizer inspection and replacement is basically the same for transmissions and transaxles. Refer to the transmission synchronizer inspection explained previously in this chapter. Synchronizer hubs and sleeves are directional. Always assemble these components in their original position. Prior to disassembly always mark the synchronizer sleeve and hub so these components are assembled in their original location in relation to each other.*

Task 10   **Inspect and replace reverse idler gear, shaft, bearings. thrust washers, and retainers.**

40. A transaxle shifts normally into all forward gears, but it will not shift into reverse gear, and there is no evidence of noise while attempting this shift.

    Technician A says the reverse shifter fork may be broken.

    Technician B says the reverse idler gear teeth may be worn.

    Who is correct?

    A. A only

    B. B only

    C. Both A and B

    D. Neither A nor B

Hint   *The reverse idler gear teeth should be inspected for chips, pits, and cracks. Check the gear bore for roughness and scoring. The reverse idler shaft and bearings must be inspected for roughness, scoring, and pitting (Figure 3-33).*

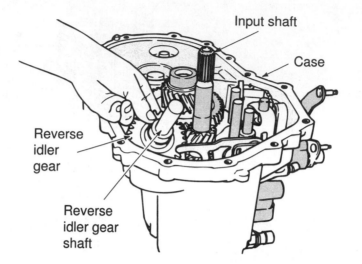

Figure 3-33   Transaxle reverse idler gear and shaft   *(Courtesy of Chrysler Corporation)*

Task 11   Inspect, repair, and replace transaxle case mating surfaces, bores, bushings, and vents.

41. Technician A says an epoxy-based sealer may be used to repair a crack in some transaxle cases.

    Technician B says some cracks in transaxle cases may be repaired with Loctite® PST. Who is correct?
    A. A only
    B. B only
    C. Both A and B
    D. Neither A nor B

Hint   *Inspect the mating surfaces of the two transaxle halves for metal burrs and scratches. Inspect the transaxle case for cracks. All bearing bores should be inspected for cracks, scoring, or gouges. Inspect all seal mounting areas for metal burrs, and gouges. Be sure the vent is not restricted. A plugged vent may result in pressure buildup in the transaxle resulting in leaks from the seals. Inspect all shift rail bushings and replace all shift rail seals.*

Transaxle case replacement often is required if the case is cracked. However, some vehicle manufacturers recommend crack repair with an epoxy-based sealer depending on the location of the crack. Always refer to the vehicle manufacturer's recommended crack repair procedure in the appropriate service manual. Damaged threads in the case may be repaired with a thread repair kit.

Task 12   Inspect and replace speedometer drive gear, driven gear, and retainers.

42. In some transaxles the speedometer drive gear is mounted on
    A. the input shaft.
    B. the transfer gear.
    C. the differential case.
    D. the drive axle inner hub.

Hint   *Some manual transaxles have a combined speedometer drive and speed-distance sensor. The speedometer driven gear should be inspected for worn teeth and looseness. Be sure the adaptor is not loose on the speed-distance sensor extension. A worn adapter may cause erratic speedometer operation. During a transmission overhaul, replace the speedometer drive O-ring. Inspect the speedometer drive gear for worn or damaged teeth and looseness on the differential case.*

Task 13   **Diagnose differential assembly noise and vibration problems; determine needed repairs.**

43. A manual transaxle chatters while driving straight ahead.
Technician A says the ring gear and pinion gear teeth may be worn and chipped.
Technician B says there may be improper preload on the differential components.
Who is correct?
A. A only
B. B only
C. Both A and B
D. Neither A nor B

Hint   *A clicking noise in all gears when the vehicle is driven straight ahead may be caused by damaged differential ring gear or drive pinion teeth. Loose ring gear bolts may cause a knocking noise when the vehicle is driven straight ahead. Differential chatter when the vehicle is driven straight ahead may result from incorrect preload on differential components. Differential gear chuckle may be caused by worn or damaged differential thrust washers, loose side gears in the case, damaged ring gear teeth, or loose ring gear bolts.*

A whining noise that changes with acceleration, deceleration, and steady throttle may be caused by worn differential ring gear and drive pinion teeth. Worn differential bearings may cause a growling noise when the vehicle is driven. If a clicking noise occurs while cornering, the differential side gears and pinion gears may have damaged teeth.

Differential vibration may be caused by worn or broken transaxle or engine mounts. Damaged or galled differential bearings, or damaged ring gear teeth, may be the cause of differential vibration. These problems also may cause the transaxle noises mentioned in the preceding paragraph.

Task 14   **Remove and replace differential assembly.**

44. While discussing differential side bearing preload in the transaxle case in Figure 3-34 Technician A says differential bearing preload is adjusted by rotating a threaded adjuster on each side of the differential bearings.
Technician B says the differential bearing preload is automatically adjusted when the case halves are reassembled.
Who is correct?
A. A only
B. B only
C. Both A and B
D. Neither A nor B

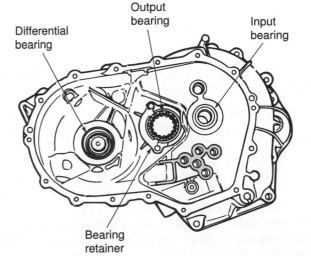

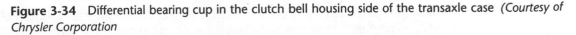

Figure 3-34 Differential bearing cup in the clutch bell housing side of the transaxle case *(Courtesy of Chrysler Corporation*

Hint   *In many transaxles after the halves of the case have been separated, the differential assembly may be lifted from the case (Figure 3-35). A differential bearing cup is pressed into each half of the case, and in some transaxles a shim for bearing preload adjustment is positioned behind bearing cup in the clutch bell housing side of the case.*

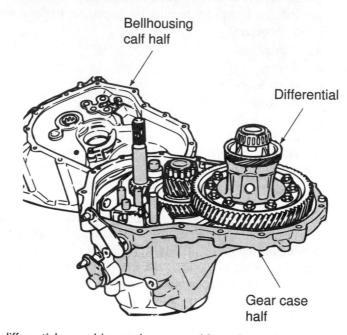

Figure 3-35   The differential assembly may be removed from the transaxle case after the case halves are separated   *(Courtesy of Chrysler Corporation)*

Task 15   Inspect, measure, adjust, and replace differential pinion gears (spiders), shaft, side gears, thrust washers, and case.

45. While determining the proper differential side gear thrust washer thickness

   A. measure the end play on one side gear to calculate the side gear spacer washer thickness.

   B. the side gear end play is measured with the thrust washers behind the gears.

   C. the correct thickness of side gear thrust washer provides the specified side gear end play.

   D. the correct thickness of side gear thrust washer provides a slight side gear preload.

Hint   *Use a hammer and punch to drive the pinion shaft roll pin from the differential case. Remove the pinion shaft, gears, side gears, and thrust washers.*

Reassemble the differential side gears, pinion gears, pinion gear thrust washers, and shaft. Do not install the side gear thrust washers. Install the roll pin to retain this shaft in the differential case. Rotate the side gears two revolutions in each direction. Install a special tool through one of the axle openings against the side gear and assemble a dial indicator so the stem rests against the tool (Figure 3-36). Move the side gear up and down and record the end play on the dial indicator. Rotate the side gear 90 degrees and record another end play reading. Turn the side gear another 90 degrees and record a third end play reading. Use the smallest end play recorded to calculate the required side gear washer thickness. The side gear end play must be 0.001 in. – 0.013 in. If the end play recorded was 0.050 in., install a 0.042 in. washer, which provides an end play of 0.008 in.

Repeat the side gear end play measurement on the opposite side gear. Disassemble the side gears and pinion gears and reassemble these components with the required side gear thrust washers.

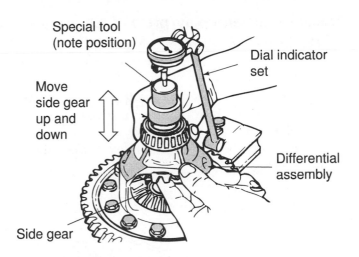

Special tool
(note position)

Dial indicator
set

Move
side gear
up and
down

Differential
assembly

Side gear

Figure 3-36   Side gear end play measurement   *(Courtesy of Chrysler Corporation)*

Task 16   Inspect and replace differential side bearings.

46. Technician A says a preload adjustment shim is positioned between both differential bearings and the differential case.

   Technician B says one of the differential bearings must be removed before the speedometer drive gear.

   Who is correct?

   A. A only
   B. B only
   C. Both A and B
   D. Neither A nor B

Hint   *On some transaxles a special puller is used to remove the bearings from the differential case. These bearings must be installed on the differential case with a special driving tool. The differential bearing cups are removed from the case with special tool (Figure 3-37). A preload shim is located behind one of these bearing cups.*

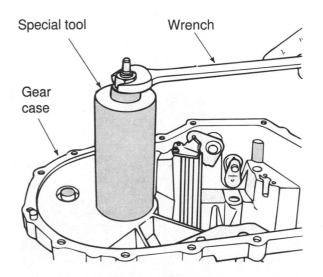

Special tool

Wrench

Gear
case

Figure 3-37   Using a special puller to remove a differential bearing cup from the case   *(Courtesy of Chrysler Corporation)*

Task 17   Measure shaft end play/preload (shim/spacer selection procedure).

47. During the measurement in Figure 3-38
   A. a new bearing cup is installed with a shim in the clutch bell housing side of the transaxle case.
   B. a medium load should be applied to the differential in the upward direction.
   C. the proper shim thickness is equal to the differential end play recorded on the dial indicator.
   D. the bolts between the transaxle case halves must be tightened to one-half the specified torque.

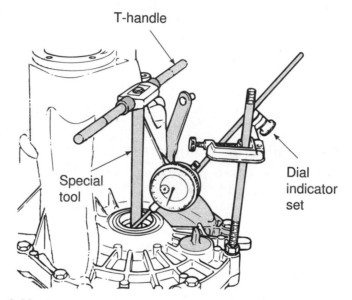

Figure 3-38   Measuring differential end play *(Courtesy of Chrysler Corporation)*

48. When the turning torque in Figure 3-39 is less than specified
   A. a thicker shim should be installed behind the side bearing cup in the bell housing side of the case.
   B. a thinner shim should be installed behind both differential side bearing cups.
   C. a thinner shim should be installed behind both differential side bearings.
   D. a thicker shim should be installed behind both differential side gears.

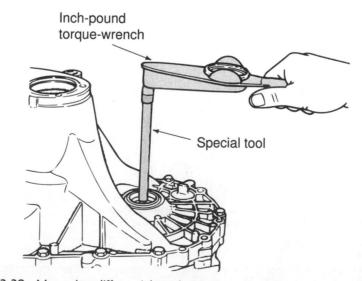

Figure 3-39   Measuring differential turning torque *(Courtesy of Chrysler Corporation)*

Hint   *After the differential bearing cup and shim have been removed from the clutch bell housing half of the transaxle case, use the proper driving tool to install a new bearing cup without the shim. Install a new bearing cup in the gear case half of the case with the proper driver. Lubricate the differential bearings with the manufacturer's specified transaxle lubricant. Install the differential assembly and reassemble the case halves. Tighten the case bolts to the specified torque.*

Install the special tool into the differential side gears and install a T-handle on top of this tool. Apply downward pressure to the T-handle and rotate this handle several times in both directions to seat the side bearings. Zero the dial indicator installed against the upper side of the differential case. Apply a medium load to the differential in the upward direction through the opposite drive axle opening. Rotate the differential back and forth with the T-handle while maintaining this upward pressure. Record the end play indicated on the dial indicator.

The proper preload shim thickness is equal to the end play plus 0.007 in. (0.18 mm) to provide side bearing preload. After the case is assembled install the special turning tool in the differential. Install an inch pound torque wrench on this tool and rotate the differential back and forth while observing the turning torque on the torque wrench. If the turning torque is more than specified, reduce shim thickness by 0.002 in. (0.5 mm). When the turning torque is less than specified, increase the shim thickness by 0.002 in. (0.5 mm).

Task 18   Inspect lubrication devices.

49. The needle bearings between the output shaft and output shaft gears are scored and blue from overheating.

    Technician A says the transaxle may have been filled with the wrong lubricant.

    Technician B says the projection is broken off on the oil feeder behind the front output shaft bearing.

    Who is correct?

    A. A only

    B. B only

    C. Both A and B

    D. Neither A nor B

Hint   *Some manual transaxles have an oil feed trough that maintains an adequate oil supply to the end bearings (Figure 3-40). Inspect the oil feed trough for damage or a bent condition. A damaged or bent oil feed trough may reduce lubricant flow to the end bearings and cause premature bearing failure. Some output shaft front bearings have an oil feeder behind the bearing to maintain the proper supply of lubricant to the bearing and shaft supported by the bearing (Figure 3-41, next page).*

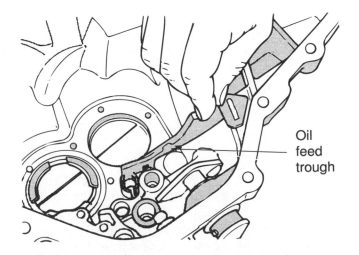

Oil feed trough

Figure 3-40 Lubricant oil trough *(Courtesy of Chrysler Corporation)*

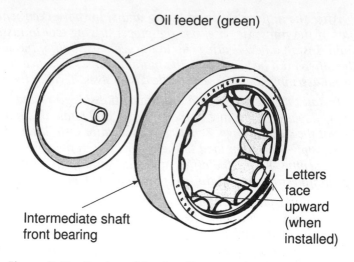

Oil feeder (green)

Letters
face
upward
(when
installed)

Intermediate shaft
front bearing

Figure 3-41   Bearing oil feeder   *(Courtesy of Chrysler Corporation)*

Drive (Half) Shaft and Universal Joint Diagnosis and Repair (Front- and Rear-Wheel-Drive)

ASE Tasks, Questions, and Related Information

Task 1   Diagnose shaft and universal/CV joint noise and vibration problems; determine needed repairs.

50.   While discussing drive shaft and universal joint diagnosis in rear wheel drive vehicles Technician A says a worn universal joint may cause a squeaking noise that decreases in relation to vehicle speed.

Technician B says a heavy vibration that only occurs during acceleration may be caused by a worn centering ball and socket on a double Cardan U-joint.

Who is correct?

A.   A only

B.   B only

C.   Both A and B

D.   Neither A nor B

Hint   *Refer to Table 5-1, page 163, in* Today's Technician Manual Transmissions and Transaxles Shop Manual *for a complete diagnosis of drive axle and CV-joint vibration and noise problems.*

Refer to Table 6-1, page 187, in Today's Technician Manual Transmissions and Transaxles Shop Manual *for a complete diagnosis of drive shaft and universal joint vibration and noise problems.*

Task 2 **Inspect, service, and replace shaft, yokes, boots, and universal/CV joints.**

51. A front-wheel-drive car has a clunking noise while decelerating.
Technician A says this noise may be caused by a worn inner drive axle joint.
Technician B says this noise may be caused by a worn front wheel bearing.
Who is correct?
   A. A only
   B. B only
   C. Both A and B
   D. Neither A nor B

Hint *Most CV joints are replaced as a complete assembly. Prior to removing the CV joint boot, mark the inner end of the boot in relation to the drive axle so the boot may be installed in the original position. When reassembling the joint always install all the grease in the joint that is provided in the repair kit.*

Prior to drive shaft removal always mark the drive shaft in relation to the differential flange so this shaft may be installed in the original position. After the universal joint retaining clips are removed from the drive shaft, a vice and the proper size of socket may be used to remove the spider from the yoke.

In a double Cardan U-joint the center yoke should be marked in relation to the ball tube yokes prior to disassembly so these components may be assembled in their original location (Figure 3-42). The bearing caps in this type of joint should be removed in the proper sequence using the same procedure as followed for a single U-joint. The centering ball and ball seats must be replaced if they are worn or scored.

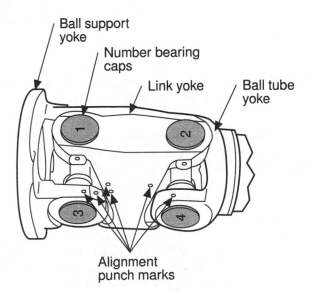

Figure 3-42 Marking the center yoke in relation to the ball tube yokes in a double Cardan U-joint

Task 3 **Inspect, service, and replace shaft center support bearings.**

52. A light-duty rear wheel drive truck has a growling noise that is not influenced by acceleration and deceleration.
Technician A says an outer rear axle wheel bearing may be rough and worn.
Technician B says the drive shaft center support bearing may be rough and worn.
Who is correct?
   A. A only
   B. B only
   C. Both A and B
   D. Neither A nor B

Hint   *When the center support bearing is removed from the drive shaft, it should be inspected for looseness and roughness. Inspect the center support bearing cushion for wear, oil soaking, and deterioration. Check for damaged or bent upper and lower center support bearing brackets. Some center support bearings have an F marked on the front side of the bearing.*

Task 4   Check and correct shaft balance.

53. The markings in Figure 3-43 are required for
    A. drive shaft removal and replacement.
    B. universal joint removal and replacement.
    C. drive shaft runout measurement.
    D. drive shaft balance testing.

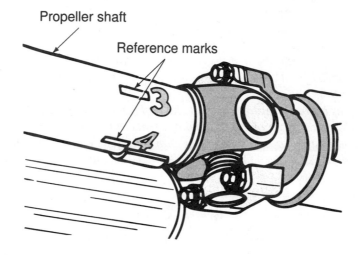

Propeller shaft

Reference marks

Figure 3-43 Chalk mark the drive shaft at four locations spaced 90° apart near the drive shaft balance weight *(Courtesy of Chevrolet Motor Division, General Motors Corporation)*

Hint   *Prior to drive shaft balancing always inspect this shaft for damage, a missing balance weight, or an accumulation of dirt or undercoating. Remove both rear wheel assemblies and install the wheel nuts on the wheel studs with the flat side toward the brake drums. Chalk mark the drive shaft at four locations 90 degrees apart just forward of the drive shaft balance weight. Fabricate a tool to hold the strobe light pickup against the rear axle housing just behind the pinion yoke. Run the vehicle in gear until the drive shaft vibration is most severe. Since 55 mph indicated on the speedometer has a possible tire speed of 110 mph, do not exceed 55 mph. Point the strobe light at the chalk marks on the drive shaft and note the position of one reference mark.*

Gently apply the brakes and shut off the engine. Rotate the drive shaft until the chalk mark is in the same position as it appeared under the strobe light. Install two screw-type hose clamps on the drive shaft near the rear of the shaft. Position these clamps so both heads are beside each other and directly opposite the number that appeared under the strobe light.

Task 5   Measure shaft runout.

54. Technician A says the dial indicator should be positioned near the front of the drive shaft to measure drive shaft runout.
    Technician B says if the drive shaft runout is excessive, the drive shaft may be straightened in a hydraulic press.
    Who is correct?
    A. A only
    B. B only
    C. Both A and B
    D. Neither A nor B

Hint   *A dial indicator must be mounted to the vehicle underbody near the center of the drive shaft to measure drive shaft runout. Position the dial indicator stem against the surface of the drive shaft. Be sure the drive shaft surface is clean and undamaged. Rotate the drive shaft one revolution to measure the runout. When the runout exceeds specifications, replace the drive shaft and recheck the runout. If the runout is still excessive, check for a bent U-joint flange or slip yoke.*

Task 6   **Measure and adjust shaft angles.**

     55. A rear-wheel-drive vehicle has a vibration that increases in relation to vehicle speed.
             Technician A says the balance pad may have fallen off the drive shaft.
             Technician B says some of the wheels may be out of balance.
             Who is correct?
             A. A only
             B. B only
             C. Both A and B
             D. Neither A nor B

Hint   *Raise the vehicle on a lift so the rear wheels are free to rotate. Clean the outer surface of the bearing caps in the front and rear U-joints. Be sure one of the rear U-joint bearing caps is facing straight down. Install the magnetic end of the inclinometer on the U-joint bearing cap facing downward, and rotate the inclinometer adjusting knob until the weighted cord is centered on the scale (Figure 3-44). Remove the inclinometer and turn the drive shaft 90 degrees. Install the inclinometer on the U-joint bearing cap facing downward, and record the degree reading where the weighted cord appears on the scale. The drive shaft angle is the difference between the zero reading and the reading when the shaft is rotated 90 degrees.*

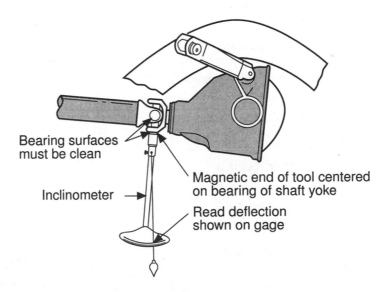

Figure 3-44 Measuring drive shaft angle *(Courtesy of Oldsmobile Division, General Motors Corporation)*

    Repeat the procedure on the front U-joint to obtain the front drive shaft angle. If the drive shaft angles are not within specifications, inspect the engine and transmission mounts for proper position, wear, looseness, breaks, and deterioration. Inspect the rear suspension mounting bushings and arms for wear, looseness, or a bent condition.

Rear-Wheel-Drive Axle Diagnosis and Repair (Ring and Pinion Gears)

ASE Tasks, Questions, and Related Information

Task 1   **Diagnose noise, vibration, and fluid leakage problems; determine needed repairs.**

56. While driving a vehicle straight ahead, the differential produces a whining noise.
    Technician A says the differential side gears are damaged.
    Technician B says the ring gear and pinion adjustments may be incorrect.
    Who is correct?
    A. A only
    B. B only
    C. Both A and B
    D. Neither A nor B
    Refer to Table 7-1, page 253, in Today's Technician Manual Transmissions and Transaxles, *by Delmar Publishers, for diagnosis of rear-wheel-drive axle problems.*

Task 2   **Inspect and replace companion flange and pinion seal; measure companion flange runout.**

57. Technician A says that insufficient pinion nut torque may cause a clunking noise during acceleration or deceleration.
    Technician B says that insufficient pinion nut torque may cause a growling noise with the vehicle in motion.
    Who is correct?
    A. A only
    B. B only
    C. Both A and B
    D. Neither A nor B

Hint   *If there is evidence of fluid leakage in the pinion seal area, the pinion seal must be replaced. A special holding tool is required to hold the differential flange while loosening the pinion nut (Figure 3-45).*

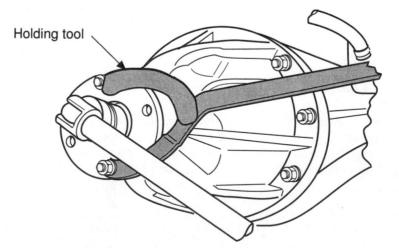

Holding tool

Figure 3-45   Differential flange holding tool *(Courtesy of Chrysler Corporation)*

Coat the outside diameter of the new seal with gasket sealer, and lubricate the seal lips with the manufacturer's specified differential lubricant. Use the proper seal driver to install the pinion seal. Install the differential flange and a new pinion nut. Since the pinion nut torque determines the pinion bearing preload, the torque on this nut is critical. Tighten the pinion nut until the manufacturer's specified turning torque is obtained with an inch-pound torque wrench and socket installed on the pinion nut. When the pinion nut is tightened to obtain the proper preload and turning torque with the original collapsible spacer, a typical turning torque is 6 in. pounds more than the specified turning torque.

Mount a dial indicator on the vehicle chassis, and position the indicator stem against the differential flange. Rotate the flange and observe the flange runout on the dial indicator. A damaged flange with excessive runout may cause a vibration while driving at a constant speed, because this problem causes incorrect drive shaft angles.

Task 3 **Inspect and replace ring gear.**

58. All of these statements about ring and pinion gears are true EXCEPT
    A. hunting-type ring and pinion gear sets must be timed.
    B. loose ring gear bolts may cause gear chuckle or knocking noise while the vehicle is in motion.
    C. damaged ring gear and pinion gear teeth may cause a ticking noise while driving the vehicle.
    D. the grooved, painted tooth on the pinion gear must be meshed with the painted notched ring gear teeth on some gear sets.

Hint *If the ring gear teeth are damaged, the ring gear and pinion gear must be replaced as a set. When installing a new ring gear on the case, be sure the bolt holes are aligned in these components. Pilot studs may be used to ensure bolt hole alignment.*

Many ring gear and pinion gear sets have timing marks that must be aligned when assembling the differential. On some gear sets one pinion gear tooth is grooved and painted, and the ring gear has a notch between two painted teeth. When the ring and pinion gears are meshed together, the grooved and painted pinion gear tooth must fit between the notched and painted ring gear teeth. Some ring gear and pinion gear sets do not have timing marks. These gear sets are referred to as hunting gears.

Task 4 **Measure ring gear runout; determine needed repairs.**

59. Excessive runout on the dial indicator in Figure 3-46 may be caused by excessive
    A. differential case runout.
    B. side bearing preload.
    C. side gear end play.
    D. ring gear bolt torque.

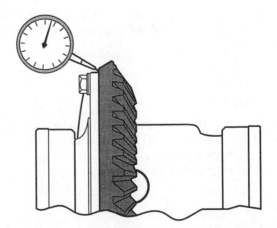

Figure 3-46 Measuring ring gear runout *(Courtesy of Chrysler Corporation)*

Hint   *The ring gear runout should be measured prior to disassembling the differential. Mount the dial indicator on the housing assembly, and position the dial indicator stem at a 90 degree angle against the back of the ring gear. Zero the dial indicator and rotate the ring gear one revolution. The difference between the highest and lowest dial indicator reading is the ring gear runout.*

To determine if excessive ring gear runout is caused by the ring gear or case, remove the ring gear and case and remove the ring gear from the case. Install the case assembly without the ring gear, and be sure the side bearings are in good condition and properly torqued. Position the dial indicator against the back side of the case, and rotate the case for one revolution to measure the case runout. If the case runout is normal but the ring gear runout is excessive, replace the ring gear and pinion. When the case runout is excessive, replace the case.

Task 5   **Inspect and replace drive pinion gear, collapsible spacers, sleeve, and bearings.**

60. Technician A says the collapsible pinion shaft spacer may be reused if the differential is disassembled and overhauled.

Technician B says the ABS toothed ring on the pinion shaft may be reinstalled if it is removed from the shaft.

Who is correct?

A. A only

B. B only

C. Both A and B

D. Neither A nor B

Hint   *If the pinion gear requires replacement, always replace the pinion gear and ring gear as a set. When the pinion bearing must be replaced, use a press plate and press to remove the rear pinion bearing. A toothed ring for the ABS system is pressed onto the shaft in some differentials. This toothed ring must be removed with a press plate and press prior to removing the rear pinion bearing. Since removal of this toothed ring may destroy its friction fit on the pinion shaft, most manufacturers recommend replacement of this ring if it is removed.*

Before pressing the new rear pinion bearing onto the pinion shaft, be sure the proper spacer washer is installed behind the pinion bearing. This shim determines the pinion depth. Press a new ABS toothed ring onto the pinion shaft and install a new collapsible sleeve on the shaft. When the differential has been disassembled and overhauled, a new collapsible spacer and pinion nut must be installed. Tighten the pinion nut a small amount at a time until the specified turning torque is obtained with an inch-pound torque wrench and socket installed on the pinion nut. Never loosen the pinion nut to obtain the correct turning torque.

Task 6   **Measure and adjust drive pinion depth.**

61. In Figure 3-47 a 0.063 in. shim that fits between the gauge block and the gauge tube with a light drag, and the pinion gear is marked +3. The proper pinion depth shim is

A. 0.060 in.

B. 0.063 in.

C. 0.066 in.

D. 0.069 in.

62. In Figure 3-48 after the dial indicator is rotated to the zero position with the stem on the gauge plate, when the dial indicator stem is moved off the gauge plate the dial indicator pointer moves 0.057 in. counterclockwise and the pinion gear is marked minus 4. The proper pinion depth shim is:

A. 0.039 in.

B. 0.041 in.

C. 0.042 in.

D. 0.043 in.

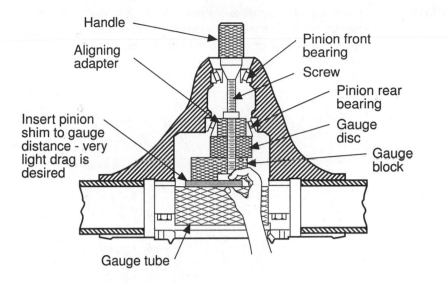

Figure 3-47   Selecting proper pinion depth shim with an arbor and gauge tube

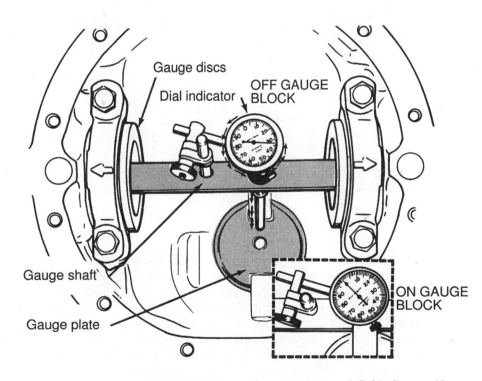

Figure 3-48   Selecting proper pinion depth shim with a gauge set and dial indicator *(Courtesy of Oldsmobile Division, General Motors Corporation)*

Hint   *Pinion depth is the distance from the nose of the pinion gear to the centerline of the axles or differential case (Figure 3-49). This depth normally is adjusted with the shim thickness behind the rear pinion bearing. Some vehicle manufacturers recommend measuring the pinion depth with an arbor and gauge tube.*

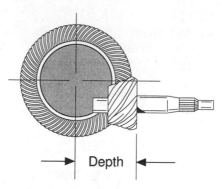

Depth

Figure 3-49   Pinion depth   *(Courtesy of Nissan Motor Co., Ltd.)*

Other manufacturers recommend measuring the pinion depth with a gauge set and dial indicator. The gauge set is installed in the pinion bearings, and a gauge shaft and discs are mounted in the side bearing openings

After the nominal pinion depth shim is determined, check the markings on the rear or side of the pinion shaft ahead of the rear pinion bearing. If the pinion is marked +3, add 0.003 (0.025 mm) to the nominal shim thickness. When the pinion is marked with a minus value subtract this amount from the nominal shim thickness.

Task 7   **Measure and adjust drive pinion bearing preload.**

63. Technician A says the pinion bearings should be lubricated when the pinion turning torque is measured.
   Technician B says the pinion nut may be loosened to obtain the specified turning torque.
   Who is correct?
   A. A only
   B. B only
   C. Both A and B
   D. Neither A nor B

Hint   *The pinion bearing preload is measured with an inch pound torque wrench and socket installed on the pinion nut. Prior to measuring the pinion turning torque with an inch pound torque wrench, the pinion shaft assembly should be installed with the bearings lubricated and a new collapsible spacer and the proper pinion depth shim. A new pinion shaft nut should be installed. Tighten the pinion nut gradually and keep measuring the turning torque. When the specified turning torque is obtained, the pinion bearing preload is correct. Never loosen the pinion nut to obtain the proper turning torque. If the pinion nut is overtightened and the turning torque is excessive, install a new collapsible spacer and repeat the procedure.*

Task 8   **Measure and adjust differential (side) bearing preload, and ring and pinion backlash (threaded cup or shim type).**

64. In Figure 3-50 the ring gear backlash and side play are zero. Right and left side-bearing adjusting nuts are determined while facing the differential from the rear. To obtain the proper ring gear backlash
    A. tighten the right and left side-bearing adjusters.
    B. loosen the left side-bearing adjuster.
    C. loosen the left side-bearing adjuster and tighten the right side-bearing adjuster.
    D. loosen the right side-bearing adjuster and tighten the left side-bearing adjuster.

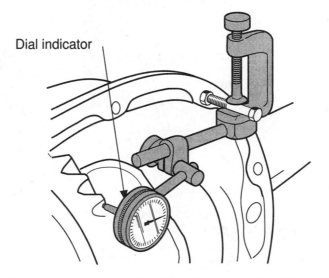

Dial indicator

Figure 3-50   Measuring ring gear backlash   *(Courtesy of Ford Motor Company)*

Hint   *The backlash between the ring gear and pinion gear is adjusted after the differential is assembled. Mount a dial indicator on the differential housing, and position the dial indicator stem against one of the ring gear teeth. Zero the dial indicator and rock the ring gear back and forth against the pinion gear teeth. The ring gear backlash is indicated on the dial indicator. Vehicle manufacturers usually recommend measuring the backlash at several locations around the ring gear.*

Side-bearing preload limits the amount of later differential case movement in the axle housing or carrier. When the differential has threaded adjusters on the outside of the side bearings, loosen the right adjuster and tighten the left adjuster to obtain zero backlash. Turn the right adjuster the specified amount to obtain the proper preload. Then rotate each adjuster the same amount in opposite directions to obtain the specified backlash.

When shims are positioned behind the side bearings to adjust backlash and preload, the differential case is pried to one side and the movement is recorded with a dial indicator or feeler gauge. Service spacers, shims, and feeler gauges are installed on each side of the side bearings to obtain zero side play and zero backlash. Then calculate the proper shim thickness to provide the specified backlash and side-bearing preload (Figure 3-51, page 172). On some differentials the proper shims have to be driven into place behind the side bearings with a special tool and a soft hammer.

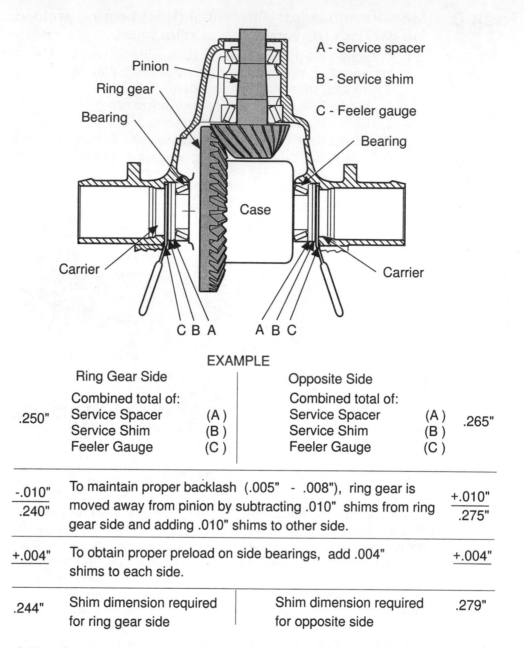

Figure 3-51 Adjusting ring gear backlash and side bearing preload with shims *(Courtesy of Cadillac Motor Car Division, General Motors Corporation)*

Task 9   Perform ring and pinion gear tooth contact pattern checks; determine needed adjustments.

65. Technician A says if the ring tooth contact pattern indicates pinion tooth contact on the toe of the pinion gear, the pinion gear should be moved toward the ring gear. Technician B says if the pinion gear teeth have low flank contact on the ring gear teeth, the pinion gear should be moved toward the ring gear.

    Who is correct?

    A. A only
    B. B only
    C. Both A and B
    D. Neither A nor B

Hint   *Observe the tooth contact pattern on the ring gear. This pattern should be centered on the drive side of the ring gear teeth. If the ring gear tooth contact pattern is not correct, the drive pinion has to be moved in relation to the ring gear, or the backlash adjusted to provide the correct pattern (Figure 3-52).*

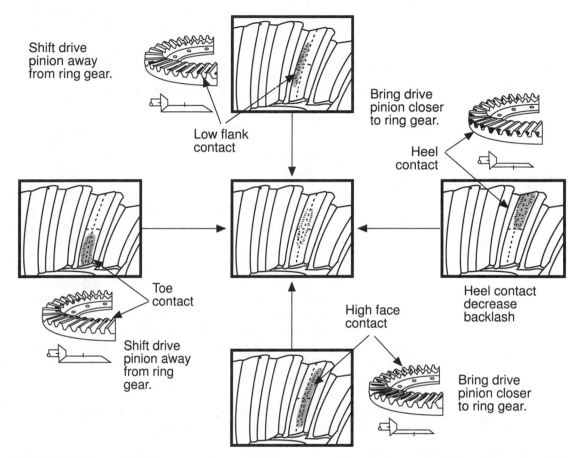

Shift drive pinion away from ring gear.

Low flank contact

Bring drive pinion closer to ring gear.

Heel contact

Toe contact

Heel contact decrease backlash

High face contact

Shift drive pinion away from ring gear.

Bring drive pinion closer to ring gear.

Figure 3-52   Correct and incorrect ring gear tooth contact patterns *(Courtesy of Nissan Motor Co., Ltd.)*

Differential Case Assembly

ASE Tasks, Questions, and Related Information

Task 1   **Diagnose differential assembly noise and vibration problems; determine needed repairs.**

66. When diagnosing a rear-wheel-drive differential that vibrates only while turning a corner
Technician A says the pinion bearings may be worn and the pinion bearing preload is less than specified.
Technician B says the bearing surfaces between the side gears or bevel pinions and the differential case may be damaged.
Who is correct?

A. A only
B. B only
C. Both A and B
D. Neither A nor B

Hint   *Refer to Table 7-1, page 253, in* Today's Technician Manual Transmissions and Transaxles Shop Manual *for a complete listing and diagnosis of differential noise and vibration problems.*

Task 2   Remove and replace differential case and ring gear.

67. All of these statements about differential case and ring gear assembly removal and replacement are true EXCEPT
A. the ring gear runout should be measured before removal of the case and ring gear assembly.
B. the case side play should be measured before removal of the case and ring gear assembly.
C. the side bearing caps should be marked in relation to the housing before removal of the case and ring gear assembly.
D. the side bearings should be clean and dry before installation of the case and ring gear assembly.

Hint   *Measure the ring gear runout and differential case side play before removing the differential case and ring gear. Be sure to mark the side bearing caps in relation to the housing prior to removal. Remove the side-bearing caps and lift the case and ring gear from the housing.*

Prior to installing the case and ring gear assembly the side bearings and bearing cups must be lubricated with the vehicle manufacturer's recommended differential lubricant. Install the case and ring gear assembly with the bearing cups and shims or adjuster nuts. Be sure the bearing cap threads are properly seated in the adjuster nut threads when the bearing caps are installed. tighten the bearing cap bolts to the specified torque.

Task 3   Inspect, measure, adjust, and replace differential pinion gears (spiders), shaft, side gears, thrust washers, and case.

Hint   *Inspection and side-play measurement of the side gears, pinion gears, thrust washers, and case in a rear-wheel-drive differential is basically the same as the inspection of these components in a transaxle differential. Refer to Task 15 in Transaxle Diagnosis and Repair for this inspection.*

Task 4   Inspect and replace differential side bearings.

Hint   *The inspection and replacement procedure for differential side bearings in a rear-wheel-drive differential is basically the same as the side-bearing inspection and replacement in a transaxle. This inspection and replacement procedure is explained in Task 16 of Transaxle Diagnosis and Repair.*

Task 5   Measure differential case runout; determine needed repairs.

68. Technician A says an accurate differential case runout measurement may be performed with scored side bearings.
Technician B says the ring gear runout should be measured before the case runout.
Who is correct?
A. A only
B. B only
C. Both A and B
D. Neither A nor B

Hint   *As mentioned previously the ring gear runout should be measured prior to removing the differential case and ring gear assembly. This procedure is explained in Task 4 of Ring and Pinion Gear Diagnosis. If the ring runout is excessive, the case runout should be measured.*

Limited Slip Differential

ASE Tasks, Questions, and Related Information

Task 1   **Diagnose limited slip differential noise, slippage, and chatter problems; determine needed repairs.**

69. All of these statements about limited slip differentials are true EXCEPT
    A. friction plates are splined to the rear axle hub.
    B. steel plates are splined to the differential case.
    C. each clutch set contains a preload spring.
    D. a special lubricant is required.

Hint   *Refer to Table 7-1, page 253, in* Today's Technician Manual Transmissions and Transaxles, *by Delmar Publishers, for diagnosis limited slip differential problems.*

Limited slip differentials have a set of multiple disc clutches behind each side gear to control differential action. The steel plates in each clutch set are splined to the case, and the friction plates between the steel plates are splined to the side gear clutch hub. Each clutch set has a preload spring that applies initial force to the clutch packs. A steel shim in each clutch set controls preload (Figure 3-53). A friction plate always is placed next to the hub.

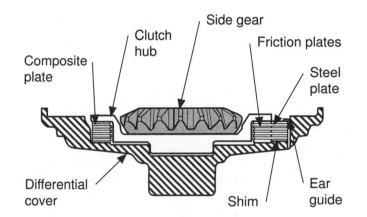

Figure 3-53   Limited slip differential clutch set *(Courtesy of Ford Motor Company)*

Limited slip differentials require a special lubricant specified by the vehicle manufacturer. The use of improper lubricant in a limited slip differential may result in differential noise such as chattering while cornering. Clicking while turning a corner may be caused by worn limited slip components such as clutches, preload springs, and shims.

Task 2   **Inspect, flush, and refill with correct lubricant.**

70. A limited slip differential chatters while cornering.
    Technician A says the differential may be filled with the wrong lubricant.
    Technician B says friction and steel plates may be worn and burned.
    Who is correct?
    A. A only
    B. B only
    C. Both A and B
    D. Neither A nor B

Hint   *Limited slip differentials should be inspected for fiber and metal cuttings in the bottom of the housing. With the differential components removed, these cuttings may be flushed out of the housing with a solvent gun and an approved cleaning solution. Wipe the bottom of the housing out with a clean shop towel. Always use new gaskets when the differential and the cover are installed in the housing. Fill the differential to the bottom of the filler plug opening with the vehicle manufacturer's specified limited slip differential lubricant.*

Task 3   **Inspect, adjust, and replace clutch (cone/plate) pack.**

71. The measurement in Figure 3-54 determines the proper
   A. friction plate thickness.
   B. steel plate thickness.
   C. shim thickness.
   D. preload spring tension.

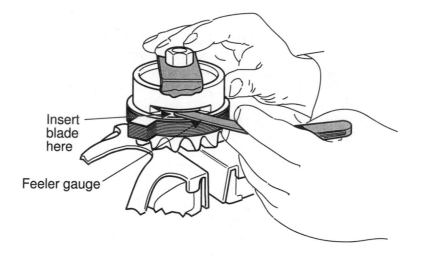

Figure 3-54   Measuring limited slip differential clutch pack to determine proper preload shim thicknes *(Courtesy of Ford Motor Company)*

Hint   *Limited slip differential clutch packs should be soaked in the specified lubricant for 30 minutes prior to installation. After each clutch pack is assembled, use the proper gauge and a feeler gauge to determine the correct shim thickness for proper preload.*

After the clutch packs are assembled in the housing, install a special tool, socket, and torque wrench in one side gear. Hold the opposite side gear and observe the torque required to rotate the side gear. If the turning torque is not within specifications, the problem is in the clutch plates, shim thickness, or preload spring.

Axle Shafts

ASE Tasks, Questions, and Related Information

Task 1 **Diagnose rear axle shaft noise, vibration, and fluid leakage problems; determine needed repairs.**

72. Technician A says an axle shaft with excessive runout may cause a vibration while driving at a constant speed of 60 mph (96 kmh).
Technician B says an axle shaft with excessive runout may cause a vibration while accelerating at low speed.
Who is correct?
   A. A only
   B. B only
   C. Both A and B
   D. Neither A nor B

Hint   *Refer to Table 7-1, page 253, in* Today's Technician Manual Transmissions and Transaxles, *by Delmar Publishers, for diagnosis of rear-wheel-drive axle problems.*

Task 2 **Inspect and replace rear axle shaft wheel studs.**

Hint   *If the threads on the axle studs are damaged, run a die over the threads. When these studs are damaged or bent they should be replaced. Remove the axle and use a hydraulic press to remove and replace the studs.*

Task 3 **Remove and replace rear axle shafts.**

Hint   *Remove the wheel and brake drum prior to axle shaft removal. Remove the bolts from the axle retaining plate. On some rear axles the C-locks on the inner end of the axles must be removed before the axles can be removed from the housing. A slide hammer-type puller attached to the axle studs may be required to remove some axles. Install a new axle seal and be sure the seal contact area on the axle shaft is in good condition before installing the axle.*

Task 4 **Inspect and replace rear axle shaft seals, bearings, and retainers.**

73. All of these statements about rear axle shaft and bearing service are true EXCEPT
   A. the retainer should be loosened by striking it with a hammer and chisel.
   B. the bearing should be pressed off the shaft with a hydraulic press.
   C. after the bearing is pressed off the axle it may be reused.
   D. press the bearing and retainer onto the axle with a hydraulic press.

Hint   *After the axle is removed, use a slide hammer-type puller to remove the axle seal from the housing. If the bearing is held on the axle shaft with a retainer, notch the retainer in several places with a hammer and chisel to loosen the retainer. Never heat the retainer during the removal or installation process. Never strike the bearing with a hammer. When the retainer is loosened, the bearing must be pressed off the axle shaft with the proper press plate and a hydraulic press. Press the new bearing and retainer onto the axle shaft.*

The axle shaft runout may be measured by placing the axle shaft in a pair of Vee blocks and positioning a dial indicator against the center of the shaft. Rotate the shaft and observe the runout on the dial indicator.

If the axle bearing is pressed into the axle housing, a slide hammer-type puller must be used to remove the bearing.

Task 5 **Measure rear axle flange runout and shaft end play; determine needed repairs.**

74. The cause of excessive runout on the dial indicator in Figure 3-55 could be
    A. a worn axle C-lock.
    B. a bent axle shaft.
    C. a bent differential housing.
    D. a worn rear axle bearing.

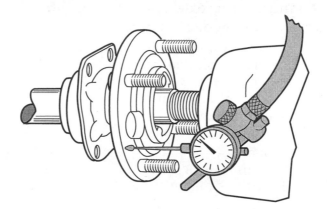

Figure 3-55 Measuring axle flange runout *(Courtesy of Chrysler Corporation)*

Hint   *Place a dial indicator stem against the axle flange to measure flange runout and end play. Maintain a slight inward pressure on the axle flange and zero the dial indicator. Observe the flange runout while rotating the axle shaft one revolution. If the runout is more than specified, replace the axle shaft.*

Move the axle inward and outward and observe the dial indicator reading to measure the axle shaft end play. If the end play exceeds specifications, inspect the bearing, retainer plate, or C-lock and C-lock groove in the axle.

Four-Wheel-Drive Component Diagnosis and Repair

ASE Tasks, Questions, and Related Information

Task 1 **Diagnose four-wheel-drive assembly noise, vibration, shifting, and steering problems; determine needed repairs.**

75. While discussing a four-wheel-drive vehicle with a vibration problem that is more noticeable while cornering.
    Technician A says the U-joints may be worn.
    Technician B says the outboard front axle joints may be worn.
    Who is correct?
    A. A only
    B. B only
    C. Both A and B
    D. Neither A nor B

Hint   *Refer to Table 8-1, page 295, in* Today's Technician Manual Transmissions and Transaxles Shop Manual *for a complete listing and diagnosis of four-wheel assembly problems.*

Task 2   **Inspect, adjust, and repair transfer case shifting mechanisms, bushings, mounts, levers, and brackets.**

76. A vacuum-shifted 4WD system does not shift into 4WD.
    Technician A says the engine vacuum may be low.
    Technician B says the vacuum motor at the front axle may be defective.
    Who is correct?
    A. A only
    B. B only
    C. Both A and B
    D. Neither A nor B

Hint   *A range control linkage adjustment is required on some transfer cases. Place the range control lever in the 2WD position, and place the specified spacer between the gate in the console and the lever. Place the outer lever on the transfer case in the 2WD position. Adjust the linkage so it is in the specified location in relation to the outer lever.*

The main components in the electric control shift system are the electronic control module, shift motor, panel-mounted switch, speed sensor, shift position sensor, and electromagnetic clutch (Figure 3-56) (Figure 3-57, next page). Use a multimeter to diagnose the system using the vehicle manufacturer's recommended procedure.

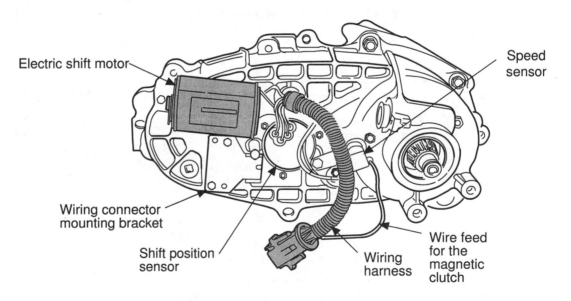

Figure 3-56   Electric transfer shift control motor and wiring harness *(Courtesy of Ford Motor Company)*

Some transfer cases are shifted by engine vacuum. A vacuum motor shifts the transfer case into 4WD, and the vacuum is then diverted to the front axle vacuum motor to lock the front axle into 4WD (Figure 3-58, next page). In these vacuum-operated systems always be sure the engine vacuum is satisfactory before proceeding with the diagnosis. Check all the vacuum hoses for leaks, kinks, and loose-fitting hoses. When the engine vacuum and hoses are satisfactory, test the vacuum motors at the transfer case and front axle.

Inspect all transfer case and engine mounts and brackets for wear, looseness, breaks, oil-soaked conditions, and deterioration. Worn or broken mounts may cause transfer case vibration.

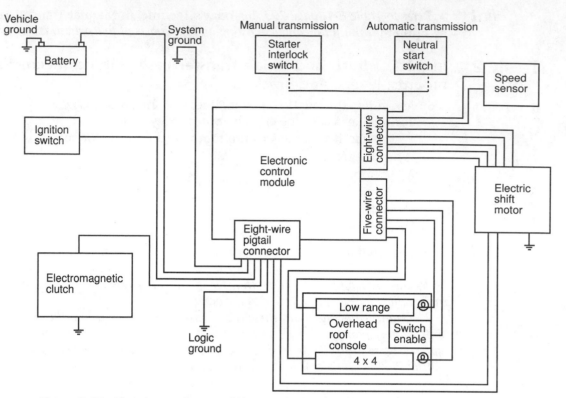

Figure 3-57 Electric transfer case shift control system *(Courtesy of Ford Motor Company)*

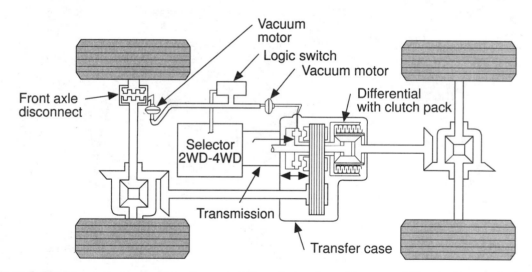

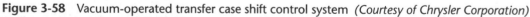

Figure 3-58 Vacuum-operated transfer case shift control system *(Courtesy of Chrysler Corporation)*

Task 3 Inspect and service transfer case components; check lube level.

77. Technician A says in 4WD low the powerflow in the transfer case is from the input shaft through the sun gear and planetary carrier to provide a gear reduction.
Technician B says in 4WD low the annulus gear in the planetary gear is rotating counterclockwise.
Who is correct?

A. A only

B. B only

C. Both A and B

D. Neither A nor B

Hint *The transfer case should be filled to the bottom edge of the filler plug opening with the vehicle on a level surface. Some manufacturers now specify automatic transmission fluid for the transfer case lubricant. After the transfer case is disassembled, clean all the components with an approved cleaning solution and inspect the condition of all transfer case components (Figure 3-59 and Figure 3-60, next page).*

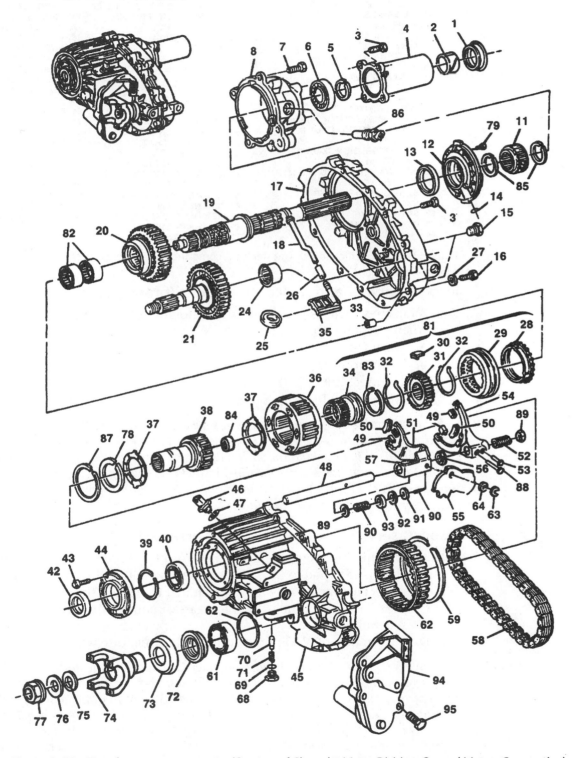

Figure 3-59 Transfer case components *(Courtesy of Chevrolet Motor Division, General Motors Corporation)*

1. SEAL, REAR OUTPUT SHAFT	35. SCREEN, PUMP PICK-UP	64. SEAL, O-RING
2. BUSHING, EXTENSION HOUSING	36. CARRIER, PLANETARY	68. PLUG, POPPET
3. BOLT, EXTENSION HOUSING/CASE HALF	37. WASHER(2), THRUST	69. SEAL, O-RING
4. EXTENSION, HOUSING	38. GEAR, INPUT	70. PLUNGER, POPPET
5. RETAINER	39. RETAINER, SNAP RING	71. SPRING, POPPET
6. BEARING, REAR OUTPUT	40. BEARING, INPUT	72. SEAL, FRONT OUTPUT SHAFT
7. BOLT, PUMP RETAINER HOUSING	42. SEAL, INPUT BEARING RETAINER	73. DEFLECTOR, FRONT OUTPUT
8. HOUSING, PUMP RETAINER	43. BOLT, INPUT BEARING RETAINER	SHAFT SEAL
11. TONE WHEEL, SPEEDOMETER	44. RETAINER, INPUT BEARING	74. YOKE, FRONT OUTPUT
12. PUMP, OIL	45. CASE HALF, FRONT	75. WASHER, RUBBER SEALING
13. SEAL, OIL PUMP	46. SWITCH, VACUUM	76. WASHER, FRONT OUTPUT YOKE
14. O-RING SEAL, PICK-UP TUBE	47. SEAL, O-RING	77. NUT, FRONT OUTPUT YOKE
15. PLUGS, DRAIN/FILL	48. RAIL, SHIFT	78. RING, CARRIER LOCK
16. BOLT, CASE HALF	49. PAD(4), RANGE AND MODE	79. SCREW, OIL PUMP
17. CASE HALF, REAR	SHIFT FORK	81. ASSEMBLY, SYNCHRONIZER
18. TUBE, OIL PICK-UP	50. PAD(2) RANGE AND MODE	82. BEARINGS, DRIVE SPROCKET
19. SHAFT, MAIN	SHIFT FORK CENTER	83. RETAINER, SNAP RING
20. SPROCKET, DRIVE	51. FORK, RANGE SHIFT	84. BEARING, MAINSHAFT PILOT
21. SHAFT, FRONT OUTPUT (DRIVEN)	52. SPRING, SHIFT FORK	85. RETAINER(2) TONE WHEEL
24. BEARING FRONT OUTPUT REAR	53. PIN, RANGE SHIFT FORK	86. SENSOR, SPEED
25. MAGNET	54. FORK, MODE SHIFT	87. RETAINER, SNAP RING
26. CONNECTOR, OIL PICK-UP TUBE	55. SECTOR, SHIFT	88. PIN, MODE FORK GUIDE
27. WASHER	56. BUSHING, SHIFT RAIL	89. CUP, SHIFT FORK SPRING
28. RING, MAIN DRIVE SYCHRONIZER	57. BRACKET, RANGE FORK	90. SPRING, SHIFT FORK
29. SLEEVE, SYCHRONIZER	58. CHAIN, DRIVE	91. WASHER, SHIFT RAIL SPRING
30. STRUT(3), SYCHRONIZER	59. RETAINER, SNAP RING	92. WASHER, SHIFT RAIL
31. HUB, SYCHRONIZER	60. RETAINER, SNAP RING	93. SPACER, SHIFT RAIL
32. SPRING(2), SYCHRONIZER STRUT	61. BEARING, FRONT OUTPUT	94. MOTOR, ENCODER
33. DOWEL, ALIGNMENT	62. GEAR, ANNULUS	95. BOLT, ENCODER MOTOR
34. HUB, RANGE SHIFT	63. RETAINER, SNAP RING	

Figure 3-60 Identification of transfer case components *(Courtesy of Chevrolet Motor Division, General Motors Corporation)*

Task 4   Inspect, service, and replace front-drive (propeller) shafts and universal joints.

78. All of these defects may cause a vibration on a 4WD vehicle that is more noticeable when changing throttle position EXCEPT
    A. worn U-joints.
    B. worn front drive axle joints.
    C. incorrect drive shaft angles.
    D. worn drive shaft slip joints.

Hint   *Rotate the drive shaft and listen for a squeaking noise from the U-joints. A squeaking noise indicates a dry, worn U-joint. Check for movement in the slip joints when up and down force is applied on the U-joint behind the slip joint. Inspect the drive shafts for impact damage.*

If the drive shaft has a center bearing, inspect this bearing for looseness and roughness. Measure the drive shaft angles with an inclinometer as explained previously in this chapter. Typical satisfactory drive shaft angles are provided in Figure 3-61.

Task 5   Inspect, service, and replace front-drive axle knuckles and driving shafts.

79. All of these statements regarding 4WD front-drive axles and joints are true EXCEPT
    A. the inner tripod joint is held on the axle shaft with a snap ring.
    B. a special swaging tool may be required to tighten the outer boot clamps.
    C. coat the new joint with the grease supplied with the joint, and discard the remaining grease.
    D. a worn outer CV joint may cause a clicking noise while cornering.

Hint   *Many 4WD front-drive axles have an outer CV joint and an inner tripod joint. Remove the outer boot clamps and slide the boot toward the center of the axle. Spread the retaining ring with a pair of snap ring pliers, and then slide the outer CV joint from the axle shaft (Figure 3-62).*

Many 4WD front axle joints have an inner tripod joint. A snap ring holds the inner spider assembly onto the shaft in this type of joint (Figure 3-63). Inspect the inner and outer joint components for looseness, wear, scoring, and damage. Check the axle boots for cracks, oil soaking, and deterioration. Replace the boots and axle joints as required. Always install all the grease supplied with the replacement joints in the joint. Always install the boot clamps using the manufacturer's recommended procedure. Some outer boot clamps must be installed with a special swaging tool (Figure 3-64).

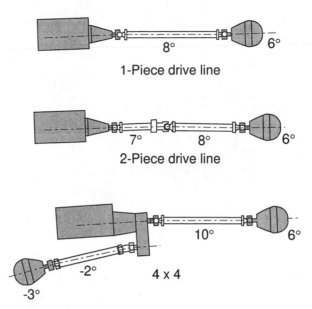

Figure 3-61   Typical drive shaft angles *(Courtesy of Ford Motor Company)*

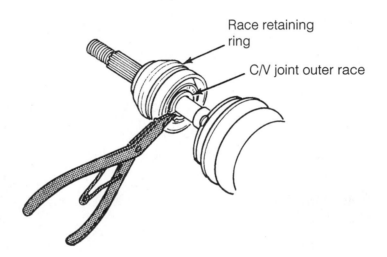

Figure 3-62   Removing outer CV joint from the axle shaft *(Courtesy of Chevrolet Motor Division, General Motors Corporation)*

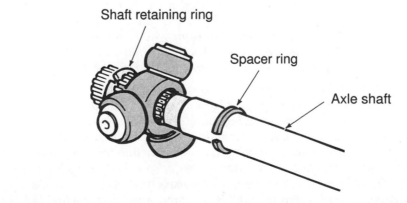

Figure 3-63 Inner tripod joint, front drive axle 4WD *(Courtesy of Chevrolet Motor Division, General Motors Corporation)*

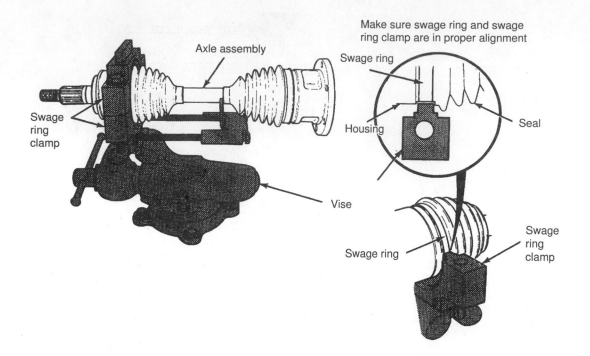

Figure 3-64 Swaging tool for outer front drive axle boot clamps 4WD *(Courtesy of Chevrolet Motor Division, General Motors Corporation)*

Task 6   Inspect, service, and replace front wheel bearings and locking hubs.

80. Technician A says automatic locking hubs should be packed with grease.

    Technician B says the cap on automatic locking hubs should be packed with grease.

    Who is correct?

    A. A only

    B. B only

    C. Both A and B

    D. Neither A nor B

Hint   *In a 4WD vehicle the front wheel bearings are serviced using the same procedure as 2WD vehicles. Since there is more load on the front wheel bearings in a 4WD vehicle, front wheel bearing adjustment is critical. After the front wheel bearings are lubricated and installed, the inner locknut is tightened to the specified torque (Figure 3-65). Some manufacturers recommend tightening this locknut until the specified front wheel turning torque is obtained with a spring scale connected to one of the wheel studs. Rotate the hub and then back off the inner locknut the specified amount, typically one-quarter turn. Install the lockwasher and tighten the outer locknut to the specified torque.*

Four-wheel-drive vehicles may have manual or automatic locking hubs (Figure 3-66 and Figure 3-67). After the wheel bearings have been adjusted and the locking screws or nut tightened to the specified torque, place some multipurpose grease on the hub inner spines. Do not pack the hub with grease. Slide the locking hub assembly into the wheel hub until it seats. Install the lock ring in the wheel hub groove. Install the lock washer and axle shaft stop on the axle bolt, and then install the bolt and tighten it to the specified torque. Place a small amount of lubricant on the cap seal, and install the cap over the body and into the wheel hub. Install the attaching bolts and tighten them to the specified torque. Rotate the locking hub control from stop to stop to make sure it operates freely. Set both controls to the same auto or lock position.

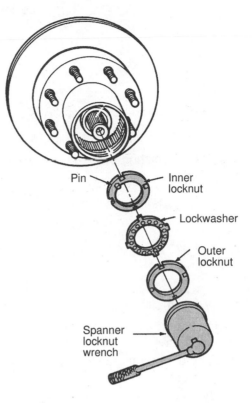

Figure 3-65 Installing and adjusting wheel bearing locknuts *(Courtesy of Ford Motor Company)*

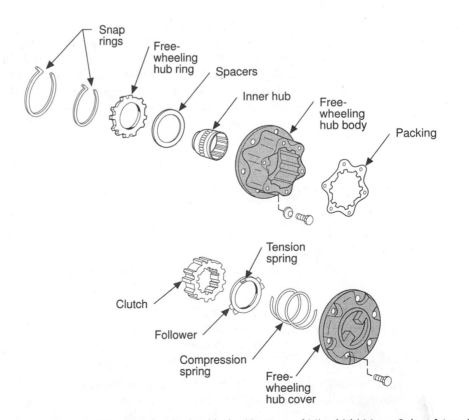

Figure 3-66 Manual locking front wheel hub *(Courtesy of Mitsubishi Motor Sales of America)*

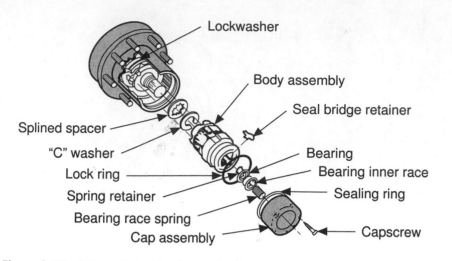

Figure 3-67 Automatic locking front wheel hub *(Courtesy of Ford Motor Company)*

Task 7   Check four-wheel-drive unit seals and remote vents.

81. Technician A says a plugged transfer case vent may cause seal leakage.
Technician B says the remote transfer case vent helps to prevent moisture from entering the transfer case when driving through water.
Who is correct?
A. A only
B. B only
C. Both A and B
D. Neither A nor B

Hint   *Inspect the front and rear output shaft flanges in the transfer case for evidence of leakage in the seal area. If there is any sign of leakage at the drive shaft flange seals, these seals must be replaced. When the flanges and seals are removed always inspect the seal contact area on the flanges for roughness and scoring. Replace the flanges if these conditions are present. Inspect the seal contact area in the case for metal burrs and gouges. Remove metal burrs with a fine-tooth file. Prior to seal installation, lube the seal lips and apply sealer to the outer diameter of the seal case. Inspect the remote venting system on the transfer case and differentials for restrictions and kinks.*

Answers and Analysis

1. D   Weak clutch plate torsional springs cause clutch chatter, but a worn clutch release bearing, pilot bearing, or excessive input shaft end play does not result in this problem. D is right.

2. D   When there is excessive free play the clutch is not released when the clutch pedal is fully depressed resulting in clutch dragging and hard shifting. If there is no clutch pedal free play, the clutch release bearing maintains pressure on the pressure plate so the clutch is not fully applied. This action results in clutch slipping. D is right.

3. C   Many self-adjusting clutch cables have no free play, no overcenter spring, and a constant running release bearing. C is right.

4. B   If the clutch pedal free play is less than specified, the clutch may not be fully engaged with the pedal released. Therefore, A is wrong.

Worn clutch facings, or a scored pressure plate may cause a slipping clutch, but these defects will not cause improper clutch disengagement. Therefore, C and D are wrong.

If there is air in the clutch hydraulic system, the slave cylinder may not push the clutch release lever enough to disengage the clutch. B is right.

5. A   A worn release bearing is noisy with the clutch pedal depressed because the release bearing is in contact with the pressure plate. Therefore, B is wrong.

The clutch pedal free-play adjustment sets the distance between the release bearing and the pressure plate fingers. A is right.

6. C   Excessive crankshaft endplay causes the pressure plate to move away from the clutch release bearing, which may result in improper clutch release. Loose engine main bearings may cause an oil leak at the rear main bearing, which contaminates the clutch facings with oil resulting in clutch slipping. Improper pressure plate to flywheel position causes engine vibrations. Therefore, A, B, and D are wrong.

A badly scored pressure plate may cause clutch chatter. C is right.

7. C   Clutch slipping may occur if the rivet heads are 0.012 in. or less below the facing surface. C is right, and A, B, and D are wrong.

8. A   When the clutch pedal is released, the clutch plate is held firmly between the flywheel and pressure plate. Under this condition the input shaft cannot rattle in a worn pilot bearing. Therefore, B, C, and D are wrong. A worn pilot bearing causes a rattling noise at low speed with the clutch pedal depressed because the clutch plate and input shaft are free to move between the pressure plate and the flywheel. A is right.

9. C   If excessive material is removed from the flywheel the torsion springs on the clutch plate are moved closer to the flywheel, and these springs may contact the flywheel. Removing excessive material from the flywheel moves the pressure plate forward away from the release bearing. This action increases free play so the slave cylinder rod may not move far enough to release the clutch. Both A and B are right, and C is the correct answer.

10. D   Excessive bell housing misalignment does not affect clutch pedal free play, noise, or vibration at high speeds. Therefore, A, B, and C are wrong.

Misalignment between the bell housing and the engine block may cause clutch chatter because the clutch disc mounted on the input shaft is not aligned properly with the pressure plate and flywheel. D is right.

11. A   The measurement shown in Figure 3-8 is for flywheel runout and crankshaft endplay. Since flywheel runout is not included in the responses, crankshaft endplay, response A is right.

12. A   Overheated clutch facings would not result in bell housing face runout. Therefore, B is wrong.

Shims may be placed between the bell housing and the engine block to correct bell housing face runout. A is right.

13. B   The dial indictor in Figure 3-9 is positioned to measure bell housing bore alignment. Bell housing bore misalignment may be corrected by turning eccentric dowels in the engine block to bell housing mounting surface. B is right.

14. A   A drive shaft vibration changes with acceleration and deceleration. Therefore, B is wrong.

Sagged engine mounts cause improper drive shaft angles in a rear-wheel-drive car. A is right.

15. C   Excessive second gear end play, or a weak detent spring on the second gear shift fork, may cause the transmission to jump out of second gear. Both A and B are right, and C is the correct answer.

16. A   Most linkage adjustments are performed with the gear shift lever in neutral. Therefore, B, C, and D are wrong, and A is right.

17. A   Excessive output shaft, or input shaft, end play results in lateral shaft movement that would not adversely affect the seal. A worn output shaft bearing would not cause premature extension housing seal failure. Therefore, B, C, and D are wrong.

> A scored drive shaft yoke results in excessive extension housing seal wear. A is right.

18. D   The drive shaft may be installed in the original position on the differential flange. The transmission weight should be supported on a transmission jack, and the engine support fixture must be installed before the transmission to engine bolts are loosened. Therefore, A, B, and C are wrong.

> A clutch plate alignment tool must be used before transmission installation. D is right.

19. C   A synchronizer hub that does not slide smoothly over the blocking ring causes the hub to jam resulting in hard shifting, and the synchronizer hub and sleeve should be marked prior to disassembly. Both A and B are right, and C is the correct answer.

20. B   A bent shift rail would not cause the transmission to jump out of gear, or result in gear clash or noise. Hard shifting may be caused by a bent shift rail, thus B is right.

21. C   A worn pilot bearing contact area on the input shaft, or a worn pilot bearing could result in noise with the clutch pedal depressed. Therefore, A and B are wrong.

> The main shaft is not turning with the engine idling and the clutch released with the transmission in neutral, so D is wrong.

> Since the input shaft is only turning with the clutch released, a rough input shaft roller bearing or needle bearings would result in a growling noise under this condition. C is right.

22. C   A worn first speed gear blocking ring, or synchronizer sleeve may cause hard shifing, but they would not result in noise while driving in first gear. Therefore, A and B are wrong. A worn, rough main shaft bearing would cause a growling noise in all gears, so D is wrong.

> Chipped and worn first speed gear teeth would cause a growling noise in first gear. C is right.

23. B   The blocking ring dog teeth tips should be pointed with smooth surfaces. Therefore, A is wrong. Clearance between the blocking ring and matching gear dog teeth is important for proper shifting, and the synchronizer sleeve must slide freely on the hub. Therefore, C and D are wrong.

> Threads in the cone area of the blocking ring must be sharp. B is right.

24. B   If the clearance between the blocking ring and the fourth-speed gear dog teeth is less than specified, the blocking ring is worn, which results in hard shifting. This problem would not result in noise while driving in fourth gear. A is wrong and B is right.

25. C   Since the counter gear is turning with the clutch pedal released in neutral and also in any gear, damaged counter gear bearings cause a growling noise under both these conditions. Both A and B are right, and C is the correct answer.

26. D   Since the reverse idler gear is only in mesh with other gears in reverse, this gear rotates only in reverse. D is right.

27. C   In fourth gear the 1-2 synchronizer is moved ahead so the synchronizer hub is meshed with the dog teeth on the fourth-speed gear on the input shaft. Excessive input shaft end play would cause the transmission to jump out of fourth gear. C is right.

28. B   A worn speedometer drive and driven gears may cause erratic speedometer operation, but this problem would not result in premature extension housing bushing wear. A is wrong.

> A plugged transmission vent may cause excessive transmission pressure, and fluid leaks, but this defect would not affect extension housing bushing wear. C is wrong.

> Excessive mainshaft end play causes lateral shaft movement, but this does not cause excessive extension housing bushing wear, so D is wrong.

Metal burrs between the extension housing and the transmission case cause misalignment of the extension housing, which forces the extension housing bushing against the drive shaft yoke in one location resulting in bushing wear. B is right.

29. A   A worn extension housing bushing may cause premature extension housing seal wear and fluid leaks, so B is wrong.

Excessive main shaft end play has no effect on speedometer operation, thus C is wrong.

The speedometer drive shown in Figure 3-20 contains a cable to drive the speedometer. Therefore, an electrical defect in the distance sensor would not affect speedometer operation, thus D is wrong.

A worn adapter bushing and speed distance sensor allows the speedometer cable to jump around causing erratic speedometer operation. A is right.

30. A   Since thicker gear oils have higher numbers, B is wrong.

ATF is much thinner compared to 90W gear oil, and therefore it does reduce friction and improve horsepower and economy. A is right.

31. B   A worn fourth speed synchronizer would only affect shifting in fourth gear, so A is wrong.

Excessive main shaft end play would not result in gear clash in all gears. A worn 3-4 shift fork may cause shifting problems in third and fourth gear, but this problem would not result in gear clash in all gears. Therefore, C and D are wrong.

A clutch disc sticking on the input shaft would cause the clutch not to release properly resulting in gear clash in all gears. B is right.

32. C   An improper shift linkage adjustment may cause hard shifting or sticking in gear. Both A and B are right, and C is the correct answer.

33. A   A worn outer drive axle joint may cause a clicking noise while cornering at low speed, but this defect would not cause repeated drive axle seal failure. B is wrong.

A plugged transaxle vent may cause excessive transaxle pressure, and repeated drive axle seal failure. A is right.

34. C   A misaligned engine and transaxle cradle may cause drive axle vibrations. Since the lower control arms are connected to this cradle, misalignment of the cradle may cause improper front suspension angles. Both A and B are right, and C is the correct answer.

35. B   After the case halves are separated, the differential may be removed first on some transaxles. Therefore, A is wrong.

During service procedures the transaxle case must be supported on a special fixture and shims, so C is wrong.

A special driving tool and a hydraulic press are required to remove the input and output shafts. D is wrong.

The input and output shafts are retained with snap rings under the transaxle cover. B is right.

36. A   Worn dog teeth on the fourth speed gear would not cause the transaxle to jump out of third gear. B is wrong.

A weak detent spring on the 3-4 shift rail may cause the transaxle to jump out of third gear, so A is right.

37. A   Worn dog teeth on the third speed gear or blocking ring may cause hard shifting or jumping out of third gear, but this defect would not cause a growling noise while driving in third gear. Therefore, B and C are wrong.

Worn threads in the third speed blocking ring may cause hard shifting, but this wear would not cause a growling noise in third gear. D is wrong.

Worn, chipped teeth on the third speed gear could result in a growling noise while driving in third gear. A is right.

38. D   Worn dog teeth on the second speed gear and blocking ring would not result in a growling noise while driving, or accelerating, in second gear. Therefore, A and B are wrong.

This wear problem would not cause hard shifting in second and third gear, so C is wrong.

Worn dog teeth on the second speed gear and blocking ring may cause the transaxle to jump out of second gear. D is right.

39. D   Synchronizer hubs are not reversible on the shaft, and synchronizer sleeves are not reversible on their hub. Both A and B are wrong, and D is the correct answer.

40. A   Worn reverse idler teeth may cause a growling noise while driving in reverse, but this problem would not cause failure to shift into reverse. Therefore, B is wrong.

A broken reverse shifter fork may cause failure of the transmission to shift into reverse without any noise. A is right.

41. A   Loctite® PST is not used to repair transaxle cases, so B is wrong. An epoxy-based sealer may be used for some transaxle case cracks. A is right.

42. C   In some transaxles the speedometer drive gear is mounted on the differential case. C is right.

43. B   Damaged ring gear teeth would cause a clicking noise while the vehicle is in motion. This defect would not cause differential chatter. A is wrong. Improper preload on differential components, such as side bearings, may cause differential chatter, and so B is right.

44. D   In the transaxle shown in Figure 3-34 differential side bearing preload is adjusted with a shim behind one of the side bearing cups. Therefore, A and B are wrong, and the correct answer is D.

45. C   The side gear end play must be measured individually on each side gear with the thrust washers removed. Therefore, A and B are wrong.

Side gears with the specified thrust washer have a slight end play but no preload. Therefore, D is wrong and C is right.

46. D   A side bearing preload adjustment shim is positioned behind one of the side bearing cups, so A is wrong.

The speedometer drive gear will slide over the side bearing so the bearing does not have to be removed first. Therefore, A and B are both wrong; D is the correct answer.

47. B   While measuring the differential side play to determine the required side bearing shim thickness, a new bearing cup is installed in the transaxle case without the shim. Therefore, A is wrong.

The proper shim thickness is equal to the differential end play plus a specified thickness for bearing preload, so C is wrong.

While measuring the differential end play the transaxle case bolts must be tightened to the specified torque. Therefore, D is wrong.

A medium load should be applied to the differential in the upward direction while measuring the differential end play. B is right.

48. A   When the differential turning torque is less than specified, the shim thickness must be increased behind the side bearing cup in the bell housing side of the transaxle case. Therefore, A is right, and B, C, and D are wrong.

49. C   Wrong transaxle lubricant may cause burned output shaft bearings. Therefore, A is right. A broken oil feeder behind the front output shaft bearing results in improper output shaft bearing lubrication and burned bearings. Therefore, both A and B are right, and the correct answer is C.

50. B   A worn U-joint may cause a squeaking noise that increases in relation to vehicle speed. Therefore, A is wrong.

        If the centering ball and socket is worn in a double Cardan U-joint, a heavy vibration may occur during acceleration. B is right.

51. A   A worn front wheel bearing usually results in a growling noise while cornering or driving straight ahead. B is wrong.

        A clunking noise while decelerating may be caused by a worn inner drive axle joint. A is right.

52. C   A worn drive shaft center bearing, or an outer rear axle bearing causes a growling noise that is not influenced by acceleration and deceleration. Both A and B are right, and C is the correct answer.

53. D   The drive shaft markings in Figure 3-43 are for drive shaft balancing. Therefore, A, B, and C are wrong. D is right.

54. D   While measuring drive shaft runout the dial indicator must be positioned near the center of the drive shaft, and the drive shaft should be replaced if the runout is excessive. Both A and B are wrong; the correct answer is D.

55. C   An out of balance drive shaft or wheel may cause a vibration that increases in relation to vehicle speed. Both A and B are right, so the correct answer is C.

56. B   Since the side gears are only turning while cornering, they do not cause a whining noise while driving straight ahead. Therefore, A is wrong. However, improper ring gear and pinion gear adjustments may cause this problem. B is right.

57. A   A loose pinion nut allows pinion shaft end play resulting in a clunking noise on acceleration and deceleration. A is right.

58. A   The question asks for the response that is not correct. Hunting-type ring and pinion gear sets do not require timing. Since the statements in B, C, and D are correct, they are not the required answer. However, A is not right, and this is the correct answer.

59. A   Excessive side bearing preload, side gear end play, or ring gear bolt torque would not cause excessive runout on the dial indicator placed against the ring gear. Therefore, B, C, and D are wrong.

        Excessive runout on the dial indicator may be caused by ring gear or case runout. A is right.

60. D   The collapsible spacer and the ABS toothed ring should be replaced when the pinion shaft has been disassembled. Therefore, both A and B are wrong, and the correct answer is D.

61. C   The marking on the pinion must be added to the shim thickness to determine the correct pinion depth shim thickness. Therefore, 0.066 in. (answer C) is right.

62. A   The reading on the dial indicator must be subtracted from 0.100 in. to obtain the nominal pinion depth shim thickness. Therefore, the nominal shim thickness is 0.43, and the pinion marking of −4 is subtracted from this figure. Therefore, 0.039 in. (answer A) is correct.

63. A   The pinion nut must never be loosened to obtain the specified pinion turning torque, so B is wrong.

        The pinion bearings should be lubricated when the pinion turning torque is measured. A is right.

64. C   The left side bearing adjusting nut must be loosened and the right side bearing adjusting nut tightened to increase ring gear backlash. C is right.

65. D   With either excessive ring gear toe contact, or low flank contact, the pinion gear must be moved away from the ring gear. Therefore, both A and B are wrong, and D is the correct answer.

66. **B**   Worn pinion bearings may cause a growling noise while driving straight ahead. Therefore, A is wrong.

Since the side gears and bevel pinions rotate only while turning a corner, rough bearing surfaces on these gears and the differential case may cause a vibration while cornering. B is right.

67. **D**   This question asks for the answer that is not correct. Ring gear runout, and case side play should be measured before removing the ring gear and case assembly, and the side bearing caps should be marked in relation to the case prior to removal. Therefore, statements A, B, and C are right, but they are not the requested answer.

The side bearings should be lubricated before installation, so statement D is wrong, and this is the correct answer.

68. **B**   The side bearings must be in good condition prior to the case runout measurement, so A is wrong.

The ring gear runout should be measured before the case runout. B is right.

69. **A**   This question asks for the statement that is not correct. In a limited slip differential the steel plates are retained in the case, each clutch set contains a preload spring, and a special lubricant is required. Therefore, statements B, C, and D are correct, but they are not the requested answer.

The friction plates are splined to the clutch hub, so statement A is not correct. A is the right answer.

70. **C**   Wrong lubricant, or worn friction and steel plates, in a limited slip differential may cause chattering while cornering. Both A and B are right, and C is the correct answer.

71. **C**   The measurement in Figure 3-54 determines the proper shim thickness. Therefore C is right, and A, B, and D are wrong.

72. **A**   Excessive axle shaft runout may cause a vibration at 60 mph (96 kmh), but this problem would not cause a vibration when accelerating at lower speed. Therefore, B is wrong, and A is right.

73. **C**   This question asks for the statement that is not correct. An axle bearing retainer should be removed by striking it in several places with a hammer and chisel. The bearing should be pressed off the axle shaft with a hydraulic press, and this press should be used to press the bearing and retainer onto the shaft. Therefore, statements A, B, and D are correct, but they are not the requested answer.

The bearing must be replaced after it is pressed off the axle. Since statement C is wrong, this is the requested answer.

74. **B**   Excessive runout on the dial indicator in Figure 3-55 could be caused by a bent axle shaft. Therefore, B is right, and A, C, and D are wrong.

75. **B**   Worn U-joints may cause a squeaking or clunking noise, and a vibration while driving straight ahead. Therefore, A is wrong.

Worn outer front drive axle joints on a 4WD vehicle may cause a vibration problem while cornering. B is right.

76. **C**   When a vacuum-operated 4WD does not shift into 4WD, the engine vacuum may be low, or the vacuum motor at the front differential may be defective. Therefore, A and B are both right, and the correct answer is C.

77. **A**   The annulus gear is locked to the case, so B is wrong.

In 4WD low the transfer case input shaft is driving the sun gear which in turn is driving the planetary carrier. A is right.

78. D   This question asks for the statement that is not correct. Worn U-joints, front axle drive joints, and incorrect drive shaft angles may cause a vibration that is more noticeable when changing throttle position. Therefore, statements A, B, and C are correct, but these statements are not the requested answer.

Worn drive shaft slip joints would not cause this type of vibration. Since statement D is wrong, it is the requested answer.

79. C   This question asks for the statement that is not correct. A snap ring holds the inner tri-pod joint on the axle shaft, and a special swaging tool may be necessary to tighten the outer boot clamp. A worn outer CV joint may cause a clicking noise while cornering. Therefore, statements A, B, and D are right, but they are not the requested answer.

All the grease supplied with the joint should be installed in the joint. Therefore, statement C is wrong, and this is the requested answer.

80. D   Neither the automatic locking hubs nor the caps should be packed in grease. Therefore, both A and B are wrong, and D is the correct answer.

81. C   A plugged transfer case vent may cause seal leakage, and the remote transfer case vent helps to keep water out of the transfer case. Both A and B are right, and C is the correct answer.

Glossary

Abrasion Wearing or rubbing away of a part.
Abrasión Desgaste o rozamiento de una pieza.

Acceleration An increase in velocity or speed.
Aceleración Aumento de velocidad o celeridad.

Adhesives Chemicals used to hold gaskets in place during the assembly of an engine. They also aid the gasket in maintaining a tight seal by filling in the small irregularities on the surfaces and by preventing the gasket from shifting due to engine vibration.
Adhesivos Productos químicos que sirven para sujetar las guarniciones en su lugar durante el montaje de un motor. También ayudan a la guarnición a mantener una junta de estanqueidad hermética al rellenar las pequeñas irregularidades en las superficies y al impedir que la guarnición se desplace debido a la vibración del motor.

Alignment An adjustment to a line or to bring into a line.
Alineación Ajuste a una línea o poner en línea.

Antifriction bearing A bearing designed to reduce friction. This type of bearing normally uses ball or roller inserts to reduce the friction.
Cojinete de antifricción Cojinete diseñado para disminuir la fricción. Normalmente a este tipo de cojinete se le insertan bolas o rodillos para disminuir la fricción.

Antiseize Thread compound designed to keep threaded connections from damage due to rust or corrosion.
Antiagarrotamiento Compuesto de filetes de un tornillo que sirve para mantener las conexiones fileteadas libres de averías causadas por el óxido o la corrosión.

Arbor press A small, hand-operated shop press used when only a light force is required against a bearing, shaft, or other part.
Prensa para calar Pequeña prensa de taller accionada a mano y utilizada solamente cuando se requiere una fuerza ligera contra un cojinete, un árbol u otra pieza.

Asbestos A material that was commonly used as a gasket material in places where temperatures are great. This material is being used less frequently today because of health hazards that are inherent to the material.
Asbesto Material que usualmente se utilizaba como material de guarnición en lugares donde las temperaturas eran muy elevadas. Hoy en día se utiliza dicho material con menos frecuencia debido a los riesgos para la salud inherentes al mismo.

Automatic locking/unlocking hubs Front wheel hubs that can engage or disengage themselves from the axles automatically.
Cubos automáticos de cierre/descerrar Cubos ubicados en las ruedas delanteras que pueden engranarse o desengranarse de los ejes automáticamente.

Automatic transmission A transmission in which gear or ratio changes are self-activated, eliminating the necessity of hand-shifting gears.
Transmisión automática Sistema en el cual los cambios de velocidades o de relación se activan automáticamente, eliminando así la necesidad de emplear la palanca del cambio de velocidades.

Axial Parallel to a shaft or bearing bore.
Axial Paralelo a un árbol o calibre de cojinete.

Axis The center line of a rotating part, a symmetrical part, or a circular bore.
Pivote Línea central de una pieza giratoria, una pieza simétrica, o un calibre circular.

Axle The shaft or shafts of a machine upon which the wheels are mounted.
Eje Árbol o árboles de una máquina sobre el cual se montan las ruedas.

Axle carrier assembly A cast-iron framework that can be removed from the rear axle housing for service and adjustment of the parts.
Conjunto portador del eje Armazón de hierro fundido que se puede remover del puente trasero para la reparación y el ajuste de las piezas.

Axle housing Designed in the removable carrier or integral carrier types to house the drive pinion, ring gear, differential, and axle shaft assemblies.
Puente trasero Diseñado en los tipos de portador desmontables o enterizos para alojar los conjuntos del piñón de mando, de la corona, del diferencial y del árbol motor.

Axle ratio The ratio between the rotational speed (rpm) of the drive shaft and that of the driven wheel; gear reduction through the differential, determined by dividing the number of teeth on the ring gear by the number of teeth on the drive pinion.
Relación del eje Relación entre las revoluciones por minuto (rpm) del árbol de mando y las de la velocidad de la rueda accionada; la desmultiplicación de engranajes por medio del diferencial, determinada al dividir el número de dientes en la corona por el número de dientes en el piñón de mando.

Axle shaft A shaft on which the road wheels are mounted.
Árbol motor Árbol sobre el cual se montan las ruedas de la carretera.

Axle shaft end thrust A force exerted on the end of an axle shaft that is most pronounced when the vehicle turns corners and curves.
Empuje longitudinal del árbol motor Fuerza ejercida sobre el extremo de un árbol motor. Dicha fuerza es más marcada cuando el vehículo dobla una esquina o una curva.

Axle shaft tubes These tubes are attached to the axle housing center section to surround the axle shaft and bearings.
Tubos del árbol motor Estos tubos se fijan a la sección central del puente trasero para rodear el árbol motor y los cojinetes.

Backlash The amount of clearance or play between two meshed gears.
Contragolpe Cantidad del espacio libre u holgura entre dos engranajes.

Balance Having equal weight distribution. The term is usually used to describe the weight distribution around the circumference and between the front and back sides of a wheel and tire assembly. Uneven weight distribution causes vibrations when the wheel is spun because the center of gravity of the wheel and tire assembly does not line up with the center line of the axle. A wheel and tire are balanced by adding weights to the wheel rim opposite the heavy section.
Equilibrio Que tiene igual distribución de peso. Usualmente este término se emplea para describir la distribución de peso alrededor de la circunferencia y entre las partes delantera y trasera de un conjunto de rueda y llanta. La distribución desigual de peso puede causar vibraciones cuando se gira la rueda porque el centro de gravedad del conjunto de rueda y llanta no está alineado con la línea central del eje. Una rueda y una llanta están en equilibrio cuando se le añade más peso a la llanta opuesta a la sección pesada.

Balanced resistance A situation where two objects, such as axle shafts, present the same resistance to driving rotation.

Resistencia equilibrada Situación en la que dos objetos, tales como los árboles motores, ofrecen la misma resistencia al mando de rotación.

Ball bearing An antifriction bearing that consists of a hardened inner and outer race with hardened steel balls that roll between the two races, and supports the load of the shaft.

Cojinete de bolas Cojinete de antifricción que consiste de un anillo interno y externo templado con bolas de acero templadas que giran entre los dos anillos, y apoya la carga del árbol.

Ball-and-trunnion universal joint A nonconstant velocity universal joint that combines the universal joint and slip joint. It uses a drivable housing connected to a shaft with a ball head through two other balls mounted on trunnions.

Junta universal de bola y muñequilla Junta universal de velocidad no constante que combina la junta universal y la junta deslizante. Utiliza un alojamiento accionable conectado a un árbol por medio de un péndulo a través de otras dos bolas montadas sobre las muñequillas.

Ball gauge A common term for a small hole gauge.

Calibrador de bolas Término común para calibrador de agujero pequeño.

Ball joint A suspension component that attaches the control arm to the steering knuckle and serves as the lower pivot point for the steering knuckle. The ball joint gets its name from its ball-and-socket design. It allows both up and down motion as well as rotation. In a MacPherson strut FWD suspension system, the two lower ball joints are nonload carrying.

Junta esférica Componente de la suspensión que fija el brazo de mando al muñón de dirección y sirve como punto de pivote inferior para el muñón de dirección. Así se le llama a la junta esférica por su diseño de rótula. Permite tanto el movimiento de ascenso y descenso, como el de rotación. En un sistema de suspensión de tracción delantera montante MacPherson, las dos juntas esféricas inferiores no tienen capacidad de carga.

Bearing The supporting part that reduces friction between a stationary and rotating part or between two moving parts.

Cojinete Pieza de soporte que disminuye la fricción entre una pieza fija y una pieza giratoria o entre dos piezas móviles.

Bearing cage A spacer that keeps the balls or rollers in a bearing in proper position between the inner and outer races.

Jaula de cojinete Espaciador que mantiene las bolas o los rodillos en posición correcta en el cojinete, entre los anillos interior y exterior.

Bearing caps In the differential, caps held in place by bolts or nuts that, in turn, hold bearings in place.

Casquillos de cojinete En el diferencial, los casquillos sujetados por tornillos o tuercas que a su vez sujetan los cojinetes en su lugar.

Bearing cone The inner race, rollers, and cage assembly of a tapered-roller bearing. Cones and cups must always be replaced in matched sets.

Cono de cojinete El anillo interior, los rodillos y el conjunto de jaula de un cojinete de rodillos cónicos. Se debe reemplazar siempre los conos y las rótulas en conjuntos que hagan juego.

Bearing cup The outer race of a tapered-roller bearing or ball bearing.

Rótula de cojinete Anillo exterior de un cojinete de rodillos cónicos o cojinete de bolas.

Bearing race The surface on which the rollers or balls of a bearing rotate. The outer race is the same thing as the cup, and the inner race is the one closest to the axle shaft.

Anillo de cojinete Superficie sobre la cual los rodillos o las bolas de un cojinete giran. El anillo exterior es lo mismo que la rótula, y el anillo interior es el que se encuentra más cerca del árbol motor.

Belleville spring A tempered spring steel cone-shaped plate used to aid the mechanical force in a pressure plate assembly.

Muelle de Belleville Lámina de muelle de acero templado en forma cónica utilizada para ayudar a la fuerza mecánica en un conjunto de placa de presión.

Bell housing A housing that fits over the clutch components and connects the engine and the transmission.

Alojamiento de campana Alojamiento que se monta sobre los componentes del embrague y que conecta el motor y la transmisión.

Bellows Rubber protective covers with accordionlike pleats used to contain lubricants and exclude contaminating dirt or water.

Fuelles Cubiertas protectivas de caucho con pliegues en forma de acordeón, utilizadas para contener los lubricantes y evitar la contaminación por polvo o agua.

Bevel spur gear Gear that has teeth with a straight center line cut on a cone.

Engranaje recto cónico Engranaje que tiene dientes con un corte de línea central recto, cortado en forma cónica.

Bleeding The process of removing air from a closed system.

Desangramiento Proceso a través del cual se remueve el aire de un sistema cerrado.

Bolt torque The turning effort required to offset resistance as the bolt is being tightened.

Par de torsión del perno Esfuerzo giratorio requerido para neutralizar la resistencia al apretarse el perno.

Boots See bellows.

Botas Véase fuelles.

Brake horsepower (bhp) Power delivered by the engine and available for driving the vehicle; bhp 5 torque 3 rpm/5,252.

Potencia en caballos indicada al freno (bhp) Energía descargada por el motor y disponible para accionar el vehículo; bhp 5 par de torsión 3 rpm/5.252.

Brinnelling Rough lines worn across a bearing race or shaft due to impact loading, vibration, or inadequate lubrication.

Acción de Brinnell Líneas toscas a través de un anillo de cojinete o un árbol, causadas por la carga de un impacto, vibración o engrase inadecuado.

Bronze An alloy of copper and tin.

Bronce Aleación de cobre y estaño.

Burnish To smooth or polish by the use of a sliding tool under pressure.

Bruñir Suavizar o pulir con una herramienta deslizante bajo presión.

Burr A feather edge of metal left on a part being cut with a file or other cutting tool.

Rebaba Bisel de metal que queda en una pieza que ha sido cortada con una lima u otra herramienta de corte.

Bushing A cylindrical lining used as a bearing assembly made of steel, brass, bronze, nylon, or plastic.

Buje Forro en forma cilíndrica utilizado como un conjunto de cojinete, hecho de acero, latón, bronce, nilón o plástico.

C-clip A C-shaped clip used to retain the drive axles in some rear axle assemblies.

Grapa-C Grapa en forma de C utilizada para retener los ejes de mando en algunos conjuntos de eje trasero.

Cage A spacer used to keep the balls or rollers in proper relation to one another. In a constant velocity joint, the cage is an open metal framework that surrounds the balls to hold them in position.
Jaula Espaciador utilizado para sujetar las bolas o los rodillos en relación correcta entre sí. En una junta de velocidad constante, la jaula es un armazón abierto de metal que rodea las bolas para que se mantengan en su posición.

Camber The amount that the center line of the wheel is titled inward or outward from the true vertical plane of the wheel.
Combadura Amplitud de la inclinación, hacia adentro o hacia afuera, de la línea central de una rueda con respecto al verdadero plano vertical de la rueda.

Canceling angles Opposing operating angles of two universal joints cancel the vibrations developed by the individual universal joint.
Angulos de supresión Los ángulos de funcionamiento opuestos de dos juntas universales cancelan las vibraciones producidas por la junta universal individual.

Carbon monoxide An odorless, colorless, and deadly gas present in the exhaust of engines.
Monóxido de carbono Gas mortífero, inodoro e incoloro presente en el escape de los motores.

Cardan universal joint A nonconstant velocity universal joint consisting of two yokes with their forked ends joined by a cross. The driven yoke changes speed twice in 360 degrees of rotation.
Junta de cardán Junta universal de velocidad no constante que consiste de dos horquillas con sus extremos unidos por una cruz. El yugo accionado cambia de velocidad dos veces en una rotación de 360°.

Carrier An object that bears, cradles, moves, or transports some other object or objects.
Portador Objeto que apoya, acojina, mueve o transporta otro u otros objetos.

Case-harden To harden the surface of steel. The carburizing method used on low carbon steel or other alloys to make the case or outer layer of the metal harder than its core.
Cementar Endurecer la superficie del acero. El método carburizante empleado en el acero de bajo contenido en carbón u otras aleaciones para hacer que la caja o capa exterior del metal sea más dura que el núcleo.

Castellate Formed to resemble a castle battlement, as in a castellated nut.
Entallar Formado para que se asemeje a la almena de un castillo, como una tuerca de corona.

Castellated nut A nut with six raised portions or notches through which a cotter pin can be inserted to secure the nut.
Tuerca de corona Tuerca con seis partes elevadas o muescas a través de las cuales se puede insertar un pasador de chaveta para sujetar la tuerca.

Caster A measurement expressed as an angle of the forward or rearward tilt of the top of the wheel spindle.
Ángulo de comba de eje Medida expresada como el ángulo de inclinación hacia adelante o hacia atrás de la parte superior del portamuela.

Center hanger bearing Ball-type bearing mounted on a vehicle cross member to support the drive shaft and provide better installation angle to the rear axle.
Silleta de suspensión central Cojinete de tipo bola montado sobre la traviesa de un vehículo para apoyar el árbol de mando y proveerle un mejor ángulo de montaje al eje trasero.

Center section The middle of the integral axle housing containing the drive pinion, ring gear, and differential assembly.
Sección central Centro del puente trasero integral que contiene el piñón de mando, la corona y el conjunto del diferencial.

Centering joint Ball socket joint placed between two Cardan universal joints to ensure that the assembly rotates on center.
Junta centradora Junta de rótula montada entre dos juntas de cardán para asegurar que el montaje gire en el centro.

Centrifugal clutch A clutch that uses centrifugal force to apply a higher force against the friction disc as the clutch spins faster.
Embrague centrífugo Embrague que utiliza fuerza centrífuga para aplicar mayor fuerza contra el disco de fricción cuando el embrague gira con más rapidez.

Centrifugal force The force acting on a rotating body that tends to move it outward and away from the center of rotation. The force increases as rotational speed increases.
Fuerza centrífuga Fuerza que acciona un cuerpo giratorio y que tiende a moverlo hacia afuera y más lejos del centro de la rotación. La fuerza aumenta cuando aumenta la velocidad de rotación.

Centimeter (cm) 0.01 meter or 0.3937 inches
Centímetro (cm) La centésima parte de un metro o 0,3937 pulgadas

Chamfer A bevel or taper at the edge of a hole or a gear tooth.
Chaflán Bisel o cono al borde de un agujero o diente de engranaje.

Chamfer face A beveled surface on a shaft or part that allows for easier assembly. The ends of FWD drive shafts are often chamfered to make installation of the CV-joints easier.
Superficie achaflanada Superficie biselada en un árbol o pieza que facilita el montaje. Los extremos de los árboles de mando de tracción delantera se achaflanan para facilitar la instalación de las juntas CV.

Chase To straighten up or repair damaged threads.
Roscar Enderezar o reparar roscas averiadas.

Chasing To clean threads with a tap.
Filetear Limpiar las roscas con un macho de roscar.

Chassis The vehicle frame, suspension, and running gear. On FWD cars, it includes the control arms, struts, springs, trailing arms, sway bars, shocks, steering knuckles, and frame. The drive shafts, constant velocity joints, and transaxle are not part of the chassis or suspension.
Chasis El armazón del vehículo, la suspensión y el tren de ruedas. En vehículos de tracción delantera, incluye los brazos de mando, los montantes, los muelles, los brazos traseros, las barras de oscilación lateral, los amortiguadores, los muñones de dirección y el armazón. Los árboles de mando, las juntas de velocidad constante y el transeje no forman parte del chasis o de la suspensión.

Cherry picker A common name for an engine hoist.
Grúa alzavagonetas Término común para montacargas del motor.

Chuckle A rattling noise that sounds much like a stick rubbing against the spokes of a bicycle wheel.
Estrépito Ruido muy grande o estruendo parecido al sonido de un palo rozando contra los rayos de la rueda de una bicicleta.

Circlip A split steel snap ring that fits into a groove to hold various parts in place. Circlips are often used on the ends of FWD drive shafts to retain the constant-velocity joints.
Grapa circular Anillo de resorte hendido, en acero, que se inserta en una ranura para sujetar varias piezas en su lugar. Con frecuencia se utilizan las grapas circulares en los extremos de árboles de mando de tracción delantera para retener las juntas de velocidad constante.

Clamp bolt Another name for a pinch bolt.
Perno de aprieto Otro nombre para perno-grapa.

Clashing Grinding sound heard when gear and shaft speeds are not the same during a gearshift operation.
Entrechoque Rechinamiento que se escucha cuando las velocidades del engranaje y del árbol no son iguales durante el cambio de velocidades.

Class A fires A type of fire in which wood, paper, and other ordinary materials are burning.
Incendio de clase A Incendio en el que se queman la madera, el papel y otros materiales comunes.

Class B fires A type of fire involving flammable liquids, such as gasoline, diesel fuel, paint, grease, oil, and other similar liquids.
Incendio de clase B Incendio en el que se queman líquidos inflamables, como por ejemplo, la gasolina, el diesel, la pintura, la grasa, el aciete, y otros líquidos similares.

Class C fires Electrical fires.
Incendio de clase C Incendios eléctricos.

Class D fires A unique type of fire, for the material burning is a metal. An example of this is a burning "mag" wheel; the magnesium used in the construction of the wheel is a flammable metal and will burn brightly when subjected to high heat.
Incendio de clase D Incendio único, porque la material que se quema es un metal. Ejemplo de esto lo será la quema de una rueda "mag"; el magnesio con el cual se fabrica la rueda es un metal inflamable que resplandece cuando es sometido a altas temperaturas.

Clearance The space allowed between two parts, such as between a journal and a bearing.
Espacio libre Espacio permitido entre dos piezas, por ejemplo, entre un muñón y un cojinete.

Close ratio A relative term for describing the gear ratios in a transmission. If the gears are numerically close, they are said to be close ratio. This design gives quicker acceleration at the expense of initial acceleration and fuel economy.
Relación próxima Término relativo que describe las relaciones de los engranajes en una transmisión. Si existe una proximidad numérica entre los engranajes, se dice que su relación es próxima. Este diseño permite una aceleración más rápida a costa de la aceleración inicial y del rendimiento de combustible.

Clunking A metallic noise most often heard when an automatic transmission is engaged into reverse or drive. It is also often heard when the throttle is applied or released. Clunking is caused by excessive backlash somewhere in the drive line and is felt or heard in the axle.
Sonido sordo Sonido metálico escuchado con más frecuencia cuando se engrana una transmisión automática en marcha atrás o en marcha adelante. Se escucha también cuando se hunde o se suelta el acelerador. El sondio metálico sordo se debe al contragolpe excesivo en alguna parte de la línea de transmisión, y se siente o escucha en el eje.

Cluster assembly A manual transmission related term applied to a group of gears of different sizes machined from one steel casting.
Conjunto desplazable Término relacionado a la transmisión manual que se aplica a un grupo de engranajes de diferentes tamaños hechos a máquina, de una pieza fundida en acero.

Clutch A device for connecting and disconnecting the engine from the transmission or for a similar purpose in other units.
Embrague Dispositivo que sirve para engranar y desengranar el motor de la transmisión o para un propósito parecido en otras unidades.

Clutch control cable A cable assembly with a flexible outer housing anchored at the upper and lower ends. Moving back and forth inside the flexible housing is a braided wire cable that transfers clutch pedal movement to the clutch release lever.
Cable de mando del embrague Conjunto de cable con un alojamiento exterior flexible sujetado a los extremos superior e inferior. Un cable trenzado de alambre que transfiere el movimiento del pedal del embrague a la palanca de desembrague se mueve de atrás para adelante dentro del alojamiento.

Clutch (friction) disc The friction material part of the clutch assembly that fits between the flywheel and pressure plate.
Disco de embrague (fricción) La pieza material de fricción del conjunto de embrague que encaja entre el volante y la placa de presión.

Clutch fork In the clutch, a Y-shaped member into which the throwout bearing is assembled.
Horquilla de embrague En el embrague, una pieza en forma de Y sobre la cual se monta el cojinete de desembrague.

Clutch housing A large aluminum or iron casting that surrounds the clutch assembly. Located between the engine and transmission, it is sometimes referred to as bell housing.
Alojamiento del embrague Pieza grande fundida en hierro o en aluminio que rodea al conjunto del embrague. Ubicado entre el motor y la transmisión, a veces se le llama alojamiento de campana.

Clutch linkage A combination of shafts, levers, or cables that transmits clutch pedal motion to the clutch assembly.
Articulación de embrague Combinación de árboles, palancas o cables que transmite el movimiento del pedal del embrague al conjunto del embrague.

Clutch packs A series of clutch discs and plates installed alternately in a housing to act as a driving or driven unit.
Paquetes del embrague Serie de discos y placas del embrague instalados por turno en un alojamiento para que funcionen como una unidad de accionamiento o accionada.

Clutch pedal A pedal in the driver's compartment that operates the clutch.
Pedal del embrague Pedal que hace funcionar el embrague; ubicado en el compartimiento del conductor.

Clutch pushrod A solid or hollow rod that transfers linear motion between movable parts; that is, the clutch release bearing and release plate.
Varilla de empuje del embrague Varilla sólida o hueca que transfiere un movimiento lineal entre piezas móviles; es decir, el cojinete de desembrague del embrague y la placa de desembrague.

Clutch safety switch See neutral start switch.
Interruptor de seguridad del embrague Véase interruptor de arranque neutro.

Clutch shaft Sometimes known as the transmission input shaft or main drive pinion. The clutch driven disc drives this shaft.
Árbol de embrague A veces llamado árbol impulsor de la transmisión o piñón principal de mando. El disco accionado del embrague acciona este árbol.

Clutch slippage Engine speed increases but increased torque is not transferred through to the driving wheels because of clutch slippage.
Deslizamiento del embrague La velocidad del motor aumenta pero el par de torsión no se transmite a las ruedas motrices a causa del deslizamiento del embrague.

Clutch teeth The locking teeth of a gear.
Dientes del embrague Dientes trabador de un engranaje.

Coefficient of friction The ratio of the force resisting motion between two surfaces in contact with the force holding the two surfaces in contact.

Coeficiente de fricción Relación de la fuerza que resiste el movimiento entre dos superficies en contacto a la fuerza que mantiene el contacto de las dos superficies.

Coil spring A heavy wirelike steel coil used to support the vehicle weight while allowing for suspension motions. On FWD cars, the front coil springs are mounted around the MacPherson struts. On the rear suspension, they may be mounted to the rear axle, to trailing arms, or around rear struts.

Muelle helicoidal Espiral grueso de acero parecido al alambre, que sirve para apoyar el peso del vehículo mientras permite el movimiento de suspensión. En vehículos de tracción delantera, los muelles helicoidales delanteros se suspenden alrededor de los montantes MacPherson. En la suspensión trasera, pueden suspenderse del eje trasero, del los brazos traseros, o alrededor de los montantes traseros.

Coil preload springs Coil springs are made of tempered steel rods formed into a spiral that resist compression; located in the pressure plate assembly.

Muelles helicoidales de carga previa Muelles helicoidales se fabrican de varillas de acero templado configuradas en forma de espiral que resisten la compresión; ubicados en el conjunto de placa de presión.

Coil spring clutch A clutch using coil springs to hold the pressure plate against the friction disc.

Embrague de muelle helicoidal Embrague que emplea muelles helicoidales para mantener la placa de presión contra el disco de fricción.

Companion flange A mounting flange that fixedly attaches a drive shaft to another drive train component.

Brida acompañante Una brida de montaje que fija un árbol de mando a otro componente del tren de mando.

Compound A mixture of two or more ingredients.

Compuesto Mezcla de dos o más ingredientes.

Concentric Two or more circles, having a common center.

Concéntrico Dos o más círculos que tienen un centro común.

Cone clutch The driving and driven parts conically shaped to connect and disconnect power flow. A clutch made from two cones, one fitting inside the other. Friction between the cones forces them to rotate together.

Embrague cónico Piezas de accionamiento y accionadas, en forma cónica, empleadas para conectar y desconectar el regulador de fuerza. Un embrague hecho de dos conos, uno se inserta dentro del otro. La fricción entre los conos les hace girar juntos.

Constant mesh Manual transmission design permits the gears to be constantly enmeshed regardless of vehicle operating circumstances.

Engrane constante Diseño de transmisión manual permite que los engranajes permanezcan siempre engranados a pesar de las condiciones del funcionamiento del vehículo.

Constant mesh transmission A transmission in which the gears are engaged at all times, and shifts are made by sliding collars, clutches, or other means to connect the gears to the output shaft.

Transmisión de engrane constante Transmisión en la que los engranajes están siempre engranados, y los cambios se llevan a cabo a través de chavetas deslizantes, embragues u otros medios para conectar los engranajes al árbol de rendimiento.

Constant velocity joint (also called CV-joint) A flexible coupling between two shafts that permits each shaft to maintain the same driving or driven speed regardless of operating angle, allowing for a smooth transfer of power. The constant velocity joint consists of an inner and outer housing with balls in between, or a tripod and yoke assembly.

Junta de velocidad constante (llamada también junta-CV) Unión flexible entre dos árboles que permite que cada árbol mantenga la misma velocidad de accionamiento o accionada, a pesar del ángulo de funcionamiento, y permite que la transferencia de fuerza sea una suave. La junta de velocidad constante consiste de un alojamiento interior y exterior entre el cual se insertan bolas, o un conjunto de trípode y yugo.

Control arm A suspension component that links the vehicle frame to the steering knuckle or axle housing and acts as a hinge to allow up and down wheel motions. The front control arms are attached to the frame with bushings and bolts and are connected to the steering knuckles with ball joints. The rear control arms attach to the frame with bushings and bolts and are welded or bolted to the rear axle or wheel hubs.

Brazo de mando Componente de suspensión que une el armazón del vehículo al muñón de dirección o al puente trasero y funciona como una bisagra para permitir el movimiento de ascenso y descenso de la rueda. Los brazos de mando delanteros se fijan al armazón con bujes y pernos y se conectan a los muñones de dirección con juntas esféricas. Se fijan los brazos de mando traseros al armazón con bujes y pernos y se sueldan o se empernan al eje trasero o a los cubos de la rueda.

Contraction A reduction in mass or dimension; the opposite of expansion.

Contracción Disminución en masa o dimensión; lo opuesto de expansión.

Corrode To eat away gradually as if by gnawing, especially by chemical action.

Corroerse Carcomer gradualmente como al roer, especialmente debido a una acción química.

Corrosion Chemical action, usually by an acid, that eats away (decomposes) a metal.

Corrosión Acción química, normalmente producida por un ácido, que carcome (descompone) un metal.

Cotter pin A type of fastener, made from soft steel in the form of a split pin, that can be inserted in a drilled hole. The split ends are spread to lock the pin in position.

Pasador de chaveta Tipo de aparato de fijación hecho de acero recocido en forma de pasador hendido que puede insertarse en un agujero barrenado. Se separan los extremos hendidos para fijar la chaveta en la posición correcta.

Counterclockwise rotation Rotating the opposite direction of the hands on a clock.

Rotación a la izquierda Rotación en el sentido inverso a la dirección de las agujas del reloj.

Counter gear assembly A cluster of gears designed on one casting with short shafts supported by antifriction bearings. Closely related to the cluster assembly.

Mecanismo contador Grupo de engranajes diseñados en una sola fundición con árboles cortos apoyados por cojinetes de antifricción. Estrechamente relacionado al conjunto desplazable.

Countershaft An intermediate shaft that receives motion from a main shaft and transmits it to a working part, sometimes called a lay shaft.

Árbol de retorno Árbol intermedio que recibe movimiento de un árbol primario y lo transmite a una pieza móvil; llamado también árbol secundario.

Coupling A connecting means for transferring movement from one part to another; may be mechanical, hydraulic, or electrical.

Acoplamiento Método de conexión para transferir movimiento de una pieza a otra; puede ser mecánico, hidráulico o eléctrico.

Coupling yoke A part of the double Cardan universal joint that connects the two universal joint assemblies.

Yugo de acoplamiento Pieza de la junta doble de cardán que conecta los dos conjuntos de junta universal.

Cover plate A stamped steel cover bolted over the service access to the manual transmission.

Cubreplaca Cubreplaca de acero estampada, empernada sobre el acceso de reparación a la transmisión manual.

Cross member A steel part of the frame structure that transverses the vehicle body to connect the longitudinal frame rails. Cross members can be welded into place or removed from the vehicle.

Traviesa Pieza de acero de la estructura del armazón que atraviesa la carrocería para conectar las barras longitudinales del armazón. Las traviesas se pueden soldar o remover del vehículo.

CV-joint See constant velocity joint.

Junta-CV Véase junta de velocidad constante.

Dead axle An axle that only supports the vehicle and does not transmit power.

Eje portante Eje que sirve sólo para apoyar el vehículo y que no transmite fuerza motriz.

Deflection Bending or movement away from normal due to loading.

Desviación Flexión o movimiento fuera de lo normal debido a la carga.

Degree A unit of measurement equal to 1/360th of a circle.

Grado Unidad de medida equivalente a cada una de las 360 partes de un círculo.

Density Compactness; relative mass of matter in a given volume.

Densidad Compacidad; masa relativa de materia en un volumen dado.

Detent A small depression in a shaft, rail, or rod into which a pawl or ball drops when the shaft, rail, or rod is moved. This provides a locking effect.

Retén Pequeña depresión en un árbol, una barra o una varilla sobre el/la cual cae un trinquete o una bola cuando se mueve el árbol, la barra o la varilla. Esto provee un efecto de blocaje.

Detent mechanism A shifting control designed to hold the manual transmission in the gear range selected.

Mecanismo de detención Control de cambio de velocidades diseñado para mantener la transmisión manual dentro del límite del engranaje elegido.

Diagnosis A systematic study of a machine or machine parts to determine the cause of improper performance or failure.

Diagnosis Estudio sistemático de una máquina o de piezas de un máquina para establecer la causa del mal funcionamiento o falla.

Dial bore gauge A commonly used name for a bore gauge.

Verificador de calibrador con indicador Nombre comúnmente utilizado para calibrador de diámetro interior.

Dial indicator A measuring instrument with the readings indicated on a dial rather than on a thimble as on a micrometer.

Indicador de cuadrante Instrumento de medida que muestra las lecturas en un cuadrante en vez de en un tambor como en el caso de un micrómetro.

Diaphragm spring A circular disc shaped like a cone, with spring tension that allows it to flex forward or backward. Often referred to as a Belleville spring.

Muelle de diafragma Disco circular en forma de cono, con tensión en el muelle que le permite moverse hacia adelante o hacia atrás. Conocido también como muelle de Belleville.

Diaphragm spring clutch A clutch in which a diaphragm spring, rather than a coil spring, applies pressure against the friction disc.

Embrague de muelle de diafragma Embrague en el que un muelle de diafragma, en vez de un muelle helicoidal, ejerce presión contra el disco de fricción.

Differential A mechanism between drive axles that permits one wheel to run at a different speed than the other while turning.

Diferencial Mecansimo entre los ejes de mando que permite que una rueda gire a una velocidad diferente que la otra.

Differential action An operational situation where one driving wheel rotates at a slower speed than the opposite driving wheel.

Acción del diferencial Situación de funcionamiento donde una rueda motriz gira más despacio que la rueda motriz opuesta.

Differential case The metal unit that encases the differential side gears and pinion gears, and to which the ring gear is attached.

Caja del diferencial Unidad metálica que reviste los engranajes laterales del diferencial y los engranajes de piñón, y sobre la cual se monta la corona.

Differential case spread Another name for preload.

Extensión de la caja del diferencial Otro nombre para carga previa.

Differential drive gear A large circular helical gear driven by the transaxle pinion gear and shaft and drives the differential assembly.

Engranaje del diferencial Engranaje helicoidal circular grande accionado por el engranaje de piñón del transeje y el árbol, que acciona el conjunto del diferencial.

Differential housing Cast-iron assembly that houses the differential unit and the drive axles. This is also called the rear axle housing.

Alojamiento del diferencial Conjunto de hierro fundido que aloja a la unidad del diferencial y a los ejes de mando. Llamado también puente trasero.

Differential pinion gears Small beveled gears located on the differential pinion shaft.

Engranajes de piñón del diferencial Pequeños engranajes biselados ubicados en el árbol de piñón del diferencial.

Differential pinion shaft A short shaft locked to the differential case. This shaft supports the differential pinion gears.

Árbol de piñón del diferencial Árbol corto fijado a la caja del diferencial. Este árbol apoya los engranajes de piñón del diferencial.

Differential ring gear A large circular hypoid-type gear enmeshed with the hypoid drive pinion gear.

Corona del diferencial Engranaje hipoide circular grande engranado con el engranaje de piñón hipoide.

Differential side gears The gears inside the differential case that are internally splined to the axle shafts, and are driven by the differential pinion gears.

Engranajes laterales del diferencial Engranajes dentro de la caja del diferencial que son ranurados internamente a los árboles motores, y accionados por los engranajes de piñón del diferencial.

Dimpling Brinelling or the presence of indentations in a normally smooth surface.

Abolladura Acción de Brinell o la formación de hendiduras en una superficie normalmente lisa.

Direct drive One turn of the input driving member compared to one complete turn of the driven member, such as when there is direct engagement between the engine and drive shaft where the engine crankshaft and the drive shaft turn at the same rpm.

Toma directa Una vuelta de la pieza de accionamiento comparada a una vuelta completa de la pieza accionada, como por ejemplo, cuando hay un engrane directo entre el motor y el árbol de mando donde el cigüeñal del motor y el árbol de mando giran a las mismas rpm.

Disengage When the operator moves the clutch pedal toward the floor to disconnect the driven clutch disc from the driving flywheel and pressure plate assembly.

Desengranar Cuando el conductor hunde el pedal del embrague para desconectar el disco de embrague accionado del volante motor y del conjunto de la placa de presión.

Distortion A warpage or change in form from the original shape.

Deformación Abarquillamiento o cambio en la forma original de la configuración original.

Dog tooth A series of gear teeth that are part of the dog clutching action in a transmission synchronizer operation. The locking teeth of a gear.

Diente de sierra Serie de dientes del engranaje que forman parte de la acción del embrague de garras durante una sincronización de transmisión. Dientes de cierre de un engranaje.

Double Cardan universal joint A near constant velocity universal joint that consists of two Cardan universal joints connected by a coupling yoke.

Junta doble de cardán Junta universal de velocidad casi constante compuesta de dos juntas de cardán conectadas por un yugo de acoplamiento.

Double-offset constant velocity joint Another name for the type of plunging, inner CV joint found on many GM, Ford, and Japanese FWD cars.

Junta de velocidad constante de desviación doble Otro nombre para el tipo de junta de velocidad constante interior de pistón tubular, instalada en muchos automóviles de tracción delantera de la GM, de la Ford y japoneses.

Double reduction axle A drive axle construction in which two sets of reduction gears are used for extreme reduction of the gear ratio.

Eje de reducción doble Eje de mando en el que se utilizan dos juegos de reductores para lograr una mayor reducción de la relación de engranajes.

Dowel A metal pin attached to one object that when inserted into a hole in another object, ensures proper alignment.

Pasador Chaveta metálica fijada a un objeto que, cuando se inserta dentro de un agujero en otro objeto, asegura una alineación correcta.

Dowel pin A pin inserted in matching holes in two parts to maintain those parts in fixed relation one to another.

Espiga de madera Chaveta insertada en agujeros parejos en dos piezas para mantener dichas piezas en una relación fija entre sí.

Downshift To shift a transmission into a lower gear.

Cambio de alta a baja velocidad Cambiar la transmisión a un engranaje de menos velocidad.

Drag link A connecting rod or link between the steering gear, Pitman arm, and the steering linkage.

Varilla de arrastre Biela o unión entre el mecanismo de dirección, el brazo pitman y el cuadrilátero de la dirección.

Drive line The universal joints, drive shaft, and other parts connecting the transmission with the driving axles.

Línea de transmisión Las juntas universales, el árbol de mando y otras piezas que conectan la transmisión a los ejes motores.

Drive line torque Relates to rear wheel drive line and is the transfer of torque between the transmission and the driving axle assembly.

Par de torsión de la línea de transmisión Relacionado a la línea de transmisión de las ruedas traseras y es la transferencia del par de torsión entre la transmisión y el conjunto del eje motor.

Drive line wrap-up A condition where axles, gears, U-joints, and other components can bind or fail if the 4WD mode is used on pavement where 2WD is more suitable.

Falla de la línea de transmisión Condición que ocurre cuando los ejes, los engranajes, las juntas universales, y otros componentes se traban o fallan si se emplea la tracción a las cuatro ruedas sobre un pavimento donde la tracción a las dos ruedas es más adecuada.

Drive pinion The gear that takes its power directly from the drive shaft or transmission and drives the ring gear.

Piñón de mando Engranajes que obtienen su fuerza motriz directamente del árbol de mando o de la transmisión y que accionan la corona.

Drive pinion flange A rim used to connect the rear of the drive shaft to the rear axle drive pinion.

Brida de piñón de mando Corona utilizada para conectar la parte trasera del árbol de mando al piñón de mando del eje trasero.

Drive pinion gear One of the two main driving gears located within the transaxle or rear driving axle housing. Together the two gears multiply engine torque.

Engranaje del piñón de mando Uno de los dos mecanismos de accionamiento principales ubicados dentro del transeje o el puente trasero. Los dos engranajes actúan juntos para multiplicar el par de torsión del motor.

Drive shaft An assembly of one or two universal joints connected to a shaft or tube; used to transmit power from the transmission to the differential. Also called the propeller shaft.

Árbol de mando Conjunto de una o dos juntas universales conectadas a un árbol o tubo; utilizado para transmitir la fuerza motriz desde la transmisión hasta el diferencial. Llamado también árbol transmisor.

Drive shaft installation angle The angle the drive shaft is mounted off the true horizontal line measured in degrees.

Ángulo de montaje del árbol de mando El ángulo al que se monta el árbol de mando fuera de la línea horizontal verdadera, medido en grados.

Driven disc The part of the clutch assembly that receives driving motion from the flywheel and pressure plate assemblies.

Disco accionado Pieza del conjunto del embrague que recibe su fuerza motriz del conjunto del volante y del conjunto de la placa de presión.

Driven gear The gear meshed directly with the driving gear to provide torque multiplication, reduction, or a change of direction.

Engranaje accionado Engranaje engranado directamente al mecanimo de accionamiento para proveer multiplicación de par de torsión, reducción o un cambio de dirección.

Driving axle A term related collectively to the rear driving axle assembly where the drive pinion, ring gear, and differential assembly are located within the driving axle housing.

Eje motor Término relacionado colectivamente al conjunto del eje motor trasero donde el piñón de mando, la corona, y el conjunto del diferencial están ubicados dentro del puente trasero.

Drop forging A piece of steel shaped between dies while hot.
Estampado Pieza de acero conformada entre troqueles mientras está caliente.

Dry-disc clutch A clutch in which the friction faces of the friction disc are dry, as opposed to a wet-disc clutch, which runs submerged in oil. The conventional type of automobile clutch.
Embrague de disco seco Embrague en el que las placas de fricción del disco de fricción están secas, lo opuesto de un disco mojado, que funciona sumergido en aceite. Tipo convencional de embrague de automóviles.

Dry friction The friction between two dry solids.
Fricción seca Fricción entre dos sólidos secos.

Dual reduction axle A drive axle construction with two sets of pinions and gears, either of which can be used.
Eje de reducción doble Conjunto de eje de mando que tiene dos juegos de piñones y engranajes. Se puede utilizar cualquiera de los dos.

Dummy shaft A shaft, shorter than the countershaft, used during disassembly and reassembly in place of the countershaft.
Árbol falso Árbol, más corto que el árbol de retorno, empleado durante el desmontaje y el remonte en vez del árbol de retorno.

Dynamic In motion.
Dinámico En movimiento.

Dynamic balance The balance of an object when it is in motion; for example, the dynamic balance of a rotating drive shaft.
Equilibrio dinámico Equilibrio de un objeto cuando está en movimiento; por ejemplo, el equilibrio dinámico de un árbol de mando giratorio.

Eccentric One circle within another circle wherein both circles do not have the same center or a circle mounted off center. On FWD cars, front-end camber adjustments are accomplished by turning an eccentric cam bolt that mounts the strut to the steering knuckle.
Excéntrico Un círculo dentro de otro donde los dos círculos no comparten el mismo centro o un círculo colocado fuera del centro. En automóviles de tracción delantera, se realizan ajustes a la combadura del tren delantero girando un perno de leva excéntrica que levanta el montante al muñón de dirección.

Eccentric washer A normal looking washer with its hole not in its center. The hole is offset from the center.
Arandela excéntrica Arandela de apariencia normal pero cuyo agujero se encuentra fuera del centro. El agujero se desvía del centro.

Efficiency The ratio between the power of an effect and the power expended to produce the effect; the ratio between an actual result and the theoretically possible result.
Rendimiento La relación entre la fuerza de un efecto y la fuerza rendida para producir tal efecto; la relación entre un resultado verdadero y un resultado teóricamente posible.

Elastomer Any rubber-like plastic or synthetic material used to make bellows, bushings, and seals.
Elastómero Cualquier material plástico o sintético parecido al caucho que se utiliza para fabricar fuelles, bujes y juntas de estanqueidad.

Emulsification When water droplets mix with grease resulting in a thicker solution than normal grease.
Emulsión Pequeñas gotas de agua que cuando se mezclan con grasa, forman un compuesto más grueso que la grasa normal.

Endplay The amount of axial or end-to-end movement in a shaft due to clearance in the bearings.
Holgadura Amplitud de movimiento axial o movimiento de extremo a extremo en un árbol debido al espacio libre en los cojinetes.

Engage When the vehicle operator moves the clutch pedal up from the floor, this engages the driving flywheel and pressure plate to rotate and drive the driven disc.
Engranar Cuando el conductor del vehículo suelta el pedal del embrague, el volante y la placa de presión se engranan para girar y accionar el disco accionado.

Engagement chatter A shaking, shuddering action that takes place as the driven disc makes contact with the driving members. Chatter is caused by a rapid grip and slip action.
Vibración de acoplamiento Movimiento de agitación y estremecimiento que ocurre cuando el disco accionado entra en contacto con las piezas de accionamiento. La causa de esta vibración es un movimiento rápido de garra y de deslizamiento.

Engine The source of power for most vehicles. It converts burned fuel energy into mechanical force.
Motor Fuente de fuerza para la mayoría de los vehículos. Convierte el combustible quemado en fuerza mecánica.

Engine torque A turning or twisting action developed by the engine measured in foot-pounds or kilogram meters.
Esfuerzo de rotación del motor Movimiento de giro o torcedura que produce el motor, medido en libras-pies o en kilográmetros.

English measuring system The USCS measuring system.
Sistema Imperial Británico Sistema de medida de USCS [Sistema Usual U.S.].

Extension housing An aluminum or iron casting of various lengths that encloses the transmission output shaft and supporting bearings.
Alojamiento de extensión Pieza fundida en aluminio o en hierro de longitudes variadas que encubre el árbol de rendimiento de la transmisión y los cojinetes soporte.

External cone clutch The external surface of one part has a tapered surface to mate with an internally tapered surface to form a cone clutch.
Embrague cónico externo La superficie externa de una pieza tiene una superficie cónica para hacer juego con una superficie internamente cónica y así formar un embrague de cono.

External gear A gear with teeth across the outside surface.
Engranaje externo Engranaje que tiene dientes a través de la superficie exterior.

Externally tabbed clutch plates Clutch plates are designed with tabs around the outside periphery to fit into grooves in a housing or drum.
Placas de embrague con orejetas externas Se diseñan placas de embrague con orejetas alrededor de la periferia exterior para que puedan ajustarse a las acanaladuras de un alojamiento o de un tambor.

Extreme pressure lubricant A special lubricant for use in hypoid gear differentials; needed because of the heavy wiping loads imposed on the gear teeth.
Lubricante para presión extrema Lubricante especial que se utiliza en diferenciales de engranaje hipoide; necesario a causa del intenso esfuerzo al que los dientes de los engranajes están sometidos.

Face The front surface of an object.
Frente Superficie frontal de un objeto.

Fatigue The buildup of natural stress forces in a metal part that eventually causes it to break. Stress results from bending and loading the material.
Fatiga Acumulación de tensiones naturales en una pieza metálica que finalmente ocasiona una ruptura. La tensión es una consecuencia de la flexión y de la carga a las cuales el material está expuesto.

Feeler gauge A metal strip or blade finished accurately with regard to thickness used for measuring the clearance between two parts; such gauges ordinarily come in a set of different blades graduated in thickness by increments of 0.001 inch.

Calibrador de espesores Lámina metálica o cuchilla acabada con precisión de acuerdo al espesor que se utiliza para medir el espacio libre entre dos piezas. Dichos calibradores normalmente están disponibles en juegos de cuchillas diferentes graduadas según el espesor, en incrementos de 0,001 pulgadas.

Fiber composites A mixture of metallic threads along with a resin form a composite offering weight and cost reduction, long-term durability, and fatigue life. Fiberglass is a fiber composite.

Compuestos de fibra Una mezcla de hilos metálicos y resina forman un compuesto que ofrece una disminución de peso y costo, mayor durabilidad y resistencia a la fatiga. La fibra de vidrio es un compuesto de fibra.

Final drive gears Main driving gears located in the axle area of the transaxle housing.

Engranajes de la transmisión final Mecanismos de accionamiento principales ubicados en la región del eje del alojamiento del transeje.

Final drive ratio The ratio between the drive pinion and ring gear.

Relación de la transmisión final Relación entre el piñón de mando y la corona.

First gear A small diameter driving helical- or spur-type gear located on the cluster gear assembly. First gear provides torque multiplication to get the vehicle moving.

Engranaje de primera velocidad Mecanismo de mando helicoide o recto de diámetro pequeño ubicado en el conjunto del tren desplazable. El engranaje de primera velocidad inicia la multiplicación de par de torsión para impulsar el vehículo.

Fit The contact between two machined surfaces.

Conexión Contacto entre dos superficies maquinadas.

Fixed-type constant velocity joint A joint that cannot telescope or plunge to compensate for suspension travel. Fixed joints are always found on the outer ends of the drive shafts of FWD cars. A fixed joint may be of either Rzeppa or tripod type.

Junta de velocidad constante de tipo fijo Junta que no puede extenderse o hundirse para compensar el movimiento de la suspensión. Siempre se encuentran juntas fijas en los extremos exteriores de los árboles de mando de vehículos de tracción delantera. Una junta fija puede ser de tipo Rzeppa o trípode.

Flange A projecting rim or collar on an object for keeping it in place.

Brida Cerco proyectado o collar en un objeto que lo mantiene en su lugar.

Flange yoke The part of the rear universal joint attached to the drive pinion.

Yugo de brida Pieza de la junta universal trasera fijada al piñón de mando.

Flexplate A lightweight flywheel used only on engines equipped with an automatic transmission. The flexplate is equipped with a starter ring gear around its outside diameter and also serves as the attachment point for the torque converter.

Placa flexible Volante liviano empleado solamente en motores equipados con transmisión automática. La placa flexible está equipada de una corona de arranque alrededor de su diámetro exterior y sirve también de punto de fijación para el convertidor del par motor.

Fluid coupling A device in the powertrain consisting of two rotating members; transmits power from the engine, through a fluid, to the transmission.

Acoplamiento fluido Mecanismo en el tren transmisor de potencia que consiste de dos piezas giratorias; transmite la fuerza desde el motor, por medio de un fluido, hasta la transmisión.

Fluid drive A drive in which there is no mechanical connection between the input and output shafts, and power is transmitted by moving oil.

Transmisión hidráulica Transmisión en la que no existe conexión mecánica alguna entre el árbol impulsor y el árbol de rendimiento. La fuerza se transmite a través del aceite motor.

Flywheel A heavy metal wheel that is attached to the crankshaft and rotates with it; helps smooth out the power surges from the engine power strokes; also serves as part of the clutch and engine-cranking system.

Volante Rueda pesada de metal que se fija al cigüeñal y que gira con él; ayuda a neutralizar las sacudidas de fuerza de las carreras motrices del motor; sirve también como parte del sistema de embrague y de arranque del motor.

Flywheel ring gear A gear, fitted around the flywheel, that is engaged by teeth on the starting-motor drive to crank the engine.

Corona del volante Engranaje ajustado alrededor del volante, engranado por dientes en el mando del motor de arranque para hacer arrancar el motor.

Foot-Pound (or ft.-lb.) This is a measure of the amount of energy or work required to lift 1 pound a distance of 1 foot.

Libra-pie Medida de la cantidad de energía o fuerza que se requiere para levantar una libra a una distancia de un pie.

Force Any push or pull exerted on an object; measured in pounds and ounces, or in newtons (N) in the metric system.

Fuerza Cualquier empuje o tirón que se ejerce sobre un objeto; medido en libras y onzas, o en newtons (N) en el sistema métrico.

Forward coast side The side of the ring gear tooth the drive pinion contacts when the vehicle is decelerating.

Cara de cabotaje delantera Cara del diente de la corona con el cual el piñón de mando entra en contacto mientras el vehículo deacelera.

Forward drive side The side of the ring gear tooth that the drive pinion contacts when accelerating or on the drive.

Cara de mando delantero Cara del diente de la corona con el cual el piñón de mando entra en contacto mientras acelera o está en marcha.

Four wheel drive (4WD) On a vehicle, driving axles at both front and rear, so that all four wheels can be driven. 4WD is the standard abbreviation for four wheel drive.

Tracción a las cuatro ruedas En un vehículo, los ejes motores se encuentran ubicados en las partes delantera y trasera, para que las cuatro ruedas se puedan accionar. 4WD es la abreviatura común para tracción a las cuatro ruedas.

Four wheel high A transfer case shift position where both front and rear drive shafts receive power and rotate at the speed of the transmission output shaft.

Alto de cuatro ruedas Posición en la caja de cambios donde los árboles de mando delantero y trasero reciben fuerza y giran a la mimsa velocidad que el árbol de rendimiento de la transmisión.

Frame The main understructure of the vehicle to which everything else is attached. Most FWD cars have only a subframe for the front suspension and drive train. The body serves as the frame for the rear suspension.

Armazón Chasis principal del vehículo al cual se fijan todas las demás piezas. La mayoría de los vehículos de tracción delantera sólo tiene un chasis que soporta la suspensión delantera y el tren de mando. La carrocería sirve como armazón para la suspensión trasera.

Free-wheeling clutch A mechanical device that will engage the driving member to impart motion to a driven member in one direction but not the other. Also known as an "overrunning clutch."

Embrague de marcha en rueda libre Mecanismo que engranará a la pieza de accionamiento para impulsar movimiento a una pieza accionada en una dirección pero no en la otra. Conocido también como "embrague de giro libre".

Freon A term commonly used for R-12.

Freón Térmimo comúnmente utilizado para R-12.

Friction The resistance to motion between two bodies in contact with each other.

Fricción Resistencia al movimiento entre dos cuerpos en contacto el uno con el otro.

Friction bearing A bearing in which there is sliding contact between the moving surfaces. Sleeve bearings, such as those used in connecting rods, are friction bearings.

Cojinete de fricción Cojinete en el cual existe un contacto deslizante entre las superficies en movimiento. Los cojinetes de manguito, como los que se utilizan en las bielas, son cojinetes de fricción.

Friction disc In the clutch a flat disc, faced on both sides with friction material and splined to the clutch shaft. It is positioned between the clutch pressure plate and the engine flywheel. Also called the clutch disc or driven disc.

Disco de fricción Disco plano del embrague, revestido en las dos caras con material de fricción y ranurado al árbol de embrague. Está ubicado entre la placa de presión del embrague y el volante de la máquina. Llamado también disco de embrague o disco accionado.

Friction facings A hard-molded or woven asbestos or paper material that is riveted or bonded to the clutch driven disc.

Revestimiento de fricción Material de moldeado duro, de asbesto tejido o de papel que es remachado o adherido al disco accionado del embrague.

Front bearing retainer An iron or aluminum circular casting fastened to the front of a transmission housing to retain the front transmission bearing assembly.

Retenedor del cojinete de rueda delantero Pieza circular fundida en hierro o en aluminio fijada a la parte delantera de un alojamiento de la transmisión para sujetar el conjunto del cojinete de transmisión delantero.

Front differential/axle assembly Like a conventional rear axle but having steerable wheels.

Conjunto de diferencial/eje delantero Igual que el puente trasero convencional, pero con ruedas orientables.

Front wheel drive (FWD) The vehicle has all drive train components located at the front.

Tracción delantera Todos los componentes del tren de mando en el vehículo se encuentran en la parte delantera.

Fulcrum rings A circular ring over which the pressure plate diaphragm spring pivots.

Anillos de fulcro Anillo circular sobre el cual gira el muelle del diafragma de la placa de presión.

Full-floating rear axle An axle that only transmits driving force to the rear wheels. The weight of the vehicle (including payload) is supported by the axle housing.

Eje trasero enteramente flotante Eje que solamente transmite la fuerza motriz a las ruedas traseras. El puente trasero soporta el peso del vehículo (incluyendo la carga útil).

Fully synchronized In a manual transmission, the synchronizer assembly operates to improve the shift quality in all forward gears.

Enteramente sincronizado En una transmisión manual, el conjunto sincronizador funciona para mejorar la calidad del desplazamiento en todos los engranajes delanteros.

Galling Wear caused by metal-to-metal contact in the absence of adequate lubrication. Metal is transferred from one surface to the other, leaving behind a pitted or scaled appearance.

Corrosión por rozamiento Desgaste causado por el contacto de un metal con otro metal debido a la ausencia de lubrificación adecuada. Se transfiere el metal de una superficie a la otra, lo cual deja un aspecto corroído o raspado.

Galvanic corrosion A type of corrosion that occurs when two dissimilar metals, such as magnesium and steel, are in contact with each other.

Corrosión galvánica Tipo de corrosión producida cuando dos metales distintos, como por ejemplo, el magnesio y el acero, entran en contacto el uno con el otro.

Gasket A layer of material, usually made of cork, paper, plastic, composition, or metal, or a combination of these, placed between two parts to make a tight seal.

Guarnición Capa de un material, normalmente hecho de corcho, papel, plástico, pasta o metal, o una combinación de éstos, ubicada entre dos piezas para crear una junta de estanqueidad hermética.

Gasket cement A liquid adhesive material, or sealer, used to install gaskets.

Cemento de guarnición Material líquido adhesivo, o de juntura, utilizado para instalar guarniciones.

Gear A wheel with external or internal teeth that serves to transmit or change motion.

Engranaje Rueda con dientes externos o internos que sirve para transmitir o cambiar movimiento.

Gear clash The noise that results when two gears are traveling at different speeds and are forced together.

Choque de engranajes Ruido producido cuando dos engranajes giran a velocidades diferentes y se unen por fuerza.

Gear lubricant A type of grease or oil blended especially to lubricate gears.

Lubrificante de engranaje Tipo de grasa o aceite mezclado especialmente para lubrificar engranajes.

Gear noise The howling or whining of the ring gear and pinion due to an improperly set gear pattern, gear damage, or improper bearing preload.

Ruido del engranaje Aullido o silbido de la corona y del piñón debido al montaje incorrecto de los engranajes, a averías en los engranajes o a carga previa incorrecta del cojinete.

Gear ratio The number of revolutions of a driving gear required to turn a driven gear through one complete revolution. For a pair of gears, the ratio is found by dividing the number of teeth on the driven gear by the number of teeth on the driving gear.

Relación de engranajes Número de revoluciones de un mecanismo de accionamiento requiridas para hacer girar un engranaje accionado una revolución completa. Para un par de engranajes, se obtiene la relación al dividir el número de dientes en el engranaje accionado por el número de dientes en el mecanismo de accionamiento.

Gear rattle A repetitive metallic impact or rapping noise that occurs when the vehicle is lugging in gear. The intensity of the noise increases with operating temperature and engine torque, and decreases with increasing vehicle speed.

Estruendo del engranaje Impacto metálico repetitivo o golpeteo que ocurre cuando el vehículo arrastra los engranajes durante la marcha. La intensidad del ruido aumenta con alta temperatura de funcionamiento y del par de torsión del motor, y disminuye al aumentar la velocidad del vehículo.

Gear reduction When a small gear drives a large gear, there is an output speed reduction and a torque increase that results in a gear reduction.

Reducción de engranajes Cuando un engranaje pequeño acciona un engranaje grande, se produce una reducción de velocidad de rendimiento y un aumento de par de torsión que resulta en una reducción de engranajes.

Gear whine A high-pitched sound developed by some types of meshing gears.

Silbido del engranaje Sonido agudo producido por algunos tipos de engranajes.

Gearshift A linkage-type mechanism by which the gears in an automobile transmission are engaged and disengaged.

Cambio de velocidades Mecanismo de tipo empalme a través del cual los engranajes en la transmisión de un vehículo se engranan y se desengranan.

Graphite Very fine carbon dust with a slippery texture used as a lubricant.

Grafito Polvo muy fino de carbón con una textura desbaladiza que se utiliza como lubricante.

Grind To finish or polish a surface by means of an abrasive wheel.

Esmerilar Acabar o pulir una superficie con una rueda abrasiva.

Half-shaft Either of the two drive shafts that connect the transaxle to the wheel hubs in FWD cars. Half-shafts have constant velocity joints attached to each end to allow for suspension motions and steering. The shafts may be of solid or tubular steel and may be of different lengths.

Semieje Cualquiera de los dos árboles de mando que conecta el transeje a los cubos de rueda en automóviles de tracción delantera. Los semiejes tienen juntas de velocidad constante fijadas a cada extremo para permitir el movimiento de suspensión y la dirección. Los semiejes pueden ser de acero sólido o tubular y sus longitudes pueden variar.

Harshness A bumpy ride caused by a stiff suspension. Can be often cured by installing softer springs or shock absorbers.

Aspereza Viaje de muchas sacudidas ocacionadas por una suspensión rígida. Puede remediarse con la instalación de muelles más flexbiles o amortigadores.

Heat treatment Heating, followed by fast cooling, to harden metal.

Tratamiento térmico Calentamiento, seguido del enfriamiento rápido, para endurecer el metal.

Heel The outside, larger half of the gear tooth.

Talón Mitad exterior más grande del diente de engranaje.

Helical Shapes like a coil spring or a screw thread.

Helicoidal Formas parecidas a un muelle helicoidal o a un filete de tornillo.

Helical gear Gears with the teeth cut at an angle to the axis of the gear.

Engranaje helicoidal Engranajes que tienen los dientes cortados a un ángulo del pivote del engranaje.

Herringbone gear A pair of helical gears designed to operate together. The angle of the pair of gears forms a V.

Engranaje bihelocoidal Par do ongranajes helicoidales diseñados para funcionar juntos. El ángulo del par de engranajes forma una V.

High pedal A clutch pedal that has an excessive amount of pedal travel.

Pedal alto Pedal del embrague con exceso de avance.

Horsepower A measure of mechanical power, or the rate at which work is done. One horsepower equals 33,000 ft.-lb. (foot-pounds) of work per minute. It is the power necessary to raise 33,000 pounds a distance of 1 foot in 1 minute.

Potencia en caballos Medida de fuerza mecánica, o velocidad a la que se realiza el trabajo. Un caballo de fuerza es equivalente a 33.000 libras-pies de trabajo por minuto. Es el esfuerzo necesario para levantar 33.000 libras a la distancia de un pie en un minuto.

Hot spots The small areas on a friction surface that are a different color, normally blue, or are harder than the rest of the surface.

Zonas de calor Zonas pequeñas sobre una superficie de rozamiento de un color diferente, normalmente azul, o más duras que el resto de la superficie.

Hotchkiss drive A type of rear suspension in which leaf springs absorb the rear axle housing torque.

Transmisión Hotchkiss Tipo de suspensión trasera en la cual muelles de láminas absorben el par de torsión del puente trasero.

Hub The center part of a wheel, to which the wheel is attached.

Cubo Parte central de una rueda, a la cual se fija la rueda.

Hydraulic clutch A clutch that is actuated by hydraulic pressure; used in cars and trucks when the engine is some distance from the driver's compartment so that it would be difficult to use mechanical linkages.

Embrague hidráulico Embrague accionado por presión hidráulica; utilizado en automóviles y camiones cuando el motor está lejos del compartimiento del conductor para dificultar la utilización de bielas motrices mecánicas.

Hydraulic fluid reservoir A part of a master cylinder assembly that holds reserve fluid.

Despósito de fluido hidráulico Parte del conjunto del cilindro primario que contiene el fluido de reserva.

Hydraulic press A piece of shop equipment that develops a heavy force by use of a hydraulic piston-and-jack assembly.

Prensa hidráulica Pieza del equipo de taller que desarrolla fuerza pesada por medio de un conjunto de gato de pistón hidráulico.

Hydraulic pressure Pressure exerted through the medium of a liquid.

Presión hidráulica Presión ejercida a través de un líquido.

Hydroscopic The property of a fluid in which it has a tendency to absorb moisture from the atmosphere.

Hidroscópico Cualidad que posee un fluido para absorber la humedad de la atmósfera.

Hypoid gear A gear that is similar in appearance to a spiral bevel gear, but the teeth are cut so that the gears match in a position where the shaft center lines do not meet; cut in a spiral form to allow the pinion to be set below the center line of the ring gear so that the car floor can be lower.

Engranaje hipoide Engranaje parecido a un engranaje cónico con dentado espiral, pero en el cual los dientes se cortan para que los engranajes se engranen en una posición donde las líneas centrales del árbol no se crucen; cortado en forma de espiral para permitir que el piñon sea colocado debajo de la línea central de la corona y que así el piso del vehículo sea más bajo.

Hypoid gear lubricant An extreme pressure lubricant designed for the severe operation of hypoid gears.

Lubricante del engranaje hipoide Lubricante de extrema presión diseñado para el funcionamiento riguroso de los engranajes hipoides.

ID Inside diameter.

ID Diámetro interior.

Idle Engine speed when the accelerator pedal is fully released and there is no load on the engine.

Marcha mínima Velocidad del motor cuando el pedal del acelerador se suelta completamente y no hay ninguna carga en el motor.

Impeller The pump or driving member in a torque converter.

Impulsor Bomba o mecanismo de accionamiento en un convertidor de torsión.

Inboard constant velocity joint The inner constant velocity joint, or the one closest to the transaxle. The inboard joint is usually a plunging-type joint that telescopes to compensate for suspension motions.

Junta de velocidad constante del interior Junta de velocidad constante interior, o la que está más cerca del transeje. La junta del interior es normalmente una junta de tipo sumergible que se extiende para compensar el movimiento de la suspensión.

Inch 25.4 mm, 2.54 cm, or 0.0254 meters

Pulgada 25,4 milímetros, 2,54 centímetros, o 0,0254 metros

Inclinometer Device designed with a spirit level and graduated scale to measure the inclination of a drive line assembly. The inclinometer connects to the drive shaft magnetically.

Inclinómetro Instrumento diseñado con nivel de burbuja de aire y escala graduada para medir la inclinación de un conjunto de la línea de transmisión. El inclinómetro se conecta magnéticamente al árbol de mando.

Increments Series of regular additions from small to large.

Incrementos Serie de aumentos regulares de lo pequeño a lo grande.

Independent rear suspension (IRS) The vehicle's rear wheels move up and down independently of each other.

Suspensión trasera independiente Las ruedas traseras del vehículo realizan un movimiento de ascenso y descenso de manera independiente la una de la otra.

Independent suspension A suspension system that allows one wheel to move up and down without affecting the opposite wheel. Provides superior handling and a smoother ride.

Suspensión independiente Sistema de suspensión que permite que una rueda realice un movimiento de ascenso y descenso sin afectar la rueda opuesta. Permite un manejo excelente y un viaje mucho más cómodo.

Index To orient two parts by marking them. During reassembly the parts are arranged so the index marks are next to each other. Used to preserve the orientation between balanced parts.

Alinear Orientar dos piezas marcándolas. Durante el remonte, se arreglan las piezas de modo que las indicaciones de alineación queden juntas. Utilizado para mantener la orientación entre piezas equilibradas.

Inner bearing race The inner part of a bearing assembly on which the rolling elements, ball or roller, rotate.

Anillo de cojinete interior Parte interior de un conjunto de cojinete sobre la cual los elementos rodantes, o sea, las bolas o los rodillos, giran.

Input shaft The shaft carrying the driving gear by which the power is applied, as to the transmission.

Árbol impulsor Árbol que soporta el mecanismo de accionamiento a través del cual se aplica la fuerza motriz; por ejemplo, a la transmisión.

Inserts One of several terms that could apply to the shift plates found in a synchronizer assembly.

Piezas insertas Uno de los varios téminos que puede aplicarse a las placas de cambio de velocidades encontradas en un conjunto sincronizador.

Insert springs Round wire springs that hold the inserts or shift plates in contact with the synchronizer sleeve. Located around the synchronizer hub.

Muelles insertos Muelles redondos de alambre que mantienen el contacto entre las piezas insertas o placas de cambio de velocidades y el manguito sincronizador. Ubicados alrededor del cubo sincronizador.

Inspection cover A removable cover that permits entrance for inspection and service work.

Cubierta de inspección Cubierta desmontable que permite la entrada para la inspección y la reparación.

Integral Built into, as part of the whole.

Integral Pieza incorporada, como parte del todo.

Integral axle housing A rear axle housing-type where the parts are serviced through an inspection cover and adjusted within and relative to the axle housing.

Puente trasero integral Tipo de puente trasero, donde se reparan las piezas a través de una cubierta de inspección y se ajustan dentro del y con relación al puente trasero.

Interference fit A press fit; for example, the inside diameter of a bore is 0.001 inch smaller than the outside diameter of a shaft, so when the shaft is fitted into the bore it must be pressed in to overcome the 0.001 inch interference fit.

Ajuste a interferencia Ajuste en prensa; por ejemplo, el diámetro interior de un calibre es un 0,001 de pulgada más pequeño que el diámetro exterior de un árbol, así que, cuando el árbol se inserta dentro del calibre, tiene que ser comprimido para compensar el ajuste a interferencia de un 0,001 de pulgada.

Interlock mechanism A mechanism in the transmission shift linkage that prevents the selection of two gears at one time.

Mecanismo de enganche Mecanismo en la biela motriz del cambio de velocidades de la transmisión que impide la selección de dos velocidades a la vez.

Intermediate bearing plate Another name for the center support plate of a transmission.

Placa intermedia del cojinete Otro nombre para la placa central de soporte de una transmisión.

Intermediate drive shaft Located between the left and right drive shafts, it equalizes drive shaft length.

Árbol de mando intermedio Ubicado entre los árboles de mando izquierdo y derecho, compensa la longitud del árbol de mando.

Intermediate plate A mechanism in the transmission shift linkage that prevents the selection of two gears at one time.

Placa intermedia Mecanismo en la biela motriz del cambio de velocidades de la transmisión que impide la selección de dos velocidades a la vez.

Internal gear A gear with teeth pointing inward, toward the hollow center of the gear.

Engranaje interno Engranaje con los dientes orientados hacia adentro, hacia el centro hueco del engranaje.

Jam nut A second nut tightened against a primary nut to prevent it from working loose. Used on inner and outer tie-rod adjustment nuts and on many pinion-bearing adjustment nuts.

Contratuerca Segunda tuerca apretada contra una tuerca principal para evitar que ésta se suelte. Utilizada en las tuercas de las barras de acoplamiento interiores y exteriores y en muchas tuercas de ajuste del cojinete del piñón.

Joint angle The angle formed by the input and output shafts of constant velocity joints. Outer joints can typically operate at angles up to 45 degrees, whereas inner joints have more restricted angles.
Ángulo de las juntas Ángulo formado por el árbol impulsor y el árbol de rendimiento de las juntas de velocidad constante. Las juntas exteriores pueden funcionar típicamente en ángulos de hasta 45 grados, mientras que las juntas interiores funcionan dentro de ángulos más limitados.

Journal A bearing with a hole in it for a shaft.
Gorrón Cojinete con un agujero para el árbol.

Key A small block inserted between the shaft and hub to prevent circumferential movement.
Chaveta Pasador pequeño insertado entre el árbol y el cubo para impedir el movimiento circunferencial.

Keyway A groove or slot cut to permit the insertion of a key.
Chavetero Ranura o hendidura cortada para que pueda insertarse una chaveta.

King pin A metal rod or pin on which steering knuckles turn.
Clavija maestra Varilla o chaveta de metal sobre la cual giran los muñones de dirección.

Knock A heavy metallic sound usually caused by a loose or worn bearing.
Golpeteo Sondio metálico pesado normalmente causado por un cojinete suelto o desgastado.

Knuckle The part of the suspension that supports the wheel hub and serves as the steering pivot. The bottom of the knuckle is attached to the lower control arm with a ball joint, and the upper portion is usually bolted to the strut.
Muñón Parte de la suspensión que apoya al cubo de la rueda y que sirve como pivote de dirección. La parte inferior del muñón se fija al brazo de mando inferior por medio de una junta esférica, y la superior normalmente se emperna al montante.

Knurl To indent or roughen a finished surface.
Estriar Endentar o poner áspera una superficie acabada.

Lapping The process of fitting one surface to another by rubbing them together with an abrasive material between the two surfaces.
Pulido Proceso de ajustar una superficie contra otra rozando la una contra la otra con un material abrasivo colocado entre las dos superficies.

Lash The amount of free motion in a gear train, between gears, or in a mechanical assembly, such as the lash in a valve train.
Juego Cantidad de movimiento libre en un tren de engranajes, entre engranajes o en un conjunto mecánico, como por ejemplo, el juego en un tren de válvulas.

Leaf spring A spring made up of a single flat steel plate or of several plates of graduated lengths assembled one on top of another; used on vehicles to absorb road shocks by bending or flexing.
Muelle de láminas Muelle compuesto de una sola placa de acero plana o de varias placas de longitudes graduadas montadas la una sobre la otra; utilizado en vehículos para absorber la aspereza de la carretera a través de la flexión.

Limited-slip differential A differential designed so that when one wheel is slipping, a major portion of the drive torque is supplied to the wheel with the better traction; also called a nonslip differential.
Diferencial de deslizamiento limitado Diferencial diseñado para que cuando una rueda se deslice, una mayor parte del par de torsión de mando llegue a la rueda que tiene mejor tracción. Llamado también diferencial antideslizante.

Line wrench A common term for a flare nut wrench.
Llave de tuerca de línea Término común para llave de tuerca abocinada.

Linkage Any series of rods, yokes, and levers, etc., used to transmit motion from one unit to another.
Biela motriz Cualquier serie de varillas, yugos y palancas, etc., utilizada para transmitir movimiento de una unidad a otra.

Live axle A shaft that transmits power from the differential to the wheels.
Eje motor Árbol que transmite fuerza motriz del diferencial a las ruedas.

Lock pin Used in some ball sockets (inner tie-rod end) to keep the connecting nuts from working loose. Also used on some lower ball joints to hold the tapered stud in the steering knuckle.
Pasador de cierre Utilizado en algunas rótulas para bolas (extremo interior de la barra de acoplamaiento) para que las tuercas de conexión no se suelten. Utilizado también en algunas juntas esféricas inferiores para mantener el espárrago cónico en el muñón de dirección.

Locked differential A differential with the side and pinion gears locked together.
Diferencial trabado Diferencial en el que el engranaje lateral y el de piñón están sujetos entre sí.

Locking A condition of a bearing caused by large particles of dirt that become trapped between a bearing and its race.
Cierre Condición de un cojinete causada por partículas de polvo que se atrapan entre un cojinete y su anillo.

Locknut A second nut turned down on a holding nut to prevent loosening.
Contratuerca Una segunda tuerca montada boca abajo sobre una tuerca de ensamble para evitar que se suelte.

Lockplates Metal tabs bent around nuts or bolt heads.
Placa de cierre Orejetas metálicas dobladas alrededor de tuercas o cabezas de perno.

Lockwasher A type of washer that, when placed under the head of a bolt or nut prevents the bolt or nut from working loose.
Arandela de muelle Tipo de arandela que cuando se coloca debajo de la cabeza de un perno o de una tuerca, evita que el perno o la tuerca se suelte.

Low speed The gearing that produces the highest torque and lowest speed of the wheels.
Baja velocidad Engranajes que producen el par de torsión más alto y la velocidad más baja de las ruedas.

Lubricant Any material, usually a petroleum product such as grease or oil, that is placed between two moving parts to reduce friction.
Lubrificante Cualquier material, normalmente un derivado de petróleo, como la grasa o el aceite, que se coloca entre dos piezas móviles para disminuir la fricción.

Lug nut The nuts that fasten the wheels to the axle hub or brake rotor. Missing lug nuts should always be replaced. Overtightening can cause warpage of the brake rotor in some cases.
Tuerca de orejetas Tuercas que sujetan las ruedas al cubo del eje o el rotor de freno. Se deben reemplazar siempre las tuercas de orejetas que se hayan perdido. En algunos casos el apretar demasiado puede causar el torcimiento del rotor de freno.

Lugging A term used to describe an operating condition in which the engine is operating at too low of an engine speed for the selected gear.
Arrastrar Término utilizado para describir una condición en la que el motor funciona a una velocidad demasiado baja para la marcha elegida.

Master cylinder The liquid-filled cylinder in the hydraulic brake system, or clutch, where hydraulic pressure is developed when the driver depresses a foot pedal.
Cilindro primario Cilindro lleno de líquido en el sistema de freno hidráulico o en el embrague, donde se produce presión hidráulica cuando el conductor oprime el pedal.

Matched gear set code Identification marks on two gears that indicate they are matched. They should not be mismatched with another gear set and placed into operation.
Código del juego de engranaje emparejado Señales de identificación en dos engranajes que indican que ambos hacen pareja. No deben emparejarse con otro juego de engranaje y ponerse en funcionamiento.

Meshing The mating, or engaging, of the teeth of two gears.
Engranar Emparejar, o endentar, los dientes de dos engranajes.

Meter (m) 39.37 inches (in.)
Metro (m) 39,37 pulgadas (pulgs.)

Micrometer A precision measuring device used to measure small bores, diameters, and thicknesses. Also called a mike.
Micrómetro Instrumento de precisión utilizado para medir calibres, espesores y diámetros pequeños. Llamado también mic.

Millimeter (mm) 0.001 meter or 0.03937 inches
Milímetro (mm) 0,001 de un metro o 0,03937 de una pulgada

Misalignment When bearings are not on the same center line.
Mal alineamiento Cuando los cojinetes no se encuentran en la misma línea central.

Mounts Made of rubber to insulate vibrations and noise while they support a powertrain part, such as engine or transmission mounts.
Monturas Hechas de caucho para aislar vibraciones y ruido mientras apoyan una pieza del tren transmisor de potencia, como por ejemplo, las monturas del motor o de la transmisión.

Multiple disc A clutch with a number of driving and driven discs as compared to a single plate clutch.
Disco múltiple Embrague con muchos discos de accionamiento y accionados, comparado con un embrague de placa simple.

Needle bearing An antifriction bearing using a great number of long, small diameter rollers. Also known as a quill bearing.
Cojinete de agujas Cojinete de antifricción que utiliza una gran cantidad de rodillos largos con diámetros pequeños. Llamado también cojinete de manguito.

Needle deflection Distance of travel from zero of the needle on a dial gauge.
Desviación de aguja Distancia a la que viaja la aguja desde el cero en un indicador de cuadrante.

Neoprene A synthetic rubber that is not affected by the various chemicals that are harmful to natural rubber.
Neopreno Caucho sintético que no es afectado por las distintas sustancias químicas nocivas para el caucho natural.

Neutral In a transmission, the setting in which all gears are disengaged and the output shaft is disconnected from the drive wheels.
Neutral En una transmisión, la regulación a la cual se desengranan todos los engranajes y se desconecta el árbol de rendimiento de las ruedas motrices.

Neutral start switch A switch wired into the ignition switch to prevent engine cranking unless the transmission shift lever is in neutral or the clutch pedal is depressed.
Interruptor de encendido neutral Interruptor conectado al botón de encendido para impedir el arranque del motor a menos que la palanca de cambios esté en neutro o se oprima el pedal de embrague.

Newton-Meter (N•m) Metric measurement of torque or twisting force.
Metro-Newton (N•m) Medida métrica de par de torsión o fuerza de torsión equivalente a libras-pies multiplicado por 1,355.

Noise Any unwanted or annoying sound.
Ruido Cualquier sonido perturbador o molesto.

Nominal shim A shim with a designated thickness.
Laminillas nominales Laminillas de un espesor específico.

Non hardening A gasket sealer that never hardens.
Antiendurecedor Junta de estanqueidad de la guarnición que nunca se endurece.

Nut A removable fastener used with a bolt to lock pieces together; made by threading a hole through the center of a piece of metal that has been shaped to a standard size.
Tuerca Aparato de fijación desmontable utilizado con un perno para que las piezas queden sujetas entre sí; se hace abriendo un hueco en el centro de una pieza de metal conformada a un tamaño estándar.

O-ring A type of sealing ring, usually made of rubber or a rubberlike material. In use, the O-ring is compressed into a groove to provide the sealing action.
Anillo-O Tipo de anillo de estanqueidad, normalmente hecho de caucho o de un material parecido al caucho. Cuando se le utiliza, el anillo-O se comprime en una ranura para proveer estanqueidad.

OD Outside diameter.
OD Diámetro exterior.

Oil seal A seal placed around a rotating shaft or other moving part to prevent leakage of oil.
Junta de aceite Junta de estanqueidad colocada alrededor de un árbol giratorio u otra pieza móvil para evitar fugas de aceite.

One-way clutch See sprag clutch.
Embrague de una sola dirección Véase embrague de horquilla.

Operating angle The difference between the drive shaft and transmission installation angles is the operating angle.
Ángulo de funcionamiento El ángulo de funcionamiento es la diferencia entre los ángulos del montaje del árbol de mando y de la transmisión.

Outboard constant velocity joint The outer constant velocity joint, or the one closest to the wheels. The outer joint is a fixed joint.
Junta de velocidad constante fuera de borde Junta de velocidad constante exterior, o la que está más cerca de las ruedas. La junta exterior es una junta fija.

Outer bearing race The outer part of a bearing assembly on which the balls or rollers rotate.
Anillo de cojinete exterior Parte exterior de un conjunto de cojinete sobre la cual giran las bolas o los rodillos.

Out-of-round Wear of a round hole or shaft that when viewed from an end will appear egg shaped.
Con defecto de circularidad Desgaste de un agujero o árbol redondo que, vistos desde uno de los extremos, parecen ovalados.

Output shaft The shaft or gear that delivers the power from a device, such as a transmission.
Árbol de rendimiento Árbol o engranaje que transmite la fuerza motriz desde un mecanismo, como por ejemplo, la transmisión.

Overcenter spring A heavy coil spring arrangement in the clutch linkage to assist the driver with disengaging the clutch and returning the clutch linkage to the full engagement position.
Muelle sobrecentro Distribución de muelles helicoidales gruesos en la biela motriz del embrague para ayudar al conductor a desengra-

nar el embrague y devolver la biela motriz del embrague a la posición de enganche total.

Overall ratio The product of the transmission gear ratio multiplied by the final drive or rear axle ratio.

Relación total Producto de la relación del engranaje de la transmisión multiplicado por la relación de la transmisión final o del eje trasero.

Overdrive Any arrangement of gearing that produces more revolutions of the driven shaft than of the driving shaft.

Sobremultiplicación Cualquier distribución de engranajes que produce más revoluciones del árbol accionado que del árbol impulsor.

Overdrive ratio Identified by the decimal point indicating less than one driving input revolution compared to one output revolution of a shaft.

Relación de sobremultiplicación Identificada por el punto decimal, lo que indica menos de una revolución impulsora de mando comparada con una revolución de rendimiento de un árbol.

Overrun coupling A free-wheeling device to permit rotation in one direction but not in the other.

Acoplamiento de giro libre Mecanismo de marcha en rueda libre que permite que se lleve a cabo la rotación en una dirección pero no en la otra.

Overrunning clutch A device consisting of a shaft or housing linked together by rollers or sprags operating between movable and fixed races. As the shaft rotates, the rollers or sprags jam between the movable and fixed races. This jamming action locks together the shaft and housing. If the fixed race should be driven at a speed greater than the movable race, the rollers or sprags will disconnect the shaft.

Embrague de rueda libre Mecanismo que consiste de un árbol o un alojamiento conectados por rodillos u horquillas que funcionan entre anillos móviles o fijos. Mientras el árbol gira, los rodillos o las horquillas se acuñan entre los anillos móviles y los anillos fijos. Este acuñamiento sujeta al árbol y al alojamiento entre sí. Si el anillo fijo debe accionarse a una velocidad más alta que el anillo móvil, los rodillos o las horquillas desconectarán el árbol.

Oxidation Burning or combustion; the combining of a material with oxygen. Rusting is slow oxidation, and combustion is rapid oxidation.

Oxidación Quema o combustión; la combinación de oxígeno con otro elemento. La corrosión es una oxidación lenta, mientras que y la combustión es una oxidación rápida.

Parallel The quality of two items being the same distance from each other at all points; usually applied to lines and, in automotive work, to machined surfaces.

Paralelo Calidad de dos objetos que se encuentran a la misma distancia el uno del otro en todos los puntos; normalmente se aplica a líneas y, en la reparación de automóviles, a superficies maquinadas.

Pawl A lever that pivots on a shaft. When lifted it swings freely and when lowered it locates in a detent or notch to hold a mechanism stationary.

Trinquete Palanca que gira sobre un árbol. Cuando se la levanta, se mueve libremente y cuando se la baja, se acuña en un retén o en una muesca para bloquear el movimiento de un mecanismo.

Pedal play The distance the clutch pedal and release bearing assembly move from the fully engaged position to the point where the release bearing contacts the pressure plate release levers.

Holgura del pedal Distancia a la que el pedal del embrague y el conjunto del cojinete de desembrague se mueven de una posición enteramente engranada a un punto donde el cojinete de desembrague entra en contacto con las palancas de desembrague de la placa de presión.

Peen To stretch or clinch over by pounding with the rounded end of a hammer.

Granallar Estirar o remachar golpeando con el extremo redondo de un martillo.

Phasing Rotational position of the universal joints on the drive shaft.

Fasaje Posición de rotación de las juntas universales sobre el árbol de mando.

Phosgene gas The gas formed when R-12 comes in contact with heat. It is poisonous and will make you sick or fatally ill.

Gas fosgeno Gas formado por la mezcla del R-12 y el calor. Sustancia venenosa que produce graves trastornos o hasta la muerte.

Pilot bearing A small bearing, such as in the center of the flywheel end of the crankshaft, which carries the forward end of the clutch shaft.

Cojinete piloto Cojinete pequeño, como por ejemplo, el ubicado en el centro del extremo del volante del cigüeñal, que soporta el extremo delantero del árbol del embrague.

Pilot bushing A plain bearing fitted in the end of a crankshaft. The primary purpose is to support the input shaft of the transmission.

Buje piloto Cojinete sencillo insertado en el extremo de un cigüeñal. El propósito principal es apoyar el árbol impulsor de la transmisión.

Pilot shaft A shaft used to align parts and that is removed before final installation of the parts; a dummy shaft.

Árbol piloto Árbol que se utiliza para alinear piezas y que se remueve antes del montaje final de las mismas; árbol falso.

Pinion gear The smaller of two meshing gears.

Engranaje de piñón El más pequeño de los dos engranajes de engrane.

Pinion carrier The mounting or bracket that retains the bearings supporting a pinion shaft.

Portador de piñón Montaje o soporte que sujeta los cojinetes que apoyan un árbol del piñón.

Pitch The number of threads per inch on any threaded part.

Paso Número de filetes de un tornillo por pulgada en cualquier pieza fileteada.

Pitman arm A short arm in the steering linkage that connects the steering gear to other steering components.

Brazo pitman Brazo corto en el cuadrilátero de la dirección que articula el mecanismo de dirección a los otros componentes de dirección.

Pivot A pin or shaft upon which another part rests or turns.

Pivote Chaveta o árbol sobre el cual se apoya o gira otra pieza.

Planet carrier In a planetary gear system, the carrier or bracket in a planetary system that contains the shafts upon which the pinions or planet gears turn.

Portador planetario En un sistema de engranaje planetario, portador o soporte en un sistema planetario que contiene los árboles sobre los cuales giran los piñones o los engranajes planetarios.

Planet gears The gears in a planetary gear set that connect the sun gear to the ring gear.

Engranajes planetarios Engranajes en un tren de engranaje planetario que conectan el engranaje principal a la corona.

Planet pinions In a planetary gear system, the gears that mesh with, and revolve about, the sun gear; they also mesh with the ring gear.

Piñones planetarios En un sistema de engranaje planetario, los engranajes que se engranan con y giran entorno del engranaje principal; se engranan también con la corona.

Planetary gear set A system of gearing modeled after the solar system. A pinion is surrounded by an internal ring gear and planet gears are in mesh between the ring gear and pinion around which all revolve.

Tren de engranaje planetario Sistema de engranaje inspirado en el sistema solar. Un piñón está rodeado por una corona interna, y los engranajes planetarios se engranan entre la corona y el piñón, alrededor de los cuales todo gira.

Plate loading Force developed by the pressure plate assembly to hold the driven disc against the flywheel.

Carga de placa Fuerza producida por el conjunto de la placa de presión para sujetar el disco accionado contra el volante.

Plunging action Telescoping action of an inner front-wheel-drive universal joint.

Acción sumergible Acción telescópica de una junta universal interior de tracción delantera.

Plunging constant velocity joint Usually the inner constant velocity joint. The joint is designed so that it can telescope slightly to compensate for suspension motions.

Junta de velocidad constante sumergible Normalmente junta de velocidad constante interior. La junta está diseñada para que pueda extenderse ligeramente y así compensar el movimiento de la suspensión.

Pneumatic tools A type of power tool that relies on compressed air for power.

Herramientas neumáticas Tipo de herramienta eléctrica que necesita aire comprimido para fucionar.

Powertrain The mechanisms that carry the power from the engine crankshaft to the drive wheels; these include the clutch, transmission, drive line, differential, and axles.

Tren transmisor de potencia Mecansimos que transmiten la potencia desde el cigüeñal del motor hasta las ruedas motrices; éstos incluyen el embrague, la transmisión, la línea de transmisión, el diferencial y los ejes.

Preload A load applied to a part during assembly so as to maintain critical tolerances when the operating load is applied later.

Carga previa Carga aplicada a una pieza durante su montaje para mantener tolerancias críticas cuando más tarde se aplique la carga de funcionamiento.

Press fit Forcing a part into an opening that is slightly smaller than the part itself to make a solid fit.

Ajuste en prensa Forzar una pieza dentro de una apertura un poco más pequeña que la pieza misma para lograr un ajuste sólido.

Pressure Force per unit area, or force divided by area. Usually measured in pounds per square inch (psi) or in kilopascals (kPa) in the metric system.

Presión Fuerza por unidad de área, o fuerza dividida por área. Normalmente se mide en libras por pulgada cuadrada (lpc) o en kilopascales (kPa) en el sistema métrico.

Pressure plate That part of the clutch that exerts force against the friction disc; it is mounted on and rotates with the flywheel. A heavy steel ring pressed against the clutch disc by spring pressure.

Placa de presión Pieza del embrague que ejerce fuerza contra el disco de fricción; se monta sobre y gira con el volante. Anillo pesado de acero, comprimido contra el disco de embrague mediante presión elástica.

Propeller shaft See drive shaft.

Árbol transmisor Véase árbol de mando.

Prussian blue A blue pigment; in solution, useful in determining the area of contact between two surfaces.

Azul de Prusia Pigmento azul; en una solución, sirve para determinar el área de contacto entre dos superficies.

psi Abbreviation for pounds per square inch; a measurement of pressure.

psi Abreviatura de libras por pulgada cuadrada; una medida de presión.

Puller Generally, a shop tool used to separate two closely fitted parts without damage. Often contains a screw, or several screws, which can be turned to apply a gradual force.

Tirador Generalmente, herramienta de taller utilizada para separar dos piezas fuertemente apretadas sin averiarlas. A menudo contiene uno o varios tornillos a los que se les puede dar vuelta para aplicar una fuerza gradual.

Pulsation To move or beat with rhythmic impulses.

Pulsación Mover o golpear con impulsos rítmicos.

Quadrant A section of a gear. A term sometimes used to identify the shift lever selector mounted on the steering column.

Cuadrante Sección de un engranaje. Término utilizado en algunas ocasiones para identificar el selector de la palanca de cambio de velocidades montado sobre la columna de dirección.

Quill shaft The term used by some manufacturers to refer to the protruding hollow shaft of the transmission's front bearing retainer.

Árbol de manguito Término utilizado por algunos fabricantes para referirse al árbol hueco proyectado del retenedor del cojinete delantero de la transmisión.

R-12 The type of refrigerant used on most cars; is commonly referred to as freon.

R-12 Tipo de refrigerante utilizado en la mayoría de los automóviles; comúnmente llamado freón.

Race A channel in the inner or outer ring of an antifriction bearing in which the balls or rollers roll.

Anillo Canal en el anillo interior o exterior de un cojinete de antifricción en el cual giran las bolas o los rodillos.

Raceway A groove or track designed into the races of a bearing or universal joint housing to guide and control the action of the balls or trunnions.

Anillo de rodadura Ranura o canal construido en el interior de los anillos de un cojinete o de junta universal para guiar y controlar el movimiento de las bolas o de las muñequillas.

Radial The direction moving straight out from the center of a circle. Perpendicular to the shaft or bearing bore.

Radial Dirección que sale directamente del centro de un círculo. Perpendicular al árbol o al calibre de cojinete.

Radial clearance (radial displacement) Clearance within the bearing and between balls and races perpendicular to the shaft.

Espacio libre radial (desplazamiento radial) Dentro del cojinete y entre las bolas y los anillos, espacio libre perpendicular al árbol.

Radial load A force perpendicular to the axis of rotation.

Carga radial Fuerza perpendicular al pivote de rotación.

Ratcheting mechanism Uses a pawl and gear arrangement to transmit motion or to lock a particular mechanism by having the pawl drop between gear teeth.

Mecanismo de trinquete Utiliza un conjunto de retén y engranaje para transmitir movimiento o para bloquear un mecanismo específico haciendo que el retén caiga entre los dientes del engranaje.

Ratio The relation or proportion that one number bears to another.

Relación Razón o proporción que existe entre un número y otro.

Ravigneaux Designer of a planetary gear system with small and large sun gears, long and short planetary pinions, planetary carriers and ring gear.

Ravigneaux Fabricante de un sistema de engranaje planetario que consiste de engranajes prinipales pequeños y grandes, piñones planetarios largos y cortos, portadores planetarios y corona.

Reamer A round metal cutting tool with a series of sharp cutting edges; enlarges a hole when turned inside it.

Escariador Herramienta redonda metálica de corte que tiene una serie de aristas agudas; ensancha un agujero cuando se le da vuelta dentro de éste.

Rear axle torque The torque received and multiplied by the rear driving axle assembly.

Torsión del eje trasero Par de torsión recibido y multiplicado por el conjunto del eje motor trasero.

Rear wheel drive (RWD) A term associated with a vehicle where the engine is mounted at the front and the driving axle and driving wheels at the rear of the vehicle.

Tracción trasera Término relacionado a un vehículo donde el motor se monta en la parte delantera, y el eje motor y las ruedas motrices en la parte trasera.

Release bearing A ball-type bearing moved by the clutch pedal linkage to contact the pressure plate release levers to either engage or disengage the driven disc with the clutch driving members.

Cojinete de desembrague Cojinete de tipo bola accionado por la biela motriz del pedal del embrague para entrar en contacto con las palancas de desembrague de la placa de presión o para engranar o desengranar el disco accionado con los mecanismos de accionamiento del embrague.

Release levers In the clutch, levers that are moved by throwout-bearing movement, causing clutch spring force to be relieved so that the clutch is disengaged, or uncoupled from the flywheel.

Palancas de desembrague En el embrague, las palancas accionadas por el movimiento del cojinete de desembrague, que hacen disminuir la fuerza del muelle del embrague para que el embrague se desengrane, o se desacople del volante.

Release plate Plate designed to release the clutch pressure plate's loading on the clutch driven disc.

Placa de desembrague Placa diseñada para desembragar la carga de la placa de presión del embrague en el disco accionado del embrague.

Removable carrier housing A type of rear axle housing from which the axle carrier assembly can be removed for parts service and adjustment.

Alojamiento portador desmontable Tipo de puente trasero del cual se puede desmontar el conjunto del portador del eje para la reparación y el ajuste de las piezas.

Retaining ring A removable fastener used as a shoulder to retain and position a round bearing in a hole.

Anillo de retención Aparato fijador desmontable utilizado como punto de apoyo para sujetar y colocar un cojinete redondo en un agujero.

Retractor clips Spring steel clips that connect the diaphragm's flexing action to the pressure plate.

Grapas retractoras Grapas de acero para muelles que conectan el movimiento flexible del diafragma a la placa de presión.

Reverse idler gear In a transmission, an additional gear that must be meshed to obtain reverse gear; a gear used only in reverse that does not transmit power when the transmission is in any other position.

Piñón de marcha atrás En una transmisión, engranaje adicional que debe engranarse para obtener un engranaje de marcha atrás; engranaje utilizado solamente durante la inversión de marcha, que no transmite fuerza cuando la transmisión se encuentra en cualquier otra posición.

Ring gear A gear that surrounds or rings the sun and planet gears in a planetary system. Also the name given to the spiral bevel gear in a differential.

Corona Engranaje que rodea los engranajes planetario y el principal en un sistema planetario. También es el nombre que se le da al engranaje cónico con dentado espiral en un diferencial.

Rivet A headed pin used for uniting two or more pieces by passing the shank through a hole in each piece, and securing it by forming a head on the opposite end.

Remanche Chaveta de cabeza utilizada para unir dos o más piezas insertando la espinilla en cada una de las piezas a través de un agujero. La espinella se asegura formando una cabeza en el extremo opuesto.

Roller bearing An inner and outer race upon which hardened steel rollers operate.

Cojinete de rodillos Anillo interior y exterior sobre el cual funcionan rodillos de acero templado.

Rollers Round steel bearings that can be used as the locking element in an overrunning clutch or as the rolling element in an antifriction bearing.

Rodillos Cojinetes redondos de acero que pueden utilizarse como el elemento de bloqueo en un embrague de rueda libre o como el elemento rodante en un cojinete de antifricción.

RPM Abbreviation for revolutions per minute, a measure of rotational speed.

RPM Abreviatura de revoluciones por minuto, una medida de velocidad de rotación.

RTV sealer Room temperature vulcanizing gasket material, which cures at room temperature; a plastic paste squeezed from a tube to form a gasket of any shape.

Junta de estanqueidad VTA Material vulcanizador de guarnición a temperatura ambiente, que se conserva a temperatura ambiente; pasta plástica que viene en tubo, utilizada para formar una guarnición de cualquier tamaño.

Rubber coupling Rubber-based disc used as a universal joint between the driving and driven shafts.

Acoplamiento de caucho Disco con base de caucho; utilizado como junta universal entre el árbol de accionamiento y el árbol accionado.

Runout Deviation of the specified normal travel of an object. The amount of deviation or wobble a shaft or wheel has as it rotates. Runout is measured with a dial indicator.

Desviación Desalineación del movimiento normal indicado de un objeto. Cantidad de desalineación o bamboleo que tiene un árbol o una rueda mientras gira. La desviación se mide con un indicador de cuadrante.

Rzeppa constant velocity joint The name given to the ball-type constant velocity joint (as opposed to the tripod-type constant velocity joint). Rzeppa joints are usually the outer joints on most FWD cars. Named after its inventor, Alfred Rzeppa, a Ford engineer.

Junta de velocidad constante Rzeppa Nombre que se le da a la junta de velocidad constante de tipo bola (en contraste con la junta de velocidad constante de tipo trípode). Las juntas Rzeppa normalmente son las juntas exteriores en la mayoría de automóviles de tracción delantera. Nombrada por su creador, Alfred Rzeppa, ingeniero de la Ford.

SAE Society of Automotive Engineers.

SAE Sociedad de Ingenieros Automotrices.

Safety stands Commonly called jack stands and are used to support a vehicle when it is raised by a jack or hoist.

Soportes de seguridad Comúnmente llamados soportes de gato. Son utilizados para sostener un vehículo cuando es levantado por un gato o montacargas.

Score A scratch, ridge, or groove marring a finished surface.

Muesca Rayado, rotura o ranura que estropea una superficie acabada.

Scuffing A type of wear in which there is a transfer of material between parts moving against each other; shows up as pits or grooves in the mating surfaces.

Frotamiento Tipo de desgaste en el cual se transfiere material entre piezas que se mueven la una contra la otra; aparece en forma de hendiduras o ranuras en las superifices emparejadas.

Seal A material, shaped around a shaft, used to close off the operating compartment of the shaft, preventing oil leakage.

Junta de estanqueidad Material conformado alrededor de un árbol, que se utiliza para sellar el compartimiento de funcionamiento del árbol y así evitar la fuga de aceite.

Sealer A thick, tacky compound, usually spread with a brush, which may be used as a gasket or sealant to seal small openings or surface irregularities.

Líquido de estanqueidad Compuesto grueso y viscoso, normalmente esparcido con una brocha, que puede emplearse como guarnición o compuesto obturador para rellenar pequeñas aperturas o irregularidades en la superficie.

Seat A surface, usually machined, upon which another part rests or seats; for example, the surface upon which a valve face rests.

Asiento Superficie, normalmente maquinada, sobre la cual se coloca o sienta otra pieza; por ejemplo, la superficie sobre la cual se coloca una cara de válvula.

Self-adjusting clutch linkage Monitors clutch pedal play through a clutch control cable and ratcheting mechanism to automatically adjust clutch pedal play.

Biela motriz del embrague de ajuste automático Controla el juego del pedal del embrague mediante un cable de mando del embrague y un mecanimso de trinquete para ajustar automáticamente el juego del pedal del embrague.

Semicentrifugal pressure plate The release levers of this pressure plate are weighted to take advantage of centrifugal force to increase plate loading resulting in reduced driven disc slip.

Placa de presión semicentrífuga A las palancas de desembrague de esta placa de presión se les añade peso para aprovechar la fuerza centrífuga y hacer que ésta aumente la carga de la placa. El resultado será un deslizamiento menor del disco accionado.

Semifloating rear axle An axle that supports the weight of the vehicle on the axle shaft in addition to transmitting driving forces to the rear wheels.

Eje trasero semi-flotante Eje que apoya el peso del vehículo sobre el árbol motor además de transmitir las fuerzas motrices a las ruedas traseras.

Separators A component in an antifriction bearing that keeps the rolling components apart.

Separadores Componente en un cojinete de antifricción que mantiene los componentes de rodamiento separados.

Shift forks Mechanisms attached to shift rails that fit into synchronizer hub for change of gears.

Horquillas de cambio de velocidades Las ranuras en el anillo sincronizador del embrague cónico deben ser afiladas para lograr la sincronización.

Shift lever The lever used to change gears in a transmission. Also, the lever on the starting motor that moves the drive pinion into or out of mesh with the flywheel teeth.

Palanca de cambio de velocidades Palanca utilizada para cambiar las velocidades en una transmisión. También es la palanca en el motor de arranque que engrana o desengrana el piñón de mando con los dientes del volante.

Shift rails Rods placed within the transmission housing that are a part of the transmission gearshift linkage.

Barras de cambio de velocidades Varillas ubicadas dentro del alojamiento de transmisión que forman parte de la biela motriz del cambio de velocidades de la transmisión.

Shim Thin sheets used as spacers between two parts, such as the two halves of a journal bearing.

Chapa de relleno Láminas delgadas utilizadas como espaciadores entre dos piezas, como por ejemplo, las dos mitades de un cojinete liso.

Shim stock Sheets of metal of accurately known thickness that can be cut into strips and used to measure or correct clearances.

Material de chapa de relleno Láminas de metal de espesor preciso que puede cortarse en tiras y utilizarse para medir el espacio libre correcto.

Shudder A shake or shiver movement.

Estremecimiento Sacudida o temblor.

SI measuring system The metric measuring system.

Sistema de medida SI Sistema métrico de medida.

Side clearance The clearance between the sides of moving parts when the sides do not serve as load-carrying surfaces.

Despojo lateral Espacio libre entre los dos lados de piezas móviles cuando éstos no sirven como superficies de carga.

Side gears Gears that are meshed with the differential pinions and splined to the axle shafts (RWD) or drive shafts (FWD).

Engranajes laterales Engranajes que se engranan con los piñones del diferencial y son ranurados a los árboles motores en vehículos de tracción trasera o a los árboles de mando en vehículos de tracción delantera.

Side thrust Longitudinal movement of two gears.

Empuje lateral Movimiento longitudinal de dos engranajes.

Slave cylinder Located at a lower part of the clutch housing. Receives fluid pressure from the master cylinder to engage or disengage the clutch.

Cilindro secundario Ubicado en la parte inferior del alojamiento del embrague. Recibe presión de fluido del cilindro primario para engranar o desengranar el embrague.

Sliding fit Where sufficient clearance has been allowed between the shaft and journal to allow free-running without overheating.

Ajuste deslizante Donde se ha permitido espacio suficiente entre el árbol y el gorrón para permitir un funcionamiento libre sin ocasionar un recalentamiento.

Sliding yoke Slides on internal and external splines to compensate for drive line length changes.

Yugo deslizante Se desliza sobre las lengüetas internas y externas para compensar los cambios de longitud de la línea de transmisión.

Sliding gear transmission A transmission in which gears are moved on their shafts to change gear ratios.

Transmisión por engranaje desplazable Transmisión en la cual los engranajes se mueven sobre sus árboles para cambiar la relación de los engranajes.

Slip fit Running or sliding fit.
Ajuste corredizo Ajuste deslizante o de marcha.

Slip joint In the powertrain, a variable-length connection that permits the drive shaft to change its effective length.
Junta corrediza En el tren transmisor de potencia, una conexión de longitud variable que le permite al árbol de mando cambiar su longitud eficaz.

Snap ring Split spring-type ring located in an internal or external groove to retain a part.
Anillo de resorte Anillo hendido de tipo muelle ubicado en una ranura interna o externa para sujetar una pieza en su lugar.

Solid axle A rear axle design that places the final drive, axles, bearings, and hubs into one housing.
Eje sólido Diseño del eje trasero que coloca la transmisión final, los ejes, los cojinetes y los cubos dentro de un solo alojamiento.

Spalling A condition of a bearing that is caused by overloading the bearing and is evident by pits on the bearings or their races.
Esquirla Condición de un cojinete provocada al éste ser sobrecargado, que se manifiesta a través de hendiduras en los cojinetes o sus anillos.

Span gauge A commonly used term for a telescoping gauge.
Indicador de extensión Término común para indicador telescópico.

Speed gears Driven gears located on the transmission output shaft. This term differentiates between the gears of the counter gear and cluster assemblies and gears on the transmission output shaft.
Engranajes de velocidades Engranajes accionados ubicados en el árbol de rendimiento de la transmisión. Este término distingue entre los engranajes de los conjuntos del mecanismo contador y de los engranajes desplazables y los engranajes sobre el árbol de rendimiento de la transmisión.

Spindle The shaft on which the wheels and wheel bearings mount.
Huso Árbol sobre el cual se montan las ruedas y los cojinetes de rueda.

Spiral bevel gear A ring gear and pinion wherein the mating teeth are curved and placed at an angle with the pinion shaft.
Engranaje cónico con dentado espiral Corona y piñón cuyos dientes emparejados son curvos y están montados en ángulo con el árbol de piñón.

Spiral gear A gear with teeth cut according to a mathematical curve on a cone. Spiral bevel gears that are not parallel have center lines that intersect.
Engranaje helocoidal Engranaje con dientes cortados en un cono según una curva matemática. Los engranajes cónicos con dentado espiral que no son paralelos tienen líneas centrales que se cruzan.

Spline Slot or groove cut in a shaft or bore; a splined shaft onto which a hub, wheel, gear, etc., with matching splines in its bore is assembled so that the two must turn together.
Lengüeta Hendidura o ranura excavada en un árbol o un calibre; árbol ranurado sobre el que se montan un cubo, una rueda, un engranaje, etc., con lengüetas que hacen pareja en su calibre, para que ambos giren juntos.

Splined hub Several keys placed radially around the inside diameter of a circular part, such as a wheel or driven disc.
Cubo ranurado Varias chavetas ubicadas de manera radial alrededor del diámetro interior de una pieza circular, como por ejemplo, una rueda o un disco accionado.

Split lip seal Typically a rope seal sometimes used to denote any two-part oil seal.
Junta de estanqueidad de reborde hendido Típicamente una junta de estanqueidad de cable utilizada en algunas ocasiones para denominar cualquier junta de aceite de dos partes.

Split pin A round split spring steel tubular pin used for locking purposes; for example, locking a gear to a shaft.
Pasador hendido Pasador tubular redondo hendido de acero para muelles utilizado para sujetar; por ejemplo, para asegurar un engranaje a un árbol.

Sprag clutch A member of the overrunning clutch family using a sprag to jam between the inner and outer races used for holding or driving action.
Embrague de horquilla Miembro de la familia del embrague de rueda libre que se sirve de una horquilla para insertarse entre los anillos interior y exterior utilizados para la retención o el funcionamiento.

Spring A device that changes shape when it is stretched or compressed, but returns to its original shape when the force is removed; the component of the automotive suspension system that absorbs road shocks by flexing and twisting.
Muelle Pieza que cambia de forma cuando se estira o comprime, pero que recobra su forma original cuando se detiene la fuerza; componente del sistema de suspensión automotriz que absorbe las sacudidas de la carretera doblándose y torciéndose.

Spring retainer A steel plate designed to hold a coil or several coil springs in place.
Retenedor de muelle Placa de acero diseñada para sujetar uno o varios muelles helicoidales en su lugar.

Spur gear Gears cut on a cylinder with teeth that are straight and parallel to the axis.
Engranaje recto Engranajes cortados en un cilindro que tienen dientes rectos y paralelos al pivote.

Squeak A high-pitched noise of short duration.
Rechinamiento Sonido agudo de corta duración.

Squeal A continuous high-pitched noise.
Chirrido Sonido agudo continuo.

Stabilizer bar Also called a sway bar. It prevents the vehicle's body from diving into turns.
Barra estabilizadora Llamada también barra de oscilación lateral. Impide que la carrocería del vehículo se desestabilice durante los virajes.

Staking punch A chisel-like punch used to create a large dimple in metal. This dimple prevents the nut from self-adjusting.
Punzón de estacas Punzón parecido a un cincel, utilizado para hacer un agujero grande en un metal. Esta abolladura impide que la tuerca se ajuste automáticamente.

Stress The force to which a material, mechanism, or component is subjected.
Esfuerzo Fuerza a la que se somete un material, mecanismo o componente.

Strut assembly Refers to all the strut components, including the strut tube, shock absorber, coil spring, and upper bearing assembly.
Conjunto de montante Se refiere a todas las piezas del montante, inclusive al tubo de montante, al amortiguador, al muelle helicoidal, y al conjunto del cojinete superior.

Stub shaft A very short shaft.
Árbol corto Árbol sumamente corto.

Sun gear The central gear in a planetary gear system around which the rest of the gears rotate. The innermost gear of the planetary gear set.
Engranaje principal Engranaje central en un sistema de engranaje planetario alrededor del cual giran los demás engranajes. Es el engranaje más interior del tren de engranaje planetario.

Sway bar Also called a stabilizer bar. It prevents the vehicle's body from diving into turns.

Barra de oscilación lateral Llamada también barra estabilizadora. Impide que la carrocería del vehículo se desestabilice durante los virajes.

Synchromesh transmission Transmission gearing that aids the meshing of two gears or shift collars by matching their speed before engaging them.

Transmisión de engranaje sincronizado Engranaje transmisor que facilita el engrane de dos engranajes o collares de cambio de velocidades al igualar la velocidad de éstos antes de engranarlos.

Synchronize To cause two events to occur at the same time; for example, to bring two gears to the same speed before they are meshed to prevent gear clash.

Sincronizar Hacer que dos sucesos ocurran al mismo tiempo; por ejemplo, hacer que dos engranajes giren a la misma velocidad antes de que se engranen para evitar el choque de engranajes.

Synchronizer assemblies Device that uses cone clutches to bring two parts rotating at two speeds to the same speed. A synchronizer assembly operates between two gears; first and second gear, third and fourth gear.

Conjuntos sincronizadores Mecansimo que utiliza embragues cónicos para hacer que dos piezas que giran a dos velocidades giren a una misma velocidad. Un conjunto sincronizador funciona entre dos engranajes; engranjes de primera y segunda velocidad, y engranajes de tercera y cuarta velocidad.

Synchronizer blocker ring Usually a brass ring that acts as a clutch and causes driving and driven units to turn at the same speed before final engagement.

Anillo de bloque sincronizador Normalmente un anillo de latón que sirve de embrague y hace que las piezas de accionamiento y las accionadas giren a la misma velocidad antes del acoplamiento final.

Synchronizer hub Center part of the synchronizer assembly that is splined to the synchronizer sleeve and transmission output shaft.

Cubo sincronizador Pieza central del conjunto sincronizador ranurada al manguito sincronizador y al árbol de rendimiento de la transmisión.

Synchronizer sleeve The sliding sleeve that fits over the complete synchronizer assembly.

Manguito sincronizador Manguito deslizante que cubre todo el conjunto sincronizador.

Tail shaft A commonly used term for a transmission's extension housing.

Extremo del árbol Término comúnmente utilizado para el alojamiento de extensión de una transmisión.

Tap To cut threads in a hole with a tapered, fluted, threaded tool.

Aterrajar Cortar filetes de tornillo en un agujero con una herramienta cónica, estriada, fileteada.

Temper To change the physical characteristics of a metal by applying heat.

Templar Cambiar las características físicas de un metal aplicándole calor.

Tension Effort that elongates or "stretches" a material.

Tensión Fuerza que alarga o estira un material.

Thickness gauge Strips of metal made to an exact thickness, used to measure clearances between parts.

Calibrador de espesor Tiras de metal hechas a un espesor extacto, utilizadas para medir el espacio libre entre las piezas.

Thread chaser A device, similar to a die, that is used to clean threads.

Fileteadora de tornillo Utensilio, parecido a un troquel, utilizado para limpiar tornillo.

Threaded insert A threaded coil that is used to restore the original thread size to a hole with damaged threads.

Piezas insertas fileteadas Espiral fileteado utilizado para devolver el tamaño original del tornillo a un agujero que tiene tornillos averiados.

Three-quarter floating axle The axle housing carries the weight of the vehicle while the bearings support the wheels on the outer ends of the axle housing tubes.

Eje flotante de tres cuartos Puente trasero que soporta el peso del vehículo mientras los cojinetes soportan las ruedas en los extremos exteriores de los tubos del puente trasero.

Throwout bearing In the clutch, the bearing that can be moved inward to the release levers by clutch-pedal action to cause declutching, which disengages the engine crankshaft from the transmission.

Cojinete de desembrague En el embrague, cojinete que puede moverse hacia adentro hasta las palancas de desembrague, por medio de la acción del pedal del embrague, para lograr el desembrague. Esta acción desengrana el cigüeñal del motor de la transmisión.

Thrust load A load that pushes or reacts through the bearing in a direction parallel to the shaft.

Carga de empuje Carga que empuja o reacciona mediante el cojinete en una dirección paralela al árbol.

Thrust washer A washer designed to take up end thrust and prevent excessive endplay.

Arandela de empuje Arandela diseñada para asegurar el empuje longitudinal y prevenir un juego longitudinal excesivo.

Tie rod The linkage between the steering rack and the steering knuckle arm. The tie rod is threaded into a tie-rod end or has a threaded split member for making toe adjustments.

Barra de acoplamiento Biela mortiz entre la cremallera y el brazo del muñón de dirección. La barra de acoplamiento se filetea en un extremo de la barra de acoplamiento o tiene una pieza fileteada hendida para ajustar el tope.

Tie-rod end The fittings on the ends of the tie rods. The outer tie-rod end connects to the steering arm, and the inner one connects to the steering rack. Both ends include ball sockets to allow pivotal action, as well as up and down flexing.

Extremo de la barra de acoplamiento Conexiones en los extremos de las barras de acoplamiento. El extremo exterior de la barra de acoplamiento se conecta al brazo de dirección, y el extremo interior a la cremallera. Ambos extremos incluyen juntas de rótula para permitir tanto el movimiento giratorio como el de ascenso y descenso.

Toe A suspension dimension that reflects the difference in the distance between the extreme front and extreme rear of the tires.

Tope Dimensión de la suspensión que refleja la diferencia de la distancia entre los extremos delantero y trasero de las ruedas.

Toe-out in turns A steering system feature that allows the front wheels to be at different angles when the vehicle is turning a corner.

Divergencia Característica del sistema de dirección que permite que las ruedas delanteras se inclinen a ángulos diferentes cuando el vehículo hace un viraje.

Tolerance A permissible variation between the two extremes of a specification or dimension.

Tolerancia Variación permisible entre los dos extremos de una especificación o dimensión.

Torque A twisting motion, usually measured in ft.-lbs. (N•m).
Par de torsión Fuerza de torsión, normalmente medida en libras-pies (N•m).

Torque converter A turbine device utilizing a rotary pump, one or more reactors (stators), and a driven circular turbine or vane whereby power is transmitted from a driving to a driven member by hydraulic action. It provides varying drive ratios; with a speed reduction, it increases torque.
Convertidor de par de torsión Turbina que utiliza una bomba giratoria, uno o más reactores (estátores) y una turbina circular accionada o paleta; la fuerza se transmite del mecanismo de accionamiento al mecanismo accionado mediante acción hidráulica. Provee relaciones de accionamiento variadas; con una reducción de velocidad, aumenta el par de torsión.

Torque curve A line plotted on a chart to illustrate the torque personality of an engine. When the engine operates on its torque curve it is producing the most torque for the quantity of fuel being burned.
Curva de torsión Línea trazada en un gráfico para ilustrar las características de torsión de un motor. Cuando un motor funciona según su curva de torsión, produce mayor par de torsión por cantidad de combustible quemado.

Torque multiplication The result of meshing a small driving gear and a large driven gear to reduce speed and increase output torque.
Multiplicación de par de torsión Resultado de engranar un engranaje de accionamiento pequeño y un engranaje accionado grande para reducir la velocidad y aumentar el par de torsión de rendimiento.

Torque steer An action felt in the steering wheel as the result of increased torque.
Dirección de torsión Acción que se advierte en el volante de dirección como resultado de un aumento en el par de torsión.

Torque tube A fixed tube over the drive shaft on some cars. It helps locate the rear axle and takes torque reaction loads from the drive axle so the drive shaft will not sense them.
Tubo de eje cardán Tubo fijo sobre el árbol de mando en algunos automóviles. Ayuda a colocar el eje trasero y remueve las cargas de reacción de torsión del eje de mando para que el árbol de mando no las reciba.

Torsional springs Round, stiff coil springs placed in the driven disc to absorb the torsional disturbances between the driving flywheel and pressure plate and the driven transmission input shaft.
Muelles de torsión Muelles helicoidales redondos y rígidos ubicados en el disco accionado para absorber las alteraciones de torsión entre el volante motor y la placa de presión, y el árbol impulsor accionado de la transmisión.

Total pedal travel The total amount the pedal moves from no free-play to complete clutch disengagement.
Avance total del pedal Distancia total a la que el pedal se mueve de cero juego al desengrane total del embrague.

Total runout The sum of the maximum readings below and above the zero line on the indicator.
Desviación total Suma de las lecturas máximas bajo y sobre la línea del cero en el indicador de cuadrante.

Total travel Distance the clutch pedal and release bearing move from the fully engaged position until the clutch is fully disengaged.
Avance total Distancia a la que el pedal del embrague y el cojinete de desembrague se mueven de la posición enteramente engranada hasta que el embrague se desengrane por completo.

Traction The gripping action between the tire tread and the road's surface.
Tracción Agarrotamiento entre la banda de la llanta y la superficie de la carretera.

Transaxle Type of construction in which the transmission and differential are combined in one unit.
Transeje Tipo de construcción en la que la transmisión y el diferencial se combinan en una sola unidad.

Transaxle assembly A compact housing most often used in front-wheel-drive vehicles that houses the manual transmission, final drive gears, and differential assembly.
Conjunto del transeje Alojamiento compacto que normalmente se utiliza en vehículos de tracción delantera y que aloja a la transmisión manual, a los engranajes de transmisión final y al conjunto del diferencial.

Transfer case An auxiliary transmission mounted behind the main transmission. Used to divide engine power and transfer it to both front and rear differentials, either full time or part time.
Caja de transferencia Transmisión secundaria montada detrás de la transmisión principal. Se utiliza para separar la energía del motor y transferirla a los diferenciales delantero y trasero, a tiempo completo o a tiempo parcial.

Transmission The device in the powertrain that provides different gear ratios between the engine and drive wheels as well as reverse.
Transmisión Mecanismo en el tren transmisor de potencia que provee diferentes relaciones de engranajes tanto entre el motor y las ruedas motrices como en la marcha atrás.

Transmission case An aluminum or iron casting that encloses the manual transmission parts.
Caja de transmisión Pieza fundida en aluminio o en hierro que encubre las piezas de la transmisión manual.

Transverse Powertrain layout in a front-wheel-drive automobile extending from side to side.
Transversal Distribución del tren transmisor de potencia en un vehículo de tracción delantera, que se extiende de un lado al otro.

Tripod (also called tripot) A three-prong bearing that is the major component in tripod constant velocity joints. It has three arms (or trunnions) with needle bearings and rollers that ride in the grooves or yokes of a tulip assembly.
Trípode Cojinete de tres puntas; componente principal en juntas trípode de velocidad constante. Tiene tres brazos (o muñequillas) con cojinetes de agujas y rodillos que van montados sobre las ranuras o los yugos de un conjunto tulipán.

Tripod universal joints Universal joint consisting of a hub with three arm and roller assemblies that fit inside a casting called a tulip.
Juntas universales de trípode Juntas universales que consisten de un cubo con conjuntos de tres brazos y de rodillos que se insertan dentro de una pieza fundida llamada un tulipán.

Trunnion One of the projecting arms on a tripod or on the cross of a four-point universal joint. Each trunnion has a bearing surface that allows it to pivot within a joint or slide within a tulip assembly.
Muñequilla Uno de los brazos salientes en un trípode o en la cruz de una junta universal de cuatro puntas. La superficie del cojinete de cada muñequilla permite que la muñequilla gire dentro de una junta o se deslice dentro de un conjunto tulipán.

Tulip assembly The outer housing containing grooves or yokes in which trunnion bearings move in a tripod constant velocity joint.
Conjunto tulipán Alojamiento exterior que contiene ranuras o yugos en los cuales se mueven los cojinetes de muñequilla en una junta trípode de velocidad constante.

Two-disk clutch A clutch with two friction discs for additional holding power; used in heavy duty equipment.
Embrague de dos discos Embrague con dos discos de fricción para proporcionar más fuerza de retención; utilizado en equipo de gran potencia.

Two-speed rear axle See double-reduction differential.
Eje trasero de dos velocidades Véase diferencial de reducción doble.

U-bolt An iron rod with threads on both ends, bent into the shape of a U and fitted with a nut at each end.
Perno en U Varilla de hierro con filetes de tornillo en los dos extermos, acodada en forma de U y provista de una tuerca a cada extremo.

U-joint A four-point cross connected to two U-shaped yokes that serves as a flexible coupling between shafts.
Junta cardánica Cruz con cuatro puntas fijadas a dos yugos en forma de U, que sirve de acoplamiento flexible entre árboles.

Universal joint A mechanical device that transmits rotary motion from one shaft to another shaft at varying angles.
Junta universal Dispositivo mecánico que transmite movimiento giratorio de un árbol a otro a ángulos cambiantes.

Universal joint operating angle The difference in degrees between the drive shaft and transmission installation angles.
Ángulo de funcionamiento de la junta universal Diferencia en grados entre los ángulos del árbol de mando y del montaje de la transmisión.

Unsprung weight The weight of the tires, wheels, axles, control arms, and springs.
Peso no suspendido Peso de las ruedas, llantas, ejes, brazos de mando, y muelles.

Upshift To shift a transmission into a higher gear.
Cambio de velocidades ascendente Acción de cambiar la transmisión a un engranaje de alta multiplicación.

USCS measuring system United States Customary System is also known as the English system.
Sistema de medida de USCS Sistema Usual U.S.; conocido también como Sistema Imperial Británico.

Vibration A quivering, trembling motion felt in the vehicle at different speed ranges.
Vibración Estremecimiento y temblor que se advierte en el vehículo a diferentes gamas de velocidades.

Vehicle identification number (VIN) The number assigned to each vehicle by its manufacturer, primarily for registration and identification purposes.
Número de identificación del vehículo Número asignado a cada vehículo por el fabricante, principalmente para su registración e identificación.

Viscosity The resistance to flow exhibited by a liquid. A thick oil has greater viscosity than a thin oil.
Viscosidad Resistencia de un fluido al movimiento relativo. Un aceite pesado tiene mayor viscosidad que un aceite fluido.

Viscous Thick; tending to resist flowing.
Viscoso Espeso; que tiende a resistir el movimiento.

Viscous friction The friction between layers of a liquid.
Fricción viscosa Fricción entre las capas de un fluido.

Wet-disc clutch A clutch in which the friction disc (or discs) is operated in a bath of oil.
Embrague de disco húmedo Embrague en el que el disco de fricción (o discos) funciona bañado en aceite.

Wheel A disc or spokes with a hub at the center that revolves around an axle, and a rim around the outside for mounting the tire on.
Rueda Disco o rayos con un cubo en el centro que gira alrededor de un eje, y una llanta alrededor del exterior sobre la cual se monta la rueda.

Wheel offset The amount of the wheel assembly that is to the side of the wheel's mounting hub.
Desviación de la rueda La porción del conjunto de la rueda ubicada al lado del cubo de montaje de la rueda.

Wheel shimmy The wobble of a tire.
Baileteo de la rueda Movimiento de zigzag de una rueda.

Worm gear A gear with teeth that resemble a thread on a bolt. It is meshed with a gear that has teeth similar to a helical tooth except that it is dished to allow more contact.
Engranaje sinfín Engranaje con dientes parecidos a los filetes de tornillo en un perno. Se engrana con un engranaje cuyos dientes son parecidos a un diente helicoidal, pero se comba para permitir un mejor contacto.

Yoke In a universal joint, the drivable torque-and-motion input and output member, attached to a shaft or tube.
Yugo En una junta universal, el mecanismo accionable impulsor y de rendimiento de torsión y movimiento, que se fija a un árbol o un tubo.

Yoke bearing A U-shaped, spring-loaded bearing in the rack-and-pinion steering assembly that presses the pinion gear against the rack.
Cojinete de yugo En el conjunto de dirección de cremallera y piñón, cojinete en forma de U, con cierre automático, que sujeta el piñón contra la cremallera.

Zerk fitting A common name for grease fittings. A very small check valve that allows grease to be injected into a component part, but keeps the grease from squirting out again.
Conexión Zerk Término común para conexiones de engrase. Válvula de retención sumamente pequeña que permite inyectar la grasa en un componente, y que a la vez impide que esa grasa se derrame nuevamente.

4 Suspension and Steering

Pretest

The purpose of this pretest is to determine the amount of review that you may require prior to writing the ASE Suspension and Steering Test. If you answer all the review questions correctly, complete the questions and study the information in this chapter to prepare for the ASE Suspension and Steering Test.

If two or more of your answers to the pretest questions are incorrect, complete a study of Chapters 3 through 17 in *Today's Technician Automotive Suspension and Steering Systems Classroom and Shop Manuals*, published by Delmar Publishers, plus a study of the questions and information in this chapter.

The pretest answers are located at the end of the pretest, and these answers also are in the answer sheets supplied with this book.

1. A customer complains about steering wander while driving straight ahead.
   Technician A says the front wheel bearing adjustments may be loose.
   Technician B says the front wheels may have negative caster settings.
   Who is correct?
   A. A only
   B. B only
   C. Both A and B
   D. Neither A nor B

2. When performing a sector lash adjustment on a manual recirculating ball steering gear use an inch-pound torque wrench and socket to rotate the worm shaft through a 45 degree arc in each direction with the worm shaft positioned
   A. one turn from the full left of its travel.
   B. in the center of its travel.
   C. one-half turn from the full right of its travel.
   D. one turn to the right from the center of its travel.

3. A power rack and pinion steering gear has excessive steering effort in both directions. The cause of this problem could be
   A. the rack piston seal is leaking.
   B. the inner rack seal is leaking.
   C. the outer rack bulkhead seal is leaking.
   D. the rack bearing adjustment is too loose.

4. During a pressure test a power steering pump has less than the specified pressure.
   Technician A says the steering gear may be defective.
   Technician B says the pump pressure relief valve may be sticking.
   Who is correct?
   A. A only
   B. B only
   C. Both A and B
   D. Neither A nor B

5. A power rack and pinion steering system has poor steering wheel returnability. The cause of this problem could be
   A. a bent or damaged rack in the steering gear.
   B. excessive negative camber on both front wheels.
   C. worn inner tie rod ends in the steering gear.
   D. reduced front suspension curb riding height.

6. During a road test a technician discovers a vehicle has bump steer. The vehicle has power rack and pinion steering.
   Technician A says the steering gear mounting bushings may be worn.
   Technician B says the upper strut mounts may be worn.
   Who is correct?
   A. A only
   B. B only
   C. Both A and B
   D. Neither A nor B

7. The front suspension trim height on a computer-controlled air suspension system is less than specified.
   Technician A says the compressor vent valve may be defective.
   Technician B says the front height sensors may require adjusting.
   Who is correct?
   A. A only
   B. B only
   C. Both A and B
   D. Neither A nor B

8. While parking a car with four-wheel steering (4WS) the rear wheels
   A. steer about 5 degrees in the same direction as the front wheels.
   B. remain centered and do not steer in either direction.
   C. steer about 5 degrees in the opposite direction to the front wheels.
   D. steer about 1 degree in the same direction as the front wheels.

9. Fluid foaming occurs in a power steering pump reservoir.
   Technician A says the power steering belt may be too loose.
   Technician B says the power steering fluid may be contaminated.
   Who is correct?
   A. A only
   B. B only
   C. Both A and B
   D. Neither A nor B

10. A variable assist power steering system provides
    A. reduced pump pressure at low vehicle speeds.
    B. reduced pump pressure at low engine speeds.
    C. increased pump pressure at low vehicle speeds.
    D. increased pump pressure at high vehicle speeds.

11. A customer complains about front wheel shimmy. All of these defects could be the cause of the problem EXCEPT
    A. improper dynamic wheel balance.
    B. excessive toe-in on the front wheels.
    C. excessive positive caster on the front wheels.
    D. sagged springs in the rear suspension.

12. The steering on a vehicle pulls to the right while driving straight ahead. The cause of this problem could be
    A. more toe-out on the right rear wheel than on the left rear wheel.
    B. more positive camber on the left front wheel than on the right front wheel.
    C. more toe-out on turns on the left front wheel than on the right front wheel.
    D. more positive caster on the left front wheel than on the right front wheel.

13. When performing a shock absorber bounce test, the left side of the front bumper is pushed downward with considerable weight and then released. The chassis completes one free upward bounce before the vertical chassis movement stops. This action indicates
    A. the coil spring is weak.
    B. the stabilizer bar is weak.
    C. the control arm bushings are worn.
    D. the shock absorber is satisfactory.

14. On a vehicle with a MacPherson strut front suspension the steering pulls to the left while braking.
    Technician A says there may be excessive difference between the left and right SAI readings.
    Technician B says the left rear wheel may have excessive positive camber.
    Who is correct?
    A. A only
    B. B only
    C. Both A and B
    D. Neither A nor B

Answers to Pretest

1. C, 2. B, 3. A, 4. B, 5. A, 6. A, 7. B, 8. C, 9. B, 10. C, 11. B, 12. D, 13. D, 14. A

Steering Systems Diagnosis and Repair, Steering Columns and Manual Steering Gears

ASE Tasks, Questions, and Related Information

In this chapter each task in the Suspension and Steering category is provided followed by a question and some information related to the task. If you answer any question incorrectly, study this information very carefully until you understand the correct answer. For additional information on any task refer to *Today's Technician Automotive Suspension and Steering Systems Classroom and Shop Manuals*, published by Delmar Publishers.

Question answers and analysis are provided at the end of this chapter and in the answer sheets provided with this book.

Task 1 **Diagnose steering column noises, looseness, and binding problems (including tilt mechanisms): determine needed repairs.**

1. With the steering column mounted in the vehicle and all linkages connected, a steering wheel has 2.35 in. (60 mm) of free play.

Technician A says the flexible coupling may be worn.
Technician B says the steering gear mounting bolts may be loose.
Who is correct?
A. A only
B. B only
C. Both A and B
D. Neither A nor B

Hint   *For complete steering column diagnosis refer to Table 9-1, page 205, in* Today's Technician
Automotive Suspension and Steering Systems Shop Manual, *published by Delmar Publishers.*

Task 2   Diagnose manual steering gear (non-rack and pinion type) noises, binding, uneven turning effort, looseness, hard steering, and lubricant leakage problems; determine needed repairs.

2. Excessive steering effort on a manual steering gear (non-rack and pinion type) may
be caused by
A. a loose worm bearing preload adjustment.
B. an overfilled steering gear.
C. less than specified positive caster.
D. a tight sector lash adjustment.

Hint   *For complete diagnosis of manual steering gears (non-rack and pinion type) refer to Table 11-1, page 261, in* Today's Technician Automotive Suspension and Steering Systems Shop
Manual, *published by Delmar Publishers.*

Task 3   Diagnose manual rack and pinion gear noises, vibration, looseness, and hard steering problems; determine needed repairs.

3. The cause of poor returnability on a manual rack and pinion steering gear could be
A. a loose rack bearing adjustment.
B. insufficient or improper lubricant in the steering gear.
C. loose steering gear mounting bolts.
D. excessive positive camber on both front wheels.

Hint   *For complete diagnosis of manual rack and pinion steering gears refer to Table 12-1, page
275, in* Today's Technician Automotive Suspension and Steering Systems Shop Manual,
published by Delmar Publishers.

Task 4   Inspect and replace steering shaft U-joint (s), flexible coupling (s), collapsible columns, steering wheels, including steering wheels with air bags.

4. A mounting bolt in a collapsible steering column bracket is illustrated in Figure 4-1.
Technician A says if the bolt head is touching the bracket, the bracket should be
replaced.
Technician B says if the bolt head is touching the bracket, the shear load is too low.
Who is correct?
A. A only
B. B only
C. Both A and B
D. Neither A nor B

Hint   *For complete diagnosis of steering columns refer to Table 9-1, page 205, in* Today's Technician Automotive Suspension and Steering Systems Shop Manual, *published by Delmar Publishers.*

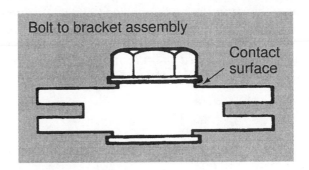

Figure 4-1 Collapsible steering column bracket *(Courtesy of Oldsmobile Division, General Motors Corporation)*

Task 5   Remove and replace manual steering gear (non-rack and pinion type)

Task 6   Adjust manual steering gear (non-rack and pinion type) worm bearing preload and sector lash.

5. In the preliminary procedure the sector lash adjuster is rotated counterclockwise until it stops, and then rotated clockwise one turn. A turning torque reading is taken with the steering gear in the center position (Figure 4-2).

Following this procedure the sector lash adjuster should be rotated in

A. a clockwise direction until the turning torque is 4 to 10 in. lbs more than in the preliminary procedure.

B. a counterclockwise direction until the turning torque is 0 in. lbs.

C. a clockwise direction until the turning torque is 20 in. lbs more than in the preliminary procedure.

D. a clockwise direction until the turning torque is 5 ft lbs more than in the preliminary procedure.

Figure 4-2 Manual recirculating ball steering gear sector lash adjustment *(Courtesy of Chevrolet Motor Division, General Motors Corporation)*

Hint   *Manual recirculating ball steering gears require worm shaft preload and sector shaft lash adjustments. The worm shaft preload adjustment must be performed before the sector shaft lash adjustment.*

Loosen the worm shaft adjuster plug locknut and rotate this plug until all the worm shaft end-play is removed. Loosen the adjuster plug one-quarter turn. With the worm shaft positioned one-half turn from the fully left or fully right position, tighten the adjuster plug until the specified turning torque is obtained on an inch-pound torque wrench installed on the steering shaft. After the turning torque is adjusted, tighten the locknut to the specified torque (Figure 4-3).

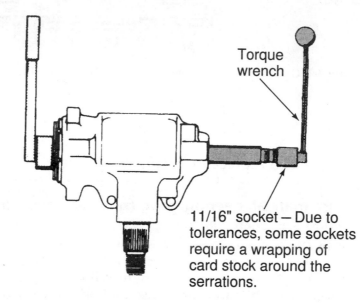

Figure 4-3 Manual recirculating ball steering gear worm shaft bearing preload adjustment *(Courtesy of Chevrolet Motor Division, General Motors Corporation)*

Prior to the sector shaft lash adjustment the lash adjuster screw is turned fully counterclockwise. This screw is then rotated one turn clockwise. With the worm shaft in the center position record the turning torque required to rotate the steering shaft. Turn the sector lash adjustment screw clockwise until the turning torque is 4 to 10 in. lbs more than the turning torque recorded with the sector lash screw backed off. Tighten the locknut on the sector shaft lash adjusting screw to the specified torque.

Task 7   Remove and replace rack and pinion steering gear.

Task 8   Adjust rack and pinion steering gear.

6. A manual rack and pinion steering gear requires excessive steering effort. Technician A says the lower ball joints on the front suspension may be worn. Technician B says the rack bearing adjustment may be too tight. Who is correct?
   A. A only
   B. B only
   C. Both A and B
   D. Neither A nor B

Hint   *During the assembly process the rack and rack teeth in a manual rack and pinion steering gear should be coated with a lithium-based grease. After the steering gear is assembled, rotate the pinion shaft fully in each direction and repeat this action. With a torque wrench and socket installed on the pinion shaft, rotate the pinion shaft back and forth while tightening the rack bearing adjuster cap. Continue tightening the adjuster cap until the specified turning torque is obtained on the torque wrench. Hold the adjuster cap and tighten the adjuster cap locknut to the specified torque.*

Task 9   Inspect and replace rack and pinion steering gear inner tie rod ends (sockets) and bellows boots.

7. During rack and pinion steering gear service Technician A says the inner tie rod end should be replaced if the articulation effort on this tie rod is less than specified. Technician B says the rack must be held while loosening the inner tie rod ends. Who is correct?
   A. A only
   B. B only
   C. Both A and B
   D. Neither A nor B

Hint   *The inner tie rods on the rack and pinion steering gear may be inspected with the steering gear installed. Loosen the clamp on the inner end of the bellows boot and move this end of the boot toward the outer tie rod end. Push inward and outward on each front wheel and watch for movement in each inner tie rod end. If any movement is present in either tie rod end, replace the tie rod end.*

With the steering gear removed connect a pull scale to the cotter key opening in the outer tie rod end. Pull upward to measure the articulation effort on the inner tie rod end. Repeat this procedure on each tie rod end. If the articulation effort is less than specified, replace the inner tie rod end.

Task 10   Inspect and replace rack and pinion steering gear mounting bushings and brackets.

8. The steering on a car with a manual rack and pinion steering gear suddenly veers in one direction when one or both front wheels hit a bump. The cause of this problem could be
   A. a loose steering gear mounting bushing.
   B. excessive positive caster on both front wheels.
   C. worn upper strut mounts.
   D. worn-out front struts.

Hint   *Rack and pinion steering gear mounting bushings may deteriorate from age, oil soaking, or heat. With the vehicle parked on the shop floor have an assistant turn the steering wheel one-half turn in both directions and watch for movement of the steering gear in the mounting bushings. If there is any movement replace the steering gear bushings. Worn, loose steering gear mounting bushings may cause unequal steering linkage heights which results in bump steer. Bump steer is the tendency of the steering to veer suddenly in one direction when one or both front wheels strike a bump.*

Power-Assisted Steering Units

ASE Tasks, Questions, and Related Information

Task 1   Diagnose power steering gear (non-rack and pinion type) noise, binding, uneven turning effort, looseness, hard steering, and fluid leakage problems; determine needed repairs.

9. Excessive looseness is experienced in a power steering gear (non-rack and pinion type). The cause of this problem could be
   A. a loose or worn power steering belt.
   B. a loose worm shaft bearing preload adjustment.
   C. a scored steering gear cylinder.
   D. a low fluid level in the power steering pump.

Hint   *For complete diagnosis of non-rack and pinion type power steering gears refer to Table 12-2, page 298, in* Today's Technician Automotive Suspension and Steering Systems Shop Manual, *published by Delmar Publishers.*

Task 2   Diagnose power rack and pinion steering gear noises, vibration, looseness, hard steering, and fluid leakage problems; determine needed repairs.

10. A power rack and pinion steering gear has a spurting oil leak at the location shown in Figure 4-4 when the rack contacts the left inner stop. The cause of this problem could be a worn
    A. inner rack seal.
    B. pinion seal.
    C. outer rack seal.
    D. input shaft seal.

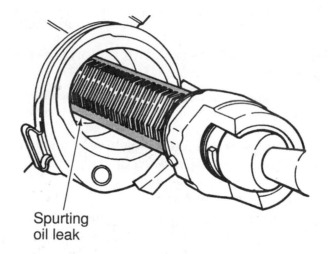

Spurting
oil leak

Figure 4-4 Power rack and pinion steering gear oil leak *(Courtesy of Chrysler Corporation)*

Hint   *For complete diagnosis of power rack and pinion steering gears refer to Table 12-2, page 298, in* Today's Technician Automotive Suspension and Steering Systems Shop Manual, *published by Delmar Publishers.*

Task 3   Inspect power steering fluid level and condition; adjust level in accordance with vehicle manufacturer's recommendations.

11. Many vehicle manufacturers recommend checking the power steering fluid level with the fluid temperature at
    A. 60°F (15.5°C).
    B. 100°F (38°C).
    C. 140°F (60°C).
    D. 175°F (80°C).

Hint   *The power steering fluid level should be checked with the power steering fluid at the normal operating temperature of 175°F (80°C). Inspect the power steering fluid for foaming, discoloration, and contamination. Foaming of the power steering fluid usually indicates a low fluid level or air in the system. If the fluid is discolored or contaminated the system should be flushed and filled with new fluid. Discolored power steering fluid may be caused by excessive heat or the wrong type of fluid. If the fluid is contaminated with metal particles, the pump or steering gear are worn.*

Task 4 Inspect, adjust tension and alignment, and replace power steering pump belt(s).

12. When checking the power steering pump belt tension, for every foot of free span the belt deflection should be
    A. 1/2 in.
    B. 1 in.
    C. 1.5 in.
    D. 1.75 in.

13. Technician A says a belt tension gauge may be used to measure the tension on a ribbed V-belt.
    Technician B says some ribbed V-belts have a spring-loaded tensioner with a belt-length scale.
    Who is correct?
    A. A only
    B. B only
    C. Both A and B
    D. Neither A nor B

Hint   *Reduced power steering pump belt tension may cause increased or erratic steering effort. Power steering pump belt tension usually is measured with a belt tension gauge. Belt tension may be checked by measuring the belt deflection. The specified deflection is usually 1/2 in. per foot of belt free span (Figure 4-5). If the power steering belt must be tightened, loosen the bracket bolt and pry on the pump ear with a pry bar to tighten the belt. Hold the pump so the belt has the proper tension, and tighten the bracket bolt.*

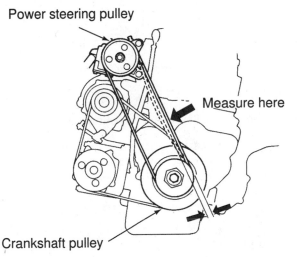

Power steering pulley

Measure here

Crankshaft pulley

Figure 4-5 Power steering pump belt deflection *(Courtesy of American Honda Motor Co., Inc.)*

Task 5 Remove and replace power steering pump; inspect pump mounts.

Task 6 Inspect and replace power steering pump seals and gaskets.

14. A power steering pump with an integral reservoir is leaking fluid between the reservoir and the pump housing.
    Technician A says the drive shaft seal may require replacement.
    Technician B says the vent in the pump cap may be restricted.
    Who is correct?

A. A only
B. B only
C. Both A and B
D. Neither A nor B

Task 7   Inspect and replace power steering pump pulley.

15. The tool in Figure 4-6 is used to
    A. remove a pressed-on power steering pump pulley.
    B. remove a bolt-on power steering pump pulley.
    C. install a pressed-on power steering pump pulley.
    D. remove the power steering pump pulley retaining nut.

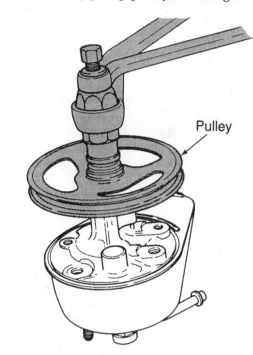

Pulley

Figure 4-6 Power steering pump service tool *(Courtesy of Chrysler Corporation)*

Hint   *A special puller must be used to remove a pressed-on power steering pump pulley. This type of pulley also requires a special installation tool. If the power steering pump pulley is retained with a nut, the pulley must be held from turning with a special tool while the nut is loosened.*

Task 8   Perform power steering pump pressure test; determine needed repairs.

16. During the power steering pump pressure test the valve shown in Figure 4-7 should be closed for
    A. 10 seconds.
    B. 15 seconds.
    C. 20 seconds.
    D. 30 seconds.

Hint   *The power steering pump pressure test must be performed with the fluid at normal operating temperature and no air in the system. A pressure gauge and valve assembly is connected in the pressure hose between the power steering pump and gear. With the engine idling, close the valve for a maximum of 10 seconds and read the pump pressure. If the power steering pump pressure is less than specified, the pump pressure relief valve may be sticking or the pump may be defective.*

If the power steering pump pressure is satisfactory, measure the steering wheel turning effort with a pull scale with the engine idling and the steering wheel centered. When the pump pressure is satisfactory, and the steering wheel turning effort is excessive, check the power steering hoses for restrictions. If the hoses are satisfactory, repair the steering gear.

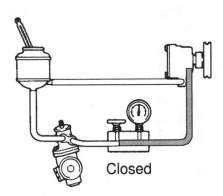

Closed

Figure 4-7 Power steering pump pressure test *(Courtesy of Toyota Motor Corporation)*

Task 9   Inspect and replace power steering hoses, fittings, and O-rings.

17. During a power steering pump pressure test the pump pressure is satisfactory, but the steering wheel turning effort is excessive.
    Technician A says the steering gear may be defective.
    Technician B says the high-pressure hose may be restricted.
    Who is correct?
    A. A only
    B. B only
    C. Both A and B
    D. Neither A nor B

Hint   *Power steering hoses should be checked for leaks, dents, restrictions, sharp bends, cracks, and contact with other components. Restricted power steering hoses reduce pressure supplied to the steering gear and increase steering effort.*

Task 10   Remove and replace power steering gear (non-rack and pinion type).

Task 11   Remove and replace power rack and pinion steering gear; inspect and replace mounting bushings and brackets.

18. When removing the steering gear on an air-bag-equipped vehicle
    Technician A says rotating the steering wheel with the steering gear disconnected may damage the clockspring electrical connector.
    Technician B says after the negative battery cable is disconnected immediately begin the steering gear removal procedure.
    Who is correct?
    A. A only
    B. B only
    C. Both A and B
    D. Neither A nor B

Hint   *On an air-bag-equipped vehicle the steering wheel should be locked in the centered position before the steering wheel or steering gear is removed. This action maintains the clockspring electrical connector in the centered position. Before starting the steering gear removal procedure on an air-bag-equipped vehicle disconnect the negative battery cable and wait for the time specified by the vehicle manufacturer. Always punch mark the lower universal joint in relation to the steering gear shaft prior to steering gear removal.*

Task 12   Adjust power steering gear (non-rack and pinion type) worm bearing preload and sector lash.

19. In the power steering gear adjustment in Figure 4-8

    Technician A says after the adjuster plug is bottomed it should be tightened to 20 ft lbs.

    Technician B says the adjuster plug should be backed off one-quarter turn from the bottomed position.

    Who is correct?

    A. A only

    B. B only

    C. Both A and B

    D. Neither A nor B

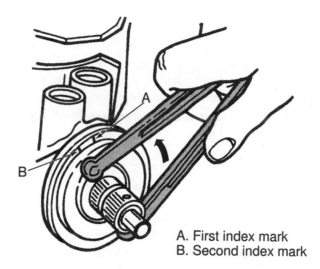

A. First index mark
B. Second index mark

Figure 4-8 Power recirculating ball steering gear adjustment *(Courtesy of Chevrolet Motor Division, General Motors Corporation)*

Hint   *When adjusting the worm shaft bearing preload on some power recirculating ball steering gears, the worm shaft bearing adjuster plug must be bottomed and then tightened to 20 ft lbs. An index mark should be placed on the housing next to or on the adjuster plug holes. Place a second index mark 0.050 in. from the original index mark in a counterclockwise direction. Rotate the adjuster plug until the hole in the adjuster plug is aligned with the second index mark.*

Task 13   Inspect and replace power steering gear (non-rack and pinion type) seals and gaskets.

20. When servicing the bearing and seal in the adjuster plug of a power recirculating ball steering gear (Figure 4-9)

    Technician A says the part number on the needle bearing should face the driving tool when installing this bearing.

    Technician B says the worm shaft bearing preload must be adjusted after the bearing and seal installation in the adjuster plug.

    Who is correct?

    A. A only

    B. B only

    C. Both A and B

    D. Neither A nor B

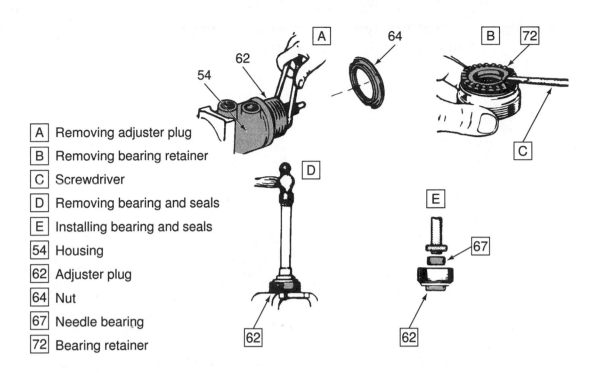

A | Removing adjuster plug
B | Removing bearing retainer
C | Screwdriver
D | Removing bearing and seals
E | Installing bearing and seals
54 | Housing
62 | Adjuster plug
64 | Nut
67 | Needle bearing
72 | Bearing retainer

Figure 4-9 Servicing adjuster plug bearing and seal, power recirculating ball steering gear *(Courtesy of Chevrolet Motor Division, General Motors Corporation)*

Hint   *Possible leak locations on a power recirculating ball steering gear include the pitman shaft seal, side cover O-ring seal, adjuster plug seal, top cover seal, and the pressure line fitting.*

Task 14   Adjust power rack and pinion steering gear.

21. Poor steering wheel returnability is experienced on a power rack and pinion steering gear.
   Technician A says the steering gear may be misaligned on the chassis.
   Technician B says the rack bearing adjustment may be too tight.
   Who is correct?
   A. A only
   B. B only
   C. Both A and B
   D. Neither A nor B

Hint   *When the rack bearing adjustment is performed a torque wrench and socket is installed on top of the pinion shaft. Turn the torque wrench back and forth to rotate the rack in both directions. Rotate the rack bearing adjuster plug until the specified turning torque is obtained on the torque wrench. Hold the adjuster plug in this position, and tighten the adjuster plug locknut to the proper torque.*

Task 15   Inspect and replace power rack and pinion steering gear inner tie rod ends (sockets), seals, gaskets, O-rings, and bellows boots.

22. In Figure 4-10 component 13 is
    A. a spacer.
    B. a rack bushing.
    C. a shock dampener.
    D. a rack seal.

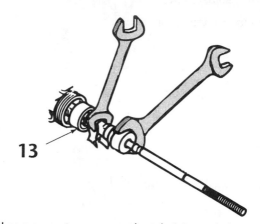

Figure 4-10 Inner tie rod components, power rack and pinion steering gear *(Courtesy of Chevrolet Motor Division, General Motors Corporation)*

Hint   *Some inner tie rod ends are threaded onto the rack. On this type of gear the rack must be held from rotating while the inner tie rods are loosened or tightened. These inner tie rod ends must be tightened to the specified torque. Some inner tie rods are staked in place after they are properly torqued. A claw washer retains the inner tie rods on other steering gears. Some inner tie rod ends are retained on the rack with a steel pin.*

Task 16   Diagnose, inspect, adjust, repair or replace components of electronically controlled steering systems.

23. When an electronically controlled four-wheel steering (4WS) system enters the fail-safe mode
    A. voltage is maintained to the rear steering unit.
    B. the control unit deenergizes the damper relay.
    C. the rear steering unit moves rapidly to the center position.
    D. the 4WS light is illuminated in the instrument panel.

Hint   *In an electronically controlled 4WS system if the control unit senses an electrical defect in the system, the control unit enters the fail-safe mode. In this mode the control unit illuminates the 4WS light in the instrument panel and shuts off the voltage supplied to the rear-wheel steering unit. Under this condition if the rear wheels are not centered, a return spring in the rear-wheel steering unit begins moving the rear wheels to the center position. When this action is taken, rear steering shaft movement rotates the armature in the rear steering unit. The control unit also energizes a damper relay, and voltage is supplied through these relay contacts to the armature in the rear steering unit. Energizing the armature under this condition causes the armature to act as a magnetic brake. This action causes the rear steering shaft to slowly return to the center position.*

Task 17   Flush, fill, and bleed power steering systems.

24. While bleeding air from a power steering system
    Technician A says if foaming is present in the reservoir after the bleeding process, the bleeding procedure should be repeated.
    Technician B says each time the steering wheel is rotated fully right or left it should be held in this position for two or three seconds.
    Who is correct?

A. A only
B. B only
C. Both A and B
D. Neither A nor B

Hint   *A power steering system should be drained through the return hose. If the system has a remote reservoir, disconnect the return hose from this reservoir to the steering gear and plug the reservoir outlet. Place the disconnected return hose in a drain pan. Flush the fluid out of the system by starting the engine and allowing it to idle. Shut the engine off and add more fluid to the reservoir to continue flushing the system.*

With the reservoir filled to the proper level, operate the engine at 1,000 rpm and rotate the steering wheel fully right and left three or four times to bleed air from the system. Hold the steering wheel in the fully right and left position for two or three seconds. Be sure the reservoir is filled to the proper level. If foaming is still present in the reservoir, repeat the bleeding procedure.

Task 18   **Diagnose, inspect, repair or replace components of variable-assist steering systems.**

25. In the variable-assist power steering system in Figure 4-11
     Technician A says the power steering assist is increased at speeds above 50 mph.
     Technician B says the power steering assist is increased when the steering wheel rotation exceeds 15 rpm.
     Who is correct?
     A. A only
     B. B only
     C. Both A and B
     D. Neither A nor B

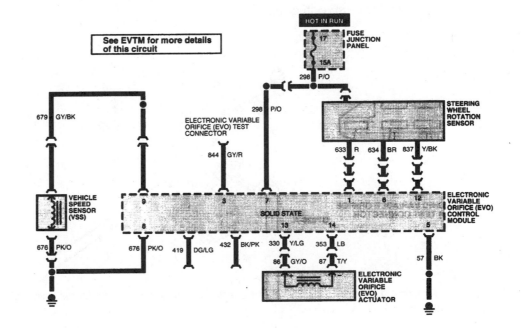

Figure 4-11   Variable assist power steering system *(Courtesy of Ford Motor Company)*

Hint   *The control module in some variable-assist power steering systems receives input signals from the vehicle speed sensor (VSS) and the steering wheel rotation sensor. In response to these inputs the control module operates an actuator in the power steering pump to control pump pressure and steering assist.*

When the vehicle speed is below 10 mph the control unit positions the actuator to provide full power steering assist. If the vehicle speed is between 10 mph and 25 mph, the control unit gradually reduces the power steering pump pressure and steering assist. Above 25 mph the control unit provides a reduced power steering assist.

When the steering wheel rotation is above 15 rpm, the control unit provides increased power steering pump pressure and steering assist. If the steering wheel rotation is below 15 rpm the control unit supplies a reduced power steering assist. Some variable assist power steering systems do not have a steering wheel rotation sensor, and these systems are only vehicle speed sensitive.

Steering Linkage

ASE Tasks, Questions, and Related Information

Task 1   **Inspect and adjust (where applicable) front and rear steering linkage geometry including attitude and parallelism.**

26. The measurement in Figure 4-12 is more than specified. This problem may result in

    A. front wheel shimmy.
    B. steering pull to the left.
    C. steering pull to the right.
    D. excessive steering effort.

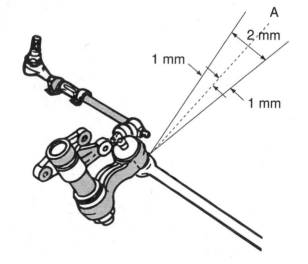

Figure 4-12   Idler arm movement check   *(Courtesy of Chevrolet Motor Division, General Motors Corporation)*

Hint   *The steering arms must be parallel to the lower control arms. In a parallelogram steering linkage the idler arm and pitman arm maintain the proper steering arm position. The steering gear mounting positions the steering arms parallel to the lower control arms in a rack and pinion steering system. If the steering arms are not parallel to the lower control arms, a condition called bump steer may occur. Bump steer may be defined as the tendency of the steering to veer suddenly in one direction when one or both front wheels strike a bump. A worn idler arm may cause wheel shimmy, bump steer, and steering looseness.*

Task 2   Inspect and replace pitman arm.

27. Bump steer is experienced during a road test for steering diagnosis.
    Technician A says the steering gear may require adjustment.
    Technician B says the pitman arm may be bent.
    Who is correct?
    A. A only
    B. B only
    C. Both A and B
    D. Neither A nor B

Hint   *Bent steering linkage components must be replaced. Never attempt to heat or straighten bent steering components such as pitman arms.*

Task 3   Inspect and replace relay (center link/intermediate) rod.

28. During a front suspension inspection the technician discovers a bent relay rod.
    Technician A says this problem changes the front wheel toe setting.
    Technician B says this problem may cause front tire wear on the inside edges.
    Who is correct?
    A. A only
    B. B only
    C. Both A and B
    D. Neither A nor B

Hint   *A bent relay rod moves the front wheel toe setting to a toe-out position. This condition may cause feathered wear across the front tire treads.*

Task 4   Inspect, adjust (where applicable), and replace idler arm and mountings.

29. A vehicle requires excessive steering effort. The power steering belt is tight and the reservoir is filled to the specified level. The cause of this problem could be
    A. worn lower ball joints.
    B. weak front springs.
    C. a seized idler arm.
    D. a weak stabilizer bar.

Hint   *Excessive steering effort may be caused by a slipping power steering belt, low fluid level, air in the power steering system, tight steering gear adjustments, worn steering gear, or a seized steering linkage component such as an idler arm.*

Task 5   Inspect, replace, and adjust tie rods, tie rod sleeves, clamps, and tie rod ends (sockets).

30. While discussing tie rod sleeve adjustments
    Technician A says the tie rod sleeves may be rotated to adjust front wheel toe.
    Technician B says the tie rod sleeves may be rotated to center the steering wheel.
    Who is correct?
    A. A only
    B. B only
    C. Both A and B
    D. Neither A nor B

Hint   *A special tie rod sleeve adjusting tool must be used to rotate the tie rod sleeves. Never rotate these sleeves with vise grips or a pipe wrench. Loosen the tie rod sleeve clamp bolts before rotating the sleeves. These sleeves may be rotated to adjust the front wheel toe or center the steering wheel. After the tie rod sleeves are rotated, the sleeve clamps must be positioned with the clamp open-*

ings away from the slots in the sleeve. After the clamp bolts are tightened to the specified torque, the outer clamp ends may touch, but some gap must be present between the clamp ends next to the sleeve (Figure 4-13).

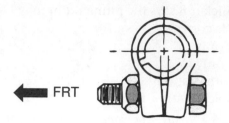

←— FRT

Figure 4-13 Proper clamp position on a tie rod sleeve *(Courtesy of Chevrolet Motor Division, General Motors Corporation)*

Task 6   Inspect and replace steering linkage damper.

31. A 4WD vehicle has excessive road shock on the steering wheel while driving on irregular road surfaces.

    Technician A says the power steering gear cylinder may be scored.

    Technician B says the steering damper may be worn out.

    Who is correct?

    A. A only

    B. B only

    C. Both A and B

    D. Neither A nor B

Hint   *The steering damper usually is connected from relay rod to the chassis. The steering damper acts like a shock absorber to reduce the transfer of road shock from the steering linkage to the steering wheel.*

Suspension Systems Diagnosis and Repair, Front Suspensions

ASE Tasks, Questions, and Related Information

Task 1   Diagnose front suspension system noises, body sway, and ride height problems; determine needed repairs.

32. The front suspension ride height is less than specified on a MacPherson strut suspension system. The cause of this problem could be

    A. worn out struts.

    B. worn upper strut mounts.

    C. worn steering gear mounting bushings.

    D. worn lower control arm bushings.

Hint   *Refer to Chapter 6, page 111, in* Today's Technician Automotive Suspension and Steering Systems Shop Manual *for complete diagnosis of front suspension systems.*

Task 2   Inspect and replace upper and lower control arms, bushings, shafts, and rebound bumpers.

33. Tool A in Figure 4-14 is used to
    A. compress the coil spring.
    B. position the shock absorber rod.
    C. measure ball joint movement.
    D. measure lower control arm bushing wear.

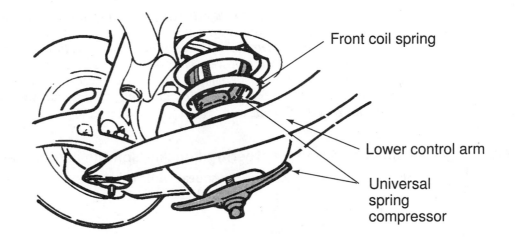

Front coil spring

Lower control arm

Universal spring compressor

Figure 4-14 Front suspension service tool *(Courtesy of Chevrolet Motor Division, General Motors Corpora tion)*

Hint   *Worn upper or lower control arm bushings on a front suspension may cause improper camber or caster angles. These improper angles may result in steering pull and/or tire tread wear. A rattling or squeaking noise on irregular road surfaces may be caused by worn control arm bushings.*

Task 3   Inspect, adjust, and replace strut rods/radius arm (compression/tension) and bushings.

34. The steering on a vehicle pulls to the right while braking, and all the brake components are in satisfactory condition.
    Technician A says the right front strut rod bushing may be worn.
    Technician B says there may be excessive negative camber on the right front wheel.
    Who is correct?
    A. A only
    B. B only
    C. Both A and B
    D. Neither A nor B

Hint   *The strut rods are connected from the lower control arm to the chassis. A large rubber bushing is mounted between the front of the strut rod and the chassis. The strut rods prevent fore-and-aft lower control arm movement. A worn strut rod bushing may allow the outer end of the lower control arm to move rearward while braking. This action provides reduced positive caster on that side of the front suspension. Since the steering pulls to the side with the least positive caster, a worn strut rod bushing may cause steering pull while braking.*

Task 4   Inspect and replace upper and lower ball joints.

35. When unloading the spring tension and vehicle weight from the ball joints prior to ball joint wear measurement on a long-and-short arm front suspension system with the coil spring positioned between the lower control arm and the chassis

    A. a jack must be placed under the chassis.
    B. the shock absorber must be disconnected.
    C. the ride height must be within specifications.
    D. a jack must be placed under the lower control arm.

Hint   *On a long-and-short arm suspension system the spring tension and vehicle weight must be unloaded from the ball joints before the ball joint wear is measured. To unload the ball joints on a long-and-short arm suspension system with the coil spring between the lower control arm and the chassis place a jack under the lower control arm. Operate the jack until the front tire is lifted off the shop floor.*

    When the coil spring is mounted between the upper control arm and the chassis, place a steel spacer between the upper control arm and the chassis. When this spacer is in place, position a jack under the chassis. Operate the jack until the front tire is lifted off the floor.

    To measure vertical ball joint wear position a dial indicator stem against the top of the ball joint stud, and use a pry bar to apply a lifting force under the tire. Place a dial indicator stem against the inside of the front wheel rim, and pull the tire and wheel inward and outward to measure horizontal ball joint wear.

Task 5   Inspect and replace steering knuckle/spindle assemblies.

36. A vehicle has excessive tire squeal while cornering.
    The cause of this problem could be
    A. a bent steering arm.
    B. excessive negative caster.
    C. worn stabilizer bushings.
    D. worn out front struts.

Hint   *A bent steering arm affects the toe-out-on turns angle. Improper toe-out-on turns, or an incorrect toe setting may cause excessive tire squeal while cornering. A bent steering knuckle may change the front wheel camber setting. In many suspension systems the steering knuckle and arm are combined in one component. Bent steering arms or knuckles must be replaced.*

Task 6   Inspect and replace front suspension system coil springs and spring insulators (silencers).

37. The coil spring in Figure 4-15 is
    A. a linear-rate spring.
    B. a variable-rate spring.
    C. a heavy-duty spring.
    D. a conventional spring.

Figure 4-15 Coil spring *(Courtesy of Sealed Power Corporation)*

38. While using a spring compressor to remove a coil spring from a strut (Figure 4-16) Technician A says the spring should be taped in the areas where the compressor contacts the spring.

   Technician B says all the spring tension must be removed from the upper strut mount before loosening the strut rod nut.

   Who is correct?

   A. A only
   B. B only
   C. Both A and B
   D. Neither A nor B

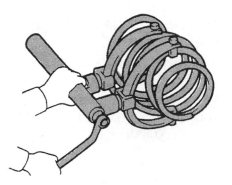

Figure 4-16 Coil spring compressor *(Courtesy of Toyota Motor Corporation)*

Hint   *Linear-rate or conventional coil springs have equal spacing between the coils and one basic shape with consistent wire diameter. Many variable rate springs have a cylindrical shape and unequally spaced coils with consistent wire diameter. Heavy-duty springs are designed to carry 3 percent to 5 percent greater loads compared to conventional springs. The wire diameter may be up to 0.100 in. greater in a heavy-duty spring than in a conventional spring.*

When removing a coil spring from a strut, the spring should be taped in the areas contacted by the spring compressor. If the coating on a coil spring is chipped, the spring may break. Before loosening the strut rod, the spring compressor must be operated until all spring tension is removed from the upper strut mount. Coil springs must be replaced in pairs.

Task 7   **Inspect and replace front suspension system leaf spring(s), leaf spring insulators (silencers), shackles, brackets, bushing, and mounts.**

39. The steering pulls to the right while driving straight ahead on a truck with a long-and-short arm front suspension and a leaf spring rear suspension. All of these defects could be the cause of the problem EXCEPT

   A. a broken center bolt in the left rear spring.
   B. more positive caster on the left front wheel than right front wheel.
   C. more positive camber on the right front wheel than the left front wheel.
   D. excessive toe-in on the front wheels.

Hint   *Worn leaf spring silencers may cause creaking and squawking noises while driving on road irregularities. A broken center bolt in a rear leaf spring may allow the differential housing to move rearward on one side. This defect affects vehicle tracking and causes steering pull.*

Task 8   **Inspect, replace, and adjust front suspension system torsion bars; inspect mounts.**

40. On a torsion bar front suspension system the ride height is below specifications on the right front side of the chassis. The ride height is satisfactory on the left front of the chassis.

Technician A says the lower control arm bushing may be worn.

Technician B says the right front torsion bar anchor bolt may need adjusting.

Who is correct?

A. A only

B. B only

C. Both A and B

D. Neither A nor B

Hint   *Torsions bars replace the coil springs in some suspension systems. Weak torsion bars or worn torsion bar mounting bushings may cause reduced chassis riding height. On many torsion bar suspension systems the torsion bar anchor bolt may be adjusted to change the ride height.*

Task 9   Inspect and replace stabilizer bar (sway bar) bushings, bracket, and links.

41. A vehicle has excessive body sway while cornering. All of these defects could be the cause of the problem EXCEPT

A. worn strut rod bushing.

B. weak stabilizer bar.

C. worn stabilizer bar bushings.

D. a broken stabilizer link.

Hint   *The stabilizer bar prevents excessive body sway while cornering. The stabilizer bar is mounted to the chassis and the control arms with heavy rubber bushings.*

Task 10   Inspect and replace MacPherson strut cartridge or assembly.

42. In Figure 4-17 the technician is

A. adjusting front wheel camber to correct tire wear.

B. loosening the strut rod nut prior to strut removal.

C. pushing the strut rod downward prior to cartridge removal.

D. aligning the strut rod during strut installation.

Hint   *A new cartridge may be installed in some front struts with the strut installed in the vehicle. Other struts must be removed to allow cartridge installation. On some struts there is no provision for cartridge installation. Prior to strut removal from the vehicle, the upper strut mounting nuts and the strut to steering knuckle bolts must be removed. A spring compressor must be used to compress the spring before the spring is removed from the strut.*

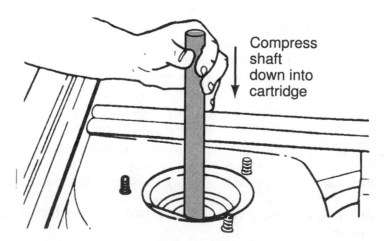

Compress
shaft
down into
cartridge

Figure 4-17 Front suspension service procedure *(Courtesy of Delco Products Division of General Motors Corporation)*

Task 11 Inspect and replace MacPherson strut upper bearing and mount.

43. A customer complains about steering chatter while cornering on a MacPherson strut front suspension system. With the vehicle parked on the shop floor the technician can feel a binding and releasing action on the left-front spring as the steering wheel is turned.

Technician A says the upper strut mount may be defective.

Technician B says the lower ball joint may have excessive wear.

Who is correct?

A. A only
B. B only
C. Both A and B
D. Neither A nor B

Hint *A worn upper strut mount may cause steering chatter while cornering and poor steering wheel return. A defective upper strut mount may result in improper camber or caster angles on the front suspension.*

Rear Suspensions

ASE Tasks, Questions, and Related Information

Task 1 Diagnose suspension system noises, body sway, and ride height problems; determine needed repairs.

44. A customer complains about harsh riding and bottoming of the rear suspension.

Technician A says the rear struts may be defective.

Technician A says the rear suspension ride height may be less than specified.

Who is correct?

A. A only
B. B only
C. Both A and B
D. Neither A nor B

Hint *A squeaking noise in the rear suspension may be caused by suspension bushings, defective struts or shock absorbers, or broken springs or spring insulators. Excessive rear suspension lateral movement may be caused by a worn track bar or bushings. Harsh riding may be caused by reduced rear suspension ride height, and defective struts or shock absorbers. Since rear suspension ride height affects some of the front suspension angles, the ride height must be measured before an alignment procedure.*

Task 2 Inspect and replace rear suspension system coil springs and spring insulators (silencers).

45. A vehicle has excessive rear suspension oscillations.

Technician A says the rear struts may be defective.

Technician B says the rear coil springs may be weak.

Who is correct?

A. A only
B. B only
C. Both A and B
D. Neither A nor B

Hint   *Weak coil springs cause harsh riding and reduced curb riding height. Broken springs or spring insulators cause a rattling noise while driving on irregular road surfaces. Worn-out struts or shock absorbers result in chassis oscillations and harsh riding.*

Task 3 **Inspect and replace rear suspension system transverse links (track bars), control arms, stabilizer bars (sway bars), bushings, and mounts.**

46. While servicing the rear suspension system in Figure 4-18
Technician A says the ball joint nut may be loosened to align the cotter key hole with the nut castellations.
Technician B says the lower control arm retaining bolts should be torqued with the vehicle weight on the suspension.
Who is correct?
A. A only
B. B only
C. Both A and B
D. Neither A nor B

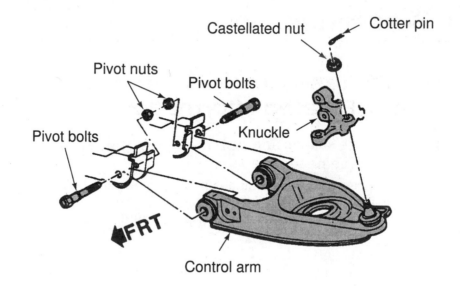

Figure 4-18 Rear suspension system with lower control arm and ball joint *(Courtesy of Oldsmobile Division, General Motors Corporation)*

Hint   *Worn track bar bushings or mounts cause excessive lateral chassis movement. Excessive rear body sway and harsh riding may be caused by a weak stabilizer bar or worn bushings, links, and mounts. Worn lower control arm bushings or ball joint may cause improper rear wheel toe and camber settings. This condition may result in tire wear or steering pull.*

Task 4 **Inspect and replace rear suspension system leaf spring(s), leaf spring insulators (silencers), shackles, brackets, bushings, and mounts.**

47. On a rear suspension with two longitudinally mounted leaf springs, the left rear spring is sagged and the left rear chassis curb riding height is less than specified. This problem could result in
A. steering pull to the right while driving straight ahead.
B. excessive left rear tire tread wear.
C. excessive steering wheel freeplay.
D. excessive left front tire tread wear.

Hint   *Sagged rear springs increase positive caster on the front suspension. When both rear springs are sagged, positive caster is increased on both front wheels. Under this condition steering effort, and steering wheel returning force after a turn, are increased. Increased positive caster may cause harsh riding.*

When one rear spring is sagged the positive caster is increased on that side of the front suspension. Since the steering tends to pull to the side with the least positive caster, this condition may cause the steering to pull to the opposite side from the sagged spring.

Task 5   Inspect and replace rear MacPherson strut cartridge or assembly, and upper mount assembly.

48. All of these problems may cause a rattling noise in the rear suspension EXCEPT
    A. a broken coil spring.
    B a broken coil spring insulator.
    C. a bent rear strut.
    D. a worn track bar bushing.

Hint   *A similar procedure is used to remove and replace front or rear struts. A spring compressor must be used to remove all the spring tension from the upper strut mount before loosening the strut rod nut. A strut cartridge may be installed in some rear struts after they are removed from the chassis.*

Task 6   Inspect nonindependent rear axle assembly for bending, warpage, and misalignment.

49. A nonindependent rear axle is offset as indicated in Figure 4-19. The result of this problem could be
    A. steering wander while driving straight ahead.
    B. steering pull to the right while driving straight ahead.
    C. poor steering wheel returnability.
    D. steering pull to the left during hard acceleration.

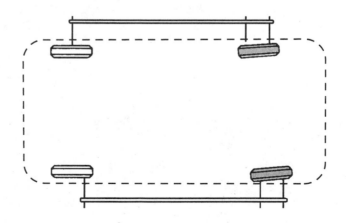

Figure 4-19 Rear axle offset

Hint   *A nonindependent rear axle may be checked for bending, warpage, and misalignment by measuring the rear wheel tracking. This operation may be performed with a track bar or a computer wheel aligner with four-wheel capabilities. A track bar measures the position of the rear wheels in relation to the front wheels. A computer wheel aligner displays the thrust angle which is the difference between the vehicle thrust line and geometric centerline of the vehicle.*

Task 7   Inspect and replace rear ball joints and tie rod assemblies.

50. While discussing the ball joint in Figure 4-20
Technician A says the ball joint is worn and should be replaced.
Technician B says the ball joint may cause an improper rear wheel camber setting.
Who is correct?

   A. A only
   B. B only
   C. Both A and B
   D. Neither A nor B

Figure 4-20 Rear ball joint *(Courtesy of Oldsmobile Division, General Motors Corporation)*

51. The left rear wheel tie rod in Figure 4-21 is longer than specified. This problem could cause

   A. excessive toe-out.
   B. excessive positive camber.
   C. excessive wear on the inside edge of the tire tread.
   D. steering pull to the left.

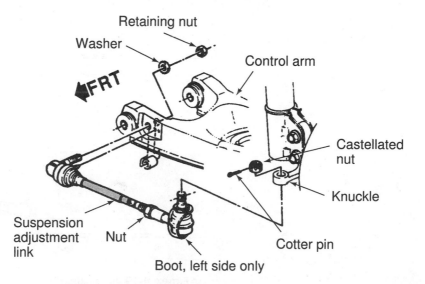

Figure 4-21 Rear suspension tie rod *(Courtesy of Oldsmobile Division, General Motors Corporation)*

Hint *Rear ball joints should be inspected in much the same way as front ball joints. Some rear suspension tie rods are connected from the knuckle to the lower control arm. The length of the tie rod determines the rear wheel toe setting.*

Miscellaneous Service

ASE Tasks, Questions, and Related Information

Task 1   **Inspect and replace shock absorbers.**

52. When one side of the front or rear bumper is pushed downward with considerable weight and then released, the bumper makes two free upward bounces before the vertical chassis movement stops. This action indicates
   A. a defective shock absorber.
   B. a weak coil spring.
   C. a broken spring insulator.
   D. worn stabilizer bushings.

Hint   *When one side of the bumper is pushed downward with considerable weight and then released, a satisfactory shock absorber or strut should complete one free upward bounce before the vertical chassis movement stops. If the bumper completes more than one free upward bounce, the shock absorber is defective, or the shock absorber mountings are loose. This is commonly called the shock absorber bounce test.*

A manual test may be performed on shock absorbers. During this test the lower end of the shock absorber is disconnected, and hand pressure is used to extend and compress the shock. A satisfactory shock absorber should provide a strong, steady resistance to movement on the compression and rebound strokes. The resistance may be different on the compression and rebound strokes.

Task 2   **Inspect and replace air shock absorbers, lines, and fittings.**

53. While discussing air shock absorbers
   Technician A says some air shock absorbers may be pressurized with the shop air hose.
   Technician B says if an air shock absorber slowly loses its air pressure, shock absorber replacement is probably necessary.
   Who is correct?
   A. A only
   B. B only
   C. Both A and B
   D. Neither A nor B

Hint   *Before removing an air shock absorber, relieve the air pressure in the shock. If air shock absorbers slowly lose their air pressure and reduce curb riding height, shock replacement is necessary.*

Task 3   **Diagnose and service front and/or rear wheel bearings.**

54. After servicing and lubricating the rear wheel bearings in a front-wheel-drive car the bearing adjusting nut is tightened to 20 ft lbs, and loosened one-half turn. The next step in the bearing adjusting procedure is to
   A. install the nut retainer and cotter key.
   B. tighten the adjusting nut so the cotter key hole lines up properly.
   C. tighten the adjusting nut to 10 to 15 ft lbs.
   D. check the wheel bearing for free play.

Hint   *Loose front wheel bearings may cause steering wander. When two tapered roller bearings are mounted in the front or rear wheel hub, a typical bearing adjustment procedure is the following:*
   • *Tighten the bearing adjustment nut to 17 to 25 ft lbs.*
   • *Back off the adjustment nut one-half turn.*
   • *Tighten the bearing adjustment nut to 10 to 15 ft lbs.*
   • *Install the nut retainer and cotter key.*

Task 4   Diagnose, inspect, adjust, repair, or replace components of electronically controlled suspension systems.

55. The electronic suspension switch in Figure 4-22 must be in the off position under all the following conditions EXCEPT
    A. diagnosing the system with a scan tester.
    B. jacking the vehicle to change a tire.
    C. hoisting the vehicle for under-car service.
    D. towing the vehicle with a tow truck.

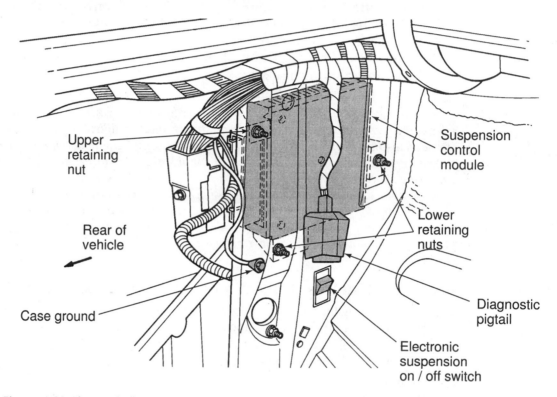

Figure 4-22 Electronically-controlled suspension system components mounted in the trunk *(Courtesy of Ford Motor Company)*

Hint   *There are several different types of electronically controlled suspension systems on cars today. Some computer controlled air suspension system have an air spring in place of each coil spring. On these systems the computer operates a compressor that supplies air into each air spring until the specified curb riding height is obtained. Most of these systems have an air suspension switch in the trunk that must be off while jacking, towing, or hoisting the vehicle. In the off position this switch shuts off the system voltage to the air suspension computer to make the system inoperative.*

Wheel Alignment Diagnosis, Adjustment, and Repair

ASE Tasks, Questions, and Related Information

Task 1   **Diagnose wheel wander, drift, pull, hard steering, bump steer, memory steer, torque steer, and steering return problems; determine needed repairs.**

    56. The steering on a vehicle pulls to the left while driving straight ahead. The cause of this problem could be
        A. more positive camber on the left front wheel than the right front wheel.
        B. more positive caster on the left front wheel than the right front wheel.
        C. improper steering axis inclination on the left front wheel.
        D. improper toe out-on-turns on the right front wheel.

Hint   *Refer to Figure 15-3, page 357 in* Today's Technician Automotive Suspension and Steering Systems *for diagnosis of steering problems.*

Task 2   **Measure vehicle ride height; determine needed repairs.**

    57. A customer complains about increased steering effort and rapid steering wheel return after turning a corner.
    Technician A says both front wheels may have a negative camber setting.
    Technician B says the rear suspension curb riding height may be reduced.
    Who is correct?
        A. A only
        B. B only
        C. Both A and B
        D. Neither A nor B

Hint   *Reduced curb riding height affects many of the front suspension angles and causes harsh riding. For example, reduced curb riding height on the rear suspension causes excessive positive caster on the front suspension. This problem results in excessive steering effort and rapid steering wheel return.*

Task 3   **Check and adjust front and rear wheel camber on suspension systems with a camber adjustment.**

    58. A long-and-short arm front suspension system has adjustment shims for adjustment of front suspension angles (Figure 4-23, next page). When shims of equal thickness are added on both upper control arm mounting bolts
        A. the camber is moved toward a negative position.
        B. the caster becomes more positive.
        C. the caster is moved toward a negative position.
        D. the steering axis inclination angle decreases.

Hint   *When adjustment shims are positioned between the upper control arm mounting and the inside of the frame, increasing the shim thickness equally on both mounting bolts moves the camber toward a more negative position. Increasing the shim thickness on the rear bolt and decreasing the shim thickness on the front bolt moves the caster toward a positive position.*

    If the adjustment shims are positioned between the upper control arm and the outside of the frame, increasing the shim thickness equally on both bolts increases the positive camber. When the shim thickness is increased on the rear bolt and decreased on the front bolt, the caster is moved toward a negative position.

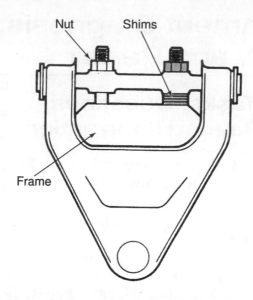

Figure 4-23 Long-and-short arm front suspension system with adjustment shims *(Courtesy of Hunter Engineering Company)*

Task 4   Check front and rear wheel camber on nonadjustable suspension systems; determine needed repairs.

59. The rear suspension system in Figure 4-24 requires a camber adjustment.

   Technician A says the lower strut to knuckle bolt hole in the strut may be elongated with a file to adjust camber.

   Technician B says the upper strut mount may be moved in elongated slots to correct the camber setting.

   Who is correct?

   A. A only
   B. B only
   C. Both A and B
   D. Neither A nor B

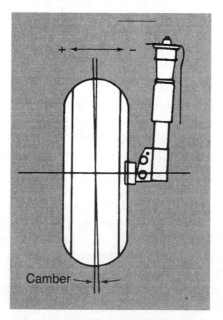

Figure 4-24 Rear suspension system *(Courtesy of Oldsmobile Division, General Motors Corporation)*

Hint   *When a vehicle manufacturer does not provide a camber adjustment, replacement components such as upper strut mounts with adjustment capabilities are often available from automotive parts suppliers.*

Task 5   Check and adjust caster on suspension system with a caster adjustment.

60. A customer complains about harsh riding on the front suspension, excessive steering effort, and rapid steering wheel return.

Technician A says the front suspension may have excessive positive caster.

Technician B says the included angle may be more than specified on the front suspension.

Who is correct?

A. A only

B. B only

C. Both A and B

D. Neither A nor B

Hint   *Excessive positive caster may cause harsh riding, rapid steering wheel return, and excessive steering effort. Negative caster causes reduced directional stability while driving straight ahead.*

Task 6   Check caster on nonadjustable suspension systems; determine needed repairs.

61. All of these statements about caster adjustment are true EXCEPT

A. the caster angle is measured with the front wheels straight ahead.

B. the front wheels are turned 20 degrees outward and then 20 degrees inward to read the caster angle.

C. the brakes must be applied with a brake pedal jack before reading the caster angle.

D. the front suspension should be jounced several times before reading the caster angle.

Hint   *On front suspension systems without a caster adjustment, some vehicle manufacturers recommend inspecting suspension components for wear or damage to determine the cause of the improper caster angle. Damaged or worn components such as the lower control arm, upper strut mount, or engine cradle may cause an improper caster angle. The damaged or worn component should be replaced. Some parts manufacturers supply suspension components that provide caster adjustment capabilities.*

Task 7   Check and adjust front wheel toe.

62. All of these statements about front wheel toe adjustment are true EXCEPT

A. the front wheels must be in the straight-ahead position when measuring front wheel toe.

B. the front wheel toe should be adjusted before the caster angle on the front suspension.

C. after the toe adjustment the steering wheel must be centered with the front wheels straight ahead.

D. while adjusting front wheel toe a tie rod sleeve rotating tool must be used to turn the tie rod sleeves.

Hint   *On parallelogram steering linkages the tie rod sleeves must be rotated to adjust front wheel toe. On a rack and pinion steering gear loosen the tie rod end locknut and the outer bellows boot clamp. Then rotate the tie rod to adjust the front wheel toe. After the toe adjustment, the steering wheel must be centered with the front wheels in the straight-ahead position.*

Task 8   Center steering wheel.

63. After the front wheel toe is adjusted, the left spoke on a steering wheel is 2 in. low on the left side when driving the vehicle straight ahead.
    Technician A says the left tie rod should be lengthened and the right tie rod should be shortened.
    Technician B says the each tie rod sleeve should be rotated 2 turns.
    Who is correct?
    A. A only
    B. B only
    C. Both A and B
    D. Neither A nor B

Hint   *When the steering wheel spoke is low on the left side, rotate the tie rod sleeves to shorten the left tie rod and lengthen the right tie rod. One quarter turn on the tie rod sleeves moves the steering wheel approximately 1 inch.*

Task 9   Check toe-out-on turns (turning radius); determine needed repairs.

64. The turning radius on the right front wheel is not within specifications. The cause of this problem could be
    A. a worn lower right ball joint.
    B. a loose outer right tie rod end.
    C. a worn lower control arm bushing.
    D. a worn right stabilizer bushing.

Hint   *On many steering systems the turning radius on the wheel on the inside of the turn is 2 degrees more than the turning radius on the wheel on the outside of the turn. If the steering radius is not within specifications, the steering arms are bent or steering linkage components such as tie rod ends are worn.*

Task 10   Check included angle; determine needed repairs.

65. A front-wheel-drive vehicle has 3 degrees difference in the included angles on the front wheels. This problem could be caused by
    A. a bent engine cradle.
    B. worn steering gear mounting bushings.
    C. loose front wheel bearings.
    D. loose inner tie rod ends.

Hint   *When the front wheel camber is positive, the camber is added to the steering axis inclination (SAI) to obtain the included angle. If the front wheel camber is negative, the camber is subtracted from the SAI to calculate the included angle. The difference between the included angle on the front wheels should not exceed 1 1/2 degrees. SAI is not considered adjustable. Improper SAI angles may be caused by a strut tower out of position, a bent lower control arm or center crossmember, or an engine cradle out of position.*

Task 11   Check rear wheel toe; determine needed repairs or adjustments.

66. A front-wheel-drive car has excessive toe out on the left rear wheel. Adjustment shims are positioned between the rear spindles and the spindle mounting surfaces.
    Technician A says a thicker shim should be installed on the front bolts in the left rear spindle.
    Technician B says this problem may cause the steering to pull to the right.
    Who is correct?
    A. A only
    B. B only
    C. Both A and B
    D. Neither A nor B

Hint   *On some rear suspension systems rear wheel toe is adjusted with shims between the spindle and spindle mounting surface. A thicker shim on the front spindle mounting bolts increases toe out. On other rear suspension systems rear wheel toe is adjusted by shifting the position of the inner end on the lower control arm. The tie rod length may be adjusted to correct the toe on some rear wheels.*

Task 12   Check rear wheel thrust angle; determine needed repairs or adjustments.

67. The thrust angle on a front-wheel-drive vehicle is more than specified, and the thrust line is positioned to the left of the geometric centerline. This problem could be caused by excessive
    A. toe out on the left rear wheel.
    B. toe out on the right rear wheel.
    C. positive camber on the left rear wheel.
    D. wear in the left lower ball joint.

Hint   *The thrust angle is the angle between the thrust line and the geometric centerline of the vehicle. If the thrust angle is not within specifications, the steering pulls to one side. For example, if the thrust line is positioned to the left of the geometric centerline, the steering pulls to the right. When the thrust angle is not within specifications on a front-wheel-drive car, the rear wheel toe may require adjusting. On a rear-wheel-drive car with a nonindependent rear axle assembly, improper rear wheel toe may be caused by a bent rear axle assembly. On this type of rear axle the complete rear axle assembly may be turned slightly in either direction causing the thrust angle to be more than specified. This condition is called rear axle offset.*

Task 13   Check front wheel setback; determine needed repairs or adjustments.

68. While discussing front wheel setback
    Technician A says front wheel setback usually is caused by worn suspension components.
    Technician B says a slight front wheel setback causes steering pull.
    Who is correct?
    A. A only
    B. B only
    C. Both A and B
    D. Neither A nor B

Hint   *Setback occurs when one front wheel is driven rearward in relation to the opposite front wheel. Setback usually is caused by collision damage, and it does not affect steering unless it is extreme.*

Task 14   Check front cradle (subframe) alignment, determine needed repairs or adjustments.

69. While discussing front cradle alignment
    Technician A says the cradle may be measured at various locations to determine if it is bent.
    Technician B says on some cradles an alignment hole in the cradle must be aligned with a hole in the chassis.
    Who is correct?
    A. A only
    B. B only
    C. Both A and B
    D. Neither A nor B

Hint   *A bent or improperly positioned front cradle may affect camber, caster, or setback angles on the front suspension. The cradle may be measured at various locations to determine if it is bent. If the cradle is properly positioned on some vehicles, an alignment hole in the cradle must be aligned with a hole in the chassis.*

Wheel and Tire Diagnosis and Repair

ASE Tasks, Questions, and Related Information

Task 1   **Diagnose tire wear patterns; determine needed repairs.**

70. The tire tread wear shown in Figure 4-25 could be caused by
    A. excessive positive camber.
    B. excessive positive caster.
    C. excessive setback
    D. improper toe adjustment.

Feathered edge

Figure 4-25 Excessive tire tread wear *(Courtesy of Chrysler Corporation)*

Hint   *Feathered wear across the tire tread usually indicates an improper toe adjustment. Wear on one side of the tire tread usually indicates an improper camber setting. Cupped tread wear may indicate improper wheel balance.*

Task 2   **Inspect tires; check and adjust air pressure.**

71. All of these statements about tire inflation are true EXCEPT
    A. overinflation causes excessive wear on the center of the tread.
    B underinflation causes excessive wear on both edges of the tread.
    C. tire pressure should be adjusted when the tires are hot.
    D. underinflation may cause wheel damage.

Hint   *Tire inflation pressures should be corrected when the tires are cool. Overinflation causes harsh riding, damage to the tire carcass, and excessive center tread wear. Underinflation causes hard steering, wheel damage, excessive wear on the tread edges, and tire overheating.*

Task 3   **Diagnose wheel/tire vibration, shimmy, and noise problems; determine needed repairs.**

72. Front wheel shimmy may be caused by
    A. excessive toe out.
    B. improper dynamic wheel balance.
    C. excessive front wheel setback.
    D. excessive positive camber.

Hint   *Tire vibration and thumping may be caused by cupped tire treads, excessive tire radial runout, heavy spots in the tire, and improper wheel balance. Front wheel shimmy may be caused by improper dynamic wheel balance, excessive positive caster, or worn steering linkage components.*

Task 4 Rotate tires/wheels according to manufacturer's recommendations.

73. The tire rotation procedure in Figure 4-26 may be used on a vehicle with
    A. radial tires and a compact spare.
    B. radial tires and a conventional spare.
    C. radial or bias ply tires.
    D. bias ply tires with a conventional spare.

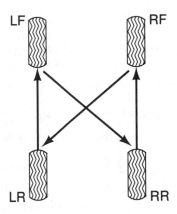

Figure 4-26 Tire rotation *(Courtesy of Chevrolet Motor Division, General Motors Corporation)*

Hint   *The vehicle manufacturer's tire rotation procedure varies depending on the type of tires and the type of spare. A different rotation is used for radial or bias ply tires. Since the compact spare cannot be used for extended mileage, this spare is not rotated with the other tires.*

Task 5 Measure wheel, tire, axle, and hub runout; determine needed repairs.

74. A vehicle has rear chassis waddle (Figure 4-27).
    Technician A says this problem may be caused by a steel belt in a tire that is not straight.
    Technician B says this problem may be caused by a bent rear hub flange.
    Who is correct?
    A. A only
    B. B only
    C. Both A and B
    D. Neither A nor B

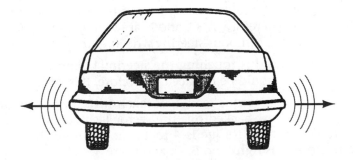

Figure 4-27 Rear suspension waddle *(Courtesy of Oldsmobile Division, General Motors Corporation)*

Hint   *Excessive lateral tire runout may cause chassis waddle. This problem may be caused by a steel belt that is not straight in a tire, and bent wheel or hub.*

Task 6   Diagnose tire pull (lead) problems; determine corrective actions.

75. A customer complains about steering pull to the left.
    Technician A says the two front tires may have different tread designs.
    Technician B says one of the front tires may have a conicity problem.
    Who is correct?
    A. A only
    B. B only
    C. Both A and B
    D. Neither A nor B

Hint   *Tires of different types, sizes, tread designs, or inflation pressures may cause steering pull. Tire conicity is caused by a tire belt that is wound off center during the manufacturing process. This problem also causes steering pull. Tire conicity is diagnosed with a special tire rotation procedure.*

Task 7   Balance wheel and tire assembly (static and/or dynamic).

76. Dynamic wheel unbalance on a front wheel may cause all of these problems EXCEPT
    A. excessive wear on the center of the tire tread.
    B. cupped tire tread wear around the tire tread.
    C. front wheel shimmy while driving at higher speeds.
    D. excessive wear on steering linkage components.

Hint   *Static wheel balance refers to proper balance of a wheel at rest. Dynamic wheel balance refers to the balance of a wheel in motion. Electronic wheel balancers spin the wheel at high speed and indicate the proper position for wheel weight installation to provide proper static and dynamic balance.*

Answers and Analysis

1. C   A worn flexible coupling or loose steering gear mounting bolts may cause excessive steering wheel freeplay. Therefore both A and B are correct, and C is the right answer.

2. D   A loose worm bearing preload adjustment causes excessive steering wheel freeplay. An overfilled steering gear does not affect steering effort. When the positive caster is less than specified, steering effort is reduced. Therefore, A, B, and C are wrong. A tight sector lash adjustment increases steering effort. D is the right answer.

3. B   A loose rack bearing adjustment or loose steering gear mounting bolts cause excessive steering wheel freeplay. Excessive positive camber causes wear on the outside edge of the tire treads. Therefore, A, C, and D are wrong. Insufficient or improper steering gear lubricant may result in poor steering wheel returnability. B is correct.

4. A   If the bolt head is touching the steering column bracket the shear load is too high and the bracket should be replaced. Therefore, B is wrong and A is correct.

5. A   After a turning torque reading is taken with the steering gear centered and the sector lash adjusting screw backed off, the sector lash adjusting screw should be turned clockwise until the turning torque is 4 to 10 in. lbs more than it was with this screw backed off. Therefore, A is right and B, C, and D are wrong.

6. B   Worn lower ball joints have little effect on steering effort. A tight rack bearing adjustment can cause excessive steering effort. A is wrong and B is right.

7. C   The inner tie rod end should be replaced if the articulation effort is less than specified, and the rack must be held while loosening the inner tie rod ends. Therefore, both A and B are correct, and C is the right answer.

8. A   Excessive positive caster may cause excessive steering effort, rapid steering wheel return, and front wheel shimmy. Worn upper strut mounts may cause chatter while turning the steering wheel, or improper camber and caster angles. Worn out struts can result in excessive vertical chassis oscillations. Therefore, B, C, and D are wrong.

A loose steering gear mounting bushing may cause a nonparallel condition between the tie rods and the lower control arms. This condition may result in the steering suddenly veering to one side. This condition is commonly called bump steer. A is correct.

9. B   A loose or worn power steering belt, or low power steering fluid level, may cause increased or erratic steering effort. A scored steering gear cylinder can also result in increased steering effort. Therefore, A, C, and D are wrong.

A loose worm shaft bearing preload adjustment may cause steering looseness. B is correct.

10. A   If the power rack and pinion steering gear has an oil leak at the left end of the rack when the rack contacts the left inner stop, the inner rack seal is leaking. When a leak at this location is not influenced by rack position, the pinion seal is leaking. Therefore B, C, and D are wrong, and A is right.

11. D   Many vehicle manufacturers recommend checking the power steering fluid level with the fluid temperature at 175°F (80°C). A, B, and C are wrong, and D is correct.

12. A   When checking power steering belt tension, for every foot of free span the belt should have 1/2 in. deflection. Therefore, B, C, and D are wrong, and A is right.

13. C   A belt tension gauge may be used to measure the belt tension on a ribbed V-belt, and some of these belts have a spring-loaded tensioner with a belt length scale. Both A and B are right, and C is the correct answer.

14. D   When a power steering pump with an integral reservoir is leaking fluid between the reservoir and the pump housing, the large housing O-ring is leaking. Both A and B are wrong, and D is the correct answer.

15. C   The tool in Figure 4-6 is used to install a pressed-on power steering pump pulley. Therefore, A, B, and D are wrong and C is right.

16. A   During the power steering pump pressure test, the pressure gauge valve should be closed for a maximum of 10 seconds. Therefore, B, C, and D are wrong, and A is right.

17. C   When the power steering pump pressure test indicates satisfactory pump pressure, but the steering wheel turning effort is excessive, the steering gear may be defective, or the high pressure hose may be restricted. Both A and B are correct, and C is the right answer.

18. A   When removing the steering wheel on an air-bag-equipped vehicle, the negative battery cable should be disconnected and the technician should wait the time specified by the car manufacturer before working on the vehicle. This time is usually one or two minutes. Therefore, B is wrong. Rotating the steering wheel with the steering gear disconnected may damage the clockspring electrical connector, thus A is correct.

19. A   When adjusting the worm shaft bearing preload on some steering gears, the adjuster plug should be bottomed and tightened to 20 ft lbs. After this procedure the adjuster plug is backed off 0.050 in. Therefore, B is wrong and A is right.

20. C   The part number on the needle bearing must face the driving tool, and the worm shaft bearing preload should be adjusted after the bearing and seal are installed. Therefore, both A and B are correct and C is the right answer.

21. B   Steering wheel returnability is not affected by a misaligned steering gear. Therefore, A is wrong. A tight rack bearing adjustment increases steering effort and causes poor steering wheel returnability. B is correct.

22. C   The component in Figure 4-10 is a shock damper. Therefore, A, B, and D are wrong, and C is correct.

23. D   When an electronically controlled four-wheel steering system enters the fail-safe mode, the voltage is shut off to the rear steering unit. In this mode the control unit energizes the damper relay, and the rear steering unit moves slowly to the center position. Therefore, A, B, and C are wrong. In the fail-safe mode the control unit illuminates the 4WS light in the instrument panel. D is correct.

24. C   When bleeding a power steering system, the steering wheel should be held in the fully right or fully left positions for two to three seconds. If foaming is still present in the reservoir after the bleeding process, the bleeding procedure should be repeated. Therefore, both A and B are correct and C is the right answer.

25. B   In variable-assist steering systems with a steering wheel rotation sensor, the power steering assist is increased when steering wheel rotation exceeds 15 rpm. Power steering assist also is increased at low vehicle speeds and decreased at higher speeds. Therefore A is wrong, and B is correct.

26. A   Excessive idler arm movement may cause front wheel shimmy, steering looseness, improper tie rod alignment with the lower control arm, and bump steer. Therefore, B, C, and D are wrong, and A is correct.

27. B   Bump steer occurs when the tie rods are not parallel to the lower control arms. This condition may be caused by a bent pitman arm or a loose, improperly adjusted idler arm. Therefore, A is wrong and B is correct.

28. A   A bent relay rod may cause improper front wheel toe. This condition causes feathered wear across the tire treads. Therefore, B is wrong and A is right.

29. C   Worn ball joints, weak front springs, or a weak stabilizer bar would not cause excessive steering effort. However, this problem may be caused by a seized idler arm. Therefore, A, B, and D are wrong and C is correct.

30. C   The tie rod sleeves must be rotated to adjust front wheel toe and center the steering wheel. Therefore, both A and B are correct, and C is the right answer.

31. B   A scored power steering gear cylinder increases steering effort, but it does not contribute to road shock on the steering wheel. Therefore, A is wrong. A worn-out steering damper may cause road shock on the steering wheel. B is the right answer.

32. D   Reduced front suspension ride height may be caused by weak or broken springs, or worn control arm bushings. Therefore, A, B, and C are wrong, and D is right.

33. A   The tool in Figure 4-14 is used to compress the coil spring on a long-and-short arm suspension system. Therefore, B, C, and D are wrong and A is correct.

34. A   Excessive negative camber on the right front wheel may cause steering pull to the left. Therefore, B is wrong. A worn right front strut rod bushing may result in the lower control arm moving rearward while braking. This action reduces positive caster on the right front wheel. The steering tends to pull to the side with the least positive caster. Therefore, A is right.

35. D   When the coil spring is mounted between the lower control arm and the chassis, a jack must be positioned under the lower control arm to unload the ball joints. Therefore, A, B, and C are wrong and D is right.

36. A   Excessive tire squeal while cornering may be caused by improper toe-out-on-turns. This problem can be caused by a bent steering arm. Therefore, B, C, and D are wrong and A is the right answer.

37. B   The coil spring in Figure 4-15 is a variable-rate spring. Therefore, A, C, and D are wrong and B is correct.

38. C   While using a spring compressor to remove a coil spring from a strut, the spring should be taped in the spring compressor contact areas, and all the spring tension must be removed from the upper strut mount before loosening the strut rod nut. Therefore, both A and B are correct and C is the right answer.

39. D In this question the requested answer does not result in steering pull to the right. A broken center bolt in the left rear spring causes rear axle offset, and this problem may cause steering pull to the right. Since the steering pulls to the side with the least positive caster, excessive positive caster on the left front wheel may cause steering pull to the right. More positive camber on the right front wheel compared to the left front wheel may cause steering pull to the right. Therefore, A, B, and C may cause steering pull to the right, but none of these are the requested answer. Excessive front wheel toe-in causes feathered tire wear, but this problem does not affect steering pull. Therefore, D is right.

40. C A worn lower control arm bushing or an improperly adjusted front torsion bar anchor bolt may cause reduced ride height. Therefore, both A and B are correct and C is the right answer.

41. A In this question the requested answer does not result in body sway while cornering. A weak stabilizer bar, worn stabilizer bar bushing, or a broken stabilizer link may cause excessive body sway while cornering. Therefore, B, C, and D may cause body sway while cornering, but none of these are the requested answer.

A worn strut rod bushing does not cause this problem. Therefore, A is right.

42. C In Figure 4-17 the technician is pushing the strut rod downward before removing the strut cartridge. Therefore, A, B, and D are wrong, and C is correct.

43. A A defective upper strut mount may result in strut chatter while cornering, but a worn lower ball joint does not cause this problem. Therefore, A is right, and B is wrong.

44. C Defective struts or reduced suspension ride height may cause harsh riding. Therefore, both A and B are correct and C is the right answer.

45. A Excessive rear suspension oscillations may be caused by defective struts, but weak coil springs do not cause this problem. Therefore, B is wrong and A is the right answer.

46. B The ball joint nuts should never be loosened to align the cotter key hole with the nut castellations. Therefore, A is wrong.

On some rear suspension systems, such as the one in Figure 4-18, the lower control arm bolts must be torqued with the vehicle weight on the suspension. Therefore, B is right.

47. A A sagged left rear leaf spring lowers the left rear ride height, and increases the positive caster on the left front wheel. Since the steering pulls to the side with the least positive caster, this problem may cause steering pull to the right. Therefore, B, C, and D are wrong, and A is the right answer.

48. C In this question the requested answer does not cause a rattling noise in the rear suspension. Since a broken spring, spring insulator, or a worn track bar bushing may cause a rattling noise in the rear suspension, these responses are not the requested answer. Therefore, A, B, and D are wrong.

A bent rear strut does not cause a ratting noise in the rear suspension. Therefore, C is correct.

49. B Rear axle offset as illustrated in Figure 4-19 may cause steering pull to the right. Therefore, A, C, and D are wrong and B is right.

50. C Many ball joints have a wear indicator. In these ball joints the shoulder of the grease fitting must extend a specific distance from the ball joint housing. If this distance is less than specified, the ball joint must be replaced. Since there is no clearance between the grease fitting shoulder and the ball joint housing, A is correct.

A worn ball joint may cause improper position of the lower end of the rear knuckle, wheel hub, and wheel. This action may result in improper rear wheel camber. Therefore, B is also correct and C is the right answer.

51. A If the tie rod in Figure 4-21 is longer than specified, the rear wheel toe-out is excessive. Therefore, A is right, and B, C, and D are wrong.

52. A When one side of the bumper is pushed downward with considerable force and then released, the bumper should only complete one free upward bounce if the shock absorber or strut is satisfactory. More than one free upward bounce of the bumper indicates defective shock absorbers or struts. Therefore, B, C, and D are wrong, and A is right.

53. C Some air shock absorbers may be pressurized with the shop air hose. Therefore, A is right. If air shock absorbers slowly lose their pressure, and reduce ride height, shock absorber replacement is probably necessary. Therefore, B is also right and C is the correct answer.

54. C After tightening the rear wheel bearing adjusting nut to 20 ft lbs, and backing the nut off one-half turn this nut should be tightened to 10 to 15 ft lbs. Therefore, A, B, and D are wrong and C is correct.

55. A In this question we are asked for the condition when the electronic suspension switch should not be in the off position. The electronic suspension switch should be off when hoisting, towing, or jacking the vehicle. Therefore, B, C, and D are not the requested answers.

The electronic suspension switch must be on while diagnosing the suspension system with a scan tool. Therefore, A is the right answer.

56. A The steering pulls to the side with the least positive caster. Therefore, B is wrong. SAI or toe-out-on turns do not affect steering pull while driving straight ahead, and thus C and D are wrong. The steering tends to pull to the side with the most positive camber. Therefore, A is the right answer.

57. B A negative camber setting does not cause rapid steering wheel return and increased steering effort. When the rear suspension ride height is reduced, the positive caster is increased on the front suspension. Since excessive positive caster causes rapid steering wheel return and increased steering effort, B is correct.

58. A When shims of equal thickness are added on both upper control arm bolts in Figure 4-23 the top of the tire is moved inward and the camber is moved toward a negative position. Therefore, B, C, and D are wrong and A is the right answer.

59. A On the rear suspension system in Figure 4-24 the lower strut to knuckle bolt hole in the strut must be filed to adjust camber. Therefore, B is wrong and A is right.

60. A Harsh riding, excessive steering effort, and rapid steering wheel return may be caused by excessive positive caster. Therefore, B is wrong and A is right.

61. A This question asks for the statement that is not true. The front suspension should be jounced several times before reading any suspension angle. The brakes should be applied with a brake jack before reading caster. The front wheels are turned outward 20 degrees and then inward 20 degrees to read the caster angle. Therefore, statements B, C, and D are correct, but none of these are the requested answer.

The caster angle is not read with the front wheels straight ahead. Therefore, statement A is not true, and this is the requested answer.

62. B When front wheel toe is measured, the front wheels must be straight ahead. After the toe adjustment is completed, the steering wheel must be centered with the front wheels straight ahead. A tie rod rotating tool must be used to rotate the tie rod sleeves. Therefore, statements A, C, and D are true, and none of these is the requested answer.

The front wheel toe should be measured after the caster measurement and adjustment. Therefore, statement B is not true, and this is the requested answer.

63. D If the steering wheel spoke is 2 in. low on the left side while driving the vehicle straight ahead, the left tie rod should be shortened, and the right tie rod lengthened. One-quarter turn on the tie rod sleeves moves the steering wheel about one inch. Therefore, both A and B are wrong, and D is the right answer.

64. B When the turning radius is not within specifications, a steering arm may be bent or a tie rod end may be loose. Therefore, A, C, and D are wrong and B is the right answer.

65. **A**   When the included angles on the front wheels have 3 degrees difference, the SAI angle is probably improper on one of the front wheels. Improper SAI may be caused by an improperly positioned strut tower or engine cradle. Therefore, B, C, and D are wrong and A is correct.

66. **B**   Excessive toe-out in the left rear wheel may be corrected by installing a thinner shim on the front spindle bolts. Therefore, A is wrong. Excessive toe-out on the left rear wheel may cause the steering to pull to the right. Therefore, B is correct.

67. **A**   If the thrust angle is excessive and the thrust line is positioned to the left of the geometric centerline, the left rear wheel may have excessive toe-out. Therefore, B, C, and D are wrong and A is right.

68. **D**   Front wheel setback usually is caused by collision damage, and this problem does not cause steering pull unless it is excessive. Therefore, both A and B are wrong, and D is the right answer.

69. **C**   The front cradle may be measured in various locations to verify a bent condition. Some cradles have an alignment hole that must be aligned with a matching hole in the chassis. Therefore, both A and B are correct, and C is the right answer.

70. **D**   The feathered tire wear in Figure 4-25 may be caused by improper toe adjustment. Therefore, A, B, and C are wrong and D is right.

71. **C**   This question asks for the statement that is not true. Overinflation causes wear on the center of the tire tread, and underinflation causes wear on the edges of the tread. Underinflation may also cause wheel damage. Therefore, statements A, B, and D are true, but none of these are the requested answer.

Tire pressure should be adjusted when the tires are cool. Therefore, statement C is not true, and this is the requested answer.

72. **B**   Front wheel shimmy may be caused by improper dynamic wheel balance, excessive positive caster, or loose steering linkage components. Therefore, A, C, and D are wrong, and B is correct.

73. **A**   The tire rotation procedure in Figure 4-26 may be used on a vehicle with radial tires and a compact spare. Therefore, A is the right answer.

74. **C**   Excessive rear chassis waddle may be caused by a steel belt in a tire that is not straight, or a bent rear hub flange. Therefore, both A and B are correct and C is the right answer.

75. **C**   Steering pull may be caused by front tires with different tread designs, or a front tire with a conicity defect. Therefore, both A and B are correct and C is the right answer.

76. **A**   This questions asks for the problem that is not caused by dynamic wheel unbalance. Cupped tire treads, wheel shimmy, or excessive steering linkage wear may be caused by dynamic wheel unbalance. Therefore, B, C, and D are not the requested answer.

Excessive wear on the center of the tire tread is not a result of improper dynamic wheel balance. Therefore, A is the requested answer.

Glossary

Accidental air bag deployment An unintended air bag deployment caused by improper service procedures.
Despliegue accidental del Airbag Despliegue imprevisto del Airbag ocasionado por procedimientos de reparación inadecuados.

Adjustment screen A display on a computer wheel aligner that allows the technician to see the results of certain suspension adjustments.
Pantalla para la visualización de ajustes Representación visual en una computadora para alineación de ruedas que le permite al técnico ver los resultados de ciertos ajustes en la suspensión.

Air bag deployment module The air bag and deployment canister assembly that is mounted in the steering wheel for the driver's side air bag, or in the dash panel for the passenger's side air bag.
Unidad de despliegue del Airbag El conjunto del Airbag y elemento de despliegue montado en el volante de dirección para proteger al conductor, o en el tablero de instrumentos para proteger al pasajero.

Air bleeding The process of bleeding air from a hydraulic system such as the power steering system.
Muestra de aire Proceso a través del cual se extrae el aire de un sistema hidráulico, como por ejemplo un sistema de dirección hidráulica.

Air conditioning programmer (ACP) A control unit that may contain the computer, solenoids, motors, and vacuum diaphragms for air conditioning control.
Programador del acondicionamiento de aire Unidad de control que puede incluir la computadora, los solenoides, los motores y los diafragmas de vacío para el control del acondicionamiento de aire.

Alignment ramp A metal ramp positioned on the shop floor on which vehicles are placed during wheel alignment procedures.
Rampa de alineación Rampa de metal ubicada en el suelo del taller de reparación de automóviles sobre la que se colocan los vehículos durante los procedimientos de alineación de ruedas.

Anhydrous calcium grease A special lubricant used in manual rack and pinion steering gears.
Grasa de calcio anhidro Lubricante especial utilizado en mecanismos de dirección de cremallera y piñón manuales.

Anti-theft lug nuts Locking lug nuts that help to prevent wheel theft.
Tuercas de orejetas anti-robo Tuercas de orejetas autobloqueantes que ayudan a evitar el robo de las ruedas.

Anti-theft wheel covers Locking wheel covers that help to prevent wheel theft.
Cubrerruedas anti-robo Cubrerruedas autobloqueantes que ayudan a evitar el robo de las ruedas.

ASE blue seal of excellence An ASE logo displayed by automotive service shops that employ ASE certified technicians.
Sello azul de excelencia de la ASE Logotipo exhibido en talleres de reparación de automóviles donde se emplean mecánicos certificados por la ASE.

ASE technician certification ASE provides certification of automotive technicians in eight different areas of expertise.
Certificación de mecánico de la ASE Certificación de mecánico de automóviles otorgada por la ASE en ocho áreas diferentes de especialización.

Auto/manual diagnostic check A test procedure in diagnosing a Ford computer-controlled suspension system.
Revisión diagnóstica automática/manual Procedimiento de prueba llevado a cabo para diagnosticar un sistema de suspensión controlado por computadora de la Ford.

Auto test A test procedure in diagnosing a Ford computer-controlled suspension system.
Prueba automática Procedimiento de prueba llevado a cabo para diagnosticar un sistema de suspensión controlado por computadora de la Ford.

Axle pullers A special puller required for axle shaft removal.
Extractores del eje Extractor especial requerido para la remoción del árbol motor.

Backup power supply A voltage source, usually located in the air bag computer, that is used to deploy an air bag if the battery cables are disconnected in a collision.
Alimentación de reserva Fuente de tensión, por lo general localizada en la computadora del Airbag, que se utiliza para desplegar el Airbag si se desconectan los cables de la batería a consecuencia de una colisión.

Ball joint axial movement Vertical movement in a ball joint because of internal joint wear.
Movimiento axial de la junta esférica Movimiento vertical en una junta esférica ocasionado por el desgaste interno de la misma.

Ball joint radial movement Horizontal movement in a ball joint because of internal joint wear.
Movimiento radial de la junta esférica Movimiento horizontal en una junta esférica ocasionado por el desgaste interno de la misma.

Ball joint removal and pressing tools Special tools required for ball joint removal and replacement.
Herramientas para la remoción y el ajuste de la junta esférica Herramientas especiales requeridas para la remoción y el reemplazo de la junta esférica.

Ball joint unloading Removing the ball joint tension supplied by the vehicle weight prior to ball joint diagnosis.
Descarga de la junta esférica Remoción de la tensión de la junta esférica provista por la carga del vehículo antes de llevarse a cabo la diagnosis de la junta esférica.

Ball joint wear indicator A visual method of checking ball joint wear.
Indicador de desgaste de la junta esférica Método visual de revisar el desgaste interior de una junta esférica.

Bearing abrasive roller wear Fine scratches on the bearing surface.
Desgaste abrasivo del rodillo del cojinete Rayados finos en la superficie del cojinete.

Bearing abrasive step wear A fine circular wear pattern on the ends of the rollers.
Desgaste abrasivo escalonado del cojinete Desgaste circular fino en los extremos de los rodillos.

Bearing brinelling Straight line indentations on the races and rollers.
Acción de Brinell en un cojinete Hendiduras en línea recta en los anillos y los rodillos.

Bearing etching A loss of material on the bearing rollers and races.
Corrosión del cojinete Pérdida de material en los rodillos y los anillos del cojinete.

Bearing fatigue spalling Flaking of the surface metal on the rollers and races.
Escamación y fatiga del cojinete Condición que ocurre cuando el metal de la superficie de los rodillos y los anillos comienza a escamarse.

Bearing frettage A fine, corrosive wear pattern around the races and rollers, with a circular pattern on the races.
Cinceladura del cojinete Desgaste corrosivo y fino alrededor de los anillos y los rodillos que se hace evidente a través de figuras circulares en los anillos.

Bearing galling Metal smears on the ends of the rollers.
Desgaste por rozamiento en un cojinete Ralladuras metálicas en los extremos de los rodillos.

Bearing heat discoloration A dark brown or bluish discoloration of the rollers and races caused by excessive heat.
Descoloramiento del cojinete ocasionado por el calor Descoloramiento marrón oscuro o azulado de los rodillos y los anillos ocasionado por el calor excesivo.

Bearing preload A tension placed on the bearing rollers and races by an adjustment or assembly procedure.
Carga previa del cojinete Tensión aplicada a los rodillos y los anillos del cojinete a través de un procedimiento de ajuste o de montaje.

Bearing pullers Special tools designed for bearing removal.
Extractores de cojinetes Herramientas especiales diseñadas para la remoción del cojinete.

Bearing smears Metal loss from the races and rollers in a circular, blotched pattern.
Ralladuras en los cojinetes Pérdida del metal de los anillos y los rodillos que se hace evidente a través de una figura circular oxidada.

Bearing stain discoloration A light brown or black discoloration of the rollers and races caused by incorrect lubricant or moisture.
Descoloramiento del cojinete Descoloramiento marrón claro o negro de los rodillos y los anillos ocasionado por la humedad o la utilización de un lubricante incorrecto.

Bellows boots Accordion-style boots that provide a seal between the tie rods and the housing on a rack and pinion steering gear.
Botas de fuelles Botas en forma de acordeón que proveen una junta de estanqueidad entre las barras de acoplamiento y el alojamiento en un mecanismo de dirección de cremallera y piñón.

Belt tension gauge A special gauge used to measure drive belt tension.
Calibrador de tensión de la correa de transmisión Calibrador especial que se utiliza para medir la tensión de una correa de transmisión.

Body sway Excessive body movement from side to side.
Oscilación de la carrocería Movimiento lateral excesivo de la carrocería.

Brake pedal jack A special tool installed between the front seat and the brake pedal to apply the brakes during certain wheel alignment measurements.
Gato del pedal de freno Herramienta especial instalada entre el asiento delantero y el pedal de freno que se utiliza para aplicar los frenos durante ciertas medidas de alineación de ruedas.

Bump steer The tendency of the steering to veer suddenly in one direction when one or both front wheels strike a bump.
Cambio de dirección ocacionado por promontorios en el terreno Tendencia de la dirección a cambiar repentinamente de sentido cuando una o ambas ruedas delanteras golpea un promontorio.

Cable reel A conductive ribbon mounted on top of the steering column to maintain electrical contact between the air bag inflator module and the air bag electrical system. This component may be called a clock spring electrical connector.
Bobina de cable Cinta conductiva montada sobre la columna de dirección que se utiliza para mantener el contacto eléctrico entre la unidad infladora y el sistema eléctrico del Airbag. Dicho componente se conoce también como conector eléctrico de cuerda de reloj.

Camber adjustment A method of adjusting the inward or outward tilt of a front or rear wheel in relation to the true vertical centerline of the tire.
Ajuste de la combadura Método de ajustar la inclinación hacia adentro o hacia afuera de una rueda delantera o trasera con relación a la línea central vertical real del neumático.

Camber angle The inward or outward tilt of a line through the center of a front or rear tire in relation to the true vertical centerline of the tire and wheel.
Ángulo de combadura Inclinación hacia adentro o hacia afuera de una línea a través del centro de una rueda delantera o trasera con relación a la línea central vertical real del neumático y la rueda.

Carbon monoxide A poisonous gas present in vehicle exhaust in small quantities.
Monóxido de carbono Gas mortífero presente en pequeñas cantidades en el escape de los motores.

Caster adjustment A method of adjusting the forward or rearward tilt of a line through the center of the upper and lower ball joints, or lower ball joint and upper strut mount, in relation to the true vertical centerline of the tire and wheel viewed from the side.
Ajuste de comba de eje Método de ajustar la inclinación hacia adelante o hacia atrás de una línea a través del centro de las juntas esféricas superior e inferior, o la junta esférica inferior y el montaje del montante superior, con relación a la línea central vertical real del neumático y la rueda vista desde la parte lateral.

Caster angle A line through the center of the upper and lower ball joints, or lower ball joint and upper strut mount, in relation to the true vertical centerline of the tire and wheel viewed from the side.
Ángulo de comba de eje Línea a través del centro de las juntas esféricas superior e inferior, o la junta esférica inferior y el montaje del montante superior, con relación a la línea central vertical real del neumático y la rueda vista desde la parte lateral.

Circlip A round circular clip used as a locking device on components such as front drive axles.
Grapa circular Grapa circular esférica que se utiliza como dispositivo de bloqueo en algunos componentes, como por ejemplo ejes de mando de tracción delantera.

Claw washer A special locking washer used to retain the tie rods to the rack in some rack and pinion steering gears.
Arandela de garra Arandela de bloqueo especial que se utiliza para sujetar las barras de acoplamiento a la cremallera en algunos mecanismos de dirección de cremallera y piñón.

"C" locks A thick metal locking device used to lock components in place, such as the rear drive axles.
Retenedores en forma de C Dispositivo de bloqueo metálico grueso que se utiliza para ajustar componentes en su posición, como por ejemplo ejes de mando de tracción trasera.

Clock spring electrical connector A conductive ribbon in a plastic container mounted on top of the steering column that maintains electrical contact between the air bag inflator module and the air bag electrical system.
Conector eléctrico de cuerda de reloj Cinta conductiva envuelta en una cubierta plástica montada sobre la parte superior de la

columna de dirección, que mantiene el contacto eléctrico entre la unidad infladora y el sistema eléctrico del Airbag.

Coil spring compressing tool   A special tool required to compress a coil spring prior to removal of the spring from the strut.

Herramienta para la compresión del muelle helicoidal   Herramienta especial requerida para comprimir un muelle helicoidal antes de remover el muelle del montante.

Collapsible steering column   A steering column that is designed to collapse when impacted by the driver in a collision to help reduce driver injury.

Columna de dirección plegable   Columna de dirección diseñada para plegarse al ser impactada por el conductor durante una colisión con el propósito de reducir las lesiones que el conductor pueda recibir.

Computer command ride (CCR) system   A computer-controlled suspension system that controls strut or shock absorber firmness in relation to driving and road conditions.

Sistema de viaje ordenado por computadora   Sistema de suspensión controlado por computadora que controla la firmeza de los montantes o de los amortiguadores de acuerdo a las condiciones de viaje o del camino.

Computer wheel aligner   A type of wheel aligner that uses a computer to measure wheel alignment angles at the front and rear wheels.

Computadora para alineación de ruedas   Tipo de alineador de ruedas que utiliza una computadora para medir los ángulos de alineación de las ruedas delanteras y traseras.

Continual strut cycling   A test procedure for a computer command ride suspension system that allows continual cycling of the strut armatures.

Funcionamiento cíclico continuo del montante   Procedimiento de prueba para un sistema de suspensión de viaje ordenado por computadora que permite el funcionamiento cíclico continuo de las armaduras de los montantes.

Control arm bushing tools   Special tools required for control arm bushing removal and replacement.

Herramientas para el buje del brazo de mando   Herramientas especiales requeridas para la remoción y el reemplazo del buje del brazo de mando.

Cornering angle   The turning angle of one front wheel in relation to the opposite front wheel during a turn. This angle may be called turning radius.

Ángulo de viraje   Ángulo de giro de una rueda delantera con relación a la rueda delantera opuesta durante un viraje. Dicho ángulo se conoce también como radio de giro.

Crocus cloth   A very fine paper for polishing or removing small abrasions from metal.

Tela fina de esmeril   Papel sumamente fino que se utiliza para pulir o remover pequeñas abrasiones del metal.

Curb riding height   The distance between the vehicle chassis and the road surface measured at specific locations.

Altura del cotén del viaje   Distancia entre el chasis del vehículo y la superficie del camino medida en puntos específicos.

Current code   A fault code in a computer that is present at all times.

Código actual   Código de fallo siempre presente en una computadora.

Cylinder end stopper   A circular bushing in the end of a rack and pinion steering gear housing.

Tapador en el extremo del cilindro   Buje circular en el extremo del alojamiento de un mecanismo de dirección de cremallera y piñón.

Data link connector (DLC)   An electrical connector for computer system diagnosis mounted under the instrument panel or in the engine compartment.

Conector de enlace de datos   Conector eléctrico montado debajo del tablero de instrumentos o en el compartimiento del motor, que se utiliza para la diagnosis del sistema informático.

Datum line   A straight reference line such as the top of a dedicated bench system.

Línea de datos   Línea recta de referencia, como por ejemplo la parte superior de un sistema de banco dedicado.

Dedicated bench system   A heavy steel bed with special fixtures for aligning unitized bodies.

Sistema de banco dedicado   Asiento pesado de acero con aparatos especiales que se utiliza para la alineación de carrocerías unitarias.

Diagnostic drawing and text screen   A display on a computer wheel aligner that provides illustrations and written instructions for the technicians.

Pantalla para la visualización de textos y diagramas diagnósticos   Representación visual en una computadora para alineación de ruedas que le provee a los mecánicos ilustraciones e instrucciones escritas.

Dial indicator   A precision measuring device with a stem and a rotary pointer.

Indicador de cuadrante   Dispositivo para medidas precisas con vástago y aguja giratoria.

Diamond-shaped chassis   A vehicle chassis that is shaped like a diamond from collision damage.

Chasis en forma de diamante   Chasis que ha adquirido la forma de un diamante a causa del impacto recibido durante una colisión.

Diesel particulates   Carbon particles emitted in diesel engine exhaust.

Partículas de diesel   Partículas de carbón presentes en el escape de un motor diesel.

Directional stability   The tendency of a vehicle steering to remain in the straight-ahead position when driven straight ahead on a smooth, level road surface.

Estabilidad direccional   Tendencia de la dirección del vehículo a permanecer en línea recta al ser así conducido en un camino cuya superficie es lisa y nivelada.

Drive cycle diagnosis   A test procedure when diagnosing computer-controlled suspension.

Diagnosis del ciclo propulsor   Procedimiento de prueba llevado a cabo durante la diagnosis de una suspensión controlada por computadora.

Dynamic imbalance   The imbalance of a wheel in motion.

Desequilibrio dinámico   Desequilibrio de una rueda en movimiento.

Eccentric camber bolt   A bolt with an out-of-round metal cam on the bolt head that may be used to adjust camber.

Perno de combadura excéntrica   Perno con una leva metálica con defecto de circularidad en su cabeza, que puede utilizarse para ajustar la combadura.

Eccentric cams or bushings   Out-of-round metal cams mounted on a retaining bolt with the shoulder of the cam positioned against a component. When the cam is rotated, the component position is changed.

Bujes o levas excéntricas   Levas metálicas con defecto de circularidad montadas sobre un perno de retenida; el apoyo de la leva está colocado contra un componente. Al girar la leva, cambia la posición del componente.

Electronically erasable programmable read only memory (EEPROM) A chip in a computer that may be erased easily with special equipment.
Memoria de solo lectura borrable y programable electrónicamente (EEPROM) Pastilla en una computadora que puede borrarse fácilmente con equipo especial.

Electronic neutral check A check while servicing an electronically controlled four wheel steering system.
Revisión electrónica neutra Revisión llevada a cabo durante la reparación de un sistema de dirección en las cuatro ruedas controlado electrónicamente.

Fail-safe function A mode entered by a computer if the computer detects a fault in the system.
Función de autoprotección Modo adaptado por una computadora si la misma detecta un fallo en el sistema.

Fault codes Numeric codes stored in a computer memory representing specific computer system faults.
Códigos de fallo Códigos numéricos almancenados en la memoria de una computadora que representan fallos específicos en el sistema informático.

Floor jack A hydraulically operated lifting device for vehicle lifting.
Gato de pie Dispositivo activado hidráulicamente que se utiliza para levantar un vehículo.

Flow control valve A special valve that controls fluid movement in relation to system demands.
Válvula de control de flujo Válvula especial que controla el movimiento del fluido de acuerdo a las exigencias del sistema.

Frame flange The upper or lower horizontal edge on a vehicle frame.
Brida del armazón Borde horizontal superior o inferior en el armazón del vehículo.

Frame web The vertical side of a vehicle frame.
Malla del armazón Lado vertical del armazón del vehículo.

Front and rear wheel alignment angle screen A display on a computer wheel aligner that provides readings of the front and rear wheel alignment angles.
Pantalla para la visualización del ángulo de alineación de las ruedas delanteras y traseras Representación visual en una computadora para alineación de ruedas que provee lecturas de los ángulos de alineación de las ruedas delanteras y traseras.

Front main steering angle sensor An input sensor mounted in the front steering gear in an electronically controlled four wheel steering system.
Sensor principal del ángulo de la dirección delantera Sensor de entrada montado en el mecanismo de dirección delantera en un sistema de dirección en las cuatro ruedas controlado electrónicamente.

Front sub steering angle sensor An input sensor mounted in the front steering gear in an electronically controlled four wheel steering system.
Sensor auxiliar del ángulo de la dirección delantera Sensor de entrada montado en el mecanismo de dirección delantera en un sistema de dirección en las cuatro ruedas controlado electrónicamente.

Functional test 211 A diagnostic procedure in a computer-controlled suspension system.
Prueba funcional 211 Procedimiento diagnóstico en un sistema de suspensión controlado por computadora.

Gear backlash, or lash Movement between gear teeth that are meshed with each other.

Contragolpe o juego en los engranajes Movimiento entre los dientes del engranaje que están endentados entre sí.

Geometric centerline An imaginary line through the exact center of the front and rear wheels.
Línea central geométrica Línea imaginaria a través del centro exacto de las ruedas delanteras y traseras.

Hard fault code A fault code in a computer that represents a fault that is present at all times.
Código de fallo duro Código de fallo en una computadora que representa un fallo que está siempre presente.

Heavy spots A spot in a tire casing that is heavier than the rest of the tire.
Puntos pesados Punto en la cubierta del neumático más pesado que el resto del neumático.

History code A fault code in a computer that represents an intermittent defect.
Código histórico Código de fallo en una computadora que representa un defecto intermitente.

Hydraulic press A hydraulically operated device for disassembling and assembling components that have a tight press-fit.
Prensa hidráulica Dispositivo activado hidráulicamente que se utiliza para desmontar y montar componentes con un fuerte ajuste en prensa.

Instrument panel cluster (IPC) A display in the instrument panel that may include gauges, indicator lights, or digital displays to indicate system functions.
Instrumentos agrupados en el tablero de instrumentos Representación visual en el tablero de instrumentos que puede incluir calibradores, luces indicadoras, o visualizaciones digitales para indicar las funciones del sistema.

Integral reservoir A reservoir that is joined with another component such as a power steering pump.
Tanque integral Tanque unido a otro componente, como por ejemplo una bomba de la dirección hidráulica.

Intermittent fault code A fault code in a computer memory that is caused by a defect that is not always present.
Código de fallo intermitente Código de fallo en la memoria de una computadora ocasionado por un defecto que no está siempre presente.

International system (SI) A system of weights and measures.
Sistema internacional Sistema de pesos y medidas.

Jack stand A metal stand that may be used to support a vehicle.
Soporte de gato Soporte de metal que puede utilizarse para apoyar un vehículo.

Lateral movement Movement from side to side.
Movimiento lateral Movimiento de un lado a otro.

Lateral runout The variation in side-to-side movement.
Desviación lateral Variación del movimiento de un lado a otro.

Light emitting diode (LED) A special diode that emits light when current flows through the diode.
Diodo emisor de luz (LED) Diodo especial que emite luz cuando una corriente fluye a través del mismo.

Lithium-based grease A special lubricant that is used on the rack bearing in manual rack and pinion steering gears.
Grasa con base de litio Lubricante especial utilizado en el cojinete de la cremallera en mecanismos de dirección de cremallera y piñón manuales.

Machinist's rule A steel ruler used for measuring short distances. These rulers are available in USC or metric measurements.

Regla para mecánicos Regla de acero que se utiliza para medir distancias cortas. Dichas reglas están disponibles en medidas del USC o en medidas métricas.

Magnetic wheel alignment gauge A wheel alignment gauge that is held on each front hub with a magnet.

Calibrador magnético para la alineación de ruedas Calibrador para la alineación de ruedas que se fija a cada uno de los cubos delanteros con un imán.

Main menu A display on a computer wheel aligner from which the technician selects various test procedures.

Menú principal Representación visual en una computadora para alineación de ruedas de la que el mecánico puede eligir varios procedimientos de prueba.

Memory steer The tendency of the vehicle steering not to return to the straight-ahead position after a turn, but to keep steering in the direction of the turn.

Dirección de memoria Tendencia de la dirección del vehículo a no regresar a la posición de línea recta después de un viraje, sino a continuar girando en el sentido del viraje.

Molybdenum disulphide lithium-based grease A special lubricant containing molybdenum disulphide and lithium that may be used on some steering components.

Grasa de bisulfuro de molibdeno con base de litio Lubricante especial compuesto de bisulfuro de molibdeno y litio que puede utilizarse en algunos componentes de la dirección.

Multipull unitized body straightening system A hydraulically operated system that pulls in more than one location when straightening unitized bodies.

Sistema de tiro múltiple para enderezar la carrocería unitaria Sistema activado hidráulicamente que tira hacia más de una dirección al enderezar carrocerías unitarias.

National Institute for Automotive Service Excellence (ASE) An organization that provides voluntary automotive technician certification in eight areas of expertise.

Instituto Nacional para la excelencia en la reparación de automóviles Organización que provee una certificación voluntaria de mecánico de automóviles en ocho áreas diferentes de especialización.

Optical toe gauge A piece of equipment that uses a light beam to measure front wheel toe.

Calibrador óptico del tope Equipo que utiliza un rayo de luz para medir el tope de la rueda delantera.

Oversteer The tendency of a vehicle to turn sharper than the turn selected by the driver.

Sobreviraje Tendencia de un vehículo a girar más de lo que el conductor desea.

Pilot bearing A bearing that supports the end of a shaft, such as a pinion shaft in a rack and pinion steering gear.

Cojinete piloto Cojinete que apoya el extremo de un árbol, como por ejemplo el árbol de piñón en un mecanismo de dirección de cremallera y piñón.

Pinion and bearing assembly The pinion gear and supporting bearings in a rack and pinion steering gear.

Conjunto del piñón y cojinete Engranaje del piñón y cojinetes de soporte en un mecanismo de dirección de cremallera y piñón.

Pinpoint tests Specific test procedures to locate the exact cause of a fault code in a computer system.

Pruebas llevadas a cabo con precisión Procedimientos de prueba específicos que se llevan a cabo para localizar la causa exacta de un código de fallo en un sistema informático.

Pitman arm puller A special puller required to remove a pitman arm.

Extractor del brazo pitman Extractor especial requerido para la remoción de un brazo pitman.

Plumb bob A weight with a sharp, tapered point that is suspended and centered on a string.

Plomada Balanza con un extremo cónico puntiagudo, suspendida y centrada en una hilera.

Ply separation A parting of the plies in a tire casing.

Separación de estrias División de las estrias en la cubierta de un neumático.

Pneumatic tools Tools operated by air pressure.

Herramientas neumáticas Herramientas activadas por aire a presión.

Power steering pump pressure gauge A gauge and valve with connecting hoses for checking power steering pump pressure.

Calibrador de presión de la bomba de la dirección hidráulica Calibrador y válvula con mangueras de conexión utilizadas para verificar la presión de la bomba de la dirección hidráulica.

Powertrain control module (PCM) A computer that controls engine and possibly transmission functions.

Unidad de control del tren transmisor de potencia Computadora que controla las funciones del motor y posiblemente las de la transmisión.

Prealignment inspection A check of steering and suspension components prior to a wheel alignment.

Inspección antes de una alineación Verificación de los componentes de la dirección y de la suspensión antes de llevarse a cabo una alineación de ruedas.

Preliminary inspection screen A display on a computer wheel aligner that allows the technician to check and record the condition of many suspension and steering components.

Pantalla para la visualización durante una inspección preliminar Representación visual en una computadora para alineación de ruedas que le permite al mecánico revisar y anotar la condición de un gran número de componentes de la suspensión y de la dirección.

Pressure relief valve A valve designed to limit pump pressure, such as a power steering pump.

Válvula de alivio de presión Válvula diseñada para limitar la presión de una bomba, como por ejemplo una bomba de la dirección hidráulica.

Programmed ride control (PRC) system A computer-controlled suspension system in which the computer operates an actuator in each strut to control strut firmness.

Sistema de control programado del viaje Sistema de suspensión controlado por computadora en el que la computadora opera un accionador en cada uno de los montantes para controlar la firmeza de los mismos.

Rack bushing A bushing that supports the rack in the rack and pinion steering gear housing.

Buje de la cremallera Buje que apoya la cremallera en el alojamiento del mecanismo de dirección de cremallera y piñón.

Radial runout The variations in diameter of a round object such as a tire.

Desviación radial Variaciones en el diámetro de un objeto circular, como por ejemplo un neumático.

Real time damping (RTD) The real time damping module controls the road-sensing suspension system.
Amortiguamiento en tiempo real La unidad del amortiguamiento en tiempo real controla el sistema de suspensión con equipo sensor.

Rear axle offset A condition in which the complete rear axle assembly has turned so one rear wheel has moved forward and the opposite rear wheel has moved rearward.
Desviación del eje trasero Condición que ocurre cuando todo el conjunto del eje trasero ha girado de manera que una de las ruedas traseras se ha movido hacia adelante y la opuesta se ha movido hacia atrás.

Rear axle sideset A condition in which the rear axle assembly has moved sideways from its original position.
Resbalamiento lateral del eje trasero Condición que ocurre cuando el conjunto del eje trasero se ha movido lateralmente desde su posición original.

Rear main steering angle sensor An input sensor in the rear steering actuator of an electronically controlled four wheel steering system.
Sensor principal del ángulo de la dirección trasera Sensor de entrada en el accionador de la dirección trasera de un sistema de dirección en las cuatro ruedas controlado electrónicamente.

Rear steering center lock pin A special pin used to lock the rear steering during specific service procedures in an electronically controlled four wheel steering system.
Pasador de cierre central de la dirección trasera Pasador especial que se utiliza para bloquear la dirección trasera durante procedimientos de reparación específicos en un sistema de dirección en las cuatro ruedas controlado electrónicamente.

Rear sub steering angle sensor An input sensor in the rear steering actuator of a four wheel steering system.
Sensor auxiliar del ángulo de la dirección trasera Sensor de entrada en el accionador de la dirección trasera de un sistema de dirección en las cuatro ruedas.

Rear wheel tracking Refers to the position of the rear wheels in relation to the front wheels.
Encarrilamiento de las ruedas traseras Se refiere a la posición de las ruedas traseras con relación a las ruedas delanteras.

Remote reservoir A container containing fluid that is mounted separately from the fluid pump.
Tanque remoto Recipiente lleno de líquido que se monta separado de la bomba del fluido.

Ribbed V-belt A belt containing a series of small "V" grooves on the underside of the belt.
Correa nervada en V Correa con una serie de ranuras pequeñas en forma de "V" en la superficie inferior de la misma.

Ride height screen A display on a computer wheel aligner that illustrates the locations for ride height measurement.
Pantalla para la visualización de la altura del viaje Representación visual en una computadora para alineación de ruedas que muestra los puntos donde se mide la altura del viaje.

Rim clamps Special clamps designed to clamp on the wheel rims and allow the attachment of alignment equipment such as the wheel units on a computer wheel aligner.
Abrazaderas de la llanta Abrazaderas especiales diseñadas para sujetar las llantas de las ruedas y así permitir la fijación del equipo de alineación, como por ejemplo las unidades de la rueda en una computadora para alineación de ruedas.

Road feel A feeling experienced by a driver during a turn when the driver has a positive feeling that the front wheels are turning in the intended direction.

Sensación del camino Sensación experimentada por un conductor durante un viraje cuando está completamente seguro de que las ruedas delanteras están girando en la dirección correcta.

Road sensing suspension (RSS) A computer-controlled suspension system that senses road conditions and adjusts suspension firmness to match these conditions in a few milliseconds.
Suspensión con equipo sensor Sistema de suspensión controlado por computadora que advierte las condiciones actuales del camino y ajusta la firmeza de la suspensión para equiparar dichas condiciones en un período de milisegundos.

Scan tester A digital computer system tester used to read trouble codes and perform other diagnostic functions.
Verificador de exploración Verificador digital del sistema informático que se utiliza para leer códigos indicativos de problemas y llevar a cabo otras funciones diagnósticas.

Seal drivers Designed to maintain even contact with a seal case and prevent seal damage during installation.
Empujadores para sellar Diseñados para mantener un contacto uniforme con el revestimiento de la junta de estanqueidad y evitar el daño de la misma durante el montaje.

Section modulus The measurement of a frame's strength based on height, width, thickness, and the shape of the side rails.
Coeficiente de sección Medida de la resistencia de un armazón, basada en la altura, el ancho, el espesor y la forma de las vigas laterales.

Selective shim Shims of different thicknesses that are used to provide an adjustment such as bearing preload.
Laminillas selectivas Laminillas de diferentes espesores que se utilizan para proveer un ajuste, como por ejemplo la carga previa del cojinete.

Self-locking nut A special nut designed so it will not loosen because of vibration.
Tuerca de cierre automático Tuerca con un diseño especial que evita su aflojamiento a causa de la vibración.

Serpentine belt A ribbed V-belt drive system in which all the belt-driven components are on the same vertical plane.
Correa serpentina Sistema de transmisión con correa nervada en V en el que todos los componentes accionados por una correa se encuentran sobre el mismo plano vertical.

Service bay diagnostics A diagnostic system supplied by Ford Motor Company to their dealers that diagnoses vehicles and provides communication between the dealer and manufacturer.
Diagnóstico para el puesto de reparación Sistema diagnóstico que les suministra la Ford a sus distribuidores de automóviles; dicho sistema diagnostica vehículos y permite una buena comunicación entre el distribuidor y el fabricante.

Service check connector A diagnostic connector used to diagnose computer systems. A scan tester may be connected to this connector.
Conector para la revisión de reparaciones Conector diagnóstico que se utiliza para diagnosticar sistemas informáticos. A dicho conector se le puede conectar un verificador de exploración.

Setback Occurs when one front wheel is driven rearward in relation to the opposite front wheel.
Retroceso Condición que ocurre cuando una de las ruedas delanteras se mueve hacia atrás con relación a la rueda delantera opuesta.

Sheared injected plastic Is inserted into steering column shafts and gear shift tubes to allow these components to collapse if the driver is thrown against the steering column in a collision.

Plástico cortado inyectado Insertado en los árboles de la columna de dirección y en los tubos del cambio de engranajes de velocidad para permitir que estos componentes se pleguen si la columna de dirección es impactada por el conductor durante una colisión.

Shim select screen A display on a computer wheel aligner that shows the technician the thickness and location of the proper shims for rear wheel alignment.

Pantalla para la visualización durante la selección de laminillas Representación visual en una computadora para alineación de ruedas que le muestra al mecánico el espesor y la ubicación de las laminillas correctas para la alineación de las ruedas traseras.

Shock absorber manual test A test in which the lower end of the shock absorber is disconnected and the shock absorber is operated manually to determine its condition.

Prueba manual del amortiguador Prueba en la que se desconecta el extremo inferior del amortiguador con el fin de poder operarlo manualmente y lograr determinar su condición.

Shock absorber or strut bounce test A test in which the vehicle is bounced by leaning on the bumper to determine shock absorber condition.

Prueba de rebote del montante o del amortiguador Prueba en la que se ejerce presión sobre el parachoques con el fin de hacer rebotar el vehículo y lograr determinar la condición del amortiguador.

Shop layout The location of all shop facilities including service bays, equipment, safety equipment, and offices.

Arreglo del taller de reparación Ubicación de todas las instalaciones del taller, incluyendo los puestos de reparación, el equipo, el equipo de seguridad, y las oficinas.

Single-pull unitized body straightening system Hydraulically operated equipment that pulls in one location while straightening unitized bodies.

Sistema enderezador de tiro único para la carrocería unitaria Sistema activado hidráulicamente que tira hacia una dirección al enderezar carrocerías unitarias.

Soft-jaw vise A vise equipped with soft metal, such as copper, in the jaws.

Tornillo con tenacilla maleable Tornillo equipado con un metal blando, como por ejemplo cobre, en las tenacillas.

Specifications menu A display on a computer wheel aligner that allows the technician to select, enter, alter, and display vehicle specifications.

Menú de especificaciones Representación visual en una computadora para alineación de ruedas que le permite al mecánico seleccionar, introducir datos, cambiar e indicar especificaciones referentes al vehículo.

Spiral cable A conductive ribbon mounted in a plastic container on top of the steering column that maintains electrical contact between the air bag inflator module and the air bag electrical system. A spiral cable may be called a clock spring electrical connector.

Cable espiral Cinta conductiva montada en un recipiente plástico sobre la columna de dirección que mantiene el contacto eléctrico entre la unidad infladora y el sistema eléctrico del Airbag. El cable espiral se conoce también como conector eléctrico de cuerda de reloj.

Spring fill diagnostics A diagnostic procedure in a computer-controlled air suspension system.

Diagnóstico con relleno para muelles Procedimiento diagnóstico en un sistema de suspensión de aire controlado por computadora.

Spring insulators, or silencers Rings usually made from plastic and positioned on each end of a coil spring to reduce noise and vibra-

tion transfer to the chassis.

Aisladores de muelles, o silenciadores Anillos, por lo general fabricados de plástico y colocados a ambos extremos de un muelle helicoidal, que se utilizan para disminuir la transferencia de ruido y de vibración al chasis.

Spring sag Occurs when a spring becomes weak, and the curb riding height is reduced compared to the original height.

Agotamiento de los muelles Condición que ocurre cuando el muelle se debilita; la altura del cotén del viaje disminuye si se la compara con la altura orginal.

Stabilizer bar, or sway bar A round steel bar connected between the front or rear lower control arms that reduces body sway.

Barra estabilizadora o barra de oscilación lateral Barra circular de acero contectada entre los brazos de mando delantero o trasero que disminuye la oscilación lateral de la carrocería.

Static imbalance Refers to the imbalance of a wheel and tire at rest.

Desequilibrio estático Se refiere al desequilibrio de una rueda y de un neumático cuando se ha detenido la marcha del vehículo.

Steering effort The amount of effort required by the driver to turn the steering wheel.

Esfuerzo de dirección Amplitud de esfuerzo requerido por parte del conductor para girar el volante de dirección.

Steering pull The tendency of the steering to pull to the right or left when the vehicle is driven straight ahead on a smooth, straight road surface.

Tiro de la dirección Tendencia de la dirección a desviarse hacia la derecha o hacia la izquierda mientras se conduce el vehículo en línea recta en un camino cuya superficie es lisa y nivelada.

Steering wheel freeplay The amount of steering wheel movement before the front wheels begin to turn.

Juego libre del volante de dirección Amplitud de movimiento del volante de dirección antes de que las ruedas delanteras comiencen a girar.

Steering wheel locking tool A special tool used to lock the steering wheel during certain wheel alignment procedures.

Herramienta para el cierre del volante de dirección Herramienta especial que se utiliza para bloquear el volante de dirección durante ciertos procedimientos de alineación de ruedas.

Stethoscope A special tool that amplifies sound to help diagnose noise location.

Estetoscopio Herramienta especial que amplifica el sonido para ayudar a diagnosticar la procedencia de los ruidos.

Strikeout, or rebound, bumper A rubber block that prevents the control arm from striking the chassis when the wheel hits a large road irregularity.

Parachoques de rebote Bloque de caucho que evita que el brazo de mando choque contra el chasis cuando la rueda golpea una irregularidad en el camino.

Strut cartridge The inner components in a strut that may be replaced rather than replacing the complete strut.

Cartucho del montante Componentes internos de un montante que pueden ser reemplazados en vez de tener que reemplazarse todo el montante.

Strut chatter A chattering noise as the steering wheel is turned often caused by a binding upper strut mount.

Vibración del montante Rechinamiento producido mientras se gira el volante de dirección. A menudo ocasionado por el trabamiento del montaje del montante superior.

Strut rod, or radius rod A rod connected from the lower control arm to the chassis to prevent forward and rearward control arm movement.
Varilla del montante o varilla radial Varilla conectada del brazo de mando inferior al chasis que se utiliza para evitar el movimiento hacia adelante o hacia atrás del brazo de mando.

Strut tower A circular, raised, reinforced area inboard of the front fenders that supports the upper strut mount and strut assembly.
Torre del montante Área circular, elevada y reinforzada en la parte interior de los guardafangos delanteros que apoya el conjunto del montante y montaje del montante superior.

Sulfuric acid A very corrosive acid used in automotive batteries.
Ácido sulfúrico Ácido sumamente corrosivo utilizado en las baterías de automóviles.

Sunburst frame cracks Radiate outward from an opening in the vehicle frame.
Grietas del armazón en forma de rayos de sol Grietas que se proyectan hacia afuera desde una abertura en el armazón del vehículo.

Supplemental inflatable restraint (SIR) An air bag system.
Sistema de seguridad inflable suplementario Sistema del Airbag.

Tapered-head bolts A bolt with a taper on the underside of the head.
Pernos de cabeza cónica Perno con una cabeza cuya superficie inferior es cónica.

Teflon ring compressing tool A special tool used to compress the Teflon rings on the pinion prior to installation in a rack and pinion steering gear.
Herramienta para la compresión de anillos de teflón Herramienta especial que se utiliza para comprimir los anillos de teflón en el piñón antes de ser instalados en un mecanismo de dirección de cremallera y piñón.

Teflon ring expander A special tool required to expand the Teflon rings prior to installation on the pinion in a rack and pinion steering gear.
Expansor para anillos de teflón Herramienta especial requerida para expandir los anillos de teflón antes de ser instalados en un mecanismo de dirección de cremallera y piñón.

Thrust line A line positioned at a 90° angle to the rear axle and projected toward the front of the vehicle.
Línea de empuje Línea colocada a un ángulo de 90° con relación al eje trasero y proyectada hacia la parte frontal del vehículo.

Tie rod end and ball joint puller A special puller designed for removing tie rod ends and ball joints.
Extractor para la junta esférica y el extremo de la barra de acoplamiento Extractor especial diseñado para la remoción de las juntas esféricas y los extremos de barras de acoplamiento.

Tie rod sleeve adjusting tool A special tool required to rotate tie rod sleeves without damaging the sleeve.
Herramienta para el ajuste del manguito de la barra de acoplamiento Herramienta especial requerida para girar los manguitos de las barras de acoplamiento sin estropearlos.

Tire changer Equipment used to demount and mount tires on wheel rims.
Cambiador de neumáticos Equipo que se utiliza para desmontar y montar los neumáticos en las llantas de las ruedas.

Tire condition screen A display on a computer wheel aligner that allows the technician to check and enter tire condition.
Pantalla para la visualización de la condición del neumático Representación visual en una computadora para alineación de ruedas que le permite al mecánico revisar e introducir datos referentes a la condición del neumático.

Tire conicity Occurs when the tire belt is wound off-center in the manufacturing process creating a cone-shaped belt, which results in steering pull.
Conicidad del neumático Condición que ocurre cuando la correa del neumático se devana fuera del centro durante el proceso de fabricación creando así una correa en forma cónica y trayendo como resultado el tiro de la dirección.

Tire rotation Involves moving each wheel and tire to a different location on the vehicle to increase tire life.
Cambio de posición de los neumáticos Se refiere a la rotación de cada rueda y neumático a una posición diferente en el vehículo para aumentar la vida útil del neumático.

Tire thump A pounding noise as the tire and wheel rotate usually caused by improper wheel balance.
Ruido sordo del neumático Ruido similar al de un golpe pesado que se produce mientras el neumático y la rueda están girando. Por lo general este ruido lo ocasiona el desequilibrio de la rueda.

Tire tread depth gauge A special tool required to measure tire tread depth.
Calibrador de la profundidad de la huella del neumático Herramienta especial requerida para medir la profundidad de la huella del neumático.

Tire vibration Vertical or sideways tire oscillations.
Vibración del neumático Oscilaciones verticales o laterales del neumático.

Toe gauge A special tool used to measure front or rear wheel toe.
Calibrador del tope Herramienta especial que se utiliza para medir el tope de las ruedas delanteras o traseras.

Toe-in A condition in which the distance between the front edges of the tires is less than the distance between the rear edges of the tires.
Convergencia Condición que ocurre cuando la distancia entre los bordes frontales de los neumáticos es menor que la distancia entre los bordes traseros de los mismos.

Toe-out A condition in which the distance between the front edges of the tires is more than the distance between the rear edges of the tires.
Divergencia Condición que ocurre cuando la distancia entre los bordes delanteros de los neumáticos es mayor que la distancia entre los bordes traseros de los mismos.

Toe-out on turns The turning angle of the wheel on the inside of a turn compared to the turning angle of the wheel on the outside of the turn.
Divergencia durante un viraje El ángulo de giro de la rueda en el interior de un viraje comparado con el ángulo de giro de la rueda en el exterior del viraje.

Torque steer The tendency of the steering to pull to one side during hard acceleration on front wheel drive vehicles with unequal length front drive axles.
Dirección de torsión Tendencia de la dirección a desviarse hacia un lado durante una aceleración rápida en un vehículo de tracción delantera con ejes de mando desiguales.

Total toe The sum of the toe angles on both wheels.
Tope total Suma de los ángulos del tope en ambas ruedas.

Track gauge A long straight bar with adjustable pointers used to measure rear wheel tracking in relation to the front wheels.
Calibrador del encarrilamiento Barra larga y recta con agujas ajustables que se utiliza para medir el encarrilamiento de las ruedas traseras con relación a las ruedas delanteras.

Traction control system (TCS) A computer-controlled system that prevents one drive wheel from spinning more than the opposite drive wheel on acceleration.
Sistema para el control de la tracción Sistema de suspensión controlado por computadora que evita que una rueda motriz gire más que la opuesta durante la aceleración del vehículo.

Tram gauge A long, straight bar with adjustable pointers used to measure unitized bodies.
Calibrador del tram Barra larga y recta con agujas ajustables que se utiliza para medir carrocerías unitarias.

Tread wear indicators Raised portions near the bottom of the tire tread that are exposed at a specific tread wear.
Indicadores del desgaste del neumático Secciones elevadas cerca de la parte inferior de la huella del neumático que quedan expuestas cuando la huella alcanza una cantidad de desgaste específica.

Trim height The normal chassis riding height on a computer-controlled air suspension system.
Altura equilibrada Altura normal de viaje del chasis en un sistema de suspensión controlado por computadora.

Turning angle screen A display on a computer wheel aligner that displays the turning angle.
Pantalla para la visualización del ángulo de giro Representación visual en una computadora para alineación de ruedas que muestra el ángulo de giro.

Turning imbalance A condition in which more effort is required to turn the steering wheel in one direction compared to the opposite direction.
Desequilibrio de giro Condición que ocurre cuando se requiere más esfuerzo para girar el volante en una dirección que en la dirección opuesta.

Turning radius The turning angle on the front wheel on the inside of a turn compared to the front wheel turning angle on the outside of the turn.
Radio de giro, o círculo de giro Ángulo de giro de la rueda delantera en el interior de un viraje con relación al ángulo de giro de la rueda delantera en el exterior del viraje.

Turning radius gauge A gauge with a degree scale mounted on turn tables under the front wheels.
Calibrador del radio de giro Calibrador con una escala de grados montado sobre plataformas giratorias debajo de las ruedas delanteras.

Turn tables Heavy steel plates placed under the front wheels during a wheel alignment that allow front wheel turning.
Plataformas giratorias Placas pesadas de acero que se colocan debajo de las ruedas delanteras durante una alineación de ruedas para permitir el giro de las mismas.

Understeer The tendency of a vehicle not to turn as much as desired by the driver.
Dirección pobre Tendencia de un vehículo a no girar tanto como lo desea el conductor.

United States Customary (USC) A system of weights and measures patterned after the British system.
Sistema usual estadounidense (USC) Sistema de pesos y medidas desarrollado según el modelo del Sistema Imperial Británico.

Vacuum hand pump A mechanical pump with a vacuum gauge and hose used for testing vacuum-operated components.
Bomba de vacío manual Bomba mecánica con un calibrador de vacío y una manguera que se utiliza para probar componentes a depresión.

V-belt A drive belt with a V-shape.
Correa en V Correa de transmisión en forma de V.

Vehicle lift A hydraulically or air-operated mechanism for lifting vehicles.
Levantamiento del vehículo Mecanismo activado hidráulicamente o por aire que se utiliza para levantar vehículos.

Vehicle wander The tendency of the steering to pull to the right or left when the vehicle is driven straight ahead on a straight road.
Desviación de la marcha del vehículo Tendencia de la dirección a desviarse hacia la derecha o hacia la izquierda cuando se conduce el vehículo en línea recta en un camino cuya superficie es lisa.

Wear indicator A device that visually indicates ball joint wear.
Indicador de desgaste Dispositivo que indica de manera visual el desgaste de una junta esférica.

Wheel balancer Equipment for wheel balancing that is usually computer controlled.
Equilibrador de ruedas Equipo para la equilibración de ruedas que por lo general es controlado por computadora.

Wheel runout compensation screen A display on a computer wheel aligner that is shown during the wheel runout compensation procedure.
Pantalla para la visualización de la compensación de desviación de la rueda Representación visual en una computadora para alineación de ruedas que aparece durante el procedimiento de compensación de desviación de la rueda.

Woodruff key A half-moon shaped metal key used to retain a component, such as a pulley, on a shaft.
Chaveta woodruff Chaveta de metal en forma de media luna que se utiliza para sujetar un componente, como por ejemplo una roldana, en un árbol.

Yield strength A measurement of the material strength from which a frame is manufactured.
Límite para fluencia Medida de la resistencia del material del que está fabricado un armazón.

Brakes

Pretest

The purpose of this pretest is to determine the amount of review that you may require prior to writing the ASE Brakes Test. If you answer all the pretest questions correctly, complete the questions and study the information in this chapter to prepare for the ASE Brakes Test.

If two or more of your answers to the pretest questions are incorrect, complete a study of Chapters 3 through 10 in *Today's Technician Automotive Brakes Classroom and Shop Manuals*, published by Delmar Publishers, plus a study of the questions and information in this chapter.

The pretest answers are located at the end of the pretest; these answers also are in the answer sheets supplied with this book. Unless stated otherwise the pretest questions apply vehicles without antilock brake systems (ABS).

1. During a brake application a vehicle experiences wheel lockup on the right rear (R/R) wheel. The cause of this problem could be
   A. weak brake shoe return springs on the R/R wheel.
   B. less-than-specified brake pedal free play.
   C. brake linings contaminated with grease on the R/R wheel.
   D. swollen rubber cups in the master cylinder.

2. A vehicle experiences brake drag on all four wheels.
   Technician A says the master cylinder compensating ports may be plugged.
   Technician B says the rubber cups in the master cylinder may be swollen.
   Who is correct?
   A. A only
   B. B only
   C. Both A and B
   D. Neither A nor B

3. A brake hose on the left front (L/F) wheel is soft and spongy in one location, but there are no visible brake fluid leaks. This defect may cause
   A. brake pedal fade during a brake application.
   B. brake drag on the L/F wheel.
   C. premature lockup on the L/F wheel.
   D. steering pull to the left during a brake application.

4. While discussing brake fluids
   Technician A says DOT 3 brake fluid is hydroscopic.
   Technician B says DOT 4 and DOT 5 brake fluid may be mixed.
   Who is correct?
   A. A only
   B. B only
   C. Both A and B
   D. Neither A nor B

5. The brake metering valve
   A. reduces fluid pressure to the rear brakes during moderate brake applications.
   B. reduces fluid pressure to the rear brakes during hard brake applications.
   C. delays fluid pressure to the front brakes during light brake applications.
   D. reduces fluid pressure to the front brakes during moderate brake applications.

6. The red brake warning light is illuminated continually with the ignition switch on.
   Technician A says the brake pedal may be binding.
   Technician B says there may be a fluid leak at the L/R wheel.
   Who is correct?
   A. A only
   B. B only
   C. Both A and B
   D. Neither A nor B

7. While pressure bleeding the brakes on a vehicle with front disc brakes and rear drum brakes
   Technician A says the metering valve should be closed.
   Technician B says the wheels may be bled in any sequence.
   Who is correct?
   A. A only
   B. B only
   C. Both A and B
   D. Neither A nor B

8. A vehicle with front disc and rear drum brakes experiences excessive pedal fade after several hard brake applications. The cause of this problem could be
   A. the diameter of the brake drums exceeds the maximum specification.
   B. the pistons are seized in the rear wheel cylinders.
   C. there is air trapped in the rear wheel cylinders.
   D. the rear brake shoe return springs are weak.

9. When the master cylinder is removed the vacuum brake booster contains some brake fluid.
   Technician A says this problem may be caused by leaking primary piston cups.
   Technician B says this problem may be caused by a defective one-way check valve in the booster vacuum hose.
   Who is correct?
   A. A only
   B. B only
   C. Both A and B
   D. Neither A nor B

10. On a vehicle with drum brakes, grabbing occurs on the R/F wheel during a brake application. The cause of this problem could be
    A. a restricted R/F brake hose.
    B. a leaking brake line near the L/F wheel.
    C. hard spots on the R/F brake drum.
    D. swollen cups in the R/F wheel cylinder.

11. While discussing brake rotor measurement
    Technician A says brake rotor runout should be measured while rotating the wheel with a dial indicator positioned against the rotor friction surface.
    Technician B says rotor parallelism should be measured with a micrometer at six or more locations around the rotor.
    Who is correct?

    A. A only

    B. B only

    C. Both A and B

    D. Neither A nor B

12. On an integral antilock brake system (ABS) with a high-pressure accumulator poor stopping ability and a hard brake pedal may be caused by

    A. an open wheel speed sensor winding.

    B. a defective solenoid valve.

    C. low accumulator pressure.

    D. improper wheel speed sensor adjustment.

13. The amber ABS warning light is illuminated with the engine running on an integral ABS system with a high-pressure accumulator. All of these defects could be the cause of the problem EXCEPT

    A. a defective wheel speed sensor.

    B. a defective pump motor.

    C. the parking brake partially applied.

    D. a defective ABS computer.

14. While obtaining diagnostic trouble codes (DTCs) to diagnose ABS systems

    A. DTCs indicate a defect in a specific component.

    B. some ABS systems do not provide flash codes.

    C. on some systems cycle the ignition switch three times to obtain DTCs.

    D. flash codes must be obtained with the engine running.

Answers To Pretest

1. C, 2. C, 3. A, 4. A, 5. C, 6. C, 7. D, 8. A, 9. A, 10. C, 11. C, 12. C, 13. C, 14. B

Hydraulic System Diagnosis and Repair, Master Cylinders (Non-ABS)

ASE Tasks, Questions, and Related Information

In this chapter each task in the Brakes category is provided followed by a question and some information related to the task. If you answer any question incorrectly, study this information very carefully until you understand the correct answer. For additional information on any task refer to *Today's Technician Automotive Brakes Classroom and Shop Manuals*, published by Delmar Publishers.

Question answers and analysis are provided at the end of this chapter and in the answer sheets provided with this book.

Task 1   **Diagnose poor stopping or dragging caused by problems in the master cylinder; determine needed repairs.**

1. While discussing a vehicle with dragging brakes
Technician A says fluid may be leaking past the master cylinder cups.
Technician B says the stop light and cruise control switch may require adjusting.
Who is correct?

    A. A only
    B. B only
    C. Both A and B
    D. Neither A nor B

Hint   *Dragging brakes may be caused by swollen master cylinder cups, or the compensation port covered in the master cylinder. This problem also may be caused by reduced brake pedal free play, or improper adjustment of the stop light and cruise control switch. If brake dragging occurs only on one front caliper, the quick take-up valve may not be allowing fluid movement from the reservoir into the primary piston area.*

    Excessive pedal effort may be caused by swollen master cylinder cups or a corroded master cylinder.

Task 2   Diagnose poor stopping, dragging, high or low pedal, or hard pedal caused by problems in the step bore master cylinder or internal valves (e.g., volume control devices quick take-up valve, fast-fill valve, pressure regulating valve); determine needed repairs.

    2. The brake pedal is low and spongy, and all brake adjustments are completed as specified by the vehicle manufacturer. The cause of this problem could be
        A. a binding pedal linkage.
        B. dented brake lines.
        C. a plugged compensating port.
        D. a weak hydraulic brake hose.

Hint   *A low, spongy brake pedal may be caused by air in the hydraulic system, or low brake fluid level in the master cylinder. If the brake drums are machined so they are thinner than specified by the manufacturer, the brake pedal may be low and spongy. Weak hydraulic hoses that expand under pressure also may cause a low, spongy brake pedal.*

    A hard brake pedal may be caused by contaminated brake fluid, swollen master cylinder cups, dented brake lines, restricted brake hoses, or a plugged quick take-up valve. Glazed brake linings or a defective brake booster also increase pedal effort.

Task 3   Measure and adjust pedal pushrod length.

    3. A vehicle has no brake pedal free play. This problem may cause
        A. a low brake pedal.
        B. a spongy brake pedal.
        C. pressure buildup and dragging brakes.
        D. fluid leaking past the primary piston cups.

Hint   *The specified brake pedal free play is usually 1/16 to 1/4 in. Excessive brake pedal free play causes a low brake pedal. When the brake pedal free play is less than specified, the compensating ports in the master cylinder may be closed by the piston cups when the brake pedal is released. This action results in fluid pressure buildup and dragging brakes.*

Task 4   Check the master cylinder for defects by depressing the brake pedal; determine needed repairs.

    4. With the engine running and the brake pedal applied and held with medium foot pressure, the brake pedal slowly moves downward.
       Technician A says brake fluid may be leaking past the master cylinder piston cups.
       Technician B says brake fluid may be leaking from a brake hose at one of the front wheels.
       Who is correct?
        A. A only
        B. B only
        C. Both A and B
        D. Neither A nor B

Hint   *When the brakes are held in the applied position with the engine running, the pedal should feel firm. On most vehicles the brake pedal should remain 2 in. from the floor for manual brakes or 1 in. from the floor for power brakes. On a manual brake system it is not necessary to have the engine running. If the pedal is low, the brakes may require adjusting, or the self adjusters may not be operating. Excessive brake pedal free play also causes a low pedal. A spongy pedal may be caused by air in the hydraulic system or drums that have been machined until they are too thin. If the brake pedal slowly moves downward, fluid may be leaking past the master cylinder piston cups, or there may be an external leak in a brake hose or line.*

Task 5   **Diagnose master cylinder for secondary cup defects by inspecting for external fluid leakage.**

5. There are signs of paint removal on the power brake booster below the master cylinder. The cause of this problem could be
   A. a leaking secondary cup on the primary piston.
   B. a defective diaphragm in the brake booster.
   C. a leaking secondary cup on the secondary piston.
   D. reduced brake free play and fluid pressure buildup.

Hint   *If there is an indication of paint removal and fluid leakage on the brake booster below the master cylinder, the secondary cup on the primary piston is leaking. If this leakage is present there probably will be brake fluid in the brake booster when the master cylinder is removed. Use a suction gun to remove this fluid.*

Task 6   **Remove master cylinder from the vehicle.**

Task 7   **Bench bleed (check for function and remove air) all non-ABS master cylinders.**

6. The service procedure illustrated in Figure 5-1 is
   A. checking the master cylinder piston travel.
   B. measuring the master cylinder displacement volume.
   C. checking the master cylinder piston return action.
   D. bleeding air from the master cylinder.

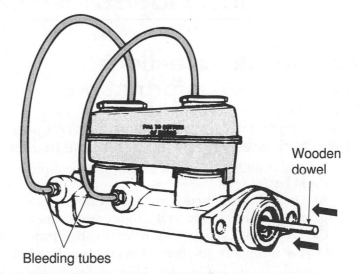

Wooden dowel

Bleeding tubes

Figure 5-1 Master cylinder service procedure *(Courtesy of Chrysler Corporation)*

Hint   *Bleeder tubes are connected from the master cylinder outlet ports to the reservoirs. The reservoir should be filled to the proper level with the specified type of brake fluid. A wooden dowel should be used to slowly depress and release the master cylinder pistons. Continue this action until there are no air bubbles coming through the bleeder tubes. Remove the bleeder tubes and plug the master cylinder outlets.*

Task 8   **Install master cylinder in vehicle; test operation of the hydraulic system.**

7. After the master cylinder is bled and installed, excessive brake pedal effort is required during brake applications.

     Technician A says the vacuum hose connected to the brake booster may be restricted.

     Technician B says the pistons may be seized in the front calipers.

     Who is correct?

     A. A only

     B. B only

     C. Both A and B

     D. Neither A nor B

Hint   *After the master cylinder is bled and installed, apply and hold the brake pedal. If the vehicle has power brakes, apply the brake pedal with the engine running. The brake pedal should be firm and remain at the proper height. Check the operation of the brakes during a road test. When the brakes are applied the steering should not pull in either direction, and the vehicle should stop with a normal amount of pedal effort. During a brake application there should be no abnormal noises such as brake squeals and rattles. While the brakes are applied, the brake pedal should remain firm without pulsating.*

Brake pull may be caused by contaminated brake linings, restricted brake lines or hoses, or seized caliper or wheel cylinder pistons.

Common causes of excessive pedal effort are glazed brake linings, seized caliper or wheel cylinder pistons, a restricted vacuum hose connected to the brake booster, or scored brake drums and rotors.

Brake squeal may be caused by improper type of brake linings, loose backing plate, brake anchor, or caliper, loose wheel bearings, and weak or broken hold-down springs.

A pulsating brake pedal may be caused by drums that are out-of-round, or rotors with excessive runout.

Fluids, Lines, and Hoses

ASE Tasks, Questions, and Related Information

Task 1   **Diagnose poor stopping, pulling, or dragging caused by problems in the brake fluid, lines, and hoses; determine needed repairs.**

8. A vehicle pulls to the left during a brake application. The cause of this problem could be

     A. the right front brake linings are contaminated with grease.

     B. the piston is seized in the right front caliper.

     C. the master cylinder pistons are swollen from contaminated fluid.

     D. the secondary compensating port is plugged in the master cylinder.

Hint   *Poor stopping ability may be caused by glazed brake linings, contaminated brake fluid and swollen master cylinder cups, restricted master cylinder tubes, or contaminated master cylinder bores.*

The most common causes of brake pull are seized caliper or wheel cylinder pistons, contaminated brake linings, scored or burned drums or rotors, and worn suspension components. If one front caliper is seized, the vehicle pulls to the opposite side from the seized caliper. When the brake linings are contaminated on a front wheel, the brake grabs on the wheel with the contaminated linings. This action causes steering pull to the side with the contaminated linings.

Brake drag may be caused by plugged master cylinder compensating ports, improper pedal free play, contaminated brake fluid and swollen master cylinder cups, or a binding brake pedal. If the brakes drag on one wheel, the brake line or hose may be restricted, or the piston may be sticking in caliper or wheel cylinder.

Task 2 Inspect brake lines and fittings for leaks, dents, kinks, rust, cracks, or wear; tighten loose fittings and supports.

9. While discussing brake lines
   Technician A says a damaged brake line may be repaired with a short piece of line and compression fittings.
   Technician B says the necessary brake line bends should be made with a tubing bending tool.
   Who is correct?
   A. A only
   B. B only
   C. Both A and B
   D. Neither A nor B

Hint *Brake lines should be inspected for leaks, cracks, rust, kinks, flattened areas, and splits. Damage brake lines should be replaced rather than repaired. A tubing bending tool should be used to complete the necessary brake line bends.*

Task 3 Inspect flexible brake hoses for leaks, kinks, cracks, bulging, or wear; tighten loose fittings and supports.

10. While discussing brake hose replacement
    Technician A says the sealing washer on the male end of a brake hose may be reused.
    Technician B says the brake line fitting should be installed and tightened in the female end of the brake hose before the male end is installed.
    Who is correct?
    A. A only
    B. B only
    C. Both A and B
    D. Neither A nor B

Hint *Flexible brake hoses allow for movement between the suspension and the chassis. These brake hoses should be inspected for cracks, leaks, twists, bulges, loose supports, and internal restrictions. Each time a brake hose is removed, the sealing washer on the male end should be replaced. When a brake hose is installed, always install and tighten the male end first.*

Task 4 Replace brake lines (double flare and ISO types), hoses, fittings, and supports.

11. The brake line flare in Figure 5-2 (next page) is
    A. a double inverted flare.
    B. an ISO flare.
    C. a single flare.
    D. an SAE flare.

Hint *Brake tubing may have double inverted flares or international standards organization (ISO) flares. Tube and hose mating surfaces should be clean and free from metal burrs.*

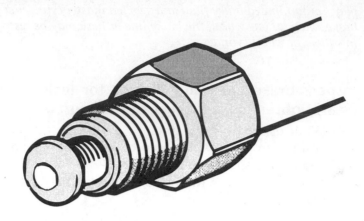

Figure 5-2 Brake tube flare *(Courtesy of Chrysler Corporation)*

Task 5   Select, handle, store, and install brake fluids (including silicone fluids).

12. While discussing brake fluids
    Technician A says silicone-based brake fluids are hydroscopic.
    Technician B says DOT 3 and DOT 4 brake fluids tend to absorb moisture.
    Who is correct?
    A.   A only
    B.   B only
    C.   Both A and B
    D.   Neither A nor B

Hint   *Brake fluid containers should be stored in a clean, dry location. When brake fluid containers are not in use, seal them tightly. Silicone-base fluid is classified as DOT 5 fluid. This type of fluid is non-hydroscopic, which means it does not absorb water. Silicone-based brake fluid has a long-term shelf life, and does not damage paint finishes.*

Commonly used nonpetroleum-based brake fluids are classified as DOT 3 or DOT 4. These brake fluids tend to absorb moisture from the air. Paint surfaces are damaged by DOT 3 or DOT 4 brake fluids. Compared to DOT 3 brake fluid, DOT 4 brake fluid has a higher boiling point and absorbs moisture more slowly.

Valves and Switches (Non-ABS)

ASE Tasks, Questions, and Related Information

Task 1   **Diagnose poor stopping, pulling, or dragging caused by problems in the hydraulic system valve(s); determine needed repairs.**

13. The purpose of the component in Figure 5-3 is to
    A. provide simultaneous front disc and rear drum brake application.
    B. prevent premature front wheel lockup during hard brake applications.
    C. prevent front brake drag when the brakes are released.
    D. reduce front brake pad temperature and provide longer pad life.

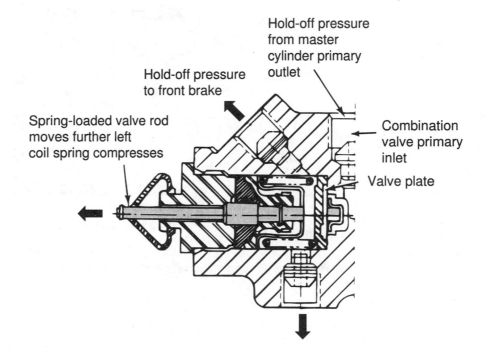

Figure 5-3 Metering valve *(Courtesy of Chrysler Corporation)*

Hint   *The metering valve is used on systems with front disc and rear drum brakes. During a brake application the brake fluid pressure to the rear brakes has to overcome the force of the brake shoe return springs and force these shoes outward. Therefore, a brief time interval is required to apply the rear brakes, whereas the front caliper pistons move the brake pads quickly against the rotors. The metering valve delays front brake fluid pressure during light brake applications so both front and rear brake applications occur at the same time. This action prevents front wheel lockup and skidding during light brake applications on slippery road surfaces. During hard brake applications the metering valve remains open and does not affect brake operation.*

Task 2 Inspect, test, and replace metering (hold-off), proportioning, pressure differential, and combination valves.

14. The pressure gauges connected to the proportioning valve inlet and outlet (Figure 5-4) should indicate
    A. higher pressure at the outlet compared to the inlet during a moderate brake application.
    B. the same pressure at the inlet and outlet during a moderate brake application.
    C. lower pressure at the outlet compared to the inlet during a moderate brake application.
    D. lower pressure at the outlet compared to the inlet during a light brake application.

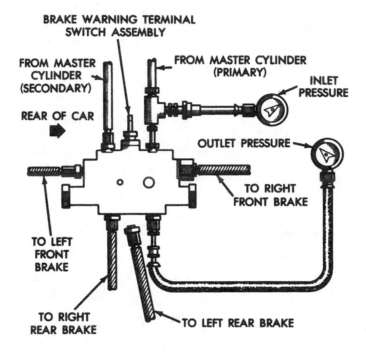

Figure 5-4 Proportioning valve test *(Courtesy of Chrysler Corporation)*

Hint   *During a moderate brake application the proportioning valve reduces pressure to the rear wheel brakes to compensate for the transfer of vehicle weight to the front wheels. This action prevents premature rear wheel lockup of the rear wheels. During a light brake application the proportioning valve does not affect rear wheel pressure. When the brakes are applied with high pedal pressure, the proportioning valve opens and allows full pressure to the rear wheels. When pressure gauges are connected to each proportioning valve inlet and outlet, the specified pressure difference between the inlet and outlet should be present during a moderate brake application.*

The metering valve stem should move slightly when the brake pedal is applied and released. If the metering valve is operating normally, a slight bump may be felt on the brake pedal when the pedal has been depressed about one inch.

Task 3 Inspect, test, replace, and adjust load or height sensing-type proportioning valve (s).

15. While discussing a load-sensing proportioning valve
    Technician A says this valve reduces brake pressure to the rear wheels as the rear suspension load is increased.
    Technician B says this valve has a linkage connected from the valve to the rear chassis.
    Who is correct?

    A. A only
    B. B only
    C. Both A and B
    D. Neither A nor B

Hint   *The height-sensing proportioning valve is mounted on the chassis, and a linkage is connected from the valve to the rear axle. When the load is light on the rear wheels the linkage positions the internal valve so brake pressure is reduced to the rear wheels during moderate brake applications. A heavy load on the rear suspension reduces rear chassis height. This action causes the linkage to move the valve in the height-sensing proportioning valve. Under this condition the proportioning valve does not reduce pressure to the rear brakes during moderate brake applications.*

Task 4   Inspect, test, and replace brake warning light, switch, and wiring.

Task 5   Reset brake pressure differential valve (if necessary).

    16. With the brake warning switch unit positioned as shown in Figure 5-5 the
        A. the brake warning light is illuminated with the ignition switch on.
        B. the pressure is higher in the secondary master cylinder section than in the primary section.
        C. the circuit is open between the switch terminal and ground.
        D. the pressure is equal in the primary and secondary sections of the master cylinder.

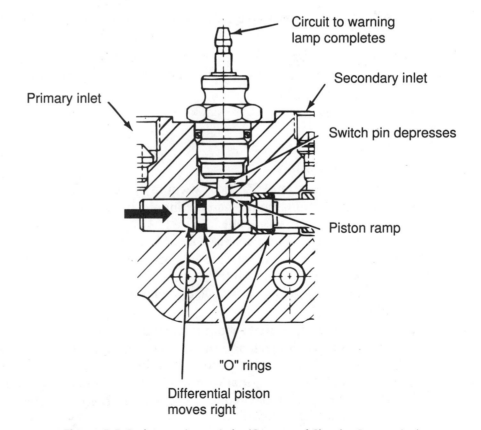

Figure 5-5 Brake warning switch *(Courtesy of Chrysler Corporation)*

Hint   *When the pressure is equal in the primary and secondary sections of the master cylinder, the warning switch piston remains centered. In this position the switch piston does not touch the switch pin. If the pressure is unequal in the primary and secondary master cylinder sections, the pressure difference moves the switch piston to one side. In this position the switch piston pushes the spring-loaded switch pin upward and closes the warning light switch. This action illuminates the brake warning light.*

The brake warning light circuit may be tested by grounding the warning switch wire with the ignition switch on. Under this condition the bulb should be illuminated. If the bulb is not illuminated check the fuse, bulb, and connecting wires.

After brake repairs are completed to restore equal pressure in the primary and secondary sections of the master cylinder, a hard brake pedal application usually centers the warning switch piston. If this action does not center the piston and put the light out, apply light brake pedal pressure and loosen a bleeder screw in the side of the brake system that had high pressure during the failure.

Bleeding, Flushing, and Leak Testing (Non-ABS Systems)

ASE Tasks, Questions, and Related Information

Task 1 **Bleed (manual, pressure, vacuum, or surge) and/or flush hydraulic system.**

17. While discussing pressure brake bleeding
Technician A says the metering valve should be closed.
Technician B says the pressure in the brake bleeder should be 20 psi.
Who is correct?
   A. A only
   B. B only
   C. Both A and B
   D. Neither A nor B

18. All of these statements about vacuum brake bleeding are true EXCEPT
   A. the vacuum pump handle should be operated two to four times before the bleeder screw is opened.
   B. the vacuum pump creates a vacuum in the tester reservoir.
   C. the bleeder screw should be opened until there is approximately 1 in. of brake fluid in the reservoir.
   D. a one-way check valve is required in the line from the bleeder screw to the reservoir.

19. All of these statements about manual brake bleeding procedure are true EXCEPT
   A   a hose is connected to the bleeder screw, and the opposite end of this hose submerged in a container of brake fluid.
   B. apply the brake pedal with moderate pressure and then open the bleeder screw.
   C. when the bleeder screw is opened and the pedal goes to the floor, release the pedal and close the screw.
   D. repeat the bleeding procedure until the fluid escaping from the bleeder hose is free of air bubbles.

20. During a surge bleeding procedure
   A. the end of the bleeder hose must be kept above the level of fluid in the container.
   B. pump the pedal quickly several times with the bleeder screw open.
   C. pump the pedal several times before opening the bleeder screw.
   D. decrease the pressure bleeder chamber pressure to 10 psi.

Hint   *During a manual brake bleeding procedure, connect a bleeder hose from a bleeder screw into a container partially filled with brake fluid. Keep the end of this hose submerged below the level of brake fluid in the container. Each wheel caliper or cylinder must be bled in the vehicle manufacturer's specified sequence in any bleeding procedure. Wheel calipers may be tapped with a soft hammer to help remove air bubbles.*

During a manual brake bleeding procedure, apply the brake pedal with moderate force and open the bleeder screw. When the brake pedal goes down to the floor, close the bleeder screw and release the pedal. Repeat the procedure until there are no air bubbles escaping from the bleeder hose.

A pressure bleeder has an adapter connected to the top of the master cylinder reservoir. A hose is connected from this adapter to the pressure bleeder fluid chamber in the top of the pressure bleeder. The pressure bleeder has an air chamber below the fluid chamber, and a diaphragm separates the air and fluid chambers. Shop air is used to pressurize the air chamber to 15 to 20 psi. If the brake system has a metering valve, this valve must be held open with a special tool. The bleeder hose is connected from each bleeder screw into a container partially filled with brake fluid. Open the bleeder screw until a clear stream of brake fluid is discharged.

The surge bleeding procedure may be used with manual or pressure bleeding. Bleed the brake system in the conventional manner. Surge bleeding involves pumping the brake pedal quickly several times with the bleeder screw open; then close the bleeder screw. After surge bleeding, use the conventional bleeding procedure to bleed the system one more time.

During the vacuum bleeding procedure the vacuum pump is connected to a sealed container. Another hose is connected from this container to the bleeder screw. A one-way check valve is connected in the hose from the bleeder screw to the container. Operate the vacuum pump handle 10 to 15 times to create vacuum in the container. Open the bleeder screw until about 1 in. of brake fluid is pulled into the container. Repeat the procedure until the fluid coming into the container is free of air bubbles.

Flushing a brake system is a continuation of the bleeding procedure. During the flushing procedure each bleeder screw is opened until all the contaminated fluid is removed. Never reuse fluid from a bleeding or flushing procedure. Discard this fluid according to environmental regulations.

Task 2   Pressure Test Brake Hydraulic System.

21. When pressure testing the hydraulic brake system equipped with a metering valve and proportioning valves
    A. during a moderate brake application the pressure at the master cylinder should exceed the pressure at the front wheel calipers.
    B. below the split point pressure the pressure at the master cylinder outlet should exceed the pressure at the rear wheel cylinders.
    C. during a moderate brake application the pressure should be equal at the front and rear wheels.
    D. above the split point pressure, the pressure at the master cylinder outlet should exceed the pressure at the rear wheel cylinders.

Hint   *A pressure test may be performed with the moderate pressure applied to the brake pedal. If the vehicle has power brakes, the engine should be running. The brake pedal should be firm at the specified pedal height. If the pedal slowly moves toward the floor, there is an internal master cylinder leak or an external leak in the lines, fittings, hoses, wheel cylinders, or calipers. When there are no external leaks, the master cylinder must be leaking internally.*

A pressure test may be performed to check the brake booster operation. Connect a pressure gauge to one of the master cylinder outlets, and apply the brake pedal with moderate pressure. Maintain this pressure and start the engine. There should be a significant increase in master cylinder pressure if the brake booster is operation is satisfactory.

A pressure gauge may be connected at each front wheel caliper. When the brakes are applied with moderate pressure, the gauge readings should be equal. When the pressure is lower at one front wheel, the line to that wheel is probably restricted.

The proportioning valves connected to the rear wheels may be tested by connecting pressure gauges on the master cylinder side and wheel cylinder side of each proportioning valve. When the brake pedal is applied with light pressure and the master cylinder pressure is below the specified split point pressure, both gauges should indicate the same pressure. If the brake pedal pressure is increased and the master cylinder pressure exceeds the split point pressure, the master cylinder pressure should exceed the wheel cylinder pressure by the specified amount.

Drum Brake Diagnosis and Repair

ASE Tasks, Questions, and Related Information

Task 1 **Diagnose poor stopping, pulling, or dragging caused by drum brake hydraulic problems; determine needed repairs.**

22. On a vehicle with drum brakes experiences poor stopping and the pedal feels springy and spongy during a brake application.
    Technician A says the vents in the master cylinder cover may be plugged.
    Technician B says there may be air in the hydraulic system.
    Who is correct?
    A. A only
    B. B only
    C. Both A and B
    D. Neither A nor B

Hint   *Poor stopping ability may be caused by hydraulic problems such as contaminated brake fluid, air in the hydraulic system, or plugged master cylinder vents.*

Brake pull may be caused by hydraulic problems such as a severely restricted brake line or hose, or wheel cylinder size that is different on opposite sides of the vehicle.

Brake drag may be caused by hydraulic problems such as contaminated brake fluid, inferior rubber cups in the master cylinder or wheel cylinders, or plugged compensating ports in the master cylinder. If one wheel is dragging the wheel cylinder rubber cups may be swollen or the wheel cylinder pistons sticking. An obstructed brake line or hose may also be the cause of brake drag.

Task 2 **Diagnose poor stopping, noise, pulling, grabbing, dragging, or pedal pulsation caused by drum brake mechanical problems; determine needed repairs.**

23. A vehicle experiences brake squeal during brake applications.
    Technician A says the drums may be distorted.
    Technician B says the backing plates may be bent.
    Who is correct?
    A. A only
    B. B only
    C. Both A and B
    D. Neither A nor B

Hint   *Reduced stopping ability may be caused by mechanical brake problems such as improper brake adjustment, incorrect, glazed, or oil-soaked linings, seized wheel cylinder pistons, or bell-mouthed barrel shaped or scored drums. Poor stopping ability combined with brake pedal fade may be caused by drums that have been machined until they are thinner than specified.*

Brake squeal may be caused by bent backing plates, distorted drums, linings loose at shoe ends, improper lining position on the shoes, weak or broken hold-down springs, or loose wheel bearings. Brake chatter may be caused by improper brake adjustment, loose backing plates, contaminated linings, out-of-round, tapered, or barrel-shaped drums, cocked or distorted shoes, or loose wheel bearings.

Brake pull may be caused by improper tire inflation, unequal tire tread wear, or different tread designs on opposite sides. Suspension and wheel alignment defects may cause brake pull. This problem also may be caused by different wheel cylinder sizes on opposite sides, seized wheel cylinder pistons, damaged or contaminated brake linings on one side, weak or broken retractor springs, or a scored brake drum on one side.

Brake grabbing may be caused by contaminated linings; shoes not centered in the drums; loose, distorted backing plates; or scored, hard-spotted, or out-of-round drums.

Dragging brakes may be caused by a binding brake pedal, swollen rubber parts caused by incorrect or contaminated brake fluid, or plugged master cylinder compensating ports. If only the rear wheels drag, the parking brake cables may be seized. When one wheel drags, that wheel may have a seized caliper or wheel cylinder piston, weak or broken shoe return springs, bent or distorted shoes, loose wheel bearings, improper shoe adjustment, or a restricted brake hose or line.

Pedal pulsations may be caused by out-of-round brake drums, or rotors with improper parallelism.

Task 3   Remove, clean, inspect, and measure brake drums; follow manufacturer's recommendations in determining need to machine or replace.

24. All of these statements are true about brake drum inside diameter measurement with a brake drum micrometer EXCEPT
    A. the drum should be cleaned before measuring the diameter.
    B. if the drum diameter is more than specified, replace the drum.
    C. the diameter should be measured at two locations around the drum.
    D. the drum diameter variation should not exceed 0.0035 in.

Hint   *Brake drums should be inspected for cracks, heat checks, out-of-round, bell-mouthed, scoring, and hard spots. The inside drum diameter should be measured with a drum micrometer. If the drum diameter exceeds the maximum limit specified by the manufacturer, replace the drum. The drum diameter should be measured every 45 degrees around the drum. The maximum allowable out-of-round specified by some manufacturers is 0.0035 in. If the out-of-round exceeds specifications, the drum may be machined if this machining does not cause the drum to exceed the maximum allowable diameter.*

Task 4   Machine drum according to manufacturer's procedures and specifications.

25. While machining a brake drum
    A. tool chatter marks may be caused by excessive damping belt tension.
    B. the tool bit depth for a rough cut should be 0.005 to 0.010 in.
    C. the tool bit depth for a rough cut should be 0.010 to 0.020 in.
    D. the tool bit depth of a finish cut should be 0.008 to 0.010 in.

Hint   *When machining a brake drum the drum must be installed securely and centered on the lathe. A dampening belt must be installed tightly around the outside of the drum to prevent the cutting tool from chattering on the drum. Many manufacturers recommend a rough cut tool depth of 0.005 to 0.010 in. and a finish cut tool depth of 0.005 in. After the machining procedure the drum should be sanded to remove minor irregularities. Many manufacturers recommend drum machining until the drum diameter is within 0.030 in. of the maximum allowable diameter. This 0.030 is necessary to allow for drum wear.*

Task 5
Using proper safety procedures, remove, clean, and inspect brake shoes/ linings, spring, pins, adjusters/self-adjusters, levers, clips, and other related brake hardware; determine needed repairs.

26. While discussing drum brake hardware
    Technician A says if item 17 in Figure 5-6 is omitted during brake hardware assembly the brake will grab on that wheel.
    Technician B says if item 8 in Figure 5-6 is omitted during brake hardware assembly there is no self-adjusting action.
    Who is correct?
    A. A only
    B. B only
    C. Both A and B
    D. Neither A nor B

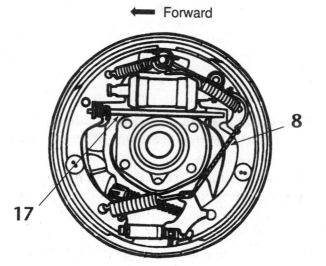

← Forward

8

17

10-inch rear brake (left side)

Figure 5-6 Drum brake hardware *(Courtesy of Ford Motor Company)*

Hint   *All brake return springs should be inspected for distortion and stretching. Brake shoes should be cleaned with a shop towel and inspected for broken welds, cracks, wear, and distortion. If the wear pattern on the brake shoes is uneven, the shoes are distorted. Check all clips and levers for wear and bending. Inspect the brake linings for contamination with oil, grease, or brake fluid. Clean and lubricate adjusting and self-adjusting mechanisms.*

Task 6
Using proper safety procedures, clean and inspect brake backing (support) plates; determine needed repairs.

27. While discussing brake backing plates
    Technician A says a bent backing plated may cause brake grabbing.
    Technician B says a loose anchor bolt may cause brake chatter.
    Who is correct?
    A. A only
    B. B only
    C. Both A and B
    D. Neither A nor B

Hint   *The backing plate should be cleaned with an approved brake cleaner that does allow the release of asbestos dust to the shop air. Inspect the backing plated for distortion, cracks, rust damage, and wear in the shoe contact areas. A distorted backing plate may cause brake grabbing. Check the anchor bolt for looseness that may result in brake chatter.*

Task 7 Disassemble and clean wheel cylinder assembly; inspect parts for wear, rust, scoring, and damage; hone cylinder (if necessary and recommended by the manufacturer); replace all cups, boots, and any damaged or worn parts; reassemble.

28. While servicing a wheel cylinder
    A. the parts should be washed in an approved cleaning solvent.
    B. if the cylinder bore is pitted or deeply scored, hone the wheel cylinder.
    C. the piston cups should be lubricated with clean brake fluid before installation.
    D. during assembly the flat side of the pistons faces the brake shoe links.

Hint *Wheel cylinders should be disassembled and cleaned with denatured alcohol or clean brake fluid. Minor cylinder bore scoring may be removed with a wheel cylinder hone. If the cylinder bore is pitted or deeply scored, replace the cylinder. During assembly lubricate the cups, pistons, and cylinder bore with clean brake fluid.*

Task 8 Lubricate brake shoe support pads on backing (support) plate, adjuster/self-adjuster mechanisms, and other brake hardware.

29. While discussing brake hardware service
    Technician A says dry shoe ledges on the backing plates may cause a squeaking noise during brake applications.
    Technician B says slightly scored shoe ledges on the backing plates may be resurfaced and lubricated with high-temperature grease.
    Who is correct?
    A. A only
    B. B only
    C. Both A and B
    D. Neither A nor B

Hint *The backing plates should be cleaned with an approved brake cleaning method. Since a warped backing plate may cause uneven shoe wear and brake grabbing, inspect the backing plates for this condition. If the shoe ledges are lightly scored, they may be resurfaced. The backing plate should be replaced if the shoe ledges are severely scored or worn. Lubricate the shoe ledges with high temperature lubricant.*

Task 9 Install brake shoes and related hardware.

30. While assembling the brake shoes and related hardware
    A. the secondary shoe faces toward the front of the vehicle.
    B. the primary and secondary shoe return springs are interchangeable.
    C. the adjuster must be installed in the proper direction.
    D. the adjuster cable usually is mounted on the primary shoe.

Hint *The secondary shoe must face toward the rear of the vehicle, and the primary shoe faces the front of the vehicle. The secondary shoe usually has a longer lining and the linings on the primary and secondary shoes usually are made of different materials. The primary and secondary shoes have different return springs. Brake adjusters must be installed in the proper direction so the star wheel is behind the backing plate adjustment opening. The adjuster cable and related mechanism usually is installed on the secondary shoe.*

Task 10

Preadjust brake shoes and parking brake before installing brake drums or drum/hub assemblies and wheel bearings.

31. Technician A says in Figure 5-7 the brake shoes are being adjusted to match the drum size.
    Technician B says in Figure 5-7 the drum and brake shoes are being measured for wear.
    Who is correct?
    A. A only
    B. B only
    C. Both A and B
    D. Neither A nor B

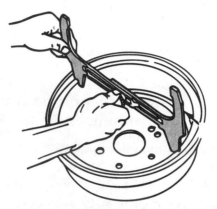

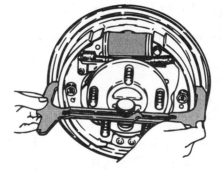

Figure 5-7 Brake shoe service *(Courtesy of Chevrolet Motor Division, General Motors Corporation)*

32. While adjusting the parking brake in Figure 5-8
    Technician A says the parking brake should be applied until a pin in hole A contacts the parking brake outer flange.
    Technician B says the parking brake shown in Figure 5-8 should be released during the adjustment.
    Who is correct?
    A. A only
    B. B only
    C. Both A and B
    D. Neither A nor B

Hint   *A special tool is used to measure the brake drum inside diameter. The other side of the tool is then installed on the outside of the brake shoes, and the shoe adjuster is turned until the shoes contact the tool. This procedure adjusts the shoes properly in relation to the drum size before installing the drum.*

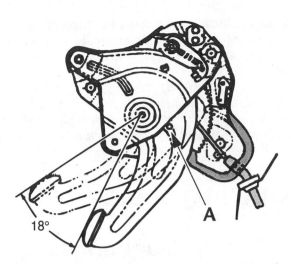

Figure 5-8 Parking brake adjustment *(Courtesy of Chevrolet Motor Division, General Motors Corporation)*

The brake shoes must be properly adjusted before the parking brake adjustment. Some manufacturers recommend adjusting the parking brake with the brake released. Tighten the parking brake cable adjusting nut until there is a light drag while rotating the rear wheels. Loosen the parking brake cable adjusting nut until the rear wheels rotate freely, and then loosen the parking brake cable adjusting nut two turns.

Task 11   Reinstall wheel, torque lug nuts, and make final checks and adjustments.

33. While discussing the brake adjustment in Figure 5-9
    Technician A says the thin tool is used to position the star wheel so it is accessible with the adjusting tool.
    Technician B says the star wheel should be rotated until there is wheel drag and then back off the star wheel until the wheel rotates freely.
    Who is correct?
    A. A only
    B. B only
    C. Both A and B
    D. Neither A nor B

Figure 5-9 Drum brake adjustment *(Courtesy of Chrysler Corporation)*

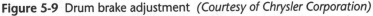

Hint   *The lug nuts must be tightened in the proper sequence with a crisscross pattern. Tighten the lug nuts to one-half the specified torque, and then tighten them to the specified torque. An improper lug nut tightening sequence, or excessive torque, may cause warped brake drums and grabbing brakes.*

If a brake adjustment is necessary, a thin tool may be inserted through the backing plate opening to release the self-adjusting mechanism. After this tool is in place, a brake adjusting tool may be used to rotate the star wheel until a slight wheel drag is felt while rotating the wheel. Rotate the star wheel in the opposite direction until the wheel rotates freely.

Disc Brake Diagnosis and Service

ASE Tasks, Questions, and Related Information

Task 1   **Diagnose poor stopping, pulling, or dragging caused by disc brake hydraulic problems; determine needed repairs.**

34. All of these defects may cause the car to pull to one side while braking EXCEPT
    A. incorrect or loose brake pads.
    B. loose caliper mounting bracket.
    C. seized master cylinder pistons.
    D. sticking caliper piston.

Hint   *Poor stopping may be caused by these hydraulic problems; restricted brake lines or hoses, sticking master cylinder pistons, or sticking caliper pistons.*

Pulling to one side while braking may be caused by a sticking caliper piston, or a restricted line or hose.

Dragging brakes may be caused by soft or swollen rubber components caused by incorrect or contaminated brake fluid. This problem also may be caused by plugged master cylinder compensating ports or a restricted hose or line.

Task 2   **Diagnose poor stopping, noise, pulling, grabbing, dragging, or pedal pulsation caused by disc brake mechanical problems; determine needed repairs.**

35. Excessive brake pedal pulsations are experienced while braking.
    Technician A says the outer drive axle joint may be worn.
    Technician B says the front wheel bearing may be worn and loose.
    Who is correct?
    A. A only
    B. B only
    C. Both A and B
    D. Neither A nor B

Hint   *Poor stopping may be caused by contaminated brake linings, defective power brake booster or low vacuum supply, or a defective power brake vacuum check valve.*

Pulling to one side while braking may be caused by the following mechanical problems: incorrect or loose pads, contaminated brake linings, loose caliper mounting, or defective suspension parts.

Dragging brakes may be caused by a binding brake pedal, loose or worn wheel bearings, a sticking caliper piston, or a restricted line or hose.

Braking noise may be caused by bent, damaged, or loose pads, worn out linings, foreign material embedded in the lining, or loose caliper mounting.

Grabbing brakes may be caused by contaminated linings, incorrect or loose pads, or loose caliper mounting.

Pedal pulsations may be caused by excessive rotor runout, and worn or loose wheel bearings.

Task 3   **Remove caliper assembly from mountings; clean and inspect for leaks and damage to caliper housing.**

    36. After a brake application the caliper piston is returned by
        A. the twisting action of the seal.
        B. a return spring.
        C. brake fluid pressure.
        D. atmospheric pressure.

Hint   *Inspect the caliper for evidence of brake fluid in and around the boot area. If brake fluid is evident, the caliper seal must be replaced. Replace the boot if it is torn, damaged, or cracked. If the caliper ears are worn, broken, or elongated, replace the caliper.*

Task 4   **Clean and inspect caliper mountings and slides for wear and damage.**

    37. On a single piston floating caliper the inside brake pad lining is worn out, but there is very little wear on the outside pad lining. The cause of this problem could be
        A. worn caliper pins and bushings.
        B. a leaking caliper piston seal.
        C. a leaking brake hose.
        D. excessive rotor lateral runout.

Hint   *Inspect the surfaces of the abutments on the caliper and anchor plate. Smooth these surfaces with emery cloth if they are rusty, burred, or corroded. Coat these surfaces with anti-seize lubricant before installing the caliper. Check for wear on caliper pins, slides, springs, and bushings. Worn components must be replaced.*

Task 5   **Remove, clean, and inspect pads and retaining hardware; determine needed repairs, adjustments, and replacements.**

    38. On a vehicle with front disc and rear drum brakes a scraping noise is present in one front wheel while driving the vehicle. The cause of this problem could be
        A. worn caliper pins and bushings.
        B. the pad wear sensor contacting the rotor.
        C. loose caliper mounting bolts.
        D. loose pad mounting in the caliper.

Hint   *Replace the brake pads if the linings are worn beyond the specified limit. Replacement also is necessary if the wear indicators are touching the rotor. The most common type of wear indicator is a metal tab attached to the edge of the pad. When the brake lining is worn to the specified limit, this metal tab contacts the rotor and produces a scraping noise while driving the vehicle.*

Task 6   **Disassemble and clean caliper assembly; inspect parts for wear, rust, scoring, and damage; replace all seals, boots, and any damaged or worn parts.**

    39. After honing a brake caliper the maximum increase in caliper bore diameter is
        A. 0.001 in.
        B. 0.002 in.
        C. 0.005 in.
        D. 0.008 in.

Hint   *Caliper pistons may be removed by pumping the brake pedal slowly with the caliper connected to the brake hose. Inspect steel pistons for worn surfaces, rust, pitting, scoring, or corrosion. Check phenolic pistons for swelling, cracks, chips, and gouges. Remove minor scratches from the caliper bore with crocus cloth. The caliper bore may be honed with the proper brake hone lubricated with brake fluid. Honing must not increase the caliper bore more than 0.001 in. Clean all parts with denatured alcohol or clean brake fluid. Replace the caliper boot and seal.*

Task 7   Reassemble caliper.

40. While reassembling a brake caliper
    A. the boot should be installed and seated followed by the seal.
    B. coat the piston seal and boot with clean brake fluid.
    C. leave the piston dry and install it through the boot and seal until it bottoms.
    D. plug the bleeder screw hole and the high pressure inlet while installing the piston.

Hint   *Be sure the caliper bore and seal groove are perfectly clean. The seal should be lubricated with clean brake fluid and seated in the groove. Lubricate the boot with clean brake fluid and install it in the proper bore groove. Coat the piston with clean brake fluid or the specified lubricant and install it through the boot and seal. Do not plug the high-pressure inlet and bleeder screw hole while installing the piston. Install the piston until it bottoms and then plug the bleeder screw hole and high-pressure inlet to keep out foreign material.*

Task 8   Clean, inspect, and measure rotor with a dial indicator and a micrometer; follow manufacturer's recommendations in determining need to machine or replace.

41. In the brake rotor measurement in Figure 5-10
    Technician A says the brake rotor is being measured for runout.
    Technician B says this measurement should be made at three locations around the rotor.
    Who is correct?
    A. A only
    B. B only
    C. Both A and B
    D. Neither A nor B

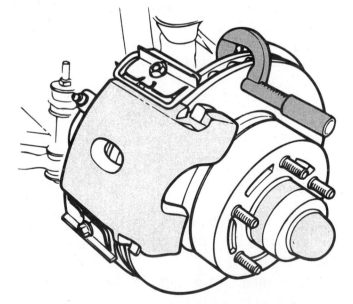

Figure 5-10   Brake rotor measurement with a micrometer *(Courtesy of Chrysler Corporation)*

Hint   *With the rotor mounted on the vehicle, position a dial indicator against the rotor friction surface and rotate the rotor for one revolution (Figure 5-11). If the runout exceeds manufacturer's specifications, replace the rotor. Excessive runout causes too much piston and pad-to-rotor clearance that results in excessive pedal travel and brake chatter.*

Use a micrometer to measure variations in rotor thickness or parallelism at six to twelve locations around the rotor. These readings should be taken near the center of the friction surface. Replace the rotor if the thickness variations exceed manufacturer's specifications. Excessive thickness variations may cause pedal pulsations and brake grabbing. Most rotors have a minimum or

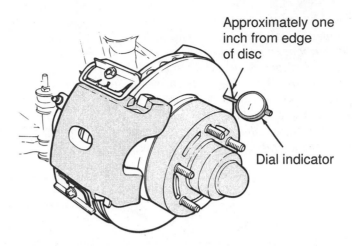

Approximately one
inch from edge
of disc

Dial indicator

Figure 5-11   Brake rotor measurement with a dial indicator *(Courtesy of Chrysler Corporation)*

discard thickness cast into them. After machining a rotor the thickness must exceed the mini-mum thickness. Some manufacturers specify the rotor thickness after machining must be at least 0.030 in. more than the minimum or discard thickness.

Task 9   Remove and replace rotor.

Task 10   Machine rotor according to manufacturer's procedures and specifications.

42. All of these statements about rotor machining are true EXCEPT
    A. a vibration damper must be placed around the outside diameter of the rotor to prevent chatter marks.
    B. with fixed caliper rotors unequal amounts of metal may be machined from each side of the rotor.
    C. machine both sides of the rotor before removing it from the rotor lathe.
    D. use a sanding pad to sand the rotor surfaces after machining is completed.

Hint   *Some rotors may be removed separately from the wheel hub. Most rotors and wheel hubs are removed as an assembly. The brake caliper must be removed prior hub and rotor removal. The dust cap, cotter key, and wheel bearing retaining nut must be removed prior to hub and rotor removal.*

The grease must be removed from the hub before mounting the hub and rotor on the lathe. Place a vibration damper around the outside diameter of the rotor, and install the hub and rotor securely on the lathe so there is no movement between the hub and lathe. Machine both sides of the rotor. On fixed caliper brake rotors, equal amounts of metal must be machined from each side of the rotor. Use a sanding pad to sand the rotor after machining. After sanding clean the rotor surface with a shop towel saturated with denatured alcohol to clean the rotor surfaces. Be sure all metal cuttings are removed from the hub.

Task 11   Install pads, calipers, and related attaching hardware; bleed system.

43. While discussing brake pad installation
    Technician A says the inboard and outboard brake pads are interchangeable in many calipers.
    Technician B says brake pad rattle may occur if there is any clearance between the pad retainer and the caliper retainer ledge.
    Who is correct?
    A. A only
    B. B only
    C. Both A and B
    D. Neither A nor B

Hint   *The manufacturer's recommended assembly procedure must be followed for each type of brake caliper. Install the inboard pad in the caliper. Install anti-rattle springs if so equipped. Position the outboard shoe in the caliper and be sure there is no clearance between the pad retainer flange and the caliper machined retainer ledge (Figure 5-12). If there is clearance at this location remove the pad and bend the pad retainer flanges as necessary. Clearance between the pad retainer flanges and the caliper retainer ledges may cause pad rattle. Be sure the pads are properly positioned and slide the caliper over the rotor. Install and tighten caliper mounting bolts. Install a new sealing washer on the brake hose and tighten the brake hose into the caliper. Install and tighten the brake line fitting in the brake hose.*

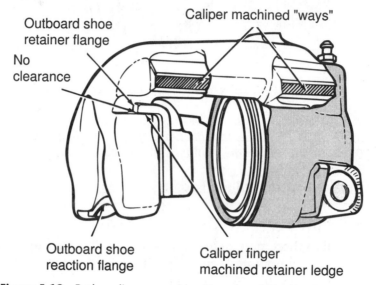

Figure 5-12   Brake caliper assembly *(Courtesy of Chrysler Corporation)*

Task 12   Adjust caliper with integrated parking brakes according to manufacturer's recommendations.

44. While discussing the rear brake caliper and parking brake mechanism in Figure 5-13
    A. during a parking brake application the caliper is moved toward the rotor surface.
    B. the parking brake should be depressed 1.5 in. before the cable adjustment check.
    C. the parking brake cable should be adjusted so the stopper pin is just touching the stop.
    D. the parking brake cable should be adjusted so there is 0.050 in. between the stopper pin and the stop.

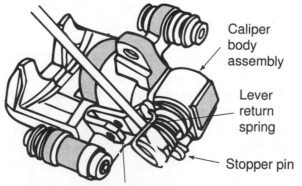

Figure 5-13   Parking brake cable in rear caliper *(Courtesy of Cadillac Motor Car Division, General Motors Corporation)*

Hint   *Before the cable adjustment the cables and the parking brake levers in the rear calipers must be working freely. With the parking brake released the cable adjuster should be rotated until the stopper pins just contact the stops in the rear calipers.*

Task 13   Fill master cylinder to proper level with recommended fluid; inspect caliper for leaks.

45. In most cast iron master cylinders the distance from the brake fluid level to the top of the reservoir casting should be
    A. 0 in.
    B. 1/4 in.
    C. 3/4 in.
    D. 1 in.

Hint   *After the brakes are bled, the master cylinder reservoir should be filled with the manufacturer's specified brake fluid to the proper level. Many cast iron master cylinder reservoirs are filled to within 1/4 in. of the reservoir top casting. Aluminum master cylinders with transparent plastic reservoirs are filled to the full mark on the reservoir.*

After several firm brake applications check the calipers for leaks in the piston and hose attachment areas.

Task 14   Reinstall wheel, torque lug nuts, and make final checks and adjustments.

46. While discussing wheel and lug nut installation
    Technician A says excessive lug nut torque may cause rotor runout.
    Technician B says an impact wrench may be used to tighten lug nuts.
    Who is correct?
    A. A only
    B. B only
    C. Both A and B
    D. Neither A nor B

Hint   *Excessive lug nut torque may distort the brake rotor and cause excessive rotor runout. Tighten the lug nuts to the specified torque following the manufacturer's recommended lug nut tightening sequence. Do not use an impact wrench to tighten the lug nuts.*

Power Assist Units Diagnosis and Repair

ASE Tasks, Questions, and Related Information

Task 1   Test pedal-free travel with and without the engine running to check power booster operation.

47. With the engine stopped the technician pumps the brake pedal several times and then holds the brake pedal on and starts the engine. The pedal moves slightly downward when the engine is started. This action indicates
    A. a restriction in the power booster vacuum hose.
    B. a defective check valve in the power booster vacuum hose.
    C. a normal vacuum supply to the power booster.
    D. a low-intake manifold vacuum.

Hint   *With the engine stopped, pump the brake pedal several times to release the vacuum stored in the brake booster. Push the brake pedal lightly downward until resistance is felt. The free play is the pedal movement from this point to the released position. Adjust the master cylinder push rod until the pedal free play is within specifications. With the engine stopped, pump the brake pedal several times, and hold the pedal in the applied position. When the engine is started the pedal should move slightly downward if the vacuum supply to the brake booster is normal. If the pedal does not move slightly downward, check the vacuum hose and one-way check valve to the brake booster.*

Task 2   Check vacuum supply (manifold or auxiliary pump) to vacuum-type power booster with a vacuum gauge.

48. While checking the vacuum supply to the power brake booster
    Technician A says the vacuum gauge should be connected between the one-way check valve and the brake booster.
    Technician B says with the engine idling the vacuum supplied to the brake booster should be 8 to 10 in Hg.
    Who is correct?
    A. A only
    B. B only
    C. Both A and B
    D. Neither A nor B

Hint   *Connect a vacuum gauge with a T connection in the hose between the one-way check valve and the brake booster. With the engine idling the vacuum should be 17 to 21 in. Hg. If the vacuum is low, connect the vacuum gauge directly to the intake manifold. When the vacuum in the intake is within specifications, check the one-way check valve and the hoses from the intake manifold to the brake booster. If the intake manifold is low, check engine compression and intake manifold vacuum leaks. When the vacuum supply to the brake booster is low on a vehicle with a vacuum pump, check the pump and hoses connected to the brake booster.*

Task 3   Inspect vacuum-type power brake booster unit for vacuum leaks; inspect the check valve for proper operation; repair, adjust, or replace parts as necessary.

49. There is evidence of engine oil in the vacuum brake booster. The cause of this problem could be
    A. a defective vacuum hose to the brake booster.
    B. a defective one-way check valve in the booster vacuum hose.
    C. a defective PCV valve with excessive restriction.
    D. a partially restricted air cleaner element.

Hint   *To check the one-way check valve in the brake booster vacuum hose operate the engine at 2,000 rpm and then allow the engine to idle. Shut the engine off and wait 90 seconds. Pump the brake pedal five or six times. The first two pedal applications should be power assisted. If the first two brake applications are not power assisted, the one-way check valve is defective in the brake booster vacuum hose. A defective one-way check valve in the brake booster vacuum hose may cause an accumulation of engine oil in the brake booster.*

To check the vacuum brake booster for air tightness operate the engine for 2 minutes and then shut off the engine. Pump the brake pedal several times with normal braking pressure. If the brake booster is operating normally, the pedal should go down normally on the first brake application. The pedal should gradually become higher with each brake application. With the engine running, apply the brakes and then shut off the engine while maintaining the pedal pressure for 30 seconds. When the pedal height does not change, the brake booster is not leaking. If the pedal height gradually rises, the brake booster is leaking.

Task 4   **Inspect and test hydro-boost system and accumulator for leaks and proper operation; repair, adjust, or replace parts as necessary.**

50. The power steering operates normally on a hydro-boost-equipped vehicle. During a hydro-boost brake test the brake pedal is pumped several times with the engine not running. When medium pressure is applied to the brake pedal and the engine is started, the brake pedal height remains unchanged.

Technician A says the power steering pump pressure may be low.

Technician B says the accumulator may be defective in the hydro-boost unit.

Who is correct?

A. A only

B. B only

C. Both A and B

D. Neither A nor B

Hint   *Pump the brake pedal several times with the engine not running to begin the hydro-boost test. Apply medium foot pressure to the brake pedal and start the engine. If the hydro-boost unit is operating normally, the brake pedal should move downward followed by a slight push back on the pedal. When the pedal does not have this action, the hydro-boost unit may be defective. Since the hydro-boost unit depends on power steering pump pressure, always be sure the power steering system is operating normally before testing or servicing the hydro-boost unit.*

To test the accumulator, operate the engine and turn the steering wheel until the wheels lightly touch their stops. Hold the steering wheel in this position for 5 seconds, and then release the steering wheel. Shut off the engine and pump the brake pedal several times. Several power-assisted brake applications should be available if the accumulator is properly charged. Start the engine and hold the steering wheel with the wheels against the stops to recharge the accumulator. A slight hissing sound should be heard while the accumulator is being charged. Shut the engine off and wait for 1 hour. If there are no leaks in the accumulator or hydro-boost system, several power-assisted brake applications should be available without starting the engine.

Miscellaneous (Wheel Bearings, Parking Brakes, Electrical, etc.) Diagnosis and Repair

ASE Tasks, Questions, and Related Information

Task 1   **Diagnose wheel bearing noises, wheel shimmy and vibration problems; determine needed repairs.**

51. A rear-wheel-drive car experiences a growling noise only on deceleration. The cause of this problem could be

A. a defective differential ring gear and pinion gear.

B. a defective rear wheel bearing.

C. a defective front wheel bearing.

D. defective differential side bearing.

Hint   *A defective wheel bearing causes a continual growling noise that is most noticeable at low speeds. Wheel bearing noise is not influenced by acceleration and deceleration. A growling noise caused by a defective front wheel bearing may be more noticeable while cornering. Wheel* *shimmy may be caused by loose front wheel bearings, improper front wheel alignment, improperly balanced front wheels.*

Task 2   Remove, clean, inspect, repack wheel bearings, or replace wheel bearings and races; replace seals; adjust wheel bearings according to manufacturer's specifications.

52. A front-wheel-drive vehicle has two tapered roller bearings in the rear wheel hubs. While adjusting the rear wheel bearings the bearing adjusting nut is tightened to 25 ft. lbs. The bearing nut should then be
    A. loosened one turn and tightened to 50 in. lbs.
    B. loosened one and one-half turns and tightened to 75 in. lbs.
    C. loosened one-half turn and tightened to 10 to 15 in. lbs.
    D. loosened one-half turn and tightened to 40 in. lbs.

Hint   *Wheel bearings should be cleaned in an approved cleaning solvent. After the bearings are cleaned they should be repacked with the manufacturer's recommended wheel bearing grease. The bearings may be repacked by hand or with a bearing repacking tool. After the bearings are installed in the hub, use the proper seal driver to install a new hub seal. Tighten the wheel bearing adjusting nut to 25 ft. lbs. while rotating the wheel. Loosen the adjusting nut one-half turn and then tighten the nut to 10 to 15 in. lbs. Install a new cotter key, and then install the dust cap.*

Task 3   Check parking brake system; inspect cables and parts for wear, rusting and corrosion; clean or replace parts as necessary; lubricate assembly.

.53. While discussing parking brake adjustment
    Technician A says the brake shoes should be properly adjusted before the parking brake adjustment.
    Technician B says after the cable adjusting nut is tightened so the rear wheels have a slight drag, loosen this nut four turns.
    Who is correct?
    A. A only
    B. B only
    C. Both A and B
    D. Neither A nor B

Hint   *The brake shoes should be adjusted properly, and the parking brake cables must be operating freely before a parking brake adjustment. Tighten the parking brake cable adjusting nut until there is a slight drag while rotating the rear wheels. Loosen the cable adjusting nut until the rear wheels rotate freely and then loosen this nut two more turns. Apply the parking brake several times and be sure the rear wheels are firmly held. After each application the rear wheels must rotate freely.*

Task 4   Adjust parking brake assembly; check operation.

54. In the measurement shown in Figure 5-14
    A. the drive shaft center support bearing wear is measured.
    B. the propeller shaft parking brake is adjusted.
    C. the brake shoes must be adjusted before this measurement.
    D. the parking brake is released during this measurement.

Hint   *During the propeller parking brake adjustment the clevis pin is removed from the parking brake lever at the brake assembly. The parking brake is applied until the specified round gauge fits in an opening in the parking brake mechanism and the gauge is contacting the parking brake outer flange. A pull-scale-type gauge is connected to the parking brake lever, and the parking brake tightening device is rotated until a specific reading is obtained on the gauge. The cable adjuster is then rotated until the clevis pin slides freely through the clevis and parking brake lever opening.*

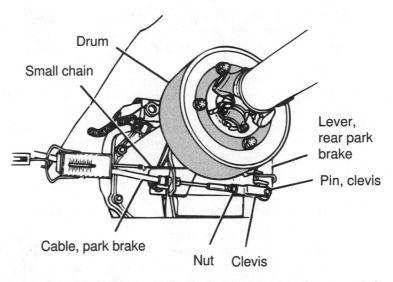

Figure 5-14 Adjustment with pull-scale on propeller shaft mechanism *(Courtesy of Chevrolet Motor Division, General Motors Corporation)*

Task 5   Test parking brake indicator light(s), switch(es), and wiring.

55. The brake warning light is illuminated continually with the ignition switch on in Figure 5-15. All of these defects may be the cause of the problem EXCEPT

   A. an open circuit in the wire to the parking brake switch.

   B. the wire to the parking brake switch touching the vehicle ground.

   C. a continually closed parking brake switch.

   D. low fluid level in the secondary section of the master cylinder reservoir.

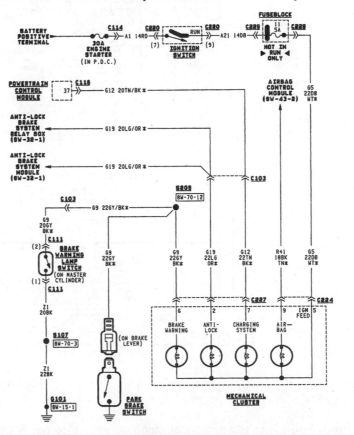

Figure 5-15 Brake warning light circuit *(Courtesy of Chrysler Corporation)*

Hint   *When the ignition switch is turned on, voltage is supplied through a fuse to one terminal on the brake warning light. When the parking brake is applied, the parking brake switch connects the other side of the brake warning bulb to ground. This action illuminates the brake warning light. The brake warning switch also may ground the brake warning light if there is unequal pressure between the primary and secondary sections of the master cylinder.*

If the brake warning light is illuminated continually with the ignition switch on, check the master cylinder brake fluid level. If the brake fluid level is satisfactory, disconnect the wire from the brake warning light switch and the parking brake switch. If the brake warning light goes out when one of these wires is disconnected, that switch is faulty. When the brake warning light is illuminated with wires disconnected from these switches, check for a grounded condition on the wires from the switches to the bulb.

Task 6   Test, adjust, repair, or replace brake stop light switch and wiring.

56. When the brake pedal is depressed, the rear stop lights are not illuminated, but the high-mounted stop lights operate normally (Figure 5-16 and Figure 5-17). The turn signals and hazard warning circuits operate normally.

    Technician A says the cause of this problem may be a blown stop-hazard fuse.

    Technician B says there may be an open circuit between junction S206 the turn signal switch.

    Who is correct?

    A. A only
    B. B only
    C. Both A and B
    D. Neither A nor B

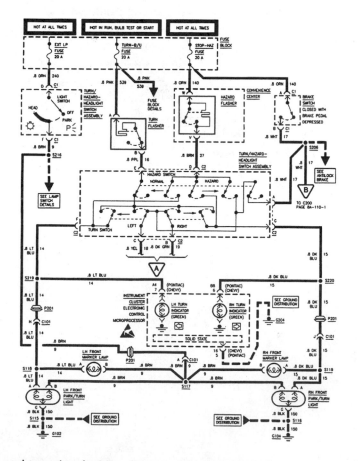

Figure 5-16 Stop and turn signal wiring diagram *(Courtesy of Chevrolet Motor Division, General Motors Corporation)*

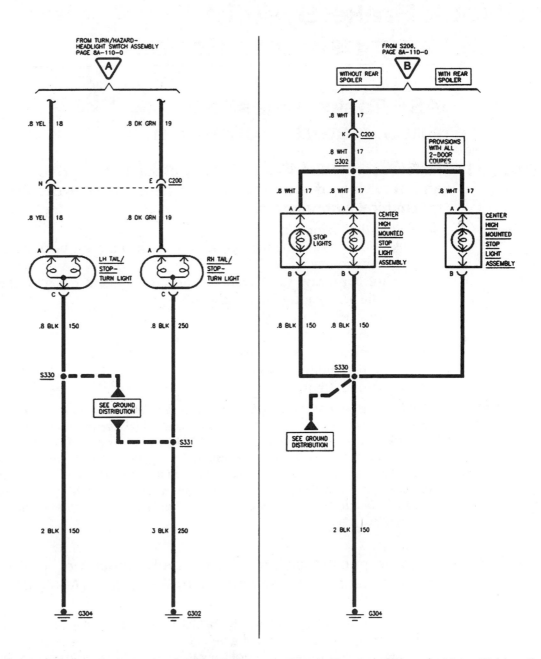

Figure 5-17   Stop and turn signal wiring diagram continued   *(Courtesy of Chevrolet Motor Division, General Motors Corporation)*

Hint   *The stop light switch may be adjusted on many vehicles. In most applications this switch should be adjusted so the stop lights are illuminated when the brake pedal is depressed 1/4 in. An improper stop light switch adjustment or a binding brake pedal may cause the stop lights to be illuminated continually with the ignition switch on. Stop light circuits vary depending on the vehicle make and model year, the technician must use a wiring diagram to diagnose each circuit. When diagnosing a stop light circuit, the stop light fuse should be tested first. With the wires disconnected from the stop light switch, connect a pair of ohmmeter leads to the switch terminals. With the brake pedal released, the ohmmeter should provide an infinite reading. When the brake pedal is depressed, the ohmmeter should indicate 0.5 ohms or less.*

Antilock Brake System (ABS) Diagnosis and Repair

ASE Tasks, Questions, and Related Information

Task 1   Follow accepted service and safety precautions during inspection, testing, and servicing of antilock brake system (ABS) hydraulic, electrical, and mechanical components.

57. While discussing ABS service
Technician A says the high pressure accumulator must be discharged before a brake line is disconnected.
Technician B says some manufacturers recommend relieving the accumulator gas pressure before accumulator disposal.
Who is correct?
A. A only
B. B only
C. Both A and B
D. Neither A nor B

Hint   *Before a brake line is removed, the brake pedal should be applied and released twenty-five to thirty times to relieve the brake fluid pressure in the high-pressure accumulator. When the amber ABS warning light is illuminated, the antilock function is cancelled because the ABS computer has sensed an electrical defect in the system. Under this condition normal power-assisted braking is available, and the car may be driven to an automotive repair facility as soon as possible so the necessary repairs can be completed. When the red brake warning light is illuminated there may be a serious braking defect, such as low brake fluid level. Under this condition the vehicle should not be driven.*

Task 2   Diagnose poor stopping, wheel lockup, pedal feel, pulsation, and noise problems caused by the antilock brake system (ABS); determine needed repairs.

58. While road testing a vehicle with ABS a clicking noise is heard for a short time when the engine is started and the vehicle starts off at low speed. During a normal stop when the ABS function is not operating, pedal pulsations are experienced.
Technician A says the clicking action may be caused by defective solenoids in the ABS hydraulic control unit.
Technician B says the pedal pulsations are normal on an ABS system.
Who is correct?
A. A only
B. B only
C. Both A and B
D. Neither A nor B

Hint   *In many ABS systems the computer enters a prove-out mode each time the ignition switch is turned on and the engine is started. During this mode the computer scans the ABS electrical system for defects. On some ABS systems the computer prove-out check includes momentarily energizing all the solenoids in the system. This computer action may cause an audible clicking action when the vehicle is driven at low speeds. If the brakes are applied during the prove-out mode, momentary pedal pulsations may be felt.*

On many ABS systems pedal pulsations are felt while the brakes are operating in the ABS mode. On some ABS systems an increase in brake pedal height may be experienced during the ABS mode. If pedal pulsations are experienced during the normal braking mode, a brake drum is out-of-round, or a rotor has excessive runout.

Poor stopping may be caused by contaminated or glazed brake linings, scored drums or rotors, or defective brake booster operation. In some ABS systems the accumulator pressure supplies power assist. In these systems low accumulator pressure causes poor stopping ability.

Brake grabbing may be caused by contaminated linings, bent backing plates, or distorted drums or rotors.

Task 3   Observe antilock brake system (ABS) warning light(s) at start-up; determine if further diagnosis is needed.

59. All of these conditions may cause illumination of the red brake warning light EXCEPT
   A. parking brake engagement.
   B. low fluid level in the master cylinder.
   C. an open wheel speed sensor winding.
   D. an accumulator pressure below 1,500 psi.

Hint   *The red brake warning light is normally illuminated while cranking the engine and for a few seconds after the engine is started. This light may be illuminated with the engine running if the parking brake is applied, the master cylinder fluid level is low, or the accumulator pressure drops below 1,500 psi.*

The amber ABS brake warning light is illuminated while cranking the engine and for a few seconds after the engine is started. When the amber ABS light is illuminated with the engine running, the computer has sensed an electrical defect in the ABS system.

Task 4   Diagnose antilock brake system (ABS) electronic control(s) and components using self diagnosis and/or recommended test equipment; determine needed repairs.

60. On ABS systems with the data link connector (DLC) shown in Figure 5-18 ABS diagnostic trouble codes (DTCs) are flashed by the amber ABS warning light when a jumper wire is connected between terminals
   A. A and B.
   B. A and H.
   C. A and K.
   D. A and D.

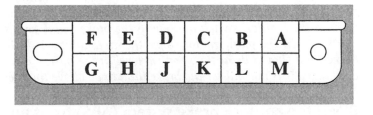

Figure 5-18 Data link connector (DLC)

Hint   *On some ABS systems a jumper wire is connected between the specified terminals in the DLC to obtain flash codes on the amber ABS warning light. On many General Motors products the jumper wire must be connected between terminals A and H in the DLC. Many ABS systems do not provide flash codes, and the DTCs must be obtained with a scan tool on these systems. A DTC indicates a defect in a certain area. In most cases voltmeter or ohmmeter tests must be performed to locate the exact cause of the problem.*

Task 5   Depressurize integral (high-pressure) components of the antilock brake system (ABS) following manufacturer's recommended safety precautions.

61. On ABS systems with a high-pressure accumulator, this accumulator is depressurized by
    A. pumping the brake pedal 25 times with the ignition switch off.
    B. pumping the brake pedal 10 times with the ignition switch on.
    C. loosening the bleeder screw on the accumulator with the ignition switch on.
    D. loosening of the front wheel bleeder screws with the ignition switch on.

Hint   *The high-pressure accumulator must be depressurized before disconnecting a brake line or hose. On many ABS systems the high pressure accumulator is depressurized by pumping the brake pedal 25 times with the ignition switch off. On other systems the accumulator pressure is relieved by connecting a bleeding tool to the bleeder screw on the hydraulic control unit and loosening the bleeder screw with the ignition switch off. When an accumulator is replaced, the gas pressure should be relieved in the old accumulator before discarding it. After the accumulator is removed, a pressure relief screw in the accumulator is loosened the specified number of turns until the gas pressure is relieved.*

Task 6   Fill the antilock brake system (ABS) master cylinder with recommended fluid to proper level following manufacturer's procedures; inspect system for leaks.

62. On an integral ABS system with a high-pressure accumulator
    Technician A says the master cylinder fluid level should be checked with a discharged accumulator.
    Technician B says the master cylinder fluid level should be checked with the engine running and the brake pedal applied.
    Who is correct?
    A. A only
    B. B only
    C. Both A and B
    D. Neither A and B

Hint   *On an integral ABS system with a high-pressure accumulator, the brake fluid level in the master cylinder should be checked with a fully charged accumulator. To fully charge the accumulator turn on the ignition switch and pump the brake pedal several times until the ABS pump motor starts. When the pump motor stops, the accumulator is fully charged. Under this condition the brake fluid should be at the specified level in the master cylinder reservoir. If the accumulator is discharged it is normal for the brake fluid to be above the specified level in the master cylinder.*

Task 7   Bleed the antilock brake system (ABS) front and rear hydraulic circuits following manufacturer's procedures.

63. All of these statements about bleeding an ABS system with a high-pressure accumulator are true EXCEPT
    A. the front and rear brakes may be bled with a pressure bleeder.
    B. be sure the ignition switch is on when using a pressure bleeder.
    C. be sure the brake pedal is released when using a pressure bleeder.
    D. the rear brakes may be bled with a fully charged accumulator.

Hint   *On an ABS system with a high-pressure accumulator, the front or rear brakes may be bled with a pressure bleeder. If a pressure bleeder is attached to the master cylinder for bleeding purposes, maintain the bleeder pressure at 35 psi (240 kPa). Be sure the ignition switch is off and the brake pedal released. Connect a hose from the bleeder screw into a plastic container, and open the bleeder screw for 10 seconds. Tighten the bleeder screw and repeat the procedure at each wheel. Be sure the bleeder pressure is maintained. Adjust the master cylinder fluid level to the max fill line when the bleeding procedure is completed.*

Since accumulator pressure is supplied to the rear wheels, the rear brakes may be bled with a fully charged accumulator.

Task 8   Perform a fluid pressure (hydraulic boost) diagnosis on the integral (high-pressure) antilock system (ABS); determine needed repairs.

64. On an integral ABS system with a high-pressure accumulator all of these statements about accumulator pressure and pump run time are true EXCEPT

   A. if the accumulator pressure is low the pump run time may be longer than specified.

   B. low accumulator pressure has no effect on brake boost power.

   C. a defective pressure switch may cause higher than specified boost pressure.

   D. a pressure gauge and adapter may be installed between the accumulator and the accumulator fluid passage.

Hint   *On an integral ABS with a high-pressure accumulator a hard brake pedal and poor stopping ability may be caused by low accumulator pressure. When the accumulator pressure is low, the pump run time may be much longer than specified, and the amber ABS warning light may be illuminated. To check the pump running time, turn off the ignition switch and pump the brake pedal at least twenty times until the pedal feels hard. Turn on the ignition switch and measure the pump run time. The accumulator should be fully charged and the pump should stop within 1 minute. When the pump run time is longer than specified, check for fluid leaks at the ABS assembly. If there are no fluid leaks, check the pump, pump relay, pressure switch, and accumulator.*

To check the accumulator pressure, pump the brake pedal twenty-five times with the ignition switch off to relieve the accumulator pressure. Remove the accumulator and install a pressure gauge between the accumulator and the accumulator fluid outlet. Install the accumulator and connect a pressure gauge to the adapter. Turn the ignition switch on and observe the pressure gauge reading when the pump stops running. If the pressure is less than specified, check the pump, pressure switch, and accumulator. When the pump pressure is more than specified check the pressure switch.

Task 9   Remove and install antilock brake system (ABS) components following manufacturer's procedures and specifications.

65. When servicing the integral ABS unit with a high-pressure accumulator in Figure 5-19 (next page)

   Technician A says the accumulator pressure should be relieved before removing the solenoid valve body.

   Technician B says the accumulator pressure should be relieved before removing the reservoir.

   Who is correct?

   A. A only

   B. B only

   C. Both A and B

   D. Neither A nor B

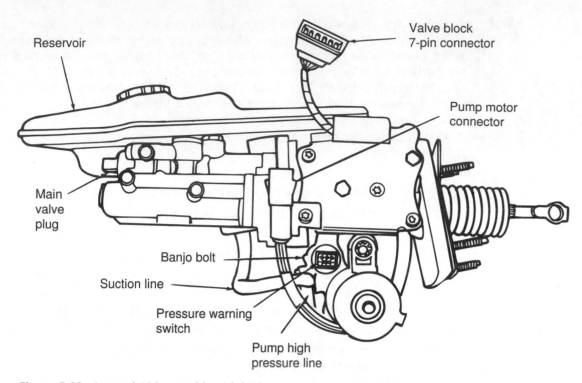

Figure 5-19   Integral ABS assembly with high pressure accumulator *(Courtesy of Ford Motor Company)*

Hint   *Always be sure the accumulator pressure is relieved before removing the ABS assembly or any component in this assembly. If the accumulator is removed, always install a new O-ring under the accumulator. Coat the new O-ring with clean brake fluid before installation. Turn off the ignition switch and disconnect the negative battery cable before disconnecting any wiring connectors in the ABS system. If the vehicle is equipped with an air bag or bags, wait the length of time specified by the manufacturer after the negative battery cable is disconnected before proceeding with the service or diagnosis procedure.*

Task 10   **Service, test, and adjust antilock brake system (ABS) speed sensors following manufacturer's recommended procedures.**

   66. A diagnostic trouble code (DTC) representing the L/R wheel speed sensor is obtained
        in the ABS system in Figure 5-20. This DTC may be caused by
        A.   an open circuit at terminal 6 on the ABS computer connector.
        B.   a grounded circuit on the #885 red wire connected to terminal 10 on the ABS
             computer.
        C.   a larger tire than specified by the manufacturer on the L/R wheel.
        D.   an open circuit in the electronic brake control relay winding.

Hint   *Some wheel speed sensors have a paper shim installed on the inner end of the sensor. If this type of sensor is removed, a new shim should be installed on the sensor. The sensor should be installed until the paper shim lightly contacts the toothed ring. Hold the sensor in this position, and tighten the sensor retaining bolt. On other sensors the clearance between the sensor tip and the toothed ring is measured with a feeler gauge. Rotate the wheel one revolution and measure the sensor clearance to be sure this clearance is uniform. If necessary, loosen the sensor retaining bolt and move the sensor to adjust the clearance.*

   If a DTC is obtained representing a wheel speed sensor, the sensor and connecting wires may be tested with an ohmmeter. Disconnect the ABS computer wires and connect the ohmmeter leads to the appropriate sensor wires to test the sensor for an open circuit. When an infinite reading is obtained, the sensor or connecting wires are open. Disconnect the wiring connector at the sensor, and connect the ohmmeter leads to the sensor terminals to determine if the open circuit is in the

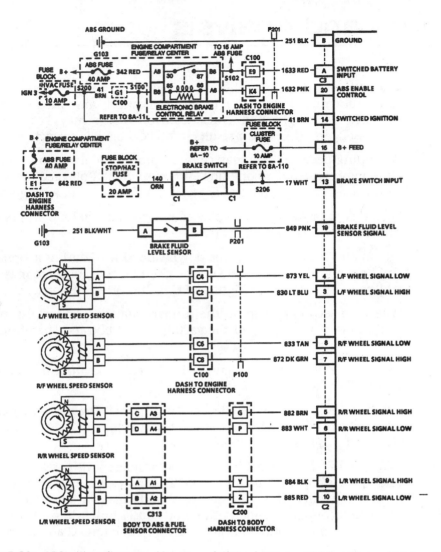

Figure 5-20 ABS wiring diagram *(Courtesy of Chevrolet Motor Division, General Motors Corporation)*

sensor winding or connecting wires. When the ohmmeter leads are connected from one of the wheel speed sensor terminals at the ABS computer connector to ground, an infinite reading should be obtained. If the ohmmeter provides a low reading, the sensor winding or connecting wires are shorted to ground.

Task 11   **Diagnose ABS problems caused by vehicle modifications (tire size, curb height, final drive ratio, etc.).**

 67. An ABS-equipped rear-wheel-drive vehicle experiences lockup on both rear wheels during the antilock brake mode.
 Technician A says larger than specified tires may be installed on the rear wheels.
 Technician B says the rear tires may not be the same size on each rear wheel.
 Who is correct?
 A. A only
 B. B only
 C. Both A and B
 D. Neither A nor B

Hint   *On an ABS-equipped vehicle the original tire size and differential ratio must be maintained. When a different tire size is installed the wheel speed is changed in relation to vehicle speed. Under this condition different wheel speed sensor signals are sent to the ABS computer. This action may reduce ABS effectiveness or cause wheel lockup in the antilock brake mode.*

Answers and Analysis

1. B   Brake fluid leaking past the master cylinder cups may cause the brake pedal to slowly move downward during a brake application, and thus A is wrong. An improperly adjusted stop light and cruise control switch may keep the brake pedal partially depressed, and this action results in brake drag. B is correct.

2. D   A binding pedal linkage may cause improper pedal return and brake drag. Dented brake lines may cause pull to one side and brake drag. A plugged master cylinder compensating port causes dragging brakes. Therefore, A, B, and C are wrong and D is right.

3. C   Insufficient brake pedal free play causes pressure buildup and brake drag. Therefore, A, B, and D are wrong, and C is correct.

4. C   When the brake pedal moves downward slowly during a brake application, the brake fluid may be leaking past the master cylinder cups, or there may be a leak in a brake hose or line. Both A and B are right and C is the correct response.

5. A   A leaking secondary cup on the primary master cylinder piston may allow fluid to leak and run down the front of the vacuum brake booster. Therefore, B, C, and D are wrong, and A is right.

6. D   The service procedure in Figure 5-1 is bleeding air from the master cylinder, and thus A, B, and C are wrong and D is correct.

7. C   When excessive brake pedal effort is required, the vacuum hose to the vacuum brake booster may be restricted or the front caliper pistons may be seized. Both A and B are correct, and C is the right response.

8. B   When both front brake linings are contaminated with grease, both front wheels may grab and pedal effort may be increased. Therefore, A is wrong. If the master cylinder piston cups are swollen the compensating ports may be covered resulting in brake drag. A plugged secondary compensating port causes brake drag on the brakes supplied from the secondary section of the master cylinder. Therefore, C and D are wrong.

   If the right front caliper piston is seized there is very little or no braking action on this wheel. With normal braking action on the left front wheel, the vehicle may pull to the left during a brake application. Therefore, B is right.

9. B   Damaged brake lines should be replaced, and so A is wrong. A tubing bender should be used to make the necessary brake tubing bends without kinking the brake tubing. Therefore, B is correct.

10. D   The male end of the brake hose should be installed and tightened before the female end. The brake hose sealing washer should be replaced when installing the brake hose. Therefore, both A and B are wrong, and D is the correct response.

11. B   The brake line flare in Figure 5-2 is an ISO flare. Therefore, A, C, and D are wrong, and B is right.

12. B   Silicone brake fluids are non-hydroscopic which means they do not absorb moisture. Since DOT 3 and DOT 4 brake fluids are both hydroscopic, they tend to absorb moisture from the atmosphere. B is the proper response.

13. A   The metering valve in Figure 5-3 delays brake fluid movement to the front wheels to provide simultaneous application of the front and rear brakes. Therefore, B, C, and D are wrong, and A is correct.

14. C   Since the proportioning valve reduces pressure to the rear brakes during moderate brake applications, the pressure gauges connected to this valve should indicate lower pressure at the valve outlet compared to the inlet. Therefore, A, B, and D are wrong and C is right.

15. D The load sensing proportioning valve allows more pressure to the rear brakes as the rear suspension load is increased. This valve is mounted on the chassis and has a linkage to the rear suspension. Therefore, both A and B are wrong, and D is the correct response.

16. A With the brake warning switch positioned as shown in Figure 5-5 the pressure is higher in the primary master cylinder section than in the secondary section. Therefore, B is wrong. The circuit is completed from the switch terminal to ground, and pressure is higher in the primary master cylinder section. Therefore, C and D are wrong.

The brake warning light is illuminated with the ignition switch on or the engine running, thus A is correct.

17. B When using a pressure bleeder the metering valve must be held open with a special tool, thus A is wrong. The pressure in the pressure bleeder should be maintained at 20 psi, so B is correct.

18. A This question asks for the statement that is not true. The vacuum pump creates a vacuum in the tester reservoir, and the bleeder screw should be opened until there is about 1 in. of brake fluid in the reservoir. A one-way check valve is required in the hose from the bleeder screw to the reservoir. Therefore, B, C, and D are right, and none of these are the requested answer.

The vacuum pump handle should be operated at least 10 times before the bleeder screw is opened. Therefore, statement A is wrong, and this is the requested answer.

19. C This question asks for the statement that is not true. A hose is connected from the bleeder screw and the other end submerged in a container of brake fluid. The brake pedal should be applied with moderate pressure and then the bleeder screw is opened. The bleeding procedure should be repeated until the fluid escaping from the bleeder hose is free of bubbles. Therefore, A, B, and D are correct, and none of these is the requested answer.

When the bleeder screw is opened and the pedal goes to the floor, the bleeder screw must be closed before releasing the brake pedal. Therefore, statement C is wrong, and this is the requested answer.

20. B When using the surge bleeding method the bleeder hose must be kept submerged in the brake fluid in the bleeder jar. The brake pedal should be pumped quickly several times with the bleeder screw open. Normal pressure should be maintained in the pressure bleeder. Therefore, A, B, and D are wrong, and B is right.

21. D During a moderate brake application the master cylinder pressure should be equal to the front wheel pressure. Therefore, A is wrong. Below the split point pressure of the proportioning valve the master cylinder pressure and the rear wheel pressure should be equal, thus B is wrong. During a moderate brake application the master cylinder pressure should be higher than the pressure at the rear wheels, so C is wrong.

Above the split point pressure of the proportioning valve, the master cylinder pressure should be higher than the rear wheel pressure. Therefore, D is correct.

22. C If the master cylinder vents are plugged, air cannot enter between the cover and the cover gasket to allow this gasket to move downward and replace the brake fluid forced from the master cylinder to the wheel cylinders or calipers. This may cause a spongy brake pedal. Air in the hydraulic brake system also may cause this problem. Therefore, both A and B are correct, and C is the proper response.

23. C Brake squeal may be caused by distorted brake drums or bent backing plates. Both A and B are correct, and C is the right response.

24. C This question asks for the statement that is not true. The drum should be cleaned before diameter measurement, and the drum should be replaced if the diameter exceeds specifications. Maximum drum diameter variation should not exceed 0.0035 in. Therefore, statements A, B, and D are right, but these are not the requested answer.

The drum diameter should be measured at eight locations around the drum. Therefore, statement C is not correct, and this is the requested answer.

25. B   If tool chatter marks occur on the brake drum friction surface, the damping belt may be too loose. The tool bit depth for a rough cut should be 0.005 to 0.010 in., and the tool bit depth for a finish cut should be 0.003 to 0.005 in. Therefore, A, C, and D are wrong, and B is right.

26. B   If item 17 in Figure 5-6 is omitted, there is no parking brake action. Therefore, A is wrong. When item 8 is omitted there is no self-adjusting action. Therefore, B is correct.

27. C   A bent backing plate may cause brake grabbing, and a loose anchor bolt may cause brake chatter. Therefore, both A and B are correct, and the right answer is C.

28. C   While servicing a wheel cylinder the parts should be washed in denatured alcohol or clean brake fluid. If the cylinder bore is pitted or deeply scored, the cylinder should be replaced. The flat side of the pistons should face the cups. Therefore, A, B, and D are wrong.

The piston cups should be lubricated with clean brake fluid before assembly. Therefore, C is right.

29. C   Dry brake shoe ledges on the backing plate may cause a squeaking noise during brake applications. Slightly scored brake shoe ledges on the backing plate may be resurfaced and lubricated with high temperature grease. Therefore, both A and B are right and C is the correct answer.

30. C   The secondary shoe faces toward the rear of the vehicle, and the primary and secondary shoe return springs are not interchangeable. The adjuster cable is usually mounted on the secondary shoe. Therefore, A, B, and D are wrong.

The adjuster must be installed in the proper direction so the star wheel is accessible through the backing plate opening, so C is the correct answer.

31. A   In Figure 5-7 the brake shoes are being adjusted to match the drum size. Therefore, B is wrong and A is correct.

32. A   While adjusting the parking brake shown in Figure 5-8, the parking brake should be applied until the pin in hole A contacts the parking brake outer flange. Therefore, B is wrong and A is correct.

33. B   In Figure 5-9 the thin tool is used to release the self-adjusting mechanism, thus A is wrong. The star wheel should be rotated until the wheel drags, and then the star wheel is rotated in the opposite direction until the wheel rotates freely, so B is right.

34. C   In this question we are asked for the statement that is not true. Incorrect or loose brake pads, a loose caliper mounting, or a sticking caliper piston may cause pull to one side while braking. Therefore, statements A, B, and D are correct, but none of these is the requested answer.

Seized master cylinder pistons do not contribute to pull while braking. Therefore, statement C is not true, and this is the requested answer.

35. B   A worn outer drive axle joint does not cause brake pedal pulsations, thus A is wrong. Brake pedal pulsations may be caused by worn or loose wheel bearings. B is correct.

36. A   After a brake application the caliper piston is returned by the twisting action of the seal. Therefore, B, C, and D are wrong and A is right.

37. A   Excessive wear on the inside brake pad lining compared to the outside lining may be caused by worn caliper pins and bushings which cause a sticking caliper. Therefore, B, C, and D are wrong and A is correct.

38. B   A scraping noise while braking may be caused by a pad wear sensor contacting the rotor. Therefore, A, C, and D are wrong and B is right.

39. A When honing a brake caliper the maximum increase in caliper bore is 0.001 in., so B, C, and D are wrong and A is correct.

40. B When reassembling a brake caliper the seal should be installed followed by the boot. Therefore, A is wrong. The piston should be lubricated with clean brake fluid before it is installed, and the bleeder screw hole or high-pressure inlet should be open when installing the piston. Therefore, C and D are wrong.

The piston seal and boot should be lubricated with clean brake fluid prior to installation, thus B is the right answer.

41. D In Figure 5-10 the rotor is measured for parallelism or thickness variation; this measurement should be made at six to twelve locations around the rotor. Therefore, both A and B are wrong and D is the correct response.

42. B In this question we are asked for the statement that is not true. During the machining process a vibration damper must be positioned around the outside diameter of the rotor. Both sides of the rotor should be machined before removing it from the lathe, and a sanding pad is used to sand the rotor surface after machining. Therefore, A, C, and D are right, but these are not the requested answer.

Equal amounts of metal must be removed from each side of the rotor on fixed caliper rotors. Therefore, B is not true, and this is the requested answer.

43. B The inboard and outboard brake pads are not interchangeable in many calipers, so A is wrong. If there is clearance between the pad retainer and the caliper retainer ledge, brake pad rattle may occur. B is right.

44. C In rear caliper in Figure 5-13 the parking brake mechanism moves the caliper piston toward the rotor surface, and the parking brake should be released during an adjustment. Therefore, A and B are wrong. The parking brake cable should be adjusted so the stopper pin just contacts the stop. Therefore, D is wrong and C is right.

45. B On many cast iron master cylinders the brake fluid level should be 1/4 in. below the top of the casting surface. Therefore, A, C, and D are wrong and B is correct.

46. A An impact wrench should not be used to tighten wheel lug nuts since this excessive torque from this procedure may distort drums and cause excessive rotor runout. Therefore, B is wrong and A is correct.

47. C The brake pedal action in this question indicates a normal vacuum supply to the power brake booster, thus A, B, and D are wrong and C is right.

48. A The vacuum gauge should be installed between the one-way check valve and the brake booster to check the vacuum supply, and with the engine idling the vacuum should be 16 to 18 in. Hg. Therefore, B is wrong and A is correct.

49. B If there is evidence of oil in the vacuum brake booster the one-way check valve in the booster vacuum hose is defective. Therefore, A, C, and D are wrong and B is right.

50. B If the brake pedal height remains unchanged as described in the question, there is no accumulator pressure to supply power assist. This may be caused by low power steering pump pressure, but this problem also reduces steering assist. Since the question informs us that power steering operation is normal, low power steering pump pressure is not the cause of the complaint. Therefore, A is wrong.

A defective accumulator may cause reduced power brake assist, so B is correct.

51. A The noise from a defective front or rear wheel bearing would not be affected by acceleration or deceleration. Therefore, B and C are wrong. Acceleration and deceleration has very little effect on defective differential side bearing noise, so D is wrong.

Noise from a defective differential ring and pinion gear usually is affected by acceleration and deceleration, so A is right.

52. C   After tightening the wheel bearing nut to 25 ft lbs, the wheel bearing adjusting nut should be loosened one-half turn and tightened to 10 to 15 in. lbs. Therefore, A, B, and D are wrong, and C is right.

53. A   On many vehicles the parking brake cable should be adjusted so there is a slight rear wheel drag, and then loosened two turns, thus B is wrong. The brake shoes should be properly adjusted before a parking brake adjustment. A is correct.

54. B   In Figure 5-14 the propeller parking brake adjustment is performed. Brake shoe adjustment does not affect this type of parking brake. During the parking brake adjustment the specified pin is installed in a hole in the parking brake pedal mechanism. The parking brake is applied until this pin contacts the outer flange on the pedal mechanism. Therefore, A, C, and D are wrong and B is right.

55. A   In this question we are asked for the answer that is not the cause of the problem. A continually illuminated brake warning light may be caused by a grounded wire to the parking brake switch, a continually closed parking brake switch, or low fluid level in one section of the master cylinder reservoir. Therefore, B, C, and D may be the cause of the problem, and none of these are the requested answer.

If an open circuit occurs in the wire to the parking brake switch, the brake warning light is not illuminated when the parking brake is applied. Therefore, A is not the cause of a continually illuminated brake warning light and this is the correct answer.

56. B   If the stop hazard fuse is blown, the stop lights and hazard warning lights are inoperative. Therefore, A is wrong. An open circuit between junction S206 and the turn signal switch causes inoperative stop lights. Since the high-mounted stop lights are connected to junction S206, they work normally. Therefore, B is correct.

57. C   The high-pressure accumulator must be discharged before a brake line is disconnected, and some manufacturers recommend relieving accumulator gas pressure before accumulator disposal. Therefore, both A and B are correct and C is the right answer.

58. D   The clicking action during initial driving is a result of the ABS computer prove-out mode in which the computer momentarily energizes the solenoids in the ABS system. Therefore, A is wrong.

On many ABS systems pedal pulsations are normal during the ABS function. However, pedal pulsations during a normal stop when the ABS function is not operating may be caused by out-of-round drums or rotors with excessive runout. Therefore, B also is wrong and the correct answer is D.

59. C   In this question we are asked for the answer that is not the cause of red brake warning light illumination. A parking brake application, low fluid level in the master cylinder, or accumulator pressure below 1,500 psi, may cause red brake warning light illumination. Therefore, A, B, and D are not the required answer.

An open wheel speed sensor winding does not cause illumination of the red brake warning light, but this problem results in illumination of the amber ABS light. Therefore, C is correct.

60. B   To obtain DTCs on a vehicle with the DLC in Figure 5-18 a jumper wire must be connected between terminals A and H. Therefore, A, C, and D are wrong and B is right.

61. A   On an ABS system with a high-pressure accumulator the accumulator is depressurized by pumping the brake pedal twenty-five times with the ignition switch off. Therefore, B, C, and D are wrong and A is right.

62. D   In an integral ABS system with a high-pressure accumulator the brake fluid level should be checked with a fully charged accumulator. Therefore, both A and B are wrong and D is correct.

63. B   In this question we are asked for the statement that is not true about bleeding an ABS system with a high pressure accumulator. On these systems the front and rear brakes may be bled with a pressure bleeder, the brake pedal should be released when using a pressure bleeder, and the rear brakes may be bled with a fully charged accumulator. Therefore, statements A, C, and D are true and none of these are the requested response.

The ignition switch should be off when using a pressure bleeder on these systems. Therefore, statement B is not true, and this is the requested answer.

64. B   In this question we are asked for the statement that is not true. On an integral ABS system with a high-pressure accumulator, low accumulator pressure causes longer pump running time, and a defective pressure switch may cause higher-than-specified boost pressure. A pressure gauge adapter may be installed between the accumulator and the accumulator fluid passage. Therefore, statements A, C, and D are true, but none of these are the requested answer.

Low accumulator pressure does reduce brake boost power. Therefore, statement B is not true and this is the requested answer.

65. C   In an integral ABS system with a high-pressure accumulator the accumulator pressure should be relieved before removing the solenoid valve body or the reservoir. Therefore, both A and B are correct, and C is the right answer.

66. B   Since terminal 6 on the ABS computer connector is connected to the R/R wheel speed sensor, an open circuit at this terminal does not cause a code representing the L/R wheel speed sensor. Therefore, A is wrong.

A larger tire than specified on the L/R wheel may affect the ABS operation, but it does not result in this type of DTC, so C is wrong. An open circuit at the electronic brake control relay winding causes an inoperative ABS system, but this problem does not result in a DTC representing the L/R wheel speed sensor. Therefore, C and D are wrong.

Since terminal 10 on the ABS computer is connected to the L/R wheel speed sensor, a grounded circuit at this location causes a DTC representing the L/R wheel. Therefore, B is correct.

67. A   Rear tires of unequal size may result in wheel lockup on one rear wheel during the ABS mode, so B is wrong. Larger-than-specified tires on both rear wheels may cause lockup on both rear wheels during the ABS mode. A is right.

Glossary

Abrasion Wearing or rubbing away of a part.
Abrasión El desgaste o consumo por rozamiento de una parte.

ABS Antilock brake system.
ABS Un sistema de frenos antideslizante.

Acceleration An increase in velocity or speed.
Aceleración Un incremento en la velocidad.

Accumulator A device used in some antilock brake systems to maintain high pressure in the system. It is normally charged with nitrogen gas.
Acumulador Un dispositivo que se usa en algunos sistemas de frenos antideslizantes para mantener una alta presión en el sistema. Normalmente cargado con gas nitrógeno.

Air bleeding The removal of air from a hydraulic system.
Purgar con aire Quitar el aire de un sistema hidráulico.

Air brakes A braking system used on heavy trucks using air pressure as the power brake source.
Frenos de aire Un sistema de frenos utilizado en los camiones pesados que usa el presión de aire para suministrar los frenos de potencia.

Alignment An adjustment to a line or to bring into a line.
Alineación Un ajuste que se efectúa en una linea o alinear.

Anchor pin Anchoring pins on which the heel of the brake shoe rotates.
Perno de anclaje Las clavijas de anclaje en las cuales gira el extremo inferior de la zapata de freno.

Antifriction bearing A bearing designed to reduce friction. This type bearing normally uses ball or roller inserts to reduce the friction.
Cojinetes de antifricción Un cojinete diseñado con el fin de disminuir la fricción. Este tipo de cojinete suele incorporar una pieza inserta esférica o de rodillos para disminuir la fricción.

Antirattle spring A spring that holds parts in clutches and disc brakes together and keeps them from rattling.
Resorte antigolpeteo Un resorte que une las partes en los embragues o en los frenos de disco y los impide de rechinar.

Anti-seize Thread compound designed to keep threaded connections from damage due to rust or corrosion.
Antiagarrotamiento Un compuesto para filetes diseñado para protejer a las conecciones fileteados de los daños de la oxidación o la corrosión.

Arbor press A small, hand-operated shop press used when only a light force is required against a bearing, shaft, or other part.
Prensa para calar Una prensa de mano pequeña del taller que se puede usar en casos que requieren una fuerza ligera contra un cojinete, una flecha u otra parte.

Asbestos A material that was commonly used as a gasket material in places where temperatures were great. This material is being used less frequently today because of health hazards that are inherent to the material.
Amianto Una materia que se usó frecuentemente como materia de empaques en sitios en los cuales las temperaturas eran extremas. Esta materia se usa menos actualmente debido a los peligros al salud que se atribuyan a esta materia.

Axial Parallel to a shaft or bearing bore.
Axial Paralelo a una flecha o al taladro del cojinete.

Axis The center line of a rotating part, a symmetrical part, or a circular bore.
Eje La linea de quilla de una parte giratoria, una parte simétrica, o un taladro circular.

Axle The shaft or shafts of a machine upon which the wheels are mounted.
Semieje El eje o los ejes de una máquina sobre los cuales se montan las ruedas.

Backing plate A plate to which drum braking mechanisms are affixed.
Placa de respaldo Una placa a la cual se afija los mecanismos de un freno de tambor.

Ball bearing An antifriction bearing consisting of hardened inner and outer races with hardened steel balls that roll between the two races and support the load of the shaft.
Rodamiento de bolas Un cojinete de antifricción que consiste de pistas endurecidas interiores e exteriores que contienen bolas de acero endurecidos que ruedan entre las dos pistas, y sostiene la carga de la flecha.

Bearing The supporting part that reduces friction between a stationary and rotating part or between two moving parts.
Cojinete La parte portadora que reduce la fricción entre una parte fija y una parte giratoria o entre dos partes que muevan.

Bearing cage A spacer that keeps the balls or rollers in a bearing in proper position between the inner and outer races.
Jaula del cojinete Un espaciador que mantiene a las bolas o a los rodillos del cojinete en la posición correcta entre las pistas interiores e exteriores.

Bearing cone The inner race, rollers, and cage assembly of a tapered roller bearing. Cones and cups must always be replaced in matched sets.
Cono del cojinete La asamblea de la pista interior, los rodillos, y el jaula de un cojinete de rodillos cónico. Se debe siempre reemplazar a ambos partes de un par de conos del cojinete y los anillos exteriores a la vez.

Bearing cup The outer race of a tapered roller bearing or ball bearing.
Anillo exterior La pista exterior de un cojinete cónico de rodillas o de bolas.

Bearing race The surface upon which the rollers or balls of a bearing rotate. The outer race is the same thing as the cup, and the inner race is the one closest to the axle shaft.
Pista del cojinete La superficie sobre la cual rueden los rodillos o las bolas de un cojinete. La pista exterior es lo mismo que un anillo exterior, y la pista interior es la más cercana a la flecha del eje.

Bleeder valve A valve on the master cylinder, caliper, or wheel cylinder that allows air and fluid to be drained from the system.
Tornillo de purga Una válvula en el cilindro maestro, en la abrazadera, o en el cilindro de la rueda que permite que el aire y el líquido se vacíen del sistema.

Bolt torque The turning effort required to offset resistance as the bolt is being tightened.
Torsión del perno El esfuerzo de torsión que se requiere para compensar la resistencia del perno mientras que esté siendo apretado.

Booster A vacuum or hydraulic device attached to the master cylinder to ease operation and/or increase the effectiveness of a brake system.
Reforzador Un dispositivo de vacío o hidráulico conectado al cilindro maestro que facilita la operación y/o hace más eficaz un sistema de frenos.

Brake anchor   Pivot pin on brake backing plate on which the shoe rests.
Anclaje de freno   Una clavija de pivote en la placa de respaldo sobre la cual descanse la zapata.

Brake drag   The continuous contact between pad or lining with the brake disc or drum.
Arrastre de los frenos   Un contacto continuo entre la almohadilla o el forro con el disco o el tambor del freno.

Brake drum   A bowl-shaped cast iron housing against which the brake shoes press to stop its rotation.
Tambor   Una caja redonda de fierro colado contra la cual se aprietan las zapatas para detener su rotación.

Brake drum micrometer   Measures the inside diameter of a brake drum.
Micrómetro del tambor   Toma la medida del diámetro interior de un tambor de freno.

Brake fade   Loss of braking effectiveness caused by excessive heat reducing the friction between the pad and disc or shoe and the drum.
Amortiguamiento del frenado   La pérdida de la eficiencia de frenar debido al calor excesivo que reduce la fricción entre la almohadilla y el disco o la zapata y el tambor.

Brake fluid   A hydraulic fluid used to transmit force through brake lines. Brake fluid must be noncorrosive to both the metal and rubber components of the brake system.
Líquido de freno   Un líquido hidráulico que se usa para transmitir la fuerza por las líneas de freno. El líquido de freno no debe corroer los componentes de metal ni de hule que se encuentran en el sistema de frenos.

Brake flushing   A procedure for cleaning a brake system by removing fluid and washing out sediment and condensation.
Purgar los frenos   Un procedimiento para limpiar el sistema de frenos en que se vacia el líquido y se quitan los sedimentos y la condensación.

Brake grab   Sudden and undesirable increase in braking force.
Amarro de freno   Un incremento del acción de frenar repentino y no deseado.

Brake horsepower (bhp)   Power delivered by the engine and available for driving the vehicle: bhp = torque 3 rpm/5,252.
Caballo indicado al freno (bhp)   Potencia que provee el motor y que es disponible para el uso del vehículo: bhp = de par motor 3 rpm/5,252.

Brake lines   Lines that carry brake fluid from the master cylinder to the wheels.
Líneas de freno   Las líneas que llevan el líquido de freno del cilíndro maestro hasta las ruedas.

Brake lining   Heat-resistant friction material that is pressed against the metal drum or disc to achieve braking force in a brake system.
Forro de frenos   Una material resistente al calor que se oprime contra el tambor o disco de metal para ejecutar una fuerza de frenar en un sistema de frenos.

Brake pads   The parts of a disc brake system that hold the linings.
Almohadilla s de freno   Las partes en un sistema de frenos de disco que sostienen los forros.

Brake pedal   Pedal pushed upon to activate the master cylinder.
Pedal de freno   El pedal que se oprime para accionar el cilíndro maestro.

Brake shoe   The metal assembly onto which the frictional lining is attached for drum brake systems.

Zapata de freno   La asamblea metálica a la cual se afija el forro de fricción para el sistema de freno de tambor.

Brinnelling   Rough lines worn across a bearing race or shaft due to impact loading, vibration, or inadequate lubrication.
Efecto brinel   Lineas ásperas que aparecen en las pistas de un cojinete o en las flechas debido al choque de carga, la vibración, o falta de lubricación.

Burnish   To smooth or polish by the use of a sliding tool under pressure.
Bruñir   Tersar o pulir por medio de una herramienta deslizando bajo presión.

Burr   A feather edge of metal left on a part being cut with a file or other cutting tool.
Rebaba   Una lima espada de metal que permanece en una parte que ha sido cortado con una lima u otro herramienta de cortar.

Bushing   A cylindrical lining used as a bearing assembly made of steel, brass, bronze, nylon, or plastic.
Buje   Un forro cilíndrico que se usa como una asamblea de cojinete que puede ser hecho del acero, del latón, del bronce, del nylon, o del plástico.

Cage   A spacer used to keep the balls or rollers in proper relation to one another. In a constant-velocity joint, the cage is an open metal framework that surrounds the balls to hold them in position.
Jaula   Una espaciador que mantiene una relación correcta entre los rodillos o las bolas. En una junta de velocidad constante, la jaula es un armazón abierto de metal que rodea a las bolas para mantenerlas en posición.

Caliper   Major component of a disc brake system. Houses the piston(s) and supports the brake pads.
Mordaza   Un componente principal del sistema de frenos de disco. Contiene el pistón (los pistones) y sostiene las almohadillas.

Castellate   Formed to resemble a castle battlement, as in a castellated nut.
Acanalado   De una forma que parece a las almenas de un castillo (véa la palabra en inglés), tal como una tuerca con entallas.

Castellated nut   A nut with six raised portions or notches through which a cotter pin can be inserted to secure the nut.
Tuerca con entallas   Una tuerca que tiene seis porciones elevadas o muescas por los cuales se puede insertar un pasador de chaveta para retener a la tuerca.

C-clip   A C-shaped clip used to retain the drive axles in some rear axle assemblies.
Grapa de C   Una grapa en forma de C que retiene a las flechas motrices en algunas asambleas de ejes traseras.

Chamfer   A bevel or taper at the edge of a hole or a gear tooth.
Chaflán   Un bisél o cono en el borde de un hoyo o un diente del engranaje.

Chase   To straighten up or repair damaged threads.
Embutir   Enderezar o reparar a los filetes dañados.

Chassis   The vehicle frame, suspension, and running gear. On FWD cars, it includes the control arms, struts, springs, trailing arms, sway bars, shocks, steering knuckles, and frame. The drive shafts, constant-velocity joints, and transaxle are not part of the chassis or suspension.
Chasis   El armazón de un vehículo, la suspensión, y el engranaje de marcha. En los coches de FWD, incluye los brazos de mando, los postes, los resortes (chapas), los brazos traseros, las estabilizadoras, las articulaciones de la dirección y el armazón. Los árboles de mando, las juntas de velocidad constante, y la flecha impulsora no son partes del chasis ni de la suspensión.

Circlip A split steel snap ring that fits into a groove to hold various parts in place. Circlips are often used on the ends of FWD drive shafts to retain the constant-velocity joints.

Grapa circular Un seguro partido circular de acero que se coloca en una ranura para posicionar a varias partes. Las grapas circulares se suelen usar en las extremidades de los árboles de mando en FWD para retener las juntas de velocidad constante.

Clearance The space allowed between two parts, such as between a journal and a bearing.

Holgura El espacio permitido entre dos partes, tal como entre un muñon y un cojinete.

Coefficient of friction The ratio of the force resisting motion between two surfaces in contact to the force holding the two surfaces in contact.

Coeficiente de la fricción La relación entre la fuerza que resiste al movimiento entre dos superficies que tocan y la fuerza que mantiene en contacto a éstas dos superficies.

Compensating port An opening in a master cylinder that permits fluid to return to the reservoir.

Abertura de compensación Un orificio en el cilindro maestro que permite que el líquido regresa al recipiente.

Compound A mixture of two or more ingredients.

Compuesto Una combinación de dos ingredientes o más.

Concentric Two or more circles having a common center.

Concéntrico Dos círculos o más que comparten un centro común.

Constant-velocity joint (also called CV joint) A flexible coupling between two shafts that permits each shaft to maintain the same driving or driven speed regardless of operating angle, allowing for a smooth transfer of power. The constant-velocity joint consists of an inner and outer housing with balls in between, or a tripod and yoke assembly.

Junta de velocidad constante (también llamado junta CV) Un acoplador flexible entre dos flechas que permite que cada flechamantenga la velocidad de propulsión o arrastre sin importar el ángulo de operación, efectuando una transferencia lisa del poder. La junta de velocidad constante consiste de una caja interior e exterior entre los cuales se encuentran bolas, o de un conjunto de trípode y yugo.

Contraction A reduction in mass or dimension: the opposite of expansion.

Contracción Una reducción en la masa o en la dimensión: el opuesto de la expansión.

Control arm A suspension component that links the vehicle frame to the steering knuckle or axle housing and acts as a hinge to allow up-and-down wheel motions. The front control arms are attached to the frame with bushings and bolts and are connected to the steering knuckles with ball joints. The rear control arms attach to the frame with bushings and bolts and are welded or bolted to the rear axle or wheel hubs.

Brazo de mando Un componente de la suspensión que une el armazón del vehículo al articulación de dirección o a la caja del eje y que se porta como una bisagra para permitir los movimientos verticales de las ruedas. Los brazos de mando delanteros se conectan al armazón por medio de pernos y bujes y se conectan al articulación de dirección por medio de los articulaciones esféricos. Los brazos de mando traseros se conectan al armazón por medio de pernos y bujes y son soldados o empernados al eje trasero o a los cubos de las ruedas.

Control valve The component beneath the master cylinder that contains the hydraulic controls for the brake system.

Válvula de control El componente en la parte inferior del cilindro maestro que contiene los controles hidráulicos para el sistema de frenos.

Corrode To eat away gradually as if by gnawing, especially by chemical action.

Corroer Roído poco a poco, primariamente por acción químico.

Corrosion Chemical action, usually by an acid, that eats away (decomposes) a metal.

Corrosión Un acción químico, regularmente de un ácido, que corroe (descompone) un metal.

Cotter pin A type of fastener made from soft steel in the form of a split pin, that can be inserted into a drilled hole. The split ends are spread to lock the pin in position.

Pasador de chaveta Un tipo de fijación hecho de acero blando en forma de una chaveta, que se puede insertar en un hueco tallado. Las extremidades partidas se despliegen para asegurar la posición de la chaveta.

Counterclockwise rotation Rotating in the opposite direction of the hands on a clock.

Rotación en sentido inverso Girando en el sentido opuesto de las manecillas de un reloj.

Coupling A connecting means for transferring movement from one part to another; may be mechanical, hydraulic, or electrical.

Acoplador Un método de conexión que transfere el movimiento de una parte a otra; puede ser mecánico, hidráulico, o eléctrico.

Deflection Bending or movement away from normal due to loading.

Desviación Curvación o movimiento fuera de lo normal debido a la carga.

Degree A unit of measurement equal to 1/360th of a circle.

Grado Una uneda de medida que iguala al 1/360 parte de un círculo.

Density Compactness; relative mass of matter in a given volume.

Densidad La firmeza; una cantidad relativa de la materia que ocupa a un volumen dado.

Depth gauge Based on a micrometer, this gauge is used to measure the depth of holes, slots, and keyways.

Galga de profundidades Principalmente como un micrómetro, esta galga se usa para medir la profundidad de los hoyos, las ranuras y las mortajas.

Diagnosis A systematic study of a machine or machine parts to determine the cause of improper performance or failure.

Diagnóstico Un estudio sistemático de una máquina o las partes de una máquina con el fín de determinar la causa de una falla o de una operación irregular.

Dial indicator A measuring instrument with the readings indicated on a dial rather than on a thimble as on a micrometer.

Indicador de carátula Un instrumento de medida cuyo indicador es en forma de muestra en contraste al casquillo de un micrómetro.

Disc brake A brake design in which the member attached to the wheel is a metal disc and braking force is applied by two brake pads that are squeezed against the disc by the caliper.

Frenos de disco Un diseño de frenos en el cual el miembro conectado a la rueda es un disco de metal y la fuerza del frenado se aplica por medio de la mordaza que aprieta dos almohadillas de freno contra el disco.

Disc runout A measurement of how much a brake disc wobbles from side to side as it rotates.

Corrimiento del disco Una medida del movimiento oscilatorio de un disco de un lado al otro mientras que gira.

Distortion A warpage or change in form from the original shape.
Distorción El abarquillamiento o un cambio en la forma original.

Dowel A metal pin attached to one object which, when inserted into a hole in another object, ensures proper alignment.
Espiga Una clavija de metal que se fija a un objeto que, al insertarla en el hoyo de otro objeto, asegura una alineación correcta.

Dowel pin A pin inserted in matching holes in two parts to maintain those parts in fixed relation to one another.
Clavija de espiga Una clavija que se inserte en los hoyos alineados de dos partes para mantener a ésos dos partes en una relación fija el uno al otro.

Drum brake A brake design in which the component attached to the wheel is shaped like a drum. Shoes press against the inside of the drum to provide braking action.
Frenos de tambor Un diseño de frenos en el cual el componente conectado a la rueda tiene la forma de un tambor. La zapatas oprimen contra la superficie interior para efectuar la acción de frenar.

Dry friction The friction between two dry solids.
Fricción seca Fricción entre dos sólidos secos.

Dual master cylinder A master cylinder consisting of two separate sections, one for the rear brakes and the other for the front brakes.
Cilindro maestro dual Un cilindro maestro que consiste de dos secciones distinctas, una para los frenos traseros y la otra para los frenos delanteros.

Dual-servo-action brakes Both brake shoes are self-energizing; that is, they tend to multiply braking forces as they are applied.
Freno duo-servo Ambas zapatas son autoenergéticas; quiere decir, suelen multiplicar la fuerza de frenar al aplicarlas.

Dynamic In motion.
Dinámico En movimiento.

Eccentric One circle within another circle wherein the circles do not have the same center or a circle is mounted off center. On FWD cars, front-end camber adjustments are accomplished by turning an eccentric cam bolt that mounts the strut to the steering knuckle.
Excéntrico Se dice de dos círculos, el uno dentro del otro, que no comparten el mismo centro o de un círculo ubicado descentrado. En los coches FWD, los ajustes de la inclinación se efectuan por medio de un perno excéntrico que fija el poste sobre el articulación de dirección.

Efficiency The ratio between the power of an effect and the power expended to produce the effect; the ratio between an actual result and the theoretically possible result.
Eficiencia La relación entre la potencia de un efecto y la potencia que se gasta para producir el efecto; la relación entre un resultado actual y el resultado que es una posibilidad teórica.

Elastomer Any rubber-like plastic or synthetic material used to make bellows, bushings, and seals.
Elastómero Cualquiera materia plástic parecida al hule o una materia sintética que se utiliza para fabricar a los fuelles, los bujes y las juntas.

Endplay The amount of axial or end-to-end movement in a shaft due to clearance in the bearings.
Juego de las extremidades La cantidad del movimiento axial o del movimiento de extremidad a extremidad en una flecha debido a la holgura que se deja en los cojinetes.

Face The front surface of an object.
Cara La superficie delantera de un objeto.

Fatigue The buildup of natural stress forces in a metal part that eventually causes it to break. Stress results from bending and loading the material.

Fatiga El incremento de tensiones e esfuerzos normales en una parte de metal que eventualmente causan una quebradura. Los esfuerzos resultan de la carga impuesta y del doblamiento de la materia.

Feeler gauge A metal strip or blade finished accurately with regard to thickness used for measuring the clearance between two parts; such gauges ordinarily come in a set of different blades graduated in thickness by increments of 0.001 inch.
Calibrador de lainas Una lámina o hoja de metal que ha sido acabado precisamente con respecto a su espesor que se usa para medir la holgura entre dos partes; estas galgas típicamente vienen en un conjunto de varias espesores graduados desde el 0.001 de una pulgada.

Fiber composites A mixture of metallic threads along with a resin form a composite offering weight and cost reduction, long-term durability, and fatigue life. Fiberglass is a fiber composite.
Compuestos de fibra Una mezcla de hilos metálicos con una resina que formen un compuesto ofreciendo una reducción en costo y peso, y una durabilidad y utilidad de largo plazo. La fibra de vidrio es una fibra compuesta.

Fit The contact between two machined surfaces.
Ajuste El contacto entre dos superficies maquinadas.

Fixed caliper A disc brake caliper that has pistons on both sides of the rotor. It is rigidly fixed to the suspension.
Mordaza fija Una mordaza de freno de disco que tiene pistones en ambos lados del rotor. Se fija rígidamente a la suspención.

Floating caliper A disc brake caliper that has one piston. The caliper is free to move on pins in response to the piston pressing against the rotor.
Mordaza flotante Una mordaza de freno de disco que tiene un pistón. La mordaza mueve libremente sobre clavijas respondiendo al pistón oprimiendo contra el rotor.

Foot-pound (or ft-lb) This is a measure of the amount of energy or work required to lift 1 pound a distance of 1 foot.
Libra-pie (o lb.p.) Una medida de la cantidad de energía o fuerza que requiere mover una libra a una distancia de un pie.

Force Any push or pull exerted on an object; measured in pounds and ounces, or in Newtons (N) in the metric system.
Fuerza Cualquier acción empujado o jalado que se efectua en un objeto; se mide en pies y onzas, o en Newtones (N) en el sistema métrico.

Four wheel drive On a vehicle, driving axles at both front and rear, so that all four wheels can be driven.
Tracción a cuatro ruedas En un vehículo, se trata de los ejes de dirección fronteras y traseras, para que cada una de las ruedas puede impulsar.

Frame The main understructure of the vehicle to which everything else is attached. Most FWD cars have only a subframe for the front suspension and drive train. The body serves as the frame for the rear suspension.
Armazón La estructura principal del vehículo al cual todo se conecta. La mayoría de los coches FWD sólo tiene un bastidor auxiliar para la suspensión delantera y el tren de propulsión. La carrocería del coche sirve de chassis par la suspensión trasera.

Friction The resistance to motion between two bodies in contact with each other.
Fricción La resistencia al movimiento entre dos cuerpos que estan en contacto.

Front wheel drive (FWD) The vehicle has all drive train components located at the front.

Tracción de las ruedas delanteras (FWD) El vehículo tiene todos los componentes del tren de propulsión en la parte delantera.

FWD Abbreviation for front wheel drive.

FWD Abreviación de tracción de las ruedas delanteras.

Galling Wear caused by metal-to-metal contact in the absence of adequate lubrication. Metal is transferred from one surface to the other, leaving behind a pitted or scaled appearance.

Desgaste por fricción El desgaste causado por el contacto de metal a metal en la ausencia de lubricación adecuada. El metal se transfere de una superficie a la otra, causando una aparencia agujerado o con depósitos.

Gasket A layer of material, usually made of cork, paper, plastic, composition, metal, or a combination of these, placed between two parts to make a tight seal.

Empaque Una capa de una materia, normalmente hecho del corcho, del papel, del plástico, de la materia compuesta, del metal, o de cualquier combinación de éstos, que se coloca entre dos partes para formar un sello impermeable.

Gasket cement A liquid adhesive material or sealer used to install gaskets.

Mastique para empaques Una substancia líquida adhesiva o una substancia impermeable que se usa para instalar a los empaques.

Glazing An extremely shiny condition of a metal surface, such as the bore of a cylinder.

Vidriado Una condición altamente luciente de una superficie de un metal, tal como el taladro de un cilindro.

Grind To finish or polish a surface by means of an abrasive wheel.

Amolar Acabar o pulir a una superficie por medio de una muela para pulverizar.

Half shaft Either of the two drive shafts that connect the transaxle to the wheel hubs in FWD cars. Half-shafts have constant-velocity joints attached to each end to allow for suspension motions and steering. The shafts may be of solid or tubular steel and may be of different lengths. Balance is not critical, as half-shafts turn at roughly one-third the speed of RWD drive shafts.

Semieje Cualquier de dos ejes o flechas de mando que conectan el transeje a los cubos de las ruedas en los coches de FWD. Los semiejes tienen juntas de velocidad continua conectado a cada extremo para permitir los movimientos de la suspención y la dirección. Los ejes pueden fabricarse del acero sólido o tubular y pueden variar en longitud. El balance no es crítico, puesto que los semiejes giran una tercera parte de la velocidad de los ejes de mando de RWD.

Hard pedal A loss in braking efficiency so that an excessive amount of pressure is needed to actuate the brakes.

Pedal de freno duro Una pérdida de la eficiencia del frenado en que se requiere una cantidad excesiva de presión para activar los frenos.

Heat treatment Heating, followed by fast cooling, to harden metal.

Tratamiento térmico Calentamiento, seguido por un enfriamiento rápido, para endurecer a un metal.

Honing A process whereby an abrasive material is used to smoothen a surface. Honing is a common operation in preparing cylinder walls for the installation of pistons.

Rectificación Un proceso por el cual una material abrasiva se usa para tersar una superficie. Rectificar es una operación comun en preparar los interiores de los cilindros para la instalación de los pistones.

Horsepower A measure of mechanical power, or the rate at which work is done. One horsepower equals 33,000 ft-lbs (foot-pounds) of work per minute. It is the power necessary to raise 33,000 pounds a distance of 1 foot in 1 minute.

Caballo de fuerza Una medida de fuerza mecánica, o el régimen en el cual se efectua el trabajo. Un caballo de fuerza iguala a 33,000 lb.p. (libras pie) de trabajo por minuto. Es la fuerza requerida para transportar a 33,000 libras una distancia de 1 pie en 1 minuto.

Hub The center part of a wheel, to which the wheel is attached.

Cubo La parte central de una rueda, a la cual se monta la rueda.

Hydraulic booster A power brake booster operated by hydraulic pressure from the power steering pump.

Hidrorreforzador Un reforzador de freno de potencia que opera con la presión hidráulica de la bomba de la dirección hidráulica.

Hydraulic fluid reservoir A part of a master cylinder assembly that holds reserve fluid.

Recipiente del líquido hidráulico Una parte de la asamblea del cilindro maestro que contiene el líquido en reserva.

Hydraulic press A piece of shop equipment that develops a heavy force by use of a hydraulic piston-and-jack assembly.

Prensa hidráulica Una herramienta del taller que provee una fuerza grande por medio de una asamblea de gato con un pistón hidráulico.

Hydraulic pressure Pressure exerted through the medium of a liquid.

Presión hidráulica La presión esforzada por medio de un líquido.

ID Inside diameter.

DI Diámetro Interior.

Idle Engine speed when the accelerator pedal is fully released and there is no load on the engine.

Marcha lenta La velocidad del motor cuando el pedal accelerador esta completamente desembragada y no hay carga en el motor.

Increments Series of regular additions from small to large.

Incremento Una serie de agregaciones regulares de pequeña a grande.

Index To orient two parts by marking them. During reassembly the parts are arranged so the index marks are next to each other. Used to preserve the orientation between balanced parts.

Índice Orientar a dos partes marcándolas. Al montarlas, las partes se colocan para que las marcas de índice estén alineadas. Se usan los índices para preservar la orientación de las partes balanceadas.

Inner bearing race The inner part of a bearing assembly on which the rolling elements, ball or roller, rotate.

Pista interior de un cojinete La parte interior de una asamblea de cojinetes en la cual ruedan las bolas o los rodillos.

Integral Built into, as part of the whole.

Íntegro Contenido, como una parte del total.

Jam nut A second nut tightened against a primary nut to prevent it from working loose. Used on inner and outer tie-rod adjustment nuts and on many pinion-bearing adjustment nuts.

Contra tuerca Una tuerca secundaria que se aprieta contra una tuerca primaria para prevenir que ésta se afloja. Se emplean en las tuercas de ajustes interiores e exteriores para las barras de acoplamiento y también en muchas de las tuercas de ajuste de portapiñones.

Key A small block inserted between the shaft and hub to prevent circumferential movement.

Chaveta Un tope pequeño que se meta entre la flecha y el cubo para prevenir un movimiento circunferencial.

Keyway A groove or slot cut to permit the insertion of a key.
Ranura de chaveta Un corte de ranura o mortaja que permite insertar una chaveta.

Knock A heavy metallic sound usually caused by a loose or worn bearing.
Golpe Un sonido metálico fuerte que suele ser causado por un cojinete suelto o gastado.

Knuckle The part of the suspension that supports the wheel hub and serves as the steering pivot. The bottom of the knuckle is attached to the lower control arm with a ball joint, and the upper portion is usually bolted to the strut.
Articulación La parte de la suspensión que sostiene al cubo de la rueda y sirve como punto pivote de dirección. La parte inferior de la articulación se une al brazo de mando inferior por medio de una articulación esférica, y la parte superior suele ser empernado al poste.

Knurl To indent or roughen a finished surface.
Moletear Indentar o desbastar a una superficie acabada.

Linkage Any series of rods, yokes, and levers, and so on, used to transmit motion from one unit to another.
Biela Cualquiera serie de barras, yugos, palancas, y todo lo demás, que se usa para transferir los movimientos de una unedad a otra.

Locknut A second nut turned down on a holding nut to prevent loosening.
Contra tuerca Una tuerca segundaria apretada contra una tuerca de sostén para prevenir que ésta se afloja.

Lock pin Used in some ball sockets (inner tie-rod end) to keep the connecting nuts from working loose. Also used on some lower ball joints to hold the tapered stud in the steering knuckle.
Clavija de cerrojo Se usan en algunas rótulas (las extremidades interiores de la barra de acoplamiento) para prevenir que se aflojan las tuercas de conexión. También se emplean en algunas juntas esféricas inferiores para retener al perno cónico en la articulación de dirección.

Lockplates Metal tabs bent around nuts or bolt heads.
Placa de cerrojo Chavetas de metal que se doblan alrededor de las tuercas o las cabezas de los pernos.

Lockwasher A type of washer that, when placed under the head of a bolt or nut, prevents the bolt or nut from working loose.
Arandela de freno Un tipo de arandela que, al colocarse bajo la cabeza de un perno, previene que el perno o la tuerca se aflojan.

Lubricant Any material, usually a petroleum product such as grease or oil, that is placed between two moving parts to reduce friction.
Lubricante Cualquier substancia, normalmente un producto de petróleo como la grasa o el aceite, que se coloca entre dos partes en movimiento para reducir la fricción.

Lug nut The nuts that fasten the wheels to the axle hub or brake rotor. Missing lug nuts should always be replaced. Overtightening can cause warpage of the brake rotor in some cases.
Tuerca de las ruedas Las tuercas que sujetan las ruedas al cubo de flecha o al rotor de los frenos. Las tuercas de las ruedas que se pierden siempre deben reemplazarse. Si se aprietan demasiado puede causar una deformación en el rotor del freno en algunos casos.

Master cylinder The liquid-filled cylinder in the hydraulic brake system or clutch where hydraulic pressure is developed when the driver depresses a foot pedal.
Cilindro maestro El cilindro lleno de líquido en el sistema de frenos hidráulico o en el embrague en el cual la presión hidráulica se presenta cuando el conductor comprime un pedal bajo su pie.

Metering valve A component that momentarily delays the application of front disc brakes until the rear drum brakes begin to move. Helps to provide balanced braking.
Válvula de medición Un componente que retrasa momentáneamente la aplicación de los frenos de disco delanteras hasta que comienzan a moverse los frenos de tambor traseros. Ayuda en proporcionar el enfrenado más equilibrado.

Micrometer A precision measuring device used to measure small bores, diameters, and thicknesses. Also called a mike.
Micrómetro Un dispositivo de medida precisa que se emplea a medir los taladros pequeños y los espesores. También se llama un mike (mayk).

Misalignment When bearings are not on the same center line.
Desalineamineto Cuando los cojinetes no comparten la misma linea central.

Mounts Made of rubber to insulate vibrations and noise while they support a power train part, such as engine or transmission mounts.
Monturas Hecho de hule para insular a las vibraciones y a los ruidos mientras que sujetan una parte del tren de propulsión, tal como las monturas del motor o las monturas de la transmisión.

Neoprene A synthetic rubber that is not affected by the various chemicals that are harmful to natural rubber.
Neoprene Un hule sintético que no se afecta por los varios productos químicos que pueden dañar al hule natural.

Newton-meter (N·m) Metric measurement of torque or twisting force.
Metro-Newton (N·m) Una medida métrica de la fuerza de torsión.

Nut A removable fastener used with a bolt to lock pieces together; made by threading a hole through the center of a piece of metal that has been shaped to a standard size.
Tuerca Un retén removable que se usa con un perno o tuerca para unir a dos piezas; se fabrica al filetear un hoyo taladrado en un pedazo de metal que se ha formado a un tamaño especificado.

Oil seal A seal placed around a rotating shaft or other moving part to prevent leakage of oil.
Empaque de aceite Un empaque que se coloca alrededor de una flecha giratoria para prevenir el goteo de aceite.

O-ring A type of sealing ring, usually made of rubber or a rubber-like material. In use, the O-ring is compressed into a groove to provide the sealing action.
Anillo en O Un tipo de sello anular, suele ser hecho de hule o de una materia parecida al hule. Al usarse, el anillo en O se comprime en una ranura para proveer un sello.

Outer bearing race The outer part of a bearing assembly on which the balls or rollers rotate.
Pista exterior de un cojinete La parte exterior de una asamblea de cojinetes en la cual ruedan las bolas o los rodillos.

Out-of-round Wear of a round hole or shaft that when viewed from an end will appear egg-shaped.
Defecto de circularidad Desgaste de un taladro o de una flecha circular, que al verse de una extremidad, tendrá una forma asimétrica, como la de un huevo.

Oxidation Burning or combustion; the combining of a material with oxygen. Rusting is slow oxidation, and combustion is rapid oxidation.
Oxidación Quemando o la combustión; la combinación de una materia con el oxígeno. El orín es una oxidación lenta, la combustión es la oxidación rápida.

Parallel The quality of two items being the same distance from each other at all points; usually applied to lines and, in automotive work, to machined surfaces.
Paralelo La calidad de dos artículos que mantienen la misma distancia el uno al otro en cada punto; suele aplicarse a las líneas y, en el trabajo automotivo, a las superficies acabadas a máquina.

Parking brake A mechanically operated brake used to hold the vehicle when it is parked.
Freno de estacionamiento (emergencia) Un freno que se opera a mano para sostener al vehículo al estacionarse.

Parking brake strut A bar between the brake shoes. When the parking lever is actuated, the parking brake strut pushes the leading brake shoe into the drum.
Poste del freno de estacionamiento Una barra entre las zapatas de freno. Al actuarse la palanca de estacionamiento, el poste del freno de estacionamiento empuje la zapata delantera contra el tambor.

Pascal's law The law of fluid motion.
Ley de Pascal La ley del movimiento del fluido.

Piston seal The seal fitted to the disc brake pistons. It provides the return motion to the piston as well as sealing in the brake fluid.
Sello del pistón El sello que se ajusta a los pistones de los frenos de disco. Provee el movimiento de regreso al pistón mientras que previene una fuga del líquido de freno.

Pitch The number of threads per inch on any threaded part.
Paso El número de filetes por pulgada de cualquier parte fileteada.

Pivot A pin or shaft upon which another part rests or turns.
Pivote Una chaveta o una flecha que sostiene a otra parte o sirve como un punto para girar.

Power booster Used to increase pedal pressure applied to a brake master cylinder.
Reforzador Se usa para incrementar la presión del pedal aplicado al cilindro maestro del freno.

Power brakes A brake system that employs vacuum or hydraulics to assist the driver in producing braking force.
Frenos de potencia Un sistema de frenos que emplea un vacío o las hidráulicas para asistir al conductor en efectuar una fuerza de enfrenado.

Preload A load applied to a part during assembly so as to maintain critical tolerances when the operating load is applied later.
Carga previa Una carga aplicada a una parte durante la asamblea para asegurar sus tolerancias críticas antes de que se le aplica la carga de la operación.

Press fit Forcing a part into an opening that is slightly smaller than the part itself to make a solid fit.
Ajustamiento a presión Forzar a una parte en una apertura que es de un tamaño más pequeño de la parte para asegurar un ajustamiento sólido.

Pressure Force per unit area, or force divided by area. Usually measured in pounds per square inch (psi) or in kilopascals (kPa) in the metric system.
Presión La fuerza por unedad de una area, o la fuerza divida por la area. Suele medirse en libras por pulgada cuadrada (lb/pulg^2) o en kilopascales (kPa) en el sistema métrico.

Pressure bleeding Pressure bleeding uses air to pressurize brake fluid in order to force air out of the hydraulic brake system.
Purga con presión En purgar con presión uno usa el aire para sobrecomprimir el líquido de freno así forzando el aire fuera del sistema hidráulico de frenos.

Pressure differential valve Used in dual brake systems to sense unequal hydraulic pressure between the front and rear brakes.
Válvula del diferencial de presión Usado en los sistemas de frenos dobles para sentir una presión desigual entre los frenos delanteros y traseros.

Primary shoe When the car is moving forward, the shoe facing the front of the car is the leading or primary shoe.
Zapata primaria Al moverse hacia frente el coche, la zapata en la dirección hacia la parte delantera del coche es la zapata de guía o primaria.

Proportioning valve This valve regulates the hydraulic pressure in the rear brake system. It is located between the inlet and outlet ports of the rear system in the control valve. It allows equal pressure to be applied to both the front and rear brakes until a particular pressure is obtained.
Válvula dosificadora Esta válvula regula la presión hidráulica en el sistema de frenos trasero. Se ubica entre las aberturas de entrada y salida del sistema trasero en la válvula de control. Permite que una presión equilibrada se aplica a ambos los frenos delanteros y traseros hasta que se obtiene una presión específica.

psi Abbreviation for pounds per square inch, a measurement of pressure.
Lb/pulg2 Una abreviación de libras por pulgada cuadrada, una medida de la presión.

Puller Generally, a shop tool used to separate two closely fitted parts without damage. Often contains a screw, or several screws, which can be turned to apply a gradual force.
Extractor Generalmente, una herramienta del taller que sirve para separar a dos partes apretadas sin incurrir daños. Suele tener una tuerca o varias tuercas, que se pueden girar para aplicar una fuerza gradual.

Pulsation To move or beat with rhythmic impulses.
Pulsación Moverse o batir con impulsos rítmicos.

Pulsing pedal A condition where the brake pedal moves up and down when it is applied. Normally due to an unparallel brake rotor.
Pedal pulsante Una condición en la cual el pedal de freno se mueve hacia arriba y abajo al aplicarse. Normalmente se debe a un rotor de freno que esta fuera de paralelo.

Race A channel in the inner or outer ring of an antifriction bearing in which the balls or rollers roll.
Pista Un canal en el anillo interior o exterior de un cojinete antifricción en el cual ruedan las bolas o los rodillos.

Ratio The relation or proportion that one number bears to another.
Relación La correlación o proporción de un número con respeto a otro.

Reamer A round metal-cutting tool with a series of sharp cutting edges; enlarges a hole when turned inside it.
Escariador Una herramienta redonda para cortar a los metales que tiene una seria de rebordes mordaces agudos; al girarse en un agujero lo agranda.

Rear wheel drive A term associated with a vehicle where the engine is mounted at the front of the vehicle and the driving axle and driving wheels are mounted at the rear of the vehicle.
Tracción trasera Un término que se asocia con un vehículo en el cual el motor se ubica en la parte delantera y el eje propulsor y las ruedas propulsores se encuentran en la parte trasera del vehículo.

Rivet A headed pin used for uniting two or more pieces by passing the shank through a hole in each piece, and securing it by forming a head on the opposite end.
Remache Una clavija con cabeza que sirve para unir a dos piezas o más al pasar el vástago por un hoyo en cada pieza y asegurarlo por formar una cabeza en el extremo opuesto.

Roller bearing   An inner and outer race upon which hardened steel rollers operate.
Cojinete de rodillos   Una pista interior y exterior en la cual operan los rodillos hecho de acero endurecido.

Rollers   Round steel bearings that can be used as the locking element in an overrunning clutch or as the rolling element in an antifriction bearing.
Rodillos   Articulaciones redondos de acero que pueden servir como un elemento de enclavamiento en un embrague de sobremarcha o como el elemento que rueda en un cojinete antifricción.

RPM   Abbreviation for revolutions per minute, a measure of rotational speed.
RPM   Abreviación de revoluciones por minuto, una medida de la velocidad rotativa.

RTV sealer   Room-temperature vulcanizing gasket material that cures at room temperature; a plastic paste squeezed from a tube to form a gasket of any shape.
Sellador RTV   Una materia vulcanizante de empaque que cura en temperaturas del ambiente; una pasta plástica exprimida de un tubo para formar un empaque de cualquiera forma.

Runout   Deviation of the specified normal travel of an object; the amount of deviation or wobble a shaft or wheel has as it rotates. Runout is measured with a dial indicator.
Corrimiento   Una desviación de la carrera normal e especificada de un objeto. La cantidad de desviación o vacilación de una flecha o una rueda mientras que gira. El corrimiento se mide con un indicador de carátula.

RWD   Abbreviation for rear wheel drive.
RWD   Abreviación de tracción trasera.

SAE   Society of Automotive Engineers.
SAE   La Sociedad de Ingenieros Automotrices.

Score   A scratch, ridge, or groove marring a finished surface.
Entalladura   Una raya, una arruga o una ranura que desfigure a una superficie acabada.

Scuffing   A type of wear in which there is a transfer of material between parts moving against each other; shows up as pits or grooves in the mating surfaces.
Erosión   Un tipo de desgaste en el cual hay una tranferencia de una materia entre las partes que estan en contacto mientras que muevan; se manifiesta como hoyitos o muescas en las superficies apareadas.

Seal   A material, shaped around a shaft, used to close off the operating compartment of the shaft, preventing oil leakage.
Sello   Una materia, formado alrededor de una flecha, que sella el compartimiento operativo de la flecha, previniendo el goteo de aceite.

Sealer   A thick, tacky compound, usually spread with a brush, that may be used as a gasket or sealant to seal small openings or surface irregularities.
Sellador   Un compuesto pegajoso y espeso, comúnmente aplicado con una brocha, que puede usarse como un empaque o un obturador para sellar a las aperturas pequeñas o a las irregularidades de la superficie.

Seat   A surface, usually machined, upon which another part rests or seats; for example, the surface upon which a valve face rests.
Asiento   Una superficie, comúnmente maquinada, sobre la cual yace o se asienta otra parte; por ejemplo, la superficie sobre la cual yace la cara de la válvula.

Secondary shoe   When the car is moving forward, the shoe facing the rear is the trailing or secondary shoe.

Zapata secundaria   Al moverse hacia frente el coche, la zapata en la dirección hacia la parte trasera del coche es la zapata seguidora o secundaria.

Self-energizing   The increase in friction contact between the toe of the brake shoe caused by the drum rotation tending to pull the shoe into the drum.
Autoenergético   El incremento del contacto frotativo entre la parte superior del freno producido por la rotación del tambor que tiene una tendencia a jalar a la zapata hacia el tambor.

Sliding fit   Where sufficient clearance has been allowed between the shaft and journal to allow free-running without overheating.
Ajuste corredera   Donde se ha dejado una holgura suficiente entre la flecha y el muñón para permitir una marcha libre sin sobrecalentamiento.

Snap ring   Split spring-type ring located in an internal or external groove to retain a part.
Anillo de seguridad   Un anillo partido tipo resorte que se coloca en una muesca interior o exterior para retener a una parte.

Spalling   A condition where the material of a bearing surface breaks away from the base metal.
Escamación   Una condición en la cual una materia de la superficie de un rodamiento se separa del metal base.

Spline   Slot or groove cut in a shaft or bore; a splined shaft onto which a hub, wheel, or gear with matching splines in its bore is assembled so that the two must turn together.
Acanaladura (espárrago)   Una muesca o ranura cortada en una flecha o en un taladro; una flecha acanalada en la cual se asambla un cubo, una rueda,o un engranaje, que tiene un acanaladura pareja en el taladro de manera de que las dos deben girar juntos.

Spongy pedal   A condition where the brake pedal does not give firm resistance to foot pressure. Normally caused by air in the hydraulic system.
Resistencia esponjosa   Una condición en la cual el pedal de freno no ofrece una resistencia firme a la presión del pie. Suele ser causado por la presencia del aire en el sistema hidráulico.

Spring   A device that changes shape when it is stretched or compressed, but returns to its original shape when the force is removed; the component of the automotive suspension system that absorbs road shocks by flexing and twisting.
Resorte   Un dispositivo que cambia de forma al ser estirado o comprimido, pero que recupera su forma original al levantarse la fuerza; es un componente del sistema de suspensión automotívo que absorba los choques del camino al doblarse y torcerse.

Spring retainer   A steel plate designed to hold a coil or several coil springs in place.
Retén de resorte   Una chapa de acero diseñado a sostener en su posición a un resorte helicoidal o más.

Squeak   A high-pitched noise of short duration.
Chillido   Un ruido agudo de poca duración.

Squeal   A continuous high-pitched noise.
Alarido   Un ruido agudo continuo.

Star-adjuster   Star-shaped rotor used as an adjustment device in drum brakes.
Ajustador de estrella y tornillo   Un rotor en forma de estrella que se usa como dispositivo de ajuste en los frenos de tambor.

Stress   The force to which a material, mechanism, or component is subjected.
Esfuerzo   La fuerza a la cual se somete a una materia, un mecanísmo o un componente.

Tap To cut threads in a hole with a tapered, fluted, threaded tool.
Roscar con macho Cortar las roscas en un agujero con una herramienta cónica, acanalada y fileteada.

Temper To change the physical characteristics of a metal by applying heat.
Templar Cambiar las características físicas de un metal mediante una aplicación del calor.

Tension Effort that elongates or "stretches" a material.
Tensión Un esfuerzo que alarga o "estira" a una materia.

Thickness gauge Strips of metal made to an exact thickness, used to measure clearances between parts.
Calibre de espesores Las tiras del metal que se han fabricado a un espesor exacto, sirven para medir las holguras entre las partes.

Thread chaser A device, similar to a die, that is used to clean threads.
Peine de roscar Un dispositivo, parecido a una terraja, que sirve para limpiar a las roscas.

Threaded insert A threaded coil that is used to restore the original thread size to a hole with damaged threads.
Pieza inserta roscada Una bobina roscada que sirve para restaurar a su tamaño original una rosca dañada.

Thrust load A load that pushes or reacts through the bearing in a direction parallel to the shaft.
Carga de empuje Una carga que empuja o reacciona por el cojinete en una dirección paralelo a la flecha.

Thrust washer A washer designed to take up end thrust and prevent excessive endplay.
Arandela de empuje Una arandela diseñada para rellenar a la holgura de la extremidad y prevenir demasiado juego en la extremidad.

Tolerance A permissible variation between the two extremes of a specification or dimension.
Tolerancia Una variación permisible entre dos extremos de una especifcación o de un dimensión.

Torque A twisting motion, usually measured in ft-lbs (N•m).
Torsión Un movimiento giratorio, suele medirse en lb.p. (N•m).

Traction The gripping action between the tire tread and the road's surface.
Tracción La acción de agarrar entre la cara de la rueda y la superficie del camino.

Vacuum Any pressure lower than atmospheric pressure.
Vacío Cualquier presión que es más baja la presión atmosférica.

Vacuum brake booster A diaphragm-type booster that uses manifold vacuum and atmospheric pressure for its power.
Reforzador de vacío Un reforzador de tipo de diafragma que usa el vacío del colector y la presión atmosférica para operar.

Vehicle identification number (VIN) The number assigned to each vehicle by its manufacturer, primarily for registration and identification purposes.
Número de identificación del vehículo (VIN) El número asignado a cada vehículo por su fabricante, primariamente con el propósito de la registración y la identificación.

Vibration A quivering, trembling motion felt in the vehicle at different speed ranges.
Vibración Un movimiento de estremecer o temblar que se siente en el vehículo en varios intervalos de velocidad.

Viscosity The resistance to flow exhibited by a liquid. A thick oil has greater viscosity than a thin oil.
Viscosidad La resistencia al flujo que manifiesta un líquido. Un aceite espeso tiene una viscosidad mayor que un aceite ligero.

Wheel A disc or spokes with a hub at the center that revolves around an axle, and a rim around the outside on which the tire is mounted.
Rueda Un disco o rayo que tiene en su centro un cubo que gira alrededor de un eje, y tiene un rim alrededor de su exterior en la cual se monta el neumático.

Wheel cylinder A mechanism located at each wheel in a drum brake system. It uses hydraulic pressure to force the brake shoes against the drum to stop the wheel from turning.
Cilindro de la rueda Un mecanismo ubicado en cada rueda de un sistema de frenos de tambor. Utiliza la presión hidráulica para forzar las zapatas contra el tambor asi previniendo que gira la rueda.

6 Electrical/Electronic Systems

Pretest

The purpose of this pretest is to determine the amount of review that you may require prior to writing the ASE Electrical/Electronic Systems Test. If you answer all the pretest questions correctly, complete the questions and study the information in this chapter to prepare for the ASE Electrical/Electronic Systems Test.

If two or more of your answers to the pretest questions are incorrect, complete a study of Chapters 2 through 13 in *Today's Technician Automotive Electricity and Electronics Classroom and Shop Manuals*, published by Delmar Publishers, plus a study of the questions and information in this chapter.

The pretest answers are located at the end of the pretest; these answers also are in the answer sheets supplied with this book.

1. In a series electrical circuit
   A. the same amount of current flows through each resistance.
   B. each resistance is a separate path for current flow.
   C. the same voltage is dropped across each resistance.
   D. a resistance increase results in more current flow.

2. A 12V light circuit has a short to ground (Figure 6-1).
   Technician A says the current flow through the lamp is higher than normal.
   Technician B says the light cannot be turned off with the switch.
   Who is correct?
   A. A only
   B. B only
   C. Both A and B
   D. Neither A nor B

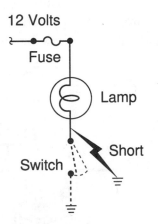

12 Volts

Fuse

Lamp

Short

Switch

Figure 6-1 Short to ground in a 12V light circuit *(Courtesy of Chrysler Corporation)*

3. A shorted condtion in an electromagnet causes
   A. an increase in coil resistance.
   B. an increase in the effective number of coil turns.
   C. an increase in coil current flow.
   D. a weaker magnetic field surrounding the coil.

4. The tester in Figure 6-2 is connected to test
   A. starting motor current draw.
   B. battery capacity.
   C. positive cable voltage drop.
   D. negative cable voltage drop.

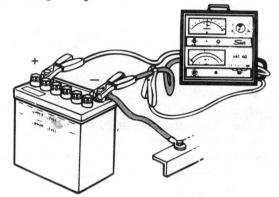

Figure 6-2 Figure 7-2 Volt-ampere tester *(Courtesy of Toyota Motor Corporation)*

5. While discussing a battery state-of-charge test on a battery with a built-in hydrometer
   Technician A says the battery may be fast charged if the hydrometer is yellow or clear.
   Technician B says the battery may be tested if the hydrometer is green.
   Who is correct?
   A. A only
   B. B only
   C. Both A and B
   D. Neither A nor B

6. The voltmeter in Figure 6-3 is connected to test
   A. battery voltage while cranking.
   B voltage drop on the positive cable.
   C. voltage drop across the starting motor.
   D. voltage drop across the solenoid contacts.

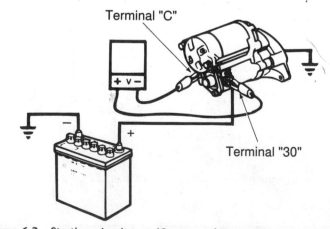

Figure 6-3 Starting circuit test *(Courtesy of Toyota Motor Corporation)*

7. When the ohmmeter is connected as illustrated in Figure 6-4, a low reading is obtained.
   Technician A says the field winding may have an open circuit.
   Technician B says the turns may be shorted together in the field coil.
   Who is correct?
   A. A only
   B. B only
   C. Both A and B
   D. Neither A nor B

Figure 6-4 Field winding test *(Courtesy of Toyota Motor Corporation)*

8. When an ohmmeter is connected to the stator leads as shown in Figure 6-5, an infinite reading is obtained. This reading indicates the stator windings
   A. have an open circuit.
   B. are shorted together.
   C. are grounded to the stator frame.
   D. have a high-resistance defect.

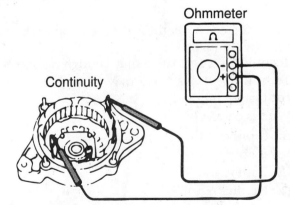

Figure 6-5 Alternator stator test *(Courtesy of Toyota Motor Corp.)*

9. A vehicle has four high-beam headlights, and two low-beam headlights. The headlights intermittently go out while driving with the lights on high beam. The problem does not occur on low beam.
   Technician A says the dimmer switch may be defective.
   Technician B says the headlight switch contacts may be defective.
   Who is correct?
   A. A only
   B. B only
   C. Both A and B
   D. Neither A nor B

10. A dome light is illuminated continually with the doors closed (Figure 6-6). The cause of this problem could be
   A. the right-hand door jam switch has a grounded condtion.
   B. the light switch internal (int) contact is grounded.
   C. the left-hand door jam switch has an open circuit.
   D. the dome light circuit is grounded between the bulb and the fuse.

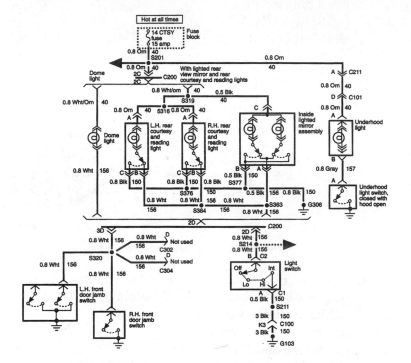

Figure 6-6   Courtesy light and dome light circuit

11. The left-side headlight is dim only on high beam (Figure 6-7). The other headlights operate normally.
   Technician A says there may be high resistance in the left-side headlight ground.
   Technician B says there may be high resistance in the dimmer switch high-beam contacts.
   Who is correct?
   A. A only
   B. B only
   C. Both A and B
   D. Neither A nor B

12. All the gauges read higher than normal in a vehicle with an instrument voltage limiter and thermal-electric gauges. The cause of this problem could be
   A. an open circuit at the instrument voltage limiter contacts.
   B. high resistance in the instrument panel ground.
   C. a grounded fuel gauge sending unit wire.
   D. a blown gauge fuse in the fuse panel.

13. A horn blows continually (Figure 6-8). All of these defects could be the cause of the problem EXCEPT
   A. a short to ground between the horn relay and the horn switch.
   B. the horn relay contacts stuck closed.
   C. a short to ground at the horn slip ring.
   D. a short to ground at the right-hand horn.

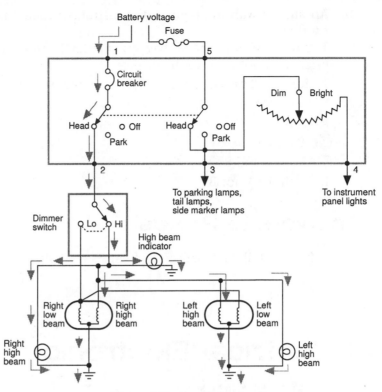

Figure 6-7   Headlight circuit

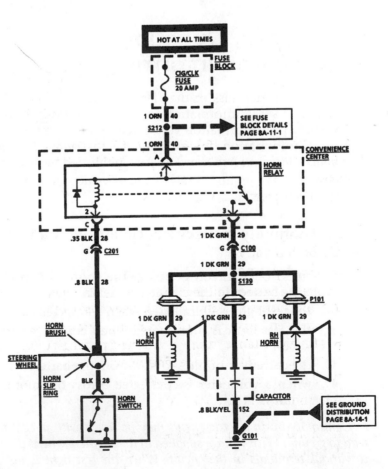

Figure 6-8   Horn circuit *(Courtesy of Pontiac Division, General Motors Corporation)*

14. An air bag warning light in the instrument panel is illuminated with the engine running.

Technician A says the air bag system may be inoperative and the customer should be advised not to drive the vehicle.

Technician B says the air bag system should be checked for diagnostic trouble codes.

Who is correct?

A. A only
B. B only
C. Both A and B
D. Neither A nor B

Answers to Pretest

1. A, 2. B, 3. C, 4. B, 5. B, 6. D, 7. D, 8. A, 9. A, 10. A, 11. D, 12. B, 13. D, 14. C

General Electrical/Electronic System Diagnosis

ASE Tasks, Questions, and Related Information

In this chapter each task in the Electrical/Electronics category is provided followed by a question and some information related to the task. If you answer any question incorrectly, study this information very carefully until you understand the correct answer. For additional information on any task refer to *Today's Technician Automotive Electricity and Electronics Classroom and Shop Manuals*, published by Delmar Publishers.

Question answers and analysis are provided at the end of this chapter and in the answer sheets provided with this book.

Task 1   **Check continuity in electrical circuits with a test light; determine needed repairs.**

1. When an open circuit occurs at the connector in Figure 6-9 and 12V is supplied from the battery to the circuit
   A. the test light is illuminated when connected as shown in Figure 6-9.
   B. the test light is illuminated when connected to the motor side of the open circuit.
   C. the current continues to flow from the battery through the motor.
   D. when a voltmeter is connected across the motor, the voltage drop across the motor is 11V.

Hint   *Continuity in an electric circuit may be tested with a 12V test lamp. Connect the test light lead to ground. With voltage supplied to the circuit, begin at the battery and connect the test light to various terminals in the circuit. When the test light is not illuminated, the open circuit is between the terminal where the test light is connected and the last terminal where the test light was illuminated.*

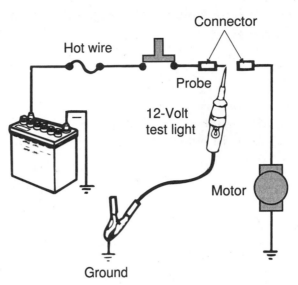

Figure 6-9 Circuit continuity diagnosis with a 12V test light *(Courtesy of Toyota Motor Corporation)*

Task 2   Check applied voltages and voltage drops in electrical/electronic circuits and components with an ammeter; determine needed repairs.

2. The battery in Figure 6-10 is fully charged and the switch is closed. The voltage drop across the light indicated on the voltmeter is 9V.

   Technician A says there may be a high resistance problem in the light.

   Technician B says the circuit may be grounded between the switch and the light.

   Who is correct?

   A. A only

   B. B only

   C. Both A and B

   D. Neither A nor B

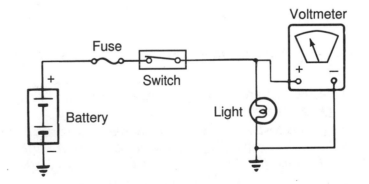

Figure 6-10 Measuring circuit voltages and voltage drops with a voltmeter *(Courtesy of Toyota Motor Corporation)*

Hint   *A voltmeter may be connected across a component in a circuit to measure the voltage drop across the component. Current must be flowing through the circuit during the voltage drop test. The amount of voltage drop depends on the resistance in the component, and the amount of current flow.*

Task 3   Check current flow in electrical/electronic circuits and components with an ammeter; determine needed repairs.

3. As indicated on the ammeter in Figure 6-11 the current flow through the light bulb is higher than specified. The cause of the high current flow could be
    A. the fuse has an open circuit.
    B. the battery voltage is low.
    C. the light bulb filament is shorted.
    D. the light bulb filament has high resistance.

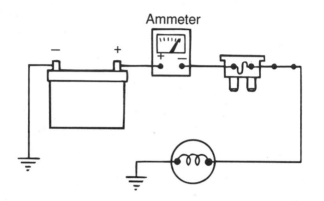

Figure 6-11   Ammeter connected to measure current flow in a circuit *(Courtesy of Chrysler Corporation)*

Hint   *An ammeter has low internal resistance, and this meter must be connected in series in a circuit. Some ammeters have an inductive clamp that fits over a wire in the circuit. These ammeters measure the current flow from the strength of the magnetic field surrounding the wire. High current flow is caused by high voltage or low resistance. Conversely, low current flow results from high resistance or low voltage.*

Task 4   Check continuity and resistances in electrical/electronic circuits and components with an ohmmeter; determine needed repairs.

4. While discussing resistance measurement with an ohmmeter
    Technician A says an ohmmeter may be connected to a circuit in which current is flowing.
    Technician B says when testing a spark plug wire with 20,000 Ω resistance, use the X100 meter scale.
    Who is correct?
    A. A only
    B. B only
    C. Both A and B
    D. Neither A nor D

Hint   *An ohmmeter has an internal power source. Meter damage may result if this meter is connected to a live circuit. The proper scale on the meter must be selected for the component being tested. For example, when testing a component with 10,000 Ω, select the X1000 scale on the meter.*

Task 5   Check electrical/electronic circuits with jumper wires; determine needed repairs.

5. The light bulb is inoperative in Figure 6-12. A jumper wire is connected from the battery positive terminal to the light bulb with the switch on, and the bulb is not illuminated. With the switch on the bulb is illuminated when a jumper wire is connected from the ground side of the light bulb to ground. The cause of the inoperative light bulb could be

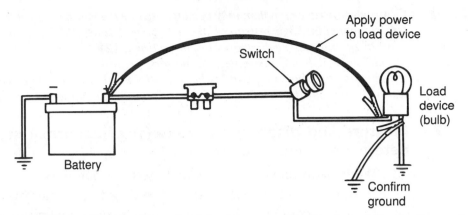

Figure 6-12   Jumper wire diagnosis of circuit components *(Courtesy of Chrysler Corporation)*

A. an open circuit in the ignition switch.
B. an open circuit in the light bulb ground.
C. an open circuit in the battery ground cable.
D. a burned out fuse.

Hint   *A jumper wire may be used to bypass a part of a circuit to locate a defect. When a component is bypassed with a jumper wire, and the circuit operation is restored to normal, the bypassed component is defective.*

Task 6   **Find shorts, grounds, opens, and high-resistance problems in electrical/electronic circuits; determine needed repairs.**

6. The light bulb in Figure 6-13 is inoperative. A 12V test light is installed in place of the fuse. When the switch is turned on the test light is on, but the light bulb remains off. When the connector near the light bulb is disconnected, the 12V test light remains illuminated.
    Technician A says the circuit may be shorted to ground between the fuse and the disconnected connector.
    Technician B says the circuit may be open between the disconnected connector and the light bulb.
    Who is correct?
    A. A only
    B. B only
    C. Both A and B
    D. Neither A nor B

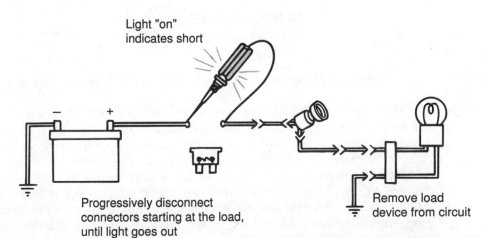

Figure 6-13   Electrical circuit diagnosis *(Courtesy of Chrysler Corporation)*

Hint   *A high-resistance problem may be diagnosed by measuring the voltage drop across various system components. High resistance in a component causes higher than specified voltage drop. A short to ground may be diagnosed by connecting a 12V test light in place of the circuit fuse. With the circuit switch on, disconnect connectors beginning at the load. When the 12V test light remains on, the short to ground is between the test light and the disconnected connector. If the test light goes out the short to ground is between the disconnected connector and the load.*

Task 7   **Measure and diagnose the cause(s) of abnormal key-off battery drain; determine needed repairs.**

7. While performing the battery drain test in Figure 6-14
   A. the switch tester switch should be closed while starting or running the engine.
   B. a battery drain of 125 milliamperes is considered normal and will not discharge the battery.
   C. the actual battery drain is recorded immediately when the switch is opened.
   D. the driver's door should be open while measuring the battery drain.

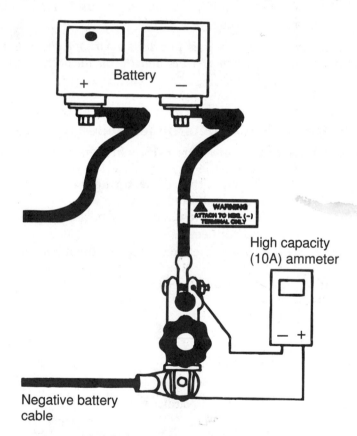

Figure 6-14   Battery drain test *(Courtesy of Pontiac Division, General Motors Corporation)*

Hint   *Many car manufacturers recommend measuring battery drain with a tester switch connected in series at the negative battery terminal. The drain test procedure must be followed in the vehicle manufacturer's service manual. A multimeter with a milliampere scale is connected parallel to the tester switch. When the tester switch is open any current drain from the battery must flow through the tester switch. Some computers require several minutes after the ignition switch is turned off before they enter the sleep mode with a reduced current drain. Therefore, after the ignition switch is turned off and the tester switch is opened, wait for the specified time before recording the milliampere reading. Some vehicle manufacturers specify a maximum battery drain of 50 milliamperes. Other vehicle manufacturers specify the battery drain is calculated by dividing the battery reserve capacity rating by 4.*

Task 8 Inspect, test, and replace fusible links, circuit breakers, and fuses.

8. A circuit breaker is removed from a power seat circuit, and an ohmmeter is connected to the circuit breaker terminals.

Technician A says the ohmmeter should provide an infinite reading if the circuit breaker is satisfactory.

Technician B says the ohmmeter current may cause the circuit breaker to open.

Who is correct?

A. A only
B. B only
C. Both A and B
D. Neither A nor B

Hint *When an ohmmeter is connected to a circuit breaker, fuse, or fuse link, the meter should read 0 ohms if the component is satisfactory. An open circuit breaker, fuse, or fuse link, causes an infinite ohmmeter reading. The current flow from an ohmmeter does not cause an automotive circuit breaker to open.*

Battery Diagnosis and Service

ASE Tasks, Questions, and Related Information

Task 1 Perform battery state-of-charge test; determine needed repairs.

9. When performing a battery hydrometer test
   A. if the battery temperature is 0°F, 0.050 should be subtracted from the hydrometer reading.
   B. if the battery temperature is 120°F, 0.020 should be subtracted from the hydrometer reading.
   C. the maximum variation in cell hydrometer readings is 0.050 specific gravity points.
   D. the battery is fully charged if all the cell hydrometer readings exceed 1.225.

Hint *When performing a battery state of charge test with a hydrometer 0.004 specific gravity points should be subtracted from the hydrometer reading for every 10°F of electrolyte temperature below 80°F. During this test 0.004 specific gravity points must be added to the hydrometer reading for every 10 degrees of electrolyte temperature above 80°F. The maximum variation is 0.050 in cell specific gravity readings. When all the cell readings exceed 1.265, the battery is fully charged.*

Task 2 Perform battery capacity (load, high-rate discharge) test; determine needed service.

10. While discussing a battery capacity test with the battery temperature at 70°F
    Technician A says the battery discharge rate is calculated by multiplying two times the battery reserve capacity rating.
    Technician B says the battery is satisfactory if the voltage remains above 9.6V.
    Who is correct?

A. A only
B. B only
C. Both A and B
D. Neither A nor B

Hint   *The battery discharge rate for a capacity test is usually one-half of the cold cranking rating. The battery is discharged at the proper rate for 15 seconds, and the battery voltage must remain above 9.6V with the battery temperature at 70°F or above.*

Task 3   Maintain or restore electronic memory functions.

11. The battery voltage is disconnected from electrical system in a vehicle with several on-board computers. This procedure may cause
    A. damage to all the computers.
    B. failure of the engine to start.
    C. erasure of the computer adaptive memories.
    D. voltage surges in the electrical system.

Hint   *If battery voltage is disconnected from a computer, the adaptive memory in the computer is erased. In the case of a powertrain control module (PCM) this action may cause erratic engine operation or erratic transmission shifting when the engine is restarted. After the vehicle is driven for 5 minutes the computer re-learns the system, and normal operation is restored. If the vehicle is equipped with memory seats or mirrors, disconnecting battery voltage also erases the memory in the computer that operates these systems. Disconnecting the battery voltage from the electrical system also erases the preprogrammed station memory in the stereo system.*

A 12V power supply from a dry-cell battery may be connected to the cigarette lighter to maintain voltage to the electrical system when the battery is disconnected.

Task 4   Inspect, clean, fill, or replace battery.

12. A maintenance-free battery is low on electrolyte, and the built-in hydrometer indicates light yellow.
    Technician A says this problem may be caused by a defective voltage regulator.
    Technician B says this problem may be caused by a loose alternator belt.
    Who is correct?
    A. A only
    B. B only
    C. Both A and B
    D. Neither A nor B

Hint   *A battery may be cleaned with a baking soda and water solution. If the built-in hydrometer indicates light yellow or clear, the electrolyte level is low, and the battery should be replaced. The low electrolyte level may be caused by a high voltage regulator setting that causes overcharging. When disconnecting battery cables always disconnect the negative battery cable first.*

Task 5   Perform slow/fast battery charge in accordance with manufacturer's recommendations.

13. While charging batteries
    A. battery charge time is the same batteries with different capacities.
    B. the battery temperature should not exceed 125°F while charging.
    C. a high charging rate may be used to charge a battery at -20°F battery temperature.
    D. the battery may be fully charged at a high rate on a fast charger.

Hint   *If the battery is charged in the vehicle, the battery cables should be disconnected during the charging procedure. The charging time depends on the battery state of charge and the battery capacity. If the battery temperature exceeds 125°F while charging, the battery may be damaged. When fast charging a battery, reduce the charging rate when the specific gravity reaches 1.225 to avoid excessive battery gassing. The battery is fully charged when the specific gravity increases to 1.265. Do not attempt to fast charge a cold battery.*

Task 6   **Inspect, clean and repair or replace battery cables, connectors, clamps, and hold downs.**

Task 7   **Jump start a vehicle with jumper cables and a booster battery or auxiliary power supply.**

14. While jump starting a vehicle with a booster battery
    Technician A says the accessories should be on in the boost vehicle while starting the vehicle being boosted.
    Technician B says the negative booster cable should be connected to an engine ground on the vehicle being boosted.
    Who is correct?
    A. A only
    B. B only
    C. Both A and B
    D. Neither A nor B

Hint   *The accessories must be off in both vehicles during the boost procedure. The negative booster cable must be connected to an engine ground in the vehicle being boosted. Always connect the positive booster cable followed by the negative booster cable, and complete the negative cable connection last on the vehicle being boosted. When disconnecting the booster cables, remove the negative booster cable first on the vehicle being boosted.*

Starting System Diagnosis and Service

ASE Tasks, Questions, and Related Information

Task 1   **Perform starter current draw test; determine needed repairs.**

15. During a starter current draw test the current draw is more than specified, and the cranking speed and battery voltage are less than specified. The cause of this problem may be
    A. worn bushings in the starting motor.
    B. high resistance in the field windings.
    C. high resistance in the battery positive cable.
    D. a burned solenoid disc and terminals.

Hint   *High starter current draw, low cranking speed, and low cranking voltage usually indicate a defective starter. This condition also may be caused by internal engine problems such as partially seized bearings.*

      Low current draw, low cranking speed, and high cranking voltage usually indicate excessive resistance in the starting circuit.

Task 2   Perform starter circuit voltage drop tests; determine needed repairs.

16. In Figure 6-15 the voltmeter is connected to test the voltage drop across
    A. the positive battery cable.
    B. the starter solenoid windings.
    C. the starter ground circuit.
    D. the starter solenoid disc and terminals.

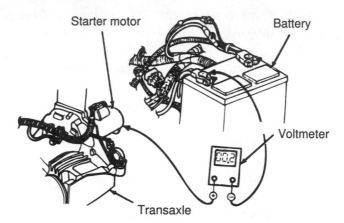

Figure 6-15   Starter circuit voltage drop test *(Courtesy of Chrysler Corporation)*

Hint   *Measure the voltage drop across each component in the starter circuit to check the resistance in that part of the circuit. The ignition and fuel systems must be disabled while making these tests. Read the voltage drop across each component while the starting motor is operating. For example, connect the voltmeter leads to the positive battery terminal and the positive cable on the starter solenoid, and crank the engine to measure the voltage drop across the positive battery cable.*

Task 3   Inspect, test, and repair or replace switches, connectors, and wires of starter control circuits.

17. In the starter circuit in Figure 6-16 the battery is fully charged and the starter relay and solenoid are completely inoperative when the ignition switch is in the start position. The cause of this problem could be
    A. a grounded circuit at terminal 85 on the starter relay.
    B. an open circuit at terminal 86 on the starter relay.
    C. a continually closed neutral safety switch.
    D. a slight resistance at terminal 87 on the starter relay.

Hint   *Relays and switches in the starting motor circuit may be tested with an ohmmeter. When an ohmmeter is connected across the relay or switch contacts, the meter should provide an infinite reading if the contacts are open. If the relay or switch contacts are closed the ohmmeter reading should be at or near zero. When the ohmmeter leads are connected across the terminals connected to the relay winding, the meter should indicate the specified resistance. A resistance below the specified value indicates a shorted winding, whereas an infinite reading proves the winding is open.*

Task 4   Inspect, test, and replace starter relays and solenoids.

18. While discussing the solenoid winding test in Figure 6-17
    Technician A says the ohmmeter is connected to test the solenoid pull-in winding.
    Technician B says if the winding is satisfactory the ohmmeter should provide an infinite reading.
    Who is correct?
    A. A only
    B. B only
    C. Both A and B
    D. Neither A nor B

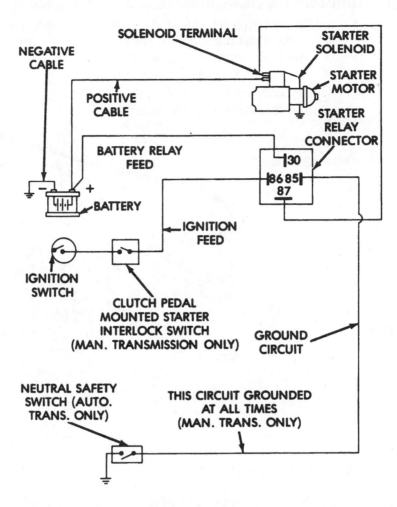

Figure 6-16   Starter relay circuit *(Courtesy of Chrysler Corporation)*

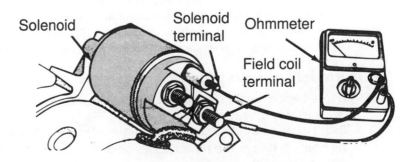

Figure 6-17   Starter solenoid test *(Courtesy of Chrysler Corporation)*

Hint   *The ohmmeter leads must be connected across the solenoid terminal and the field coil terminal to test the pull-in winding. Connect the ohmmeter leads from the solenoid terminal to ground to test the hold-in winding.*

Task 5   Remove and replace starter.

Task 6   Disassemble, clean, inspect, test, and replace starter components.

19. In Figure 6-18 a self-powered test light is used to test the starter field coils, and the test light is not illuminated. The cause of this test result could be
   A.   the fields coils are satisfactory.
   B.   the field coils have a grounded circuit.
   C.   the field coils have a shorted condition.
   D.   the field coils have an open circuit.

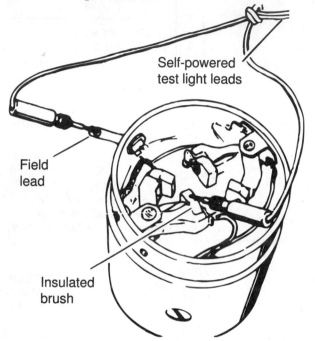

Figure 6-18   Field coil tests *(Courtesy of Pontiac Division, General Motors Corporation)*

Hint   *The test light leads are connected to the field coil lead and one of the insulated brushes to test the field coils for an open circuit. If the test light is not illuminated, the field coils have an open circuit. Connect the test light leads from the field coil lead to ground on the starter housing to test the field coils for a grounded condition. When the test light is illuminated, the field coils are grounded.*

Connect the test light leads from the commutator bars to the shaft to test the armature for a grounded condition. If the test light is illuminated, the armature is grounded. Rotate the armature in a growler with a hacksaw blade held above the core. When the blade vibrates against the core, the armature windings or commutator bars are shorted.

Task 7   Perform free-running (bench) tests; determine needed repairs.

20. While bench testing a starting motor the current draw is more than specified and the starter rpm is less than specified.
   Technician A says the starter may have a bent armature shaft.
   Technician B says the starter series field coils may have an open circuit.
   Who is correct?
   A.   A only
   B.   B only
   C.   Both A and B
   D.   Neither A nor B

Hint   *During a starter bench test if the starter current is more than specified and the starter rpm is less than specified, the field coils or armature may have an electrical defect. These test results also may be caused by a bent armature shaft or worn bushings. If the starter does not rotate and the current draw is zero, the starter has an open circuit in the field coils, armature, or brushes.*

Charging System Diagnosis and Repair

ASE Tasks, Questions, and Related Information

Task 1   **Diagnose charging system problems that cause an undercharge, a no-charge, or an overcharge condition.**

21. When discussing an alternator with zero output
Technician A says the alternator field circuit may have an open circuit.
Technician B says the fuse link may be open in the alternator to battery wire.
Who is correct?
    A. A only
    B. B only
    C. Both A and B
    D. Neither A nor B

Hint   *A low-charging voltage caused by a defective voltage regulator or alternator results in a reduced charging rate and an undercharged battery. This problem also may be caused by a loose alternator belt or excessive resistance in the wire from the alternator battery terminal to the positive battery terminal. An overcharged battery usually is caused by a defective voltage regulator that allows high charging circuit voltage. A no-charge condition may be caused by an open alternator field circuit, or an open fuse link in the wire from the alternator battery terminal to the positive battery terminal.*

Task 2   **Inspect, adjust, and replace alternator drive belts, pulleys, and fans.**

22. An alternator with a 90 ampere rating produces 45 amperes during an output test. The alternator is driven with a V-belt and the belt has the specified tension.
Technician A says the V-belt may be worn and bottomed in the pulley.
Technician B says the alternator pulley may be misaligned with the crankshaft pulley.
Who is correct?
    A. A only
    B. B only
    C. Both A and B
    D. Neither A nor B

Hint   *An undercharged battery may be caused by a slipping alternator belt. A slipping belt may be caused by insufficient belt tension or a worn, glazed, or oil-soaked belt. Belt tension may be tested with a belt tension gauge or by measuring the belt deflection in the center of the belt span. A belt should have 1/2 in. of deflection for every foot of free span. Many ribbed V-belts have an automatic spring-loaded tensioner with a belt wear scale.*

Task 3   **Perform charging system output test; determine needed repairs.**

23. While discussing a charging system output test
Technician A says the vehicle accessories should be on during the test.
Technician B says the charging system voltage should be limited to 17V.
Who is correct?
    A. A only
    B. B only
    C. Both A and B
    D. Neither A nor B

Hint   *The alternator belt tension and condition should be checked before an output test is performed. Turn off the vehicle accessories during the test. If the alternator is full-fielded during the output test, a carbon pile load in the volt-ampere tester must be used to maintain the voltage below 15V. The alternator output may be tested by lowering the voltage to the voltage specified by the vehicle manufacturer.*

Task 4 Perform alternator output test; determine needed repairs.

24. During an output test using the full-field method, a 100 ampere alternator with an integral electronic regulator produces 30 amperes. The cause of the low alternator output could be
    A. a shorted diode in the alternator.
    B. a broken brush lead wire in the alternator.
    C. an open circuit in the voltage regulator.
    D. a defective alternator capacitor.

Hint   *When the alternator output is zero during an output test, the field circuit is probably open. This problem may be caused by worn brushes or an open field winding in the rotor.*
   If the alternator output is less than specified, there is probably a defect in the diodes or stator. A high resistance in the field winding also reduces output. When the alternator is full-fielded to test output, the voltage regulator is bypassed and does not affect output.

Task 5 Inspect, test, repair, or replace voltage regulator; determine needed repairs.

25. The charging system voltage on a vehicle is 16.2V. This condition may cause all the following problems EXCEPT
    A. an overcharged battery.
    B. burned-out electrical components.
    C. electrolyte gassing in the battery.
    D. reduced headlight brilliance.

Hint   *When the alternator voltage is erratic or too low, the alternator may be full-fielded to determine the cause of the problem. When the alternator is full-fielded and the alternator current and voltage output are normal, the voltage regulator is probably defective. If the charging system voltage is higher than specified, the voltage regulator probably is defective. On a charging system with an external regulator, this problem may be caused by excessive resistance in the field circuit between the ignition switch and the regulator.*

Task 6 Perform charging circuit voltage drop tests; determine needed repairs.

26. With the ammeter and voltmeter connected to the charging system as indicated in Figure 6-19, the voltmeter indicates 2V and the ammeter reads 10 amperes.
    Technician A says this condition may cause an undercharged battery.
    Technician B says this condition may result in head lamp flare-up during acceleration.
    Who is correct?
    A. A only
    B. B only
    C. Both A and B
    D. Neither A nor B

Hint   *A voltmeter may be connected from the alternator battery wire to the positive battery terminal to measure voltage drop in the charging circuit. Many car manufacturers recommend a 10 ampere charging rate while measuring this voltage drop. When the voltage drop is more than specified, the circuit resistance is excessive. High charging circuit resistance between the alternator battery terminal and the positive battery terminal may cause an undercharged battery.*

Task 7 Inspect, repair, or replace connectors and wires of charging circuits.

Task 8 Remove and replace alternator.

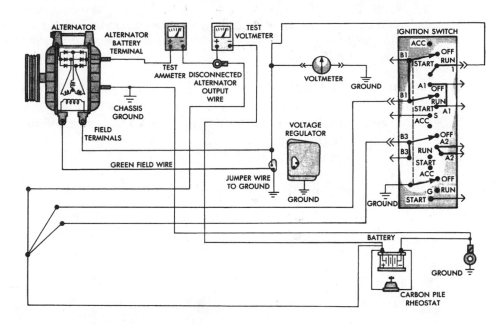

Figure 6-19 Charging system voltage test *(Courtesy of Chrysler Corporation)*

27. The charging system in Figure 6-20 has a glowing charge indicator bulb with the engine running. An alternator output test indicates satisfactory output.

    Technician A says there may be excessive resistance in the wire from the alternator battery terminal to the positive battery terminal.

    Technician B says there may be excessive resistance in the wire from the alternator L terminal to the charge indicator bulb.

    Who is correct?

    A. A only
    B. B only
    C. Both A and B
    D. Neither A nor B

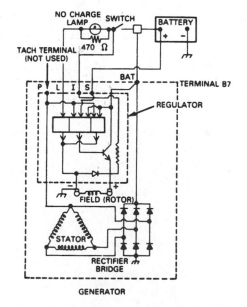

Figure 6-20 Charging system with integral electronic regulator *(Courtesy of Oldsmobile Division, General Motors Corporation)*

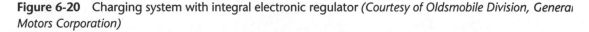

Hint   *Disconnect the negative battery terminal before removing the alternator. If the vehicle is equipped with an air bag system, wait for the time specified by the vehicle manufacturer, and then begin the alternator removal procedure.*

Task 9   Disassemble, clean, inspect, test, and replace alternator components.

28. When testing diodes connect the ohmmeter leads across each diode and then reverse the leads (Figure 6-21). A satisfactory diode provides
    A. one high meter reading and one low reading.
    B. two infinite meter readings.
    C. two low meter readings.
    D. a meter reading of 0 Ω and 25 Ω.

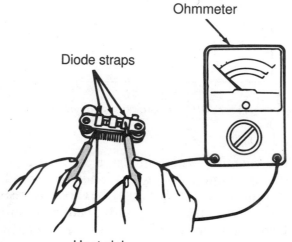

Figure 6-21 Diode testing with an ohmmeter *(Courtesy of Ford Motor Company)*

Hint   *Connect the ohmmeter leads to each pair of stator leads to test the stator for an open condition. When the ohmmeter leads are connected from one of the stator leads to the stator frame, the stator is tested for a grounded condition.*

The ohmmeter leads may be connected across the rotor slip rings to test the field winding for an open or a shorted condition. Connect the ohmmeter leads from one of the slip rings to the shaft to test the field winding and slip rings for a grounded condition.

Connect the ohmmeter leads across each diode, and then reverse the leads to test the diodes. A satisfactory diode provides one low and one high ohmmeter reading. A shorted diode is indicated by two low meter readings, and an open diode provides two infinite readings.

Lighting Systems Diagnosis and Repair, Headlights, Parking Lights, Taillights, Dash Lights, and Courtesy Lights

ASE Tasks, Questions, and Related Information

Task 1   **Diagnose the cause of brighter than normal, intermittent, dim, or no operation of headlights.**

29. The headlights on a vehicle go out intermittently and come back on in a few minutes. Technician A says this problem may be caused by an intermittent short to ground. Technician B says this problem may be caused by high charging system voltage. Who is correct?
    A. A only
    B. B only
    C. Both A and B
    D. Neither A nor B

Hint   *If the headlights are inoperative check the circuit breakers or fuses. Many headlight circuits have a circuit breaker in the headlight switch. If the headlights are dim, check for resistance in the headlight circuit or low charging system voltage. When the headlight operation is intermittent, check for an intermittent open circuit in the headlight wiring, dimmer switch, or headlight switch. Intermittent headlight operation also may be caused by a shorted condition or a short to ground. Either of these conditions cause excessive current flow in the circuit that may cause the circuit breaker to open. This action turns off the headlights. When the circuit breaker cools, the headlights come back on.*

Task 2   **Inspect, replace, and aim headlights.**

30. All of these statements about halogen headlight bulb replacement are true EXCEPT
    A. handle the bulb only by its base.
    B. do not drop or scratch the bulb.
    C. change the bulb with the headlights on.
    D. keep moisture away from the bulb.

Hint   *Always turn off the headlights and allow the bulb to cool before changing a halogen bulb. Keep moisture away from the bulb and handle the bulb by its base. Do not scratch or drop the bulb. Many manufacturers recommend using a headlight screen to align the headlights. The vehicle is parked 25 ft from the screen on a level floor (Figure 6-22, next page). Many vehicle manufacturers recommend aligning the low beams, and the high beams are considered nonadjustable. On some older vehicles the manufacturers recommended aligning the high beams.*

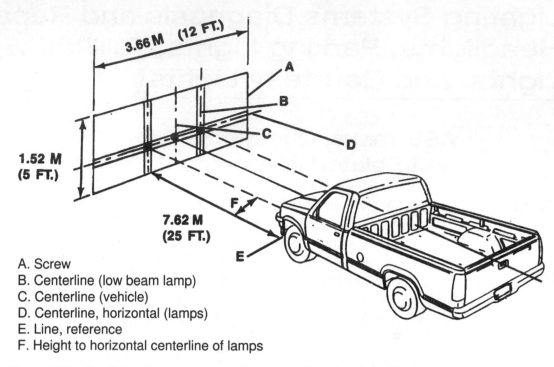

A. Screw
B. Centerline (low beam lamp)
C. Centerline (vehicle)
D. Centerline, horizontal (lamps)
E. Line, reference
F. Height to horizontal centerline of lamps

Figure 6-22 Headlight alignment screen *(Courtesy of Chevrolet Motor Division, General Motors Corporation)*

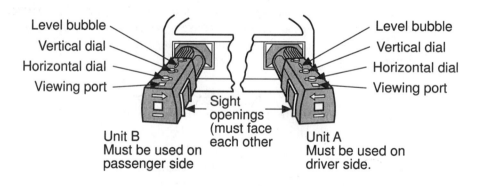

Figure 6-23 Headlight aimers *(Courtesy of Chrysler Corporation)*

On some older vehicles the manufacturers recommended measuring headlight alignment with headlight aimers (Figure 6-23). These aimers are held on the headlights with a suction cup. The headlights are adjusted until the level bubble in the aimers is in the specified position. Some vehicle manufacturers recommended measuring headlight alignment with a photoelectric aimer that measured headlight beam intensity. Headlight adjustments are provided on each headlight unit (Figure 6-24 and Figure 6-25).

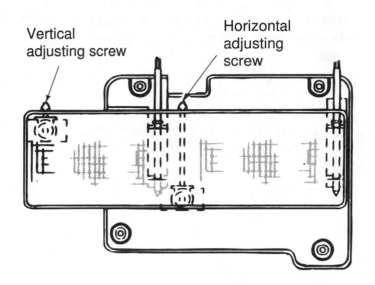

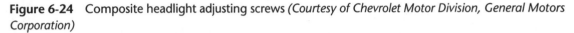

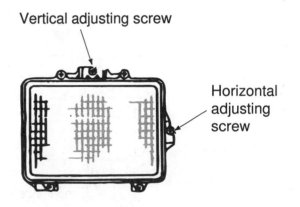

Figure 6-24   Composite headlight adjusting screws *(Courtesy of Chevrolet Motor Division, General Motors Corporation)*

Figure 6-25   Sealed beam headlight adjustment screws *(Courtesy of Chevrolet Motor Division, General Motors Corporation)*

Task 3 · Inspect, test, and repair, or replace headlight and dimmer switches, relays, control units, sensors, sockets, connectors, and wires of headlight circuits.

31. An open circuit in fuse number 12 in Figure 6-26 could result in
    A. inoperative taillights.
    B. inoperative stoplights.
    C. inoperative instrument panel lights.
    D. inoperative low-beam headlights

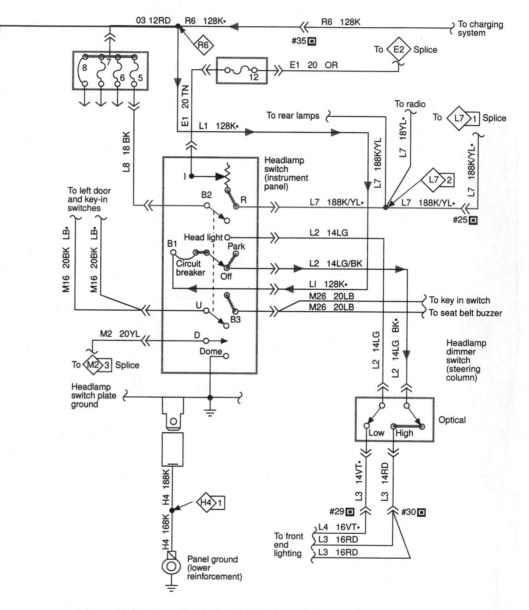

Figure 6-26 Headlight circuit *(Courtesy of Chrysler Corporation)*

Hint *Many headlight switches contain a circuit breaker that is connected in the headlight circuit. Other light circuits such as the taillights, stoplights, or instrument panel lights have separate fuses. The dimmer switch usually is part of the multifunction switch in the steering column. The dimmer switch is connected in series between the headlight switch and the headlights. When the headlights are turned on, the dimmer switch directs the current flow to the low beam or high beam headlights.*

Task 4   Diagnose the cause of intermittent, slow, or no operation of retractable headlight assembly.

Task 5   Inspect, test, and repair or replace motors, switches, relays, connectors, and wires of retractable headlight assembly circuits.

32. Terminal A has an open circuit on the top side of the headlight doors module (Figure 6-27).

    Technician A says this problem may cause both headlight doors to be inoperative.

    Technician B says this problem may damage the headlight doors module.

    Who is correct?

    A. A only
    B. B only
    C. Both A and B
    D. Neither A nor B

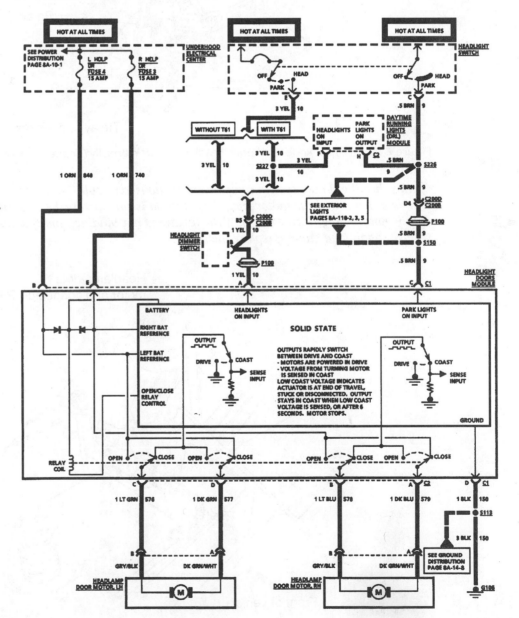

Figure 6-27   Headlight doors wiring diagram *(Courtesy of Pontiac Division, General Motors Corporation)*

Hint   When the driver turns the headlights on, voltage is supplied from the headlight switch to terminal A on the headlight doors module. In response to this signal the headlight doors module supplies voltage to both headlight door motors. This action operates both motors to open the headlight doors. If the headlights are shut off, the module reverses the motor action to close the doors.

Some headlight door motors have a manual knob on the headlight door motors (Figure 6-28). If the motors do not open the doors, this knob may be rotated to lift the doors.

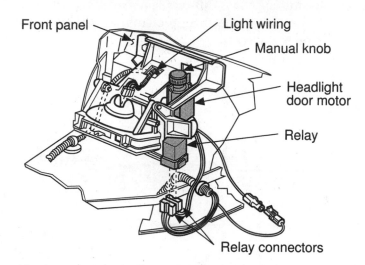

Figure 6-28   Manual knob used open headlight doors manually (*Courtesy of General Motors Corporation*)

Most headlight door system components are nonserviceable. For example, if voltage is supplied to a door motor and the motor ground connection is satisfactory, be sure the motor linkage to the door is not binding (Figure 6-29). If the linkage is not binding, and the motor has a satisfactory voltage supply and ground, replace the motor.

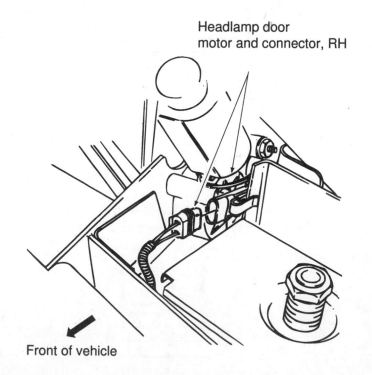

Figure 6-29   Headlight door motor (*Courtesy of Pontiac Division, General Motors Corporation*)

Task 6   Diagnose the cause of brighter than normal intermittent, dim, or no operation of parking lights and/or taillights.

Task 7   Inspect, test, and repair or replace switches, relays, bulbs, sockets, connectors, and wires of parking light and taillight circuits.

33. The rear light ground connection on the left side of Figure 6-30 has an open circuit. The ground connection on the right side of Figure 6-30 is satisfactory. This problem could result in
   A. inoperative left rear tail, stop, and side marker lights.
   B. no change in the rear light operation.
   C. inoperative left rear tail and stop lights.
   D. inoperative backup lights.

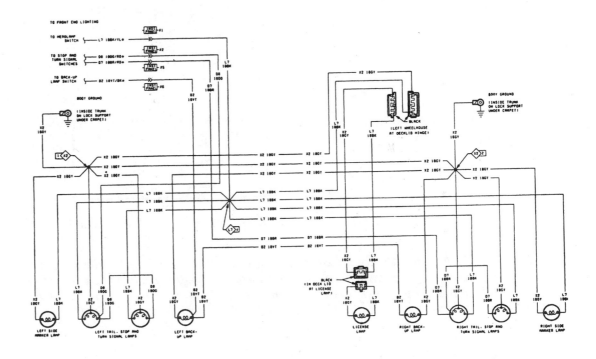

Figure 6-30   Rear lighting circuit *(Courtesy of Chrysler Corporation)*

Hint   *Many rear light bulbs are a combination bulb containing stop and taillight filaments. Backup light bulbs and side marker bulbs are single-filament. When headlight switch is turned to the park or headlight position, voltage is supplied from this switch to the taillight bulbs and side marker bulbs. Most rear lights share a common ground connection. When the brakes are applied, voltage is supplied from the brake light switch to the stop light bulbs. If the gear selector is placed in reverse, voltage is supplied from the backup light switch to the backup light bulbs.*

If any bulb is dim, there is a resistance problem in the voltage supply wire or ground wire connected to the bulb. When the engine is accelerated and all the lights are brighter than normal, the charging system voltage is higher than specified.

Task 8   Diagnose the cause of intermittent, dim, no lights, or no brightness control or instrument lighting circuits.

Task 9
Inspect, test, and repair or replace switches, relays, bulbs, sockets, connectors, wires, and printed circuit boards of instrument lighting circuits.

34. While discussing the instrument panel lights in Figure 6-31
    Technician A says an open circuit in the rheostat may cause all the bulbs to be inoperative.
    Technician B says an open circuit in one of the bulbs may cause all the bulbs to be inoperative.
    Who is correct?
    A. A only
    B. B only
    C. Both A and B
    D. Neither A nor B

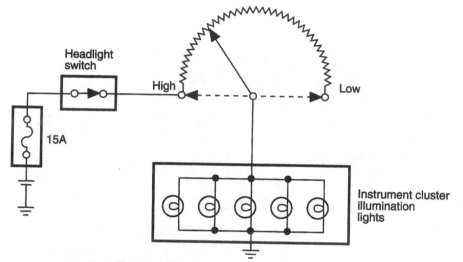

Figure 6-31   Instrument panel light circuit

Hint   *A rheostat is connected in series with the instrument panel bulbs. This rheostat is operated by the headlight switch knob or by a separate control knob. When the rheostat control knob is rotated, the voltage to the instrument panel bulbs is reduced. This action lowers the current flow and reduces the brilliance of the bulbs. The instrument panel bulbs are connected parallel to the battery. If one bulbs burns out, the other bulbs remain illuminated.*

Task 10
Diagnose the cause of intermittent, dim, or no operation of courtesy lights (dome, map, vanity).

Task 11
Inspect, test, and repair or replace switches, relays, bulbs, sockets, connectors, and wires of courtesy lights (dome, map, vanity).

35. Circuit 156 is shorted to ground at terminal S363 in Figure 6-32. This problem may cause
    A. continual operation of the courtesy lights.
    B. no operation of the courtesy lights and lighted mirror.
    C. continual operation of the underhood light.
    D. a burned out courtesy light fuse.

Hint   *Some courtesy light circuits have ground-side switches. In these circuits voltage is supplied from the positive battery terminal through a fuse to the courtesy light bulbs. When a door is opened one of the door jamb switches closes. This switch provides a ground for the courtesy light bulbs.*

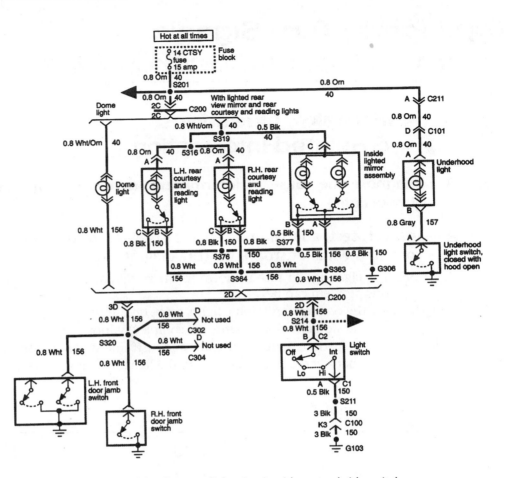

Figure 6-32 Courtesy light circuit with ground-side switches

In other courtesy light circuits the switches are connected on the insulated side of the circuit between the battery positive terminal and the courtesy light bulbs. A ground wire is connected from each bulb to ground (Figure 6-33).

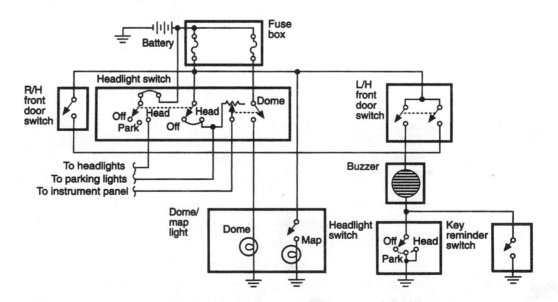

Figure 6-33 Courtesy light circuit with insulated-side switches.

Stop Lights, Turn Signals, Hazard Lights, and Backup Lights

ASE Tasks, Questions, and Related Information

Task 1   Diagnose the cause of intermittent, dim, or no operation of stop light (brake light).

Task 2   Inspect, test, adjust, and repair or replace switch, bulbs, sockets, connectors, and wires of stop light (brake light circuits).

36. The cigar lighter fuse is blown in the stop light circuit in Figure 6-34. The result of this problem can be
    A.   the courtesy and dome lights come on dimly when the cigar lighter is pushed in.
    B.   the stop and dome lights are completely inoperative.
    C.   the parking lights, taillights, and instrument panel lights are inoperative.
    D.   a battery drain occurs with all the light switches in the off position.

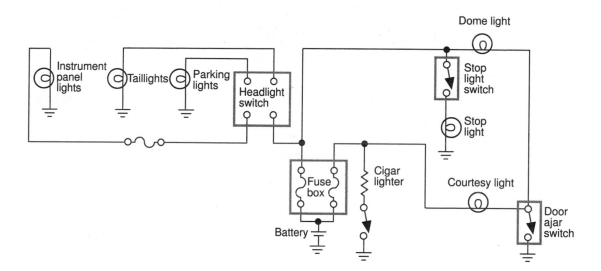

Figure 6-34   Stop light and cigar lighter circuits

Hint   *If the cigar lighter is pushed in and the cigar lighter fuse is blown, current flows through the dome light, courtesy light and cigar lighter to ground (Figure 6-35). Since these lights are now in series with the cigar lighter, the lights glow dimly.*

In many stop light circuits voltage is supplied to the brake light switch from the battery positive terminal. When the brakes are applied, brake pedal movement closes the stop light switch. This action supplies voltage to the stop lights and the collision avoidance light (Figure 6-36). In many stop light systems the stop light filaments are mounted in the same bulb as the taillight filaments.

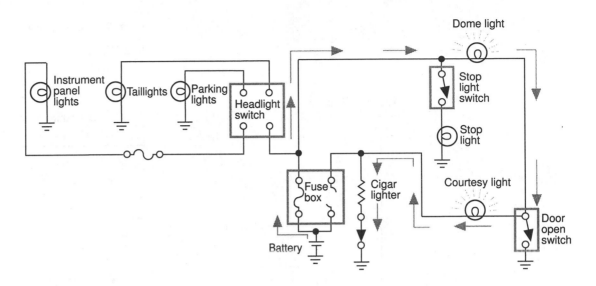

Figure 6-35   Stop light and cigar lighter circuits with blown fuse.

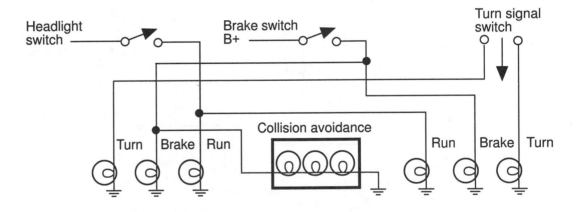

Figure 6-36   Stop light circuit with collision avoidance light.

Task 3   **Diagnose the cause of no turn signal and hazard lights, or lights with no flash on one or both sides**

Task 4   **Inspect, test, and repair or replace switches, flasher units, bulbs, sockets, connectors, and wires of turn signal and hazard light circuits.**

37. In the signal light circuit in Figure 6-37 (next page) the right rear signal light is dim, and all the other lights work normally. The cause of this problem may be
   A. high resistance in the DB 180G RD wire from the signal light switch to the rear lamp wiring.
   B. a short to ground in the DB 180G RD wire from the signal light switch to the rear lamp wiring.
   C. high resistance in the D7 18BR RD wire from the signal light switch to the rear lamp wiring.
   D. high resistance in the D2 18 RD wire from the signal light flasher to the switch.

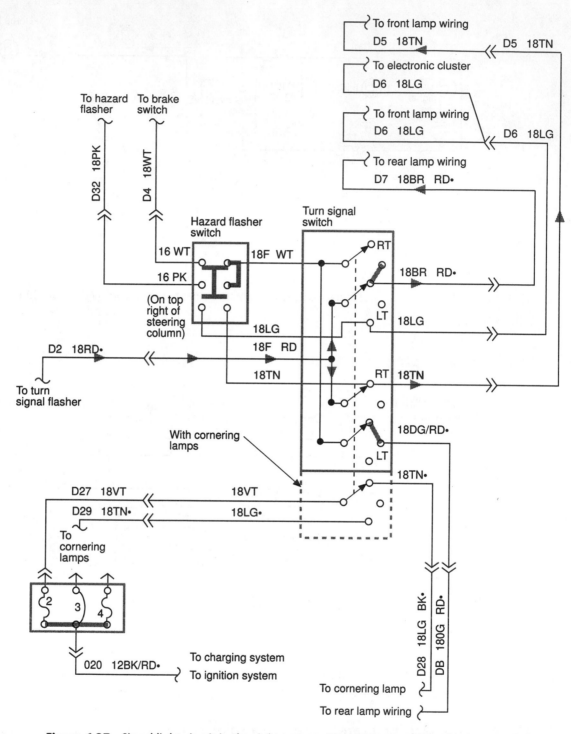

Figure 6-37 Signal light circuit in the right turn position *(Courtesy of Chrysler Corporation)*

Hint   *The signal light flasher contains a bimetallic heating strip surrounded by a heating coil. One flasher contact is mounted on the bimetallic strip, and the other contact is stationary. When the ignition switch is turned on, voltage is supplied through the flasher contacts to the signal light switch (Figure 6-38). This switch directs the voltage to the left or right signal light bulbs depending on the signal light lever position selected by the driver. When current starts flowing through the flasher, the heat from the heating coil bends the bimetallic strip and opens the flasher contacts. The bimetallic strip cools allowing the contacts to close, and this action repeats to provide a flashing action. If the brake pedal is applied during a right turn, the left brake light is illuminated.*

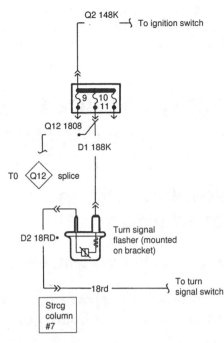

Figure 6-38 Signal light flasher *(Courtesy of Chrysler Corporation)*

When the hazard switch is pressed, voltage is supplied from the hazard flasher through the hazard switch and signal light switch to the front and rear signal lights. The hazard flasher has the same internal design as the conventional flasher.

Task 5   **Diagnose the cause of intermittent, dim, or no operation of backup light.**

Task 6   **Inspect, test, and repair or replace switch, bulbs, sockets, connections, and wires of backup light circuits.**

38. The right-hand backup light circuit is grounded on the switch side of the bulb (Figure 6-39).
    Technician A says this condition may blow the backup light fuse.
    Technician B says the left-hand backup light may work normally while the right-hand backup light is inoperative.
    Who is correct?

    A. A only
    B. B only
    C. Both A and B
    D. Neither A nor B

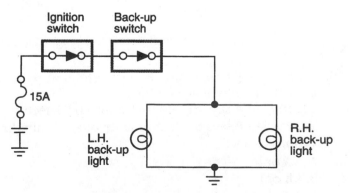

Figure 6-39 Back-up light circuit

Hint *When the ignition switch is on and the backup light switch is closed, voltage is supplied through these switches to the backup lights. The backup light switch is operated by the gear selector linkage. This switch may be mounted on the steering column or transmission.*

Gauges, Warning Devices, and Driver Information Systems Diagnosis and Repair

ASE Tasks, Questions, and Related Information

Task 1 Diagnose the cause of intermittent, high, low, or no gauge readings.

Task 2 Inspect, test and repair or replace gauges, gauge sending units, connectors, wires, and printed circuit boards of gauge circuits.

39. The fuel gauge in Figure 6-40 reads lower than the actual level of fuel in the tank. All the other gauges operate normally. The cause of this problem may be
    A. high resistance in the sending unit ground wire.
    B. high resistance between the instrument voltage limiter and the gauge.
    C. a short to ground between the gauge and the sending unit.
    D. an open circuit in the wire from the gauge to the sending unit.

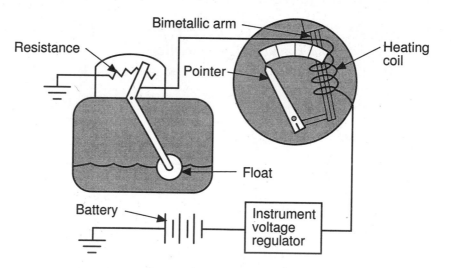

Figure 6-40 Thermal-electric fuel gauge *(Courtesy of Chrysler Corporation)*

40. All the gauges are erratic in an instrument panel with thermal-electric gauges and an instrument voltage limiter.
    Technician A says the alternator may be defective.
    Technician B says the instrument voltage limiter may be defective.
    Who is correct?
    A. A only
    B. B only
    C. Both A and B
    D. Neither A nor B

41. On a two-coil temperature gauge the gauge pointer remains in the hot position regardless of the engine temperature (Figure 6-41). The cause of this problem could be
   A. the wire may be open from the hot coil to the sending unit.
   B. the wire may be open from the cold coil to ground.
   C. the sending unit may have excessive internal resistance.
   D. an excessively high resistance between the sending unit and ground.

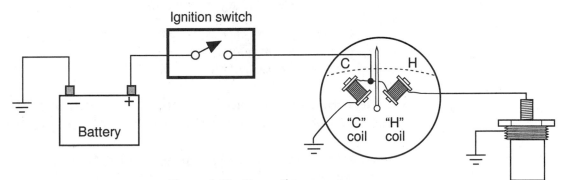

Figure 6-41   Two-coil temperature gauge

Hint   *Many vehicles are equipped with thermal-electric gauges. These gauges contain a bimetallic strip surrounded by a heating coil. The pivoted gauge pointer is connected to the bimetallic strip. The sending unit contains a variable resistor. In a fuel gauge this variable resistor is connected to a float in the fuel tank. if the tank is filled with fuel, the sending unit resistance decreases, and current flow through the bimetallic strip increases. This increased current flow heats the bimetallic strip and pushes the pointer toward the full position.*

The voltage limiter supplies about 5V to the gauges regardless of the charging system voltage. If the voltage limiter voltage is higher than specified, all the gauges have high readings. A defective voltage limiter may also cause low or erratic readings on all the gauges. The voltage limiter requires a ground connection through the instrument panel. High resistance in the instrument panel ground reduces heating coil current in the voltage limiter. This action allows the limiter contacts to remain closed longer. Under this condition voltage output from the limiter increases and gauge readings are higher.

Some gauges contain two coils, and the pointer is mounted on magnet under these coils. In a temperature gauge the sending unit is connected to the hot coil and the cold coil is grounded. If the coolant is cold the sending unit has high resistance. Under this condition current flows through the lower resistance of the cold coil. Coil magnetism around the cold coil attracts the magnet and pointer to the cold position. As the coolant temperature increases the sending unit resistance decreases. When the engine is at normal operating temperature, the current flows through the lower resistance of the hot coil and sending unit. This action attracts the magnet and pointer to the hot position.

Task 3   **Diagnose the cause(s) of intermittent, high, low, or no readings on electronic digital instrument circuits.**

Task 4   **Inspect, test, repair or replace sensors, sending units, connectors, and wires of electronic instrument circuits.**

42. When the ignition switch is turned on, most of the electronic instrument displays are brightly illuminated, but a few of the displays are not illuminated.
   Technician A says the inputs for the nonilluminated displays may be defective.
   Technician B says the electronic instrument display may be defective.
   Who is correct?

       A. A only
       B. B only
       C. Both A and B
       D. Neither A nor B

Hint   *Many electronic instrument displays provide an initial illumination of all segments when the ignition switch is turned on. This illumination proves the operation of the display segments. During this initial display all the segments in the electronic instrument displays should be brightly illuminated for a few seconds. If some of the segments are not illuminated, replace the electronic instrument cluster. When none of the segments are illuminated, check the fuses and voltage supply to the display.*

    Many electronic instrument displays have self-diagnostic capabilities. In some electronic instrument displays a specific gauge illumination or digital displays indicate certain defects in the display. Other electronic instrument displays may be diagnosed with a scan tester.

Task 5 Diagnose the cause of constant, intermittent, or no operation of warning lights, indicator lights, and other driver information systems.

Task 6 Inspect, test, and repair or replace bulbs, sockets, connectors, wires, and electronic components of warning light/driver information system circuits.

43. The door ajar light in the message center is illuminated continually with the ignition switch on (Figure 6-42). The cause of this problem may be
    A. an open circuit in wire between BCM terminal 3 and the door ajar switches.
    B. a short to ground on the BCM side of the RF door ajar switch.
    C. a defective LR door ajar switch that never moves to the closed position.
    D. an open circuit from all the door ajar switches to the chassis ground.

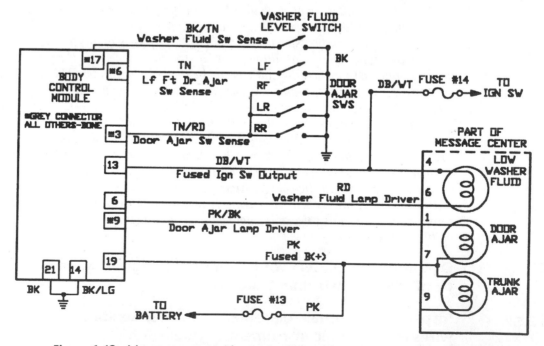

Figure 6-42 Message center with warning lights *(Courtesy of Chrysler Corporation)*

Hint   *Some warning lights are operated by the body control module (BCM). The door ajar switches and the low washer fluid switch send an input signal to the BCM if a door is ajar or the washer reservoir is low. When one of these signals is received, the BCM grounds the appropriate bulb.*

Task 7   Diagnose the cause of constant, intermittent, or no operation of audible warning devices.

44. The seat belt buzzer and the seat belt light operate continually with the ignition switch on and the driver's seat belt buckled in the seat belt and key buzzer system (Figure 6-43). The cause of this problem may be
   A. the timer contacts are stuck in the closed position.
   B. the circuit is shorted to ground at terminal 3 on the buzzer.
   C. the timer contacts and the seat belt switch are stuck closed.
   D. the circuit is open at terminal 2 on the buzzer relay.

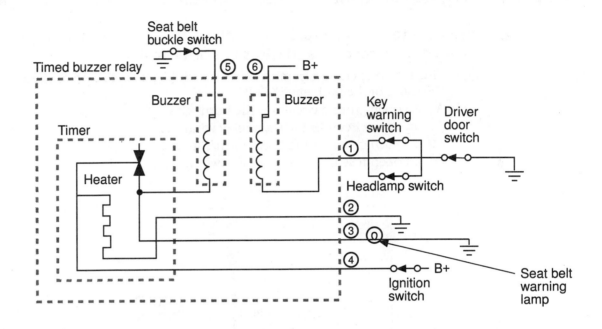

Figure 6-43   Buzzer relay circuit *(Courtesy of Chrysler Corporation)*

Hint   *Some buzzer relays contain the seat belt and the key buzzer. If the key is left in the ignition switch or the headlights are on, and the driver's door is opened, the circuit is completed from buzzer terminal 1 through the key warning switch, or headlight switch, and the driver door switch to ground. Under this condition current flows through the circuit and the buzzer is activated.*

When the ignition switch is turned on, the current flows through the timer, seat belt buzzer, and seat belt buckle switch to ground. Under this condition the buzzer is activated. Current also flows from the timer through the seat belt warning light to ground. When the driver's seat belt is buckled, the buzzer circuit is open and the buzzer is deactivated. The heater opens the timer contacts after 8 seconds and the light goes out.

Horn and Wiper/Washer Diagnosis and Repair

ASE Tasks, Questions, and Related Information

Task 1   Diagnose the cause of constant intermittent, or no operation of horn(s).

Task 2   Inspect, test, and repair or replace horn(s), horn relay, horn button (switch), connectors, and wires of horn circuits.

45. There is no operation from the horn circuit (Figure 6-44). All of these defects may be the cause of the problem EXCEPT
    A. an open ground circuit on the horn relay.
    B. an open circuit in the horn relay winding.
    C. an open circuit at the horn brush/slip ring.
    D. an open fuse link in the relay power wire.

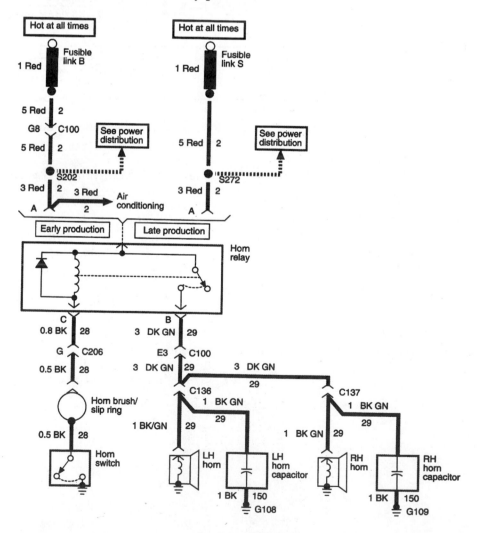

Figure 6-44   Horn circuit

Hint   *Many horn circuits contain a relay. Voltage is supplied from the positive battery terminal through a fuse link to the relay winding and contacts. When the horn switch is closed on top of the steering column, the relay winding is grounded through the switch. This action closes the relay contacts, and voltage is supplied through these contacts to the horns.*

   When the horn switch is closed in some circuits, voltage is supplied through the horn switch to the horns. A relay is not used in these circuits. Many vehicles have a low pitch and a high pitch horn. Some horns have a pitch adjustment screw.

Task 3   Diagnose the cause of wiper problems including constant, intermittent, poor speed control, parking, or no operation of wiper.

46. The wiper circuit has an open shunt coil.
    Technician A says this problem may cause the wiper motor to operate only at low speed.
    Technician B says this problem may cause the wiper motor to park with the wipers up on the windshield.
    Who is correct?
    A. A only
    B. B only
    C. Both A and B
    D. Neither A nor B

Hint   *Some wiper motors contain a series field coil, a shunt field coil, and a relay. When the wiper switch is turned on, the relay winding is grounded through one set of switch contacts. This action closes the relay contacts, and current is supplied through these contacts to the series field coil and armature. Under this condition the wiper motor starts turning. If the wiper switch is in the high-speed condition, the shunt coil is not grounded and the motor turns at high speed.*

   When the wiper switch is in the low-speed position, the shunt coil is grounded through the second set of wiper switch contacts. Under this condition current flows through the shunt coil and the wiper switch to ground. Current flow through the shunt coil creates a strong magnetic field that induces more opposing voltage in the armature windings. This opposing voltage in the armature windings reduces current flow through the series coil and armature windings to slow the armature.

   If the wiper motor fails to park, or parks in the wrong position, the parking switch or cam probably are defective.

   Some wiper motors have permanent magnets in place of the field coils. These motors have a low-speed and a high-speed brush. In some of these motors the low-speed brush is directly opposite the common brush, and the high-speed brush is positioned in between these two brushes.

Task 4   Inspect, test, and replace intermittent (pulsing) wiper speed controls.

47. An intermittent wiper system does not operate on low speed or in the intermittent mode (Figure 6-45, next page). The wipers operate normally at high speed.
    Technician A says there may be an open circuit between the low speed relay winding and the wiper module.
    Technician B says there may be an open circuit at the ground connection on the wiper module.
    Who is correct?
    A. A only
    B. B only
    C. Both A and B
    D. Neither A nor B

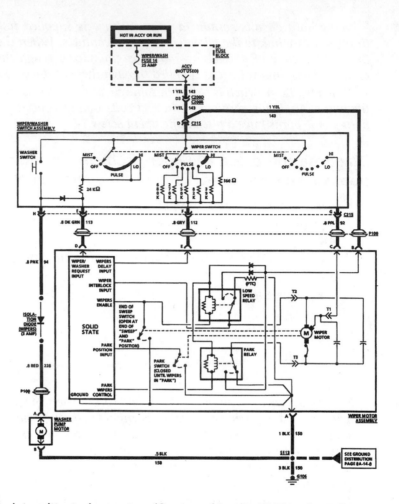

Figure 6-45   Intermittent wiper system *(Courtesy of Pontiac Division, General Motors Corporation)*

Hint   *Many vehicles are equipped with intermittent or interval wiper motor circuits. In most of these circuits a driver rotates a control knob to adjust the wiper delay interval. A variable resistor in the intermittent wiper control provides a voltage input to the intermittent wiper module. This module operates the wiper motor to provide the proper delay interval. In some intermittent wiper controls, various resistors are connected in the circuit as the switch is rotated.*

When the wiper switch is placed in the high-speed position, voltage is supplied through the high-speed switch contact to the high-speed brush in the wiper motor.

If the wiper switch is placed in the low-speed position, a voltage signal is sent from the wiper switch to the intermittent wiper module. When this signal is received, the module grounds the low-speed relay winding. This action closes the relay contacts and voltage is supplied through these contacts to the low-speed brush in the wiper motor.

If the wiper switch is placed in one of the intermittent positions, a unique voltage signal is sent to the intermittent wiper module. When this signal is received, the module opens and closes the ground circuit on the low-speed relay winding to provide the proper delay interval.

Task 5   **Diagnose the cause of constant, intermittent, or no operation of windshield washer.**

Task 6   **Inspect, test, and repair or replace washer motor, pump assembly, relays, switches, connectors, and wires of washer circuits.**

48. There is no operation from the windshield washer system (Figure 6-45). The wiper motor operation is normal.
     Technician A says the wiper/washer fuse may have an open circuit.

Technician B says the isolation diode may have an open circuit.

Who is correct?

A. A only

B. B only

C. Both A and B

D. Neither A nor B

Hint   *Many windshield washer systems have an electric pump mounted in the bottom of the washer fluid reservoir. When the washer button is pressed, voltage is supplied through the switch to the washer motor. This motor operates a pump that forces washer fluid through the hoses to the nozzles in front of the windshield.*

Accessories Diagnosis and Repair, Body

ASE Tasks, Questions, and Related Information

Task 1 **Diagnose the cause of slow, intermittent, or no operation of power side windows and power tailgate window.**

49. A power window operates normally from the master switch, but the window does not work using the window switch (Figure 6-46). The cause of this problem may be

A. an open circuit between the ignition switch and the window switch.

B. an open circuit in the window switch movable contacts.

C. an open circuit in the master switch ground wire.

D. a short to ground at the circuit breaker in the motor.

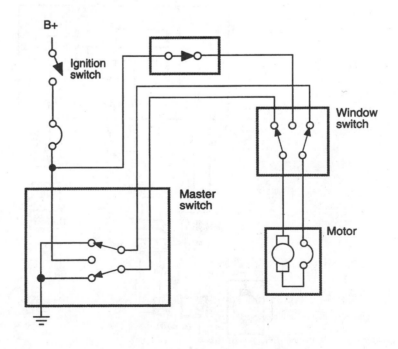

Figure 6-46 Power window circuit

Hint   *Power window circuits usually contain a master switch, individual window switches, and a window motor in each door. Some power window circuits have a window lockout switch to prevent operation of the window switches. When the master switch is placed in the down position, voltage is supplied from the center contact in this switch through the movable switch contact to the brush on the lower side of the commutator. The other brush is grounded through the master switch. Under this condition the motor moves the window to the down position.*

When the up position is selected in the master switch, current flow through the motor is reversed. Voltage is supplied from the ignition switch circuit breaker and lockout switch to the window switch. Pressing the window switch has the same effect as pressing the master switch.

Task 2   Inspect, test, and repair or replace regulators (linkages), switches, relays, motors, connectors. and wires of power side window and power tailgate window circuits.

50. There is no operation from the right rear power quarter window (Figure 6-47). The left rear power quarter window operates normally.

    Technician A says there may be an open circuit between the right rear power window switch and ground.

    Technician B says the upper contacts in the right rear power window switch may be open.

    Who is correct?

    A. A only

    B. B only

    C. Both A and B

    D. Neither A nor B.

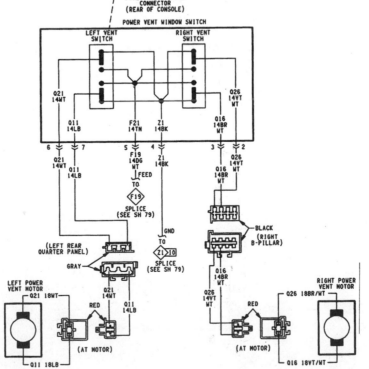

Figure 6-47 Rear power window circuit *(Courtesy of Chrysler Corporation)*

Hint   *Some vans are equipped with power windows in the rear quarter panels. In these circuits the left and right window switches share a common power supply and ground wire. When one of the switches is placed in the down position, current is supplied through the window motor in the proper direction to lower the window. Window motor current is reversed to raise the window if the switch is placed in the up position.*

Task 3   Diagnose the cause of slow, intermittent, or no operation of power seat.

Task 4   Inspect, test, adjust, and repair or replace power seat gear box, cables, slave units, switches, relays, solenoids, motors, connectors, and wires of power seat circuits.

51. A six-way power seat moves vertically at the front and rear, but there is no horizontal seat movement (Figure 6-48). All of these defects may be the cause of the problem EXCEPT
    A. a newspaper jammed in the seat track mechanism.
    B. an open circuit between the switch and the horizontal motor.
    C. an open circuit in the circuit from the switch assembly to ground.
    D. burned contacts in the horizontal seat switch.

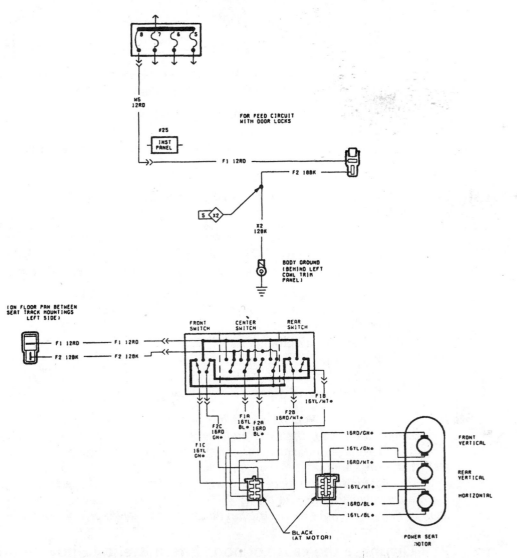

Figure 6-48   Power seat circuit *(Courtesy of Chrysler Corporation)*

Hint   *A six-way power seat moves vertically at the front and rear, and horizontally forward and rearward. This type of seat has two vertical motors and a horizontal motor. These motors are connected through gear boxes and cables to the seat track mechanisms (Figure 6-49).*

The front and rear switches have upward and downward positions, and the center switch has forward and rearward positions. When any or the switches are pressed, voltage is supplied to the appropriate motor in the proper direction. The motor moves the seat in the desired direction.

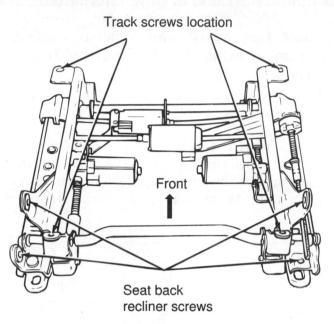

Track screws location

Front

Seat back
recliner screws

Viewed from top of track

Figure 6-49   Power seat motors and track mechanisms *(Courtesy of Chrysler Corporation)*

Task 5   **Diagnose the cause of poor, intermittent, or no operation of rear window defogger.**

Task 6   **Inspect, test, and repair or replace switches, relays, window grid, blower motors, connectors, and wires of rear window defogger circuits.**

52. When the rear defogger switch is turned on, the rear defogger light is illuminated, but there is no defogger grid operation (Figure 6-50). The cause of this problem could be
   A. an open defogger relay winding.
   B. an open circuit at the defogger relay contacts.
   C. an open circuit between the switch/timer and the grid.
   D. a defective defogger on/off switch.

Hint   *When the rear defogger switch is pressed, a signal is sent to the solid-state timer. When this signal is received, the timer grounds the relay winding. Under this condition the relay contacts supply voltage to the defogger grid. When the relay is closed current flows through the LED indicator to ground. After 10 minutes the timer opens the relay to shut off the grid current.*

The grid tracks may be tested with a 12V test light. A special compound is available to repair open circuits in the grid tracks

Some cars have an electric fan motor to circulate air past the rear window for defogging action.

Task 7   **Diagnose the cause of poor, intermittent, or no operation of electric door and hatch/trunk lock.**

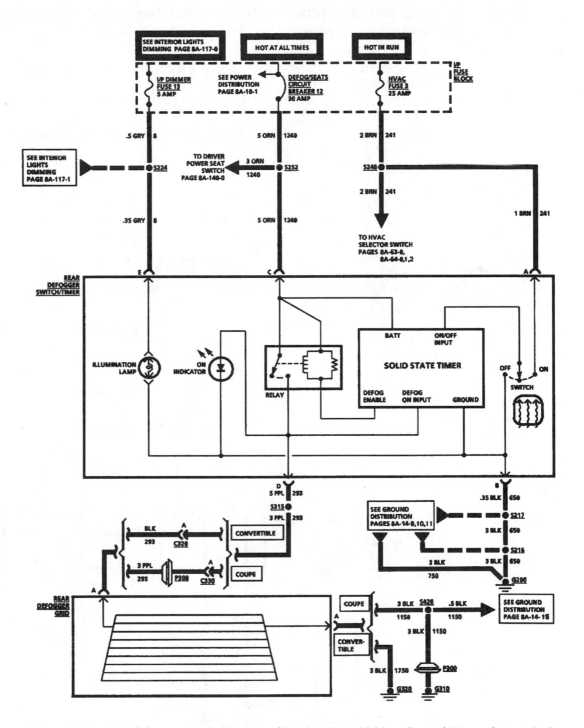

Figure 6-50   Rear defogger circuit *(Courtesy of Pontiac Motor Division, General Motors Corporation)*

Task 8   Inspect, test, and repair or replace switches, relays, controllers, actuators, connectors, and wires of electric door lock circuits.

53. When the left-hand door lock is pressed there is no operation from the power door locks (Figure 6-51). The power door locks operate normally when the right-hand door lock is pressed.

    Technician A says there may be an open circuit between junction 5500 and the left-hand door lock switch.

    Technician B says there may be an open circuit between terminal C and junction 5504 on the left-hand door lock.

    Who is correct?

    A. A only
    B. B only
    C. Both A and B
    D. Neither A nor B

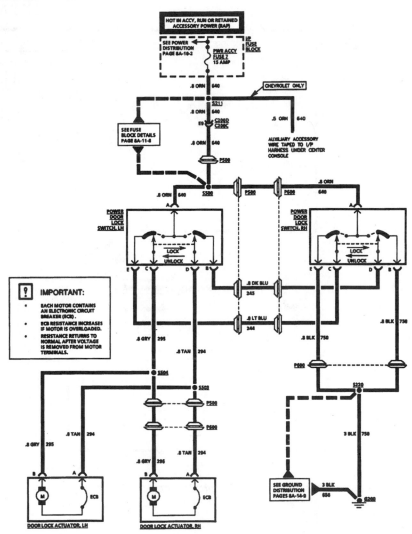

Figure 6-51   Power door lock circuit *(Courtesy of Pontiac Motor Division, General Motors Corporation)*

Hint   *Most electric door lock circuits have small electric motors to operate the door locks. When either door lock switch is pushed to the lock position, voltage is supplied to all the door lock motors in the proper direction to provide lock action. If either door lock switch is pushed to the unlock position, voltage is supplied to all the door lock motors in the opposite direction to provide unlock action.*

Task 9   Diagnose the cause of poor, intermittent, or no operation of keyless and remote lock/unlock devices.

Task 10   Inspect, test, and repair or replace components, connectors, and wires of keyless and remote lock/unlock device circuits.

54. There is no operation from the remote keyless entry system (Figure 6-52). Technician A says the battery may be discharged in the remote transmitter. Technician B says fuse number 4 may be blown in the fuse block.
Who is correct?

A. A only
B. B only
C. Both A and B
D. Neither A nor B

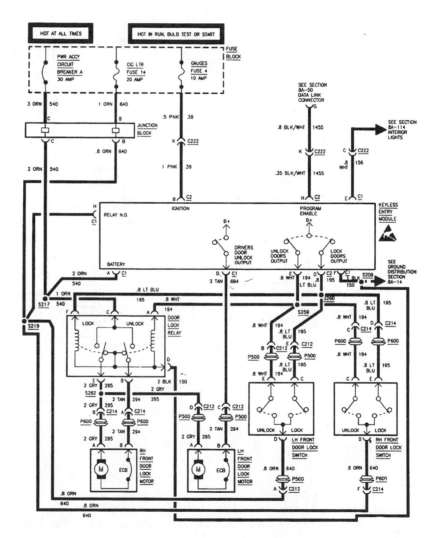

Figure 6-52   Remote keyless entry circuit (*Courtesy of Chevrolet Motor Division, General Motors Corporation*)

Hint   *The remote keyless entry module is connected to the power door lock circuit. A small remote transmitter sends lock and unlock signals to this module when the appropriate buttons are pressed on the remote control. When the handheld remote transmitter is a short distance from the truck, the module responds to the transmitter signals. When the unlock button is pressed on the remote transmitter, the module supplies voltage to the unlock relay winding to close these relay contacts and move the door lock motors to the unlock position.*

Task 11   Diagnose the cause of slow, intermittent, or no operation of electrical sunroof and convertible top.

55. The sunroof opens normally, but does not close when the close switch is pressed (Figure 6-53). All of the following defects may be the cause of the problem EXCEPT

   A. an open circuit at the close relay contacts.

   B. an open close relay winding.

   C. an open circuit at the close switch.

   D. an open circuit at terminal C on the sunroof switch.

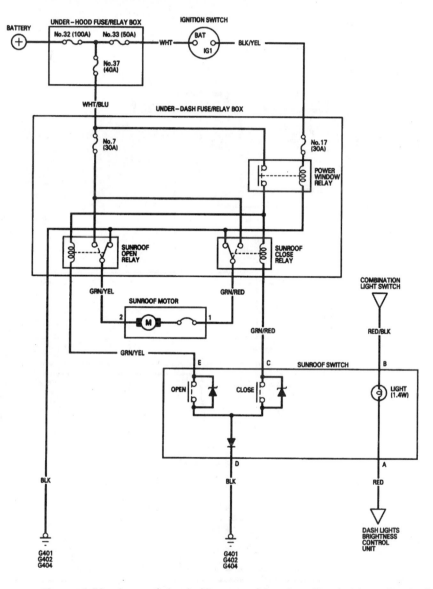

Figure 6-53   Sunroof circuit *(Courtesy of American Honda Motor Co., Inc.)*

Hint   *When the open sunroof switch is pressed, the open relay winding is grounded through the switch contacts. Under this condition the relay contacts close and supply voltage to the sunroof motor brush. The other motor brush is connected through the close relay contacts to ground. Current now flows through the motor and the motor opens the sunroof.*

If the close button is pressed, the close relay winding is grounded through the close switch contacts. Under this condition the close relay contacts supply voltage to the sunroof motor in the opposite direction to close the sunroof.

Task 12   Inspect, test, and repair or replace motors, switches, relays, connectors, and wires of electrically operated sunroof and convertible top circuits.

56. The convertible top goes down normally, but there is no motor operation when the up button is pressed (Figure 6-54). The cause of this problem may be

    A. an open circuit breaker in the convertible top motor.

    B. an open ground wire on the convertible top switch.

    C. a jammed linkage mechanism on the convertible top.

    D. an open circuit at terminal C on the convertible top switch.

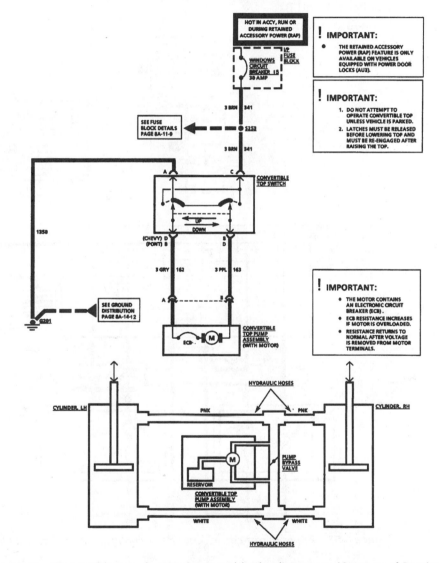

Figure 6-54   Convertible top electric circuit and hydraulic system *(Courtesy of Pontiac Division, General Motors Corporation)*

Hint   *The convertible top system contains a dual switch, pump motor, hydraulic cylinders, and linkages from these cylinders to the convertible top. When the down button is pressed, voltage is supplied through these switch contacts to a motor brush. The opposite motor brush is grounded through the up contacts. Under this condition current flows through the motor, and the motor drives the pump. With this motor rotation the pump supplies hydraulic pressure to the proper side of the cylinder pistons to move the top downward.*

If the up button is pressed, motor and pump rotation are reversed and the pump supplies hydraulic pressure to the upward side of the cylinder pistons.

Task 13   Diagnose the cause of poor, intermittent, or no operation of electrically operated/heated mirror.

57. There is no operation from the power mirror system (Figure 6-55).
    Technician A says there may be an open circuit in the mirror select switch.
    Technician B says there may be an open circuit between the mirror switch assembly and ground.
    Who is correct?
    A. A only
    B. B only
    C. Both A and B
    D. Neither A nor B

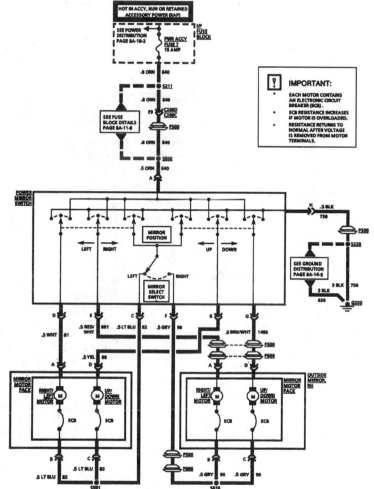

Figure 6-55   Power mirror circuit *(Courtesy of Pontiac Division, General Motors Corporation)*

Hint   *Voltage is supplied through a fuse to the power mirror switch assembly. When the mirror select switch is in the left position, it supplies voltage to the left mirror motor. When the left/right switch is pressed to the left position, a ground connection is completed from the right/left motor through the switch to ground. Under this condition the motor moves the mirror to the left.*

If the left/right switch is pressed to the right position, voltage is supplied through this switch to the right/left motor. The mirror select switch now provides a ground for the left-side mirror motor. This action reverses the current flow through the left-side mirror motor, and the mirror moves to the right.

Up/down mirror operation and operation of the right-side mirror is basically the same. The mirror select switch and the power mirror switches direct current flow through the mirror motors to supply the desired mirror movement.

Task 14

Inspect, test, and repair or replace motors, heated mirror grids, switches, relays, connectors, and wires of electrically operated/heated mirror circuits.

58. There is no operation from the heated mirror element, but the rear defogger operates normally (Figure 6-56). All of the following defects may be the cause of the problem EXCEPT
    A. an open circuit in the number 4 circuit breaker in the fuse block.
    B. a blown fuse number 1 in the relay center.
    C. an open circuit between the heated mirror element and ground.
    D. an open circuit between the timer relay and fuse number 1.

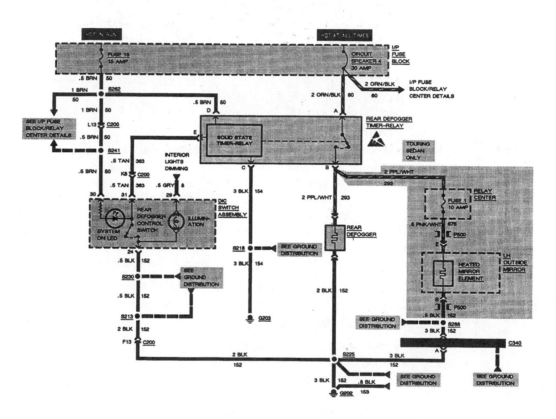

Figure 6-56   Heated mirror circuit (*Courtesy of Oldsmobile Division, General Motors Corporation*)

Hint   *Some vehicles have a heated driver's side mirror. When the rear defogger button is pressed the timer relay supplies voltage to the rear defogger grid and also to the heated mirror element. After 10 minutes the timer relay shuts off the voltage supply to the defogger grid and the heated mirror element.*

Miscellaneous

ASE Tasks, Questions, and Related Information

Task 1 Diagnose the cause of radio static and weak, intermittent, or no radio reception.

59. While discussing a radio static problem
Technician A says there may be poor metal-to-metal connection between the hood and other body components.
Technician B says the suppression coil may be defective on the instrument voltage limiter.
Who is correct?
A. A only
B. B only
C. Both A and B
D. Neither A nor B

Hint *Radio static may be caused by a defective alternator or spark plug wires. A defective ground connection between the engine and the chassis or between chassis components may result in radio static. Defective radio suppression devices, such as a suppression coil on an instrument voltage limiter, may cause static on the radio. A defective antenna with poor ground shielding may result in radio static. An open circuit in the antenna or lead-in wire may cause weak radio operation.*

Intermittent radio operation may be caused by an intermittent open circuit in the antenna or lead-in wire. An intermittent open circuit in the voltage supply wire to the radio may cause intermittent radio operation. No radio operation may be caused by a blown fuse, or an open circuit in the voltage supply wire to the radio.

Task 2 Inspect, test, and repair or replace speakers, antennas, leads, grounds, connectors, and wires of sound system circuits.

60. All of the following statements about radio antenna diagnosis with an ohmmeter are true EXCEPT
A. continuity should be present between the end of the antenna mast and the center pin on the lead-in wire.
B. continuity should be present between the ground shell of the lead in wire and the antenna mounting hardware.
C. no continuity should be present between the center pin on the lead in wire and the ground shell.
D. continuity should be present between the end of the antenna mast and the antenna-mounting hardware.

Hint *An antenna may be tested with an ohmmeter. Continuity should be present between the end of the antenna mast and the center pin on the lead-in wire. Continuity also should be present between the ground shell on the lead-in wire and the antenna-mounting hardware. No continuity should exist between the center pin on the lead-in wire and the ground shell.*

Task 3 Inspect, test, and repair or replace switches, motor, connectors, and wires of power antenna circuits.

61. The power antenna goes up when the radio is turned on, but the antenna does not go back down (Figure 6-57). The cause of this problem could be an open circuit in
A. the radio fuse.
B. the antenna fuse.
C. the down limit switch.
D. the antenna motor.

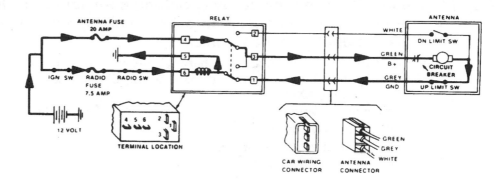

Figure 6-57   Power antenna circuit *(Courtesy of Delco Products Division of General Motors Corporation)*

Hint   *When the radio is turned on, voltage is supplied to the relay winding. This action moves the relay points to the up position, and current flows through the motor to move the antenna upward. When the antenna is fully extended, the up limit switch opens and stops current flow through the motor.*

If the radio is turned off, current flow through the relay coil stops. Under this condition the relay contacts move to the down position. This action reverses current flow through the motor and moves the antenna downward. When the antenna is fully retracted, the down limit switch opens and stops the antenna movement.

Task 4   Inspect, test, and replace noise suppression components.

62. The ohmmeter leads are connected from a radio suppression capacitor lead to the capacitor case. The ohmmeter reading is 65 Ω. This reading indicates the capacitor
   A. is satisfactory.
   B. has a high-resistance problem.
   C. has insulation leakage
   D. has reduced plate capacity.

Hint   *An electric or electronic component with a varying magnetic field may cause radio static. A radio choke coil is connected to some components to reduce radio static (Figure 6-58). In some circuits a radio suppression capacitor may be connected from the circuit to ground to reduce radio static. An ohmmeter may be connected from the capacitor lead to the case to check the capacitor for insulation leakage between the capacitor plates. When the ohmmeter is placed on the X1000 scale, the meter should provide an infinite reading. A capacitor tester may be used to test the capacitor for leakage, capacity and resistance.*

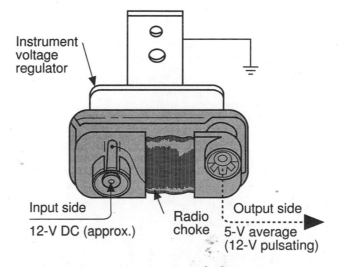

Instrument voltage regulator

Input side
12-V DC (approx.)

Radio choke

Output side
5-V average
(12-V pulsating)

Figure 6-58   Radio suppression choke

Task 5

Identify the component (unit) causing poor sound quality, noisy, erratic, intermittent, or no operation of the audio system; remove and reinstall audio system component (unit).

63. A radio has a whining noise that increases with engine speed. When the alternator field wire is disconnected, the noise stops. All of these defects may be the cause of the problem EXCEPT
    A. a defective stator.
    B. a defective diode.
    C. a defective capacitor.
    D. an open field winding.

Hint   *A radio noise locator may be made from an antenna lead-in wire. Cut the connector off the antenna end of the lead-in wire and remove a 2 in. length of the coax shield to expose the center conductor. Plug the other end of the lead-in wire into the radio, and place the exposed center conductor near any suspected noise sources. When noise is heard with the end of the lead-in wire near a component, that component is the source of the noise.*

Some audio systems have self-diagnostic capabilities. When two of the control buttons, such as the number 3 and seek down buttons, are pressed simultaneously for 3 seconds, the audio system is placed in the diagnostic mode. In this mode diagnostic trouble codes (DTCs) are shown in the audio system display.

Task 6

Inspect, test, and repair or replace case, fuse, connectors, and wires of cigar lighter circuits.

64. The cigar lighter and the dome light are both inoperative (Figure 6-59).
    Technician A says there may be an open circuit at terminal B on the cigar lighter.
    Technician B says there may be an open circuit at ground connection G202.
    Who is correct?
    A. A only
    B. B only
    C. Both A and B
    D. Neither A nor B

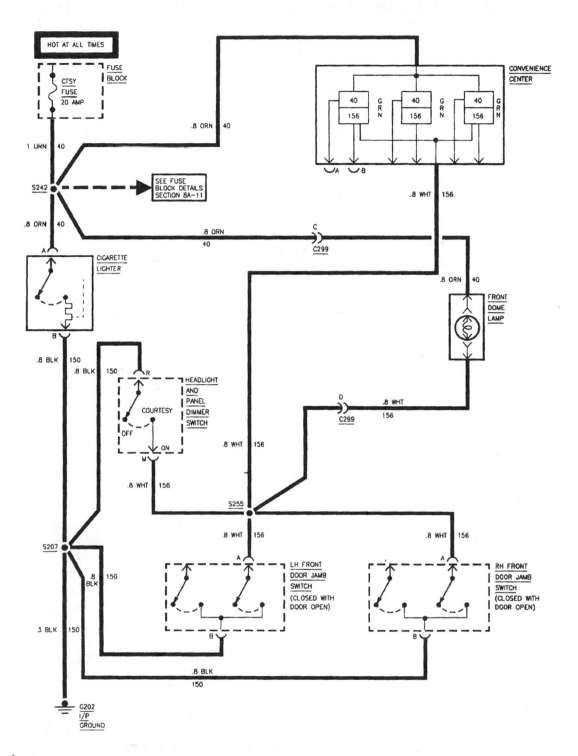

Figure 6-59   Cigar lighter and dome light circuit *(Courtesy of Chevrolet Motor Division, General Motors Corporation)*

Hint   *Voltage is supplied from the battery positive terminal through a fuse to one terminal on the cigar lighter. The other terminal on the cigar lighter is connected to ground. When the cigar lighter element is pushed inward, the circuit is completed through the lighter to ground. Current flow through the lighter heats the lighter element. When the element is hot, the lighter element moves outward and opens the circuit. On some vehicles the lighter fuse also supplies voltage to the dome light and these components share a common ground.*

Task 7   Inspect, test, and repair or replace clock, connectors, and wires of clock circuits.

65. The clock is inoperative (Figure 6-60). All of these defects may be the cause of the problem EXCEPT
    A. a grounded circuit on the SB wire at connector 51M.
    B. an open circuit at body ground B.
    C. an open circuit at terminal B in the clock connector.
    D. a grounded circuit on wire B/R at the clock connector.

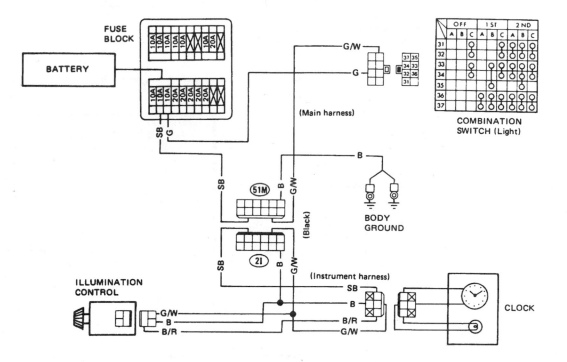

Figure 6-60   Clock circuit *(Courtesy of Nissan Motor Co., Ltd.)*

Hint   *Voltage is supplied from the positive battery terminal through a 10A fuse to the clock. The other clock terminal is grounded. The illumination control contains a variable resistor to control illumination brilliance in the clock display.*

Task 8   Diagnose the cause of unregulated, intermittent, or no operation of cruise control.

Task 9   Inspect, test, adjust, and repair or replace speedometer cables, regulator, servo hoses, switches, relays, electronic control units, speed sensors, connectors, and wires of cruise control circuits.

66. The cruise control is inoperative (Figure 6-61).
    Technician A says the vehicle speed sensor may be defective.
    Technician B says the 20 amp gauge fuse may be defective.
    Who is correct?
    A. A only
    B. B only
    C. Both A and B
    D. Neither A nor B

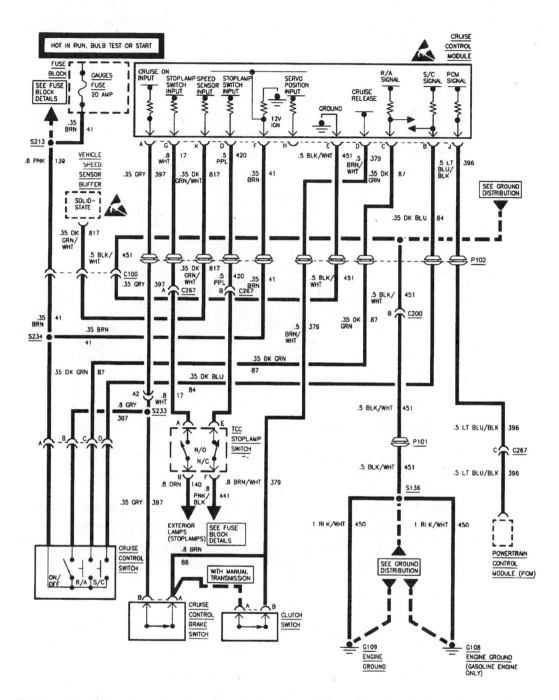

Figure 6-61 Electronic cruise control circuit (*Courtesy of Chevrolet Motor Division, General Motors Corporation*)

Hint   *Many vehicles have an electronic cruise control. In some of these systems the control module and the stepper motor are combined in one unit. A cable is connected from the stepper motor to the throttle linkage. The control unit receives inputs from the cruise control switch, brake switch, and vehicle speed sensor (VSS). The control module sends output commands to the stepper motor to provide the desired throttle opening. A defective VSS may cause erratic or no cruise control operation. A cruise control cable adjustment is required on these systems. Remove the cruise control cable from the throttle linkage. With the throttle closed and the cable pulled all the way outward, install the cable on the throttle linkage. Turn the adjuster screw on the cruise control cable to obtain 0.197 in. lash in the cable.*

Some cruise control systems have the control module mounted in the powertrain control module (PCM). The control module is connected to an external servo. This servo contains a vacuum diaphragm that is connected by a cable to the throttle linkage. The servo also contains a vent solenoid and a vacuum solenoid (Figure 6-62). The control module receives the same inputs described in the previous paragraph. In response to these inputs the control module operates the vent and vacuum solenoids to supply the proper vacuum to the servo diaphragm. Since the servo diaphragm is connected to the throttle, the vacuum supplied to this diaphragm provides the desired throttle opening. In these systems a leak in the servo diaphragm may cause erratic cruise control operation, or a gradual reduction in the cruise set speed.

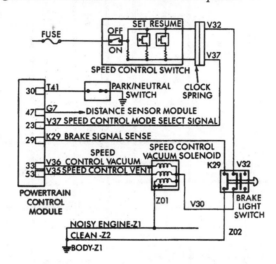

Figure 6-62 Electronic cruise control with external servo *(Courtesy of Chrysler Corporation)*

Task 10   Diagnose the cause of false, intermittent, unintended, or no operation of anti-theft system.

Task 11   Inspect, test, and repair or replace components, switches, relays, connectors, sensors, and wires of anti-theft system circuits.

67. All of these statements about an anti-theft system are true EXCEPT (Figure 6-63)
    A. the system is triggered by unauthorized operation of the doors, hood, trunk lock, or ignition.
    B. when triggered the system sounds the horn for 3 minutes, and flashes the park lights for 18 minutes.
    C. when triggered the system prevents engine starting by disabling the injectors.
    D. when triggered the system may be disarmed by disconnecting the battery.

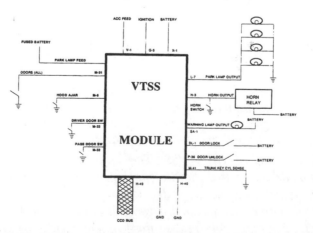

Figure 6-63 Vehicle anti-theft system *(Courtesy of Chrsyler Corporation)*

Hint   *The driver arms the anti-theft system by turning off the ignition switch, opening a door, lock the doors using the power door lock switch, and closing the door. These actions do not have to be sequential. After the arming actions are completed, the security alarm light in the instrument panel flashes for 15 seconds. When this light stops flashing, the system is armed.*

    The system may be triggered by unauthorized operation of doors, hood, trunk lock, or ignition. When the system is triggered, the horn sounds and the park lights flash for 3 minutes. After this time period if the system is not disarmed, the horn stops blowing, but the park lights continue flashing for an additional 15 minutes. The system is disarmed by unlocking either front door with the key. Disconnecting the battery does not disarm the system. Once the battery is reconnected, the alarm action continues.

Task 12   Diagnose the cause(s) of the airbag warning light staying on or flashing.

68. An air bag warning light is illuminated intermittently with the engine running.

    Technician A says the air bag system has an electrical defect.

    Technician B says this defect may cause the air bag to inflate accidentally.

    Who is correct?

    A. A only

    B. B only

    C. Both A and B

    D. Neither A nor B

Hint   *In many air bag systems the warning light is illuminated for 5 or 6 seconds after the engine starts. After this time the air bag warning light should remain off while the engine is running. If the air bag warning light is illuminated with the engine running, a fault is present in the air bag system. On Ford air bag systems the air bag warning light begins flashing a diagnostic trouble code (DTC), if an electrical defect occurs in the system. On other systems the air bag warning light is illuminated continually if an electrical defect occurs in the system. When a scan tester is connected to the air bag diagnostic connector, or data link connector (DLC), the DTCs stored in the air bag module memory are indicated on the scan tester.*

Task 13   Inspect, test, repair, or replace the air bag, air bag module, sensors, connectors, and wires of the air bag system circuit(s).

69. All of these statements about air bag system service are true EXCEPT (Figure 6-64, next page)

    A. the negative battery cable should be disconnected and manufacturer's recommended waiting period completed.

    B. safety glasses and gloves should be worn when handling deployed air bags.

    C. a 12V test light may be used to test continuity between the inflator module and the sensors.

    D. sensor operation may be affected if the sensor brackets are bent or twisted.

Hint   *When servicing an air bag system always disconnect the negative battery cable and wait for the time period specified by the vehicle manufacturer. This time usually is 1 to 2 minutes. Never use a 12V test light to diagnose an air bag system. Diagnose these systems with an ohmmeter, voltmeter, or the manufacturer's recommended equipment. Since deployed air bags may contain small quantities of sodium hydroxide, wear safety glasses and gloves when handling these components. Sensors must always be mounted in their original direction. Most sensors have a directional arrow that must face toward the front of the vehicle. Always store inflator modules face upward on the bench, and carry these components with the trim cover facing away from your body.*

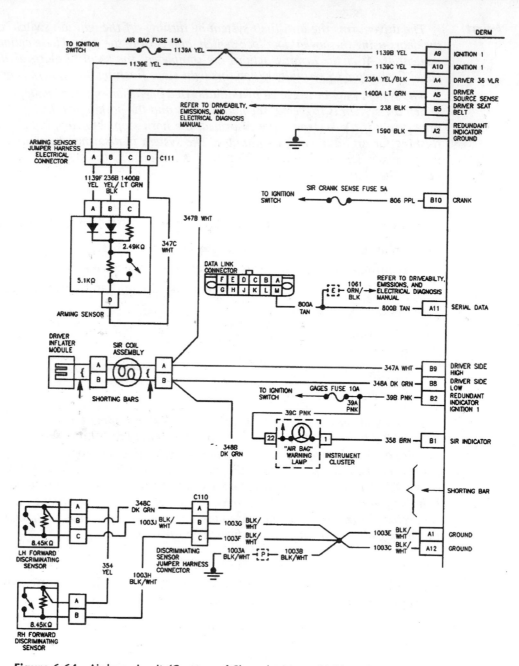

Figure 6-64   Air bag circuit *(Courtesy of Chevrolet Motor Division, General Motors Corporation)*

Answers and Analysis

1. **A**   The test light is illuminated when connected to the battery side of the open circuit, but it is not on if it is connected to the motor side of the open circuit. With an open circuit there is no current flow through the circuit, and the voltage drop across the motor is 0V. A is right.

2. **D**   The light bulb is the only resistance in the circuit and nearly all the battery voltage is dropped across this bulb. If the voltage drop across the bulb is 9V, there is high resistance

in the switch or wires. This resistance causes some voltage drop across the switch or wires and reduces the voltage drop across the bulb. Therefore, both A and B are wrong and D is the correct answer.

3. C A shorted condition causes lower resistance and increased current flow. Therefore, A, B, and D are wrong and C is the right answer.

4. D An ohmmeter should never be connected to a circuit in which current is flowing. Therefore, A is wrong. When testing a spark plug wire with 20,000 Ω, the X1000 meter scale should be used, so B is also wrong and D is the correct response.

5. B If the bulb is illuminated when a jumper wire is connected from the ground side of the light bulb to ground, there is an open circuit in the light bulb ground. Therefore, A, C, and D are wrong and B is correct.

6. A When the 12V test light remains illuminated with the connector disconnected at the light bulb, the circuit must be shorted to ground between the fuse and the disconnected connector. Therefore, B is wrong and A is right.

7. A Normal battery drain on some vehicles is 50 milliamperes. Therefore, B is wrong. Since some computers require a time period to enter the sleep mode, the drain should not be recorded immediately when the switch is opened. C is wrong. The doors should be closed and all accessories turned off during the drain test, so D is wrong. The tester switch must be closed when starting or running the engine to carry the starter or alternator current. Therefore, A is the correct response.

8. D If the circuit breaker is satisfactory, the ohmmeter reading should be zero ohms. Therefore, A is wrong. The ohmmeter current is not high enough to open the circuit breaker, so B also is wrong. D is the right answer.

9. C At 0°F battery temperature, 0.032 point should be subtracted from the hydrometer reading, so A is wrong. If the battery temperature is 120°F, 0.016 point should be added to the hydrometer reading. B also is wrong. The battery is fully charged when the hydrometer reading is 1.265, thus D is wrong. The maximum variation in battery cells is 0.050 specific gravity points. The correct answer is C.

10. B The battery discharge rate is one-half the cold cranking rating, so A is wrong. The battery is satisfactory if the voltage remains above 9.6V at 70°F. Therefore, B is correct.

11. C Disconnecting the battery from the electrical system erases the computer adaptive memories. The correct answer is C.

12. A If a built-in hydrometer indicates yellow, the battery electrolyte level is low. This problem may be caused by high charging voltage resulting from a defective voltage regulator. A loose alternator belt causes an undercharged battery. Therefore, B is wrong and A is correct.

13. B The battery charging time varies for batteries with different capacities, so A is wrong. A high charging rate should not be used on a cold battery. The charging rate should be reduced when the battery reaches 1.225 specific gravity. Therefore, C and D are wrong. During the charging process the battery temperature should not exceed 125°F. Therefore, B is the correct answer.

14. B When using a booster battery, the accessories should be off on both vehicles. Therefore, A is wrong. The negative booster cable should be connected to an engine ground on the vehicle being boosted. B is the correct response.

15. A If the field windings or positive battery cable have high resistance, starter current draw and cranking speed are reduced. Therefore, B and C are wrong. A burned solenoid disc and terminals may cause a clicking action from the starter solenoid, so D is wrong. Worn starter bushings may cause high current draw and reduced cranking speed. A is the correct answer.

16. C In Figure 6-15 the voltmeter is connected to test the voltage drop in the starter ground circuit. Therefore, A, B, and D are wrong and C is correct.

17. B   A grounded circuit at relay terminal 85, or a continually closed neutral safety switch cause the starting motor to operate in any gear selector position. Therefore, A and C are wrong. A slight resistance at relay terminal 86 may reduce the voltage at the solenoid windings resulting in a clicking action from the solenoid. Therefore, D is wrong.

An open circuit at terminal 86 on the starter relay stops the current flow through the relay winding. This problem causes a completely inoperative starter relay and solenoid. Therefore, B is correct.

18. A   In Figure 6-17 the ohmmeter is connected to test the pull-in winding. A satisfactory winding provides a low ohmmeter reading. Therefore, B is wrong and A is correct.

19. D   If the self-powered test light in Figure 6-18 is not illuminated, the field coils have an open circuit. Therefore, A, B, and C are wrong, and D is the right answer.

20. A   Open series field coils would result in an inoperative starting motor, thus B is wrong. A bent armature shaft may cause reduced starter rpm and high current draw. A is the correct response.

21. C   An open fuse link between the alternator battery wire causes zero alternator output. Therefore, B is correct. If the field circuit is open, there is no magnetic field around the rotor. Therefore, alternator output is zero, and both A and B are correct. C is the correct answer.

22. A   If the alternator belt is bottomed in the pulley, the belt may slip and alternator output is reduced. A misaligned belt may cause rapid belt wear, but the belt should not slip. Therefore, B is wrong and A is correct.

23. D   When performing an alternator output test the vehicle accessories should be off and the voltage should be limited to 15V. Therefore, both A and B are wrong, and D is the right answer.

24. A   A broken brush may cause zero field current and alternator output, so B is wrong. When the alternator is full-fielded, the voltage regulator is bypassed. Therefore, a defective regulator does not reduce output and C is wrong. Since a defective capacitor usually does not affect output, D is wrong.

A shorted diode reduces alternator output, thus A is correct.

25. D   This question asks for the statement that is not a result of the problem. A high charging voltage may cause an overcharged battery, burned out electrical components, or electrolyte gassing from the battery. Therefore, A, B, and C are a result of the problem and none of these responses are right.

A high charging voltage does not reduce headlight brilliance, so D is not a result of the problem, and this is the requested answer.

26. A   In Figure 6-19 the voltmeter is connected to measure the voltage drop from the alternator battery wire to the positive battery cable. A 2V drop in this circuit indicates excessive resistance. This resistance reduces charging current, and causes an undercharged battery. When this problem is present, the headlights may be dimmer than normal. Therefore, B is wrong and A is correct.

27. C   When the engine is running the charge indicator light remains off because there is equal voltage supplied to each side of the bulb. A glowing charge indicator bulb may be caused by unequal voltage on each side of the bulb.

High resistance in the alternator to battery wire reduces the voltage on the ignition switch side of the charge indicator bulb, while full voltage is supplied to the L terminal side of the bulb. High resistance in the wire from the L terminal to the bulb reduces the voltage on the L terminal side of the bulb. Therefore, both A and B are correct and C is right.

28. A   A satisfactory diode provides one high and one low ohmmeter reading during a diode test. Therefore, B, C, and D are wrong and A is right.

29. A   A short to ground may cause excessive current flow in the light circuit. This excessive current flow may cause the circuit breaker to open the circuit and shut off the lights. If the charging system voltage is high enough to open the headlight circuit breaker, many other problems would occur such as damaged electronic components and an overcharged battery. Therefore, B is wrong and A is correct.

30. C   This question asks for the statement that is not true. When replacing halogen bulbs handle the bulb by the base, do not scratch or drop the bulb, and avoid moisture contact with the bulb. Therefore, statements A, B, and D are true, and these are not the requested answer.

The bulb should be changed with the headlights off, and thus statement C is not true. C is the requested answer.

31. C   Since fuse number 12 is connected only to the instrument panel lights, a blown fuse causes these lights to be inoperative. Therefore, A, B, and D are wrong and C is right.

32. A   An open circuit at terminal A on the headlight doors module causes both headlight doors to be inoperative, but this problem does not damage the module. Therefore, B is wrong and A is correct.

33. B   Since the right-side and left-side ground connections are connected together, an open circuit at one ground connection does not affect rear light operation. Therefore, A, C, and D are wrong and B is right.

34. A   An open circuit in the rheostat causes all the instrument panel bulbs to be inoperative. Since these bulbs are connected in parallel, an open circuit in one bulb does not affect the other bulbs. Therefore, B is wrong and A is the correct response.

35. A   A short to ground at terminal S 363 does not affect the underhood light, so C is wrong. Since current still flows through the courtesy lights and the short to ground, this problem does not increase current flow and blow a fuse. Therefore, D is wrong. This short to ground causes continual operation of the courtesy lights, thus B is wrong and A is correct.

36. A   If the cigar lighter fuse is blown, the courtesy and dome lights glow dimly when the cigar lighter is pushed in because the cigar lighter is now in series with these lights. Therefore, B, C, and D are wrong and A is right.

37. C   The D7 18BR RD wire is connected from the signal light switch to the right rear signal light. When this is the only dim light, the problem is in this wire. Therefore, A, B, and D are wrong and C is right.

38. A   If the right-hand backup light circuit is grounded on the switch side of the bulb, excessive current blows the backup light fuse. This action makes both backup lights inoperative. Therefore, B is wrong and A is the correct answer.

39. A   High resistance between the voltage limiter and the gauge would probably affect the other gauges, so B is wrong. A short to ground between the gauge and the sending unit causes a high gauge reading, so C is wrong. An open circuit between the gauge and the sending unit causes an empty gauge reading, thus D is wrong

High resistance in the sending unit ground wire reduces gauge current flow and provides a lower gauge reading. A is the right answer.

40. B   Since the voltage limiter produces a constant 5V regardless of charging system voltage, a defective alternator does not affect the gauge reading. Therefore, A is wrong.

A defective instrument voltage limiter may cause erratic gauge operation, so B is right.

41. B   An open wire from the hot coil to the sending unit causes the current to flow through the cold coil. This action results in a low gauge reading, so A is wrong. Excessive resistance in the sending unit or sending unit ground reduces hot coil current, and increases cold coil current. Under this condition the gauge reading is lower. Therefore, C and D are wrong.

An open wire from the cold coil to ground causes all the gauge current to flow throught the hot coil. This action causes a continual hot gauge reading, so B is right.

42. B If some of the electronic instrument panel display segments are not illuminated when the ignition switch is turned on, the display is defective and must be replaced. Therefore, A is wrong and B is correct.

43. B An open circuit in the wire between the BCM terminal 3 and the door ajar switches causes an inoperative door ajar light. An open circuit from the door ajar switches to ground also causes an inopertive door ajar light. A continually open LR door ajar switch causes no door ajar light operation when the LR door is opened. Therefore, A, C, and D are wrong.

A short to ground on the BCM side of the RF door ajar switch causes continual door ajar light operation. B is the right answer.

44. C If the timer contacts are stuck closed, the seat belt buzzer would go off once the driver's seat belt is buckled. Therefore, A is wrong. If the circuit is shorted to ground at buzzer terminal three the seat belt light is inoperative, and the higher current flow may damage the buzzer contacts. Therefore, B is wrong. An open circuit at terminal two causes zero current flow in the timer heater circuit and continually closed contacts on the timer. However, the seat belt buzzer still goes off when the driver's seat belt is buckled, so D is wrong.

If the timer contacts and seat belt switch are stuck closed, the buzzer and seat belt warning light operate continually. C is right.

45. A In this question we are asked for the statement that is not the cause of the problem. An open circuit in the horn relay winding, or in the brush/slip ring would cause no horn operation. An open fuse link in the relay power wire also causes this problem. Therefore, B, C, and D may cause the problem, and these are not the requested answer.

Since the relay does not require a ground, an open circuit in the relay ground has no effect and is not a cause of the problem. A is right.

46. D When current flows through the shunt coil, the shunt coil magnetic field produces more opposing voltage in the armature windings. This action reduces armature and wiper speed. An open shunt coil causes high armature speed, and this problem does not affect wiper parking. Therefore, A and B are both wrong and D is the correct response.

47. C An open circuit between the low-speed relay winding and the module, or an open circuit in the module ground would cause no low speed or intermittent wiper operation. Therefore, both A and B are correct and C is the right answer.

48. B An open circuit in the wiper/washer fuse causes inoperative wipers and washers, so A is wrong. Since the isolation diode is only connected in the washer circuit, an open diode causes no operation of the washer. Therefore, B is the right answer.

49. A An open circuit in the window switch movable contacts causes no window operation from the master switch, so B is wrong. An open circuit in the master switch ground wire also causes no window operation from the master switch, thus C is wrong. A short to ground at the motor circuit breaker causes high current flow in the down position and a blown fuse. Therefore, D is wrong.

If there is an open circuit between the ignition switch and the window switch, there is no voltage supplied to the window switch and the window does not work from this switch. A is the correct response.

50. B Since the left and right rear window switches both use the same ground, an open circuit in this ground circuit causes no operation of either windows. Therefore, A is wrong.

An open circuit at the upper contacts on the right rear power window switch causes no operation of this window and B is correct.

51. C   This question asks for the statement that is not the cause of the problem. A newspaper jammed in the horizontal seat track may cause no horizontal seat operation. An open circuit between the horizontal switch and motor may be the cause of no horizontal seat operation. Burned horizontal switch contacts also may cause this problem. Since A, B, and D may cause the problem, they are not the requested answer.

An open circuit from the switch assembly to ground causes no operation of the seat in any direction. Therefore, this is not the cause of the problem, and C is the requested answer.

52. C   The defogger relay contacts must close to illuminate the indicator light. An open defogger relay winding, an open circuit at the relay contacts, or a defective on/off switch would not allow the relay contacts to close. Therefore, A, B, and D are wrong.

An open circuit between the switch/timer and the grid causes the indicator to be on with no grid operation. The correct answer is C.

53. A   An open circuit between terminal C and junction 5504 on the left-hand door lock switch prevents operation of the door locks from either switch. Therefore, B is wrong.

An open circuit between junction 5500 and the left-hand door lock switch prevents operation of the door locks from the left-hand switch. Therefore, A is the right answer.

54. C   Discharged remote control batteries, or blown number 4 fuse prevent operation of the remote keyless entry system. Therefore, both A and B are correct and C is the right answer.

55. A   This question asks for the statement that is not the cause of the problem. An open circuit in the close relay winding at the close switch, or at terminal C in the sunroof switch, prevents closure of the sunroof. Therefore, B, C, and D may be causes of the problem and these are not the requested answer.

An open circuit at the close relay contacts prevents sunroof opening, so A is not a cause of the problem and this is the requested answer.

56. C   An open circuit breaker in the motor, an open switch ground wire, or an open circuit at switch terminal C prevent downward operation of the convertible top. Therefore, A, B, and D are wrong.

A jammed linkage mechanism on the convertible top may allow downward top operation but no upward movement. Therefore, C is correct.

57. C   An open circuit in the mirror select switch, or an open circuit between the mirror switch and ground may cause no mirror operation. Therefore, both A and B are correct and C is the right answer.

58. A   This question asks for the statement that is not the cause of the problem. A blown fuse number 1 in the relay center, an open circuit between the heated mirror element and ground, or an open circuit between the timer relay and fuse number 1 may prevent operation of the heated mirror element and allow rear defogger operation. Since statements B, C, and D may cause the problem, these are not the requested answer.

An open circuit in the number 4 circuit breaker in the fuse block prevents operation of the rear defogger and the heated mirror element. Therefore, A is not a cause of the problem and this is the requested answer.

59. C   Poor metal-to-metal connection between the hood and other body components may cause radio static. A defective suppression coil on the instrument voltage limiter also results in this problem. Therefore, both A and B are correct and C is right.

60. D   This question asks for the statement that is not true. There should be continuity between the end of the antenna mast and the center pin in the lead-in wire. Continuity also should be present between the ground shell of the lead-in wire and the antenna hardware. No continuity should be present between the center pin in the lead-in wire and the ground shell. Therefore, statements A, B, and C are correct and these are not the requested answer.

Continuity should not be present between the end of the antenna mast and the mounting hardware. Therefore, statement D is not true and this is the requested answer.

61. C   An open circuit in the radio fuse, antenna fuse, or antenna motor would cause no operation from the antenna. Therefore, A, B, and D are wrong. Since the down limit switch only has current flow through it during antenna down operation, an open down limit switch allows upward antenna movement but no downward action. C is correct.

62. C   When the ohmmeter leads are connected from the capacitor lead to the case and a low reading is obtained, the capacitor has insulation leakage between the capacitor plates. Therefore, A, B, and D are wrong, and C is the right answer.

63. D   This question asks for the statement that is not the cause of the problem. A defective stator, diode, or capacitor may cause radio noise from the alternator. Therefore, A, B, and C may cause the problem, and these are not the requested answer.

An open field winding reduces alternator output to zero, so this problem would not cause radio static from the alternator. Therefore, D is the requested answer.

64. B   An open circuit at terminal B on the cigar lighter causes an inoperative cigar lighter, and normal dome light operation. Therefore, A is wrong. Since the cigar lighter and the dome light share a common ground connection at terminal G202, an open circuit at this location causes both of these components to be inoperative.

65. D   This question asks for the statement that is not the cause of the problem. A grounded circuit on the SB wire causes excessive current flow and this results in a blown clock fuse and inoperative clock. Since body ground B is the clock ground, an open circuit at this location causes an inoperative clock. The clock also is inoperative if there is an open circuit at terminal B in the clock connector. Therefore, A, B, and C may be the cause of the problem, and these are not the requested answer.

Since wire B/R at the clock connector is only in the clock illumination circuit, a grounded circuit at this location does not affect clock operation. Therefore, D is the requested answer.

66. C   Since the 20A gauges fuse supplies voltage to the cruise control switch, a blown fuse results in an inoperative cruise control. Therefore, B is correct. A defective vehicle speed sensor also causes an inoperative cruise control. Both A and B are correct, and C is the right answer.

67. D   This question asks for the statement that is not true. The anti-theft system is triggered by unauthorized operation of the doors, hood, trunk lock, or ignition. When triggered, the system sounds the horn for 3 minutes, flashes the lights for 18 minutes, and disables the injectors. Therefore, statements A, B, and C are correct, and these are not the requested answer.

When triggered, the system can be disarmed by operating one of the front door locks with the key. Disconnecting the battery does not disarm the system. Therefore, D is not true and this is the requested answer.

68. A   If the air bag warning light is on with the engine running, an electrical defect is present in the system. Since the sensors have to close to deploy the air bag, it is very unlikely that an electrical defect in the system will deploy the air bag. Therefore, B is wrong and A is correct.

69. C   This question asks for the statement that is not true. Before servicing an air bag system, the negative battery cable should be disconnected and the technician should wait for the time period specified by the vehicle manufacturer. Safety glasses and gloves should be worn when handling deployed air bags. Sensor operation may be affected if the brackets are bent or twisted. Therefore, statements A, B, and D are true, and these are not the requested answer.

A 12V test light must not be used to diagnose an air bag system. Therefore, statement C is not true and this is the requested answer.

Glossary

Accumulator A gas-filled pressure chamber that provides hydraulic pressure for ABS operation.
Acumulador Cámara de presión que contiene gas y que proporciona gran presión hidráulica para una función ABS.

A circuit A generator circuit that uses an external grounded field circuit. The regulator is on the ground side of the field coil.
Circuito A Circuito regulador del generador que utiliza un circuito inductor externo puesto a tierra. En el circuito A, el regulador se encuentra en el lado a tierra de la bobina inductora.

Actuators Devices that perform the actual work commanded by the computer. They can be in the form of a motor, relay, switch, or solenoid.
Accionadores Dispositivos que realizan el trabajo efectivo que ordena la computadora. Dichos dispositivos pueden ser un motor, un relé, un conmutador o un solenoide.

Air bag module Composed of the air bag and inflator assembly that is packaged into a single module.
Unidad del Airbag Formada por el conjunto del Airbag y el inflador. Este conjunto se empaqueta en una sola unidad.

Air bag system A supplemental restraint that will deploy a bag out of the steering wheel or passenger side dash panel to provide additional protection against head and face injuries during an accident.
Sistema de Airbag Resguardo complementario que expulsa una bolsa del volante o del panel de instrumentos del lado del pasajero para proveer protección adicional contra lesiones a la cabeza y a la cara en caso de un accidente.

Ambient temperature The temperature of the outside air.
Temperatura ambiente Temperatura del aire ambiente.

Ambient temperature sensor Thermistor used to measure the temperature of the air entering the vehicle.
Sensor de temperatura ambiente Termistor utilizado para medir la temperatura del aire que entra al vehículo.

Ammeter A test meter used to measure current draw.
Amperímetro Instrumento de prueba utilizado para medir la intensidad de una corriente.

Amperes See current.
Amperios Véase corriente.

Analog A voltage signal that is infinitely variable or can be changed within a given range.
Señal analógica Señal continua y variable que debe traducirse a valores numéricos discontinuos para poder ser tratada por una computadora.

Antilock brakes (ABS) A brake system that automatically pulsates the brakes to prevent wheel lock-up under panic stop and poor traction conditions.
Frenos antibloqueo Sistema de frenos que pulsa los frenos automáticamente para impedir el bloqueo de las ruedas en casos de emergencia y de tracción pobre.

A-pillar The pillar in front of the driver or passenger that supports the windshield.
Soporte A Soporte enfrente del conductor o del pasajero que sostiene el parabrisas.

Aspirator Tubular device that uses a venturi effect to draw air from the passenger compartment over the in-car sensor. Some manufacturers use a suction motor to draw the air over the sensor.

Aspirador Dispositivo tubular que utiliza un efecto venturi para extraer aire del compartimiento del pasajero sobre el sensor dentro del vehículo. Algunos fabricantes utilizan un motor de succion para extraer el aire sobre el sensor.

Back probe A term used to mean that a test is being performed on the circuit while the connector is still connected to the component. The test probes are inserted into the back of the wire connector.
Sonda exploradora de retorno Término utilizado para expresar que se está llevando a cabo una prueba del circuito mientras el conectador sigue conectado al componente. Las sondas de prueba se insertan a la parte posterior del conectador de corriente.

Battery leakage test Used to determine if current is discharging across the top of the battery case.
Prueba de pérdida de corriente de la batería Prueba utilizada para determinar si se está descargando corriente a través de la parte superior de la caja de la batería.

Battery terminal test Checks for poor electrical connections between the battery cables and terminals. Use a voltmeter to measure voltage drop across the cables and terminals.
Prueba del borne de la batería Verifica si existen conexiones eléctricas pobres entre los cables y los bornes de la batería. Utiliza un voltímetro para medir caídas de tensión entre los cables y los bornes.

B circuit A generator regulator circuit that is internally grounded. In the B circuit, the voltage regulator controls the power side of the field circuit.
Circuito B Circuito regulador del generador puesto internamente a tierra. En el circuito B, el regulador de tensión controla el lado de potencia del circuito inductor.

Bench test A term used to indicate that the unit is to be removed from the vehicle and tested.
Prueba de banco Término utilizado para indicar que la unidad será removida del vehículo para ser examinada.

B-pillar The pillar located over the shoulder of the driver or passenger.
Soporte B Soporte ubicado sobre el hombro del conductor o del pasajero.

Brushes Electrically conductive sliding contacts, usually made of copper and carbon.
Escobillas Contactos deslizantes de conducción eléctrica, por lo general hechos de cobre y de carbono.

Bus bar A common electrical connection to which all of the fuses in the fuse box are attached. The bus bar is connected to battery voltage.
Barra colectora Conexión eléctrica común a la que se conectan todos los fusibles de la caja de fusibles. La barra colectora se conecta a la tensión de la batería.

Capacity test The part of the battery test series that checks the battery's ability to perform when loaded.
Prueba de capacidad Parte de la serie de prueba de la batería que verifica la capacidad de funcionamiento de la batería cuando está cargada.

Carbon monoxide An odorless, colorless, and toxic gas that is produced as a result of combustion.
Monóxido de carbono Gas inodoro, incoloro y tóxico producido como resultado de la combustión.

Cartridge fuses See maxi-fuse.
Fusibles cartucho Véase maxifusible.

Charging system requirement test Diagnóstic test used to determine the total electrical demand of the vehicle's electrical system.
Prueba del requisito del sistema de carga Prueba diagnóstica utilizada para determinar la exigencia eléctrica total del sistema eléctrico del vehículo.

Circuit The path of electron flow consisting of the voltage source, conductors, load component, and return path to the voltage source.
Circuito Trayectoria del flujo de electrones, compuesto de la fuente de tensión, los conductores, el componente de carga y la trayectoria de regreso a la fuente de tensión.

Clock spring Maintains a continuous electrical contact between the wiring harness and the air bag module.
Muelle de reloj Mantiene un contacto eléctrico continuo entre el cableado preformado y la unidad del Airbag.

Closed circuit A circuit that has no breaks in the path and allows current to flow.
Circuito cerrado Circuito de trayectoria ininterrumpida que permite un flujo continuo de corriente.

Cold cranking amps (CCA) Rating indicates the battery's ability to deliver a specified amount of current to start an engine at low ambient temperatures.
Amperios de arranque en frío Tasa indicativa de la capacidad de la batería para producir una cantidad específica de corriente para arrancar un motor a bajas temperaturas ambiente.

Color codes Used to assist in tracing the wires. In most color codes, the first group of letters designates the base color of the insulation and the second group of letters indicates the color of the tracer.
Códigos de colores Utilizados para facilitar la identificación de los alambres. Típicamente, el primer alfabeto representa el color base del aislamiento y el segundo representa el color del indicador.

Commutator A series of conducting segments located around one end of the armature.
Conmutador Serie de segmentos conductores ubicados alrededor de un extremo de la armadura.

Component locator Service manual used to find where a component is installed in the vehicle. The component locator uses both drawings and text to lead the technician to the desired component.
Manual para indicar los elementos componentes Manual de servico utilizado para localizar dónde se ha instalado un componente en el vehículo. En dicho manual figuran dibujos y texto para guiar al mecánico al componente deseado.

Computer An electronic device that stores and processes data and is capable of operating other devices.
Computadora Dispositivo electrónico que almacena y procesa datos y que es capaz de ordenar a otros dispositivos.

Conductor A substance that is capable of supporting the flow of electricity through it.
Conductor Sustancia capaz de conducir corriente eléctrica.

Continuity Refers to the circuit being continuous with no opens.
Continuidad Se refiere al circuito ininterrumpido, sin aberturas.

Crimping The process of bending, or deforming by pinching, a connector so that the wire connection is securely held in place.
Engarzado Proceso a través del cual se curva o deforma un conectador mediante un pellizco para que la conexión de alambre se mantenga firme en su lugar.

Curb height The height of the vehicle when it has no passengers or loads, and normal fluid levels and tire pressure.
Altura del contén La altura del vehículo cuando no lleva pasajeros ni cargas, y los niveles de los fluidos y de la presión de las llantas son normales.

Current The aggregate flow of electrons through a wire. One ampere represents the movement of 6.25 billion billion electrons (or one coulomb) past one point in a conductor in one second.
Corriente Flujo combinado de electrones a través de un alambre. Un amperio representa el movimiento de 6,25 mil millones de mil millones de electrones (o un colombio) que sobrepasa un punto en un conductor en un segundo.

Current draw test Diagnostic test used to measure the amount of current that the starter draws when actuated. It determines the electrical and mechanical condition of the starting system.
Prueba de la intensidad de una corriente Prueba diagnóstica utilizada para medir la cantidad de corriente que el arrancador tira cuando es accionado. Determina las condiciones eléctricas y mecánicas del sistema de arranque.

Current output testing Diagnostic test used to determine the maximum output of the ac generator.
Prueba de la salida de una corriente Prueba diagnóstica utilizada para determinar la salida máxima del generador de corriente alterna.

d'Arsonval gauge A gauge design that uses the interaction of a permanent magnet and an electromagnet, and the total field effect to cause needle movement.
Calibrador d'Arsonval Calibrador diseñado para utilizar la interacción de un imán permanente y de un electroimán, y el efecto inductor total para generar el movimiento de la aguja.

Deep cycling Discharging the battery completely before recharging it.
Operacion cíclica completa La descarga completa de la batería previo al recargo.

Digital A voltage signal is either on-off, yes-no, or high-low.
Digital Una señal de tensión está Encendida-Apagada, es Sí-No o Alta-Baja.

Dimmer switch A switch in the headlight circuit that provides the means for the driver to select either high beam or low beam operation, and to switch between the two. The dimmer switch is connected in series within the headlight circuit and controls the current path for high and low beams.
Conmutador reductor Conmutador en el circuito para faros delanteros que le permite al conductor elegir la luz larga o la luz corta, y conmutar entre las dos. El conmutador reductor se conecta en serie dentro del circuito para faros delanteros y controla la trayectoria de la corriente para la luz larga y la luz corta.

Diode An electrical one-way check valve that will allow current to flow in one direction only.
Diodo Válvula eléctrica de retención, de una vía, que permite que la corriente fluya en una sola dirección.

Diode rectifier bridge A series of diodes that are used to provide a reasonably constant dc voltage to the vehicle's electrical system and battery.
Puente rectificador de diodo Serie de diodos utilizados para proveerles una tensión de corriente continua bastante constante al sistema eléctrico y a la batería del vehículo.

Diode trio Used by some manufacturers to rectify the stator of an ACgenerator current so that it can be used to create the magnetic field in the field coil of the rotor.
Trío de diodos Utilizado por algunos fabricantes para rectificar el estátor de la corriente de un generador de corriente alterna y poder

así utilizarlo para crear el campo magnético en la bobina inductora del rotor.

Duty-cycle The percentage of on time to total cycle time.
Ciclo de trabajo Porcentaje del trabajo efectivo a tiempo total del ciclo.

Eddy currents Small induced currents.
Corriente de Foucault Pequeñas corrientes inducidas.

Electrical load The working device of the circuit.
Carga eléctrica Dispositivo de trabajo del circuito.

Electrochemical The chemical action of two dissimilar materials in a chemical solution.
Electroquímico Acción química de dos materiales distintos en una solución química.

Electrolyte A solution of 64% water and 36% sulfuric acid.
Electrolito Solucion de un 64% de agua y un 36% de ácido sulfúrico.

Electromagnetic gauge Gauge that produces needle movement by magnetic forces.
Calibrador electromagnético Calibrador que genera el movimiento de la aguja mediante fuerzas magnéticas.

Electromagnetic induction The production of voltage and current within a conductor as a result of relative motion within a magnetic field.
Inducción electrómagnética Producción de tension y de corriente dentro de un conductor como resultado del movimiento relativo dentro de un campo magnético.

Electromagnetic interference (EMI) An undesirable creation of electromagnetism whenever current is switched on and off.
Interferencia electromagnética Fenómeno de electromagnetismo no deseable que resulta cuando se conecta y se desconecta la corriente.

Electromagnetism A form of magnetism that occurs when current flows through a conductor.
Electromagnetismo Forma de magnetismo que ocurre cuando la corriente fluye a través de un conductor.

Electromotive force (EMF) See voltage.
Fuerza electromotriz Véase tensión.

Equivalent series load (equivalent resistance) The total resistance of a parallel circuit. It is equivalent to the resistance of a single load in series with the voltage source.
Carga en serie equivalente (resistencia equivalente) Resistencia total de un circuito en paralelo, equivalente a la resistencia de una sola carga en serie con la fuente de tensión.

Excitation current Current that magnetically excites the field circuit of the ac generator.
Corriente de excitación Corriente que excita magnéticamente al circuito inductor del generador de corriente alterna.

Failsoft Computer substitution of a fixed input value if a sensor circuit should fail. This provides for system operation, but at a limited function.
Falla activa Sustitución por la computadora de un valor fijo de entrada en caso de que ocurra una falla en el circuito de un sensor. Esto asegura el funcionamiento del sistema, pero a una capacidad limitada.

Fast charging Battery charging using a high amperage for a short period of time.
Carga rápida Carga de la batería que utiliza un amperaje máximo por un corto espacio de tiempo.

Feedback 1. Data concerning the effects of the computer's commands are fed back to the computer as an input signal. Used to determine if the desired result has been achieved. 2. A condition that can occur when electricity seeks a path of lower resistance, but the alternate path operates another component than that intended. Feedback can be classified as a short.
Realimentación 1. Datos referentes a los efectos de las órdenes de la computadora se suministran a la misma como señal de entrada. La realimentación se utiliza para determinar si se ha logrado el resultado deseado. 2. Condición que puede ocurrir cuando la electricidad busca una trayectoria de menos resistencia, pero la trayectoria alterna opera otro componente que aquel deseado. La realimentación puede clasificarse como un cortocircuito.

Fiber optics A medium of transmitting for the transmission of light through polymethylmethacrylate plastic that keeps the light rays parallel even if there are extreme bends in the plastic.
Transmisión por fibra óptica Técnica de transmisión de luz por medio de un plástico de polimetacrilato de metilo que mantiene los rayos de luz paralelos aunque el plástico esté sumamente torcido.

Field current draw test Diagnostic test that determines if there is current available to the field windings.
Prueba de la intensidad de una corriente inductora Prueba diagnóstica que determina si se está generando corriente a los devanados inductores.

Fire extinguisher A portable apparatus that contains chemicals, water, foam, or special gas that can be discharged to extinguish a small fire.
Extinctor de incendios Aparato portátil que contiene elementos químicos, agua, espuma o gas especial que pueden descargarse para extinguir un incendio peque:o.

Floor jack A portable hydraulic tool used to raise and lower a vehicle.
Gato de pie Herramienta hidráulica portátil utilizada para levantar y bajar un vehículo.

Forward-bias A positive voltage that is applied to the P-type material and negative voltage to the N-type material of a semiconductor.
Polarización directa Tensión positiva aplicada al material P y tensión negativa aplicada al material N de un semiconductor.

Free speed test Diagnostic test that determines the free rotational speed of the armature. This test is also referred to as the no-load test.
Prueba de velocidad libre Prueba diagnóstica que determina la velocidad giratoria libre de la armadura. A dicha prueba se le llama prueba sin carga.

Full field Field windings that are constantly energized with full battery current. Full fielding will produce maximum ac generator output.
Campo completo Devanados inductores que se excitan constantemente con corriente total de la batería. EL campo completo producirá la salida máxima de un generador de corriente alterna.

Full field test Diagnostic test used to isolate if the detected problem lies in the ac generator or the regulator.
Prueba de campo completo Prueba diagnóstica utilizada para determinar si el problema descubierto se encuentra en el generador de corriente alterna o en el regulador.

Fuse A replaceable circuit protection device that will melt should the current passing through it exceed its rating.
Fusible Dispositivo reemplazable de protección del circuito que se fundirá si la corriente que fluye por el mismo excede su valor determinado.

Fuse box A term used that indicates the central location of the fuses contained in a single holding fixture.
Caja de fusibles Término utilizado para indicar la ubicación central de los fusibles contenidos en un solo elemento permanente.

Fusible link A wire made of meltable material with a special heat-resistant insulation. When there is an overload in the circuit, the link melts and opens the circuit.

Cartucho de fusible Alambre hecho de material fusible con aislamiento especial resistente al calor. Cuando ocurre una sobrecarga en el circuito, el cartucho se funde y abre el circuito.

Gauge 1. A device that displays the measurement of a monitored system by the use of a needle or pointer that moves along a calibrated scale. 2. The number that is assigned to a wire to indicate its size. The larger the number the smaller the diameter of the conductor.

Calibrador 1. Dispositivo que muestra la medida de un sistema regulado por medio de una aguja o indicador que se mueve a través de una escala calibrada. 2. El número asignado a un alambre indica su tamaño. Mientras mayor sea el número, más pequeño será el diámetro del conductor.

Gauss gauge A meter that is sensitive to the magnetic field surrounding a wire conducting current. The gauge needle will fluctuate over the portion of the circuit that has current flowing through it. Once the ground has been passed, the needle will stop fluctuating.

Calibrador gauss Instrumento sensible al campo magnético que rodea un alambre conductor de corriente. La aguja del calibrador se moverá sobre la parte del circuito a través del cual fluye la corriente. Una vez se pasa a tierra, la aguja dejará de moverse.

Ground The common negative connection of the electrical system that is the point of lowest voltage.

Tierra Conexión negativa común del sistema eléctrico. Es el punto de tensión más baja.

Ground circuit test A diagnostic test performed to measure the voltage drop in the ground side of the circuit.

Prueba del circuito a tierra Prueba diagnóstica llevada a cabo para medir la caída de tensión en el lado a tierra del circuito.

Grounded circuit An electrical defect that allows current to return to ground before it has reached the intended load component.

Circuito puesto a tierra Falla eléctrica que permite el regreso de la corriente a tierra antes de alcanzar el componente de carga deseado.

Ground side The portion of the circuit that is from the load component to the negative side of the source.

Lado a tierra Parte del circuito que va del componente de carga al lado negativo de la fuente.

Growler Test equipment used to test starter armatures for shorts and grounds. It produces a very strong magnetic field that is capable of inducing a current flow and magnetism in a conductor.

Indicador de cortocircuitos Equipo de prueba utilizado para localizar cortociruitos y tierra en armaduras de arranque. Genera un campo magnético sumamente fuerte, capaz de inducir flujo de corriente y magnetismo en un conductor.

Hall-effect switch A sensor that operates on the principle that if a current is allowed to flow through thin conducting material being exposed to a magnetic field, another voltage is produced.

Conmutador de efecto Hall Sensor que funciona basado en el principio de que si se permite el flujo de corriente a través de un material conductor delgado que ha sido expuesto a un campo magnético, se produce otra tensión.

Halogen The term used to identify a group of chemically related nonmetallic elements. These elements include chlorine, fluorine, and iodine.

Halógeno Término utilizado para identificar un grupo de elementos no metálicos relacionados químicamente. Dichos elementos incluyen el cloro, el flúor y el yodo.

Hard-shell connector An electrical connector that has a hard plastic shell that holds the connecting terminals of separate wires.

Conectador de casco duro Conectador eléctrico con casco duro de plástico que sostiene separados los bornes conectadores de alambres individuales.

Heat-shrink tubing A hollow insulation material that shrinks to an airtight fit over a connection when exposed to heat.

Tubería contraída térmicamente Material aislante hueco que se contrae para acomodarse herméticamente sobre una conexión cuando se encuentra expuesto al calor.

Hoist A lift that is used to raise the entire vehicle.

Elevador Montacargas utilizado para elevar el vehículo en su totalidad.

Hydrometer A test instrument used to check the specific gravity of the electrolyte to determine the battery's state of charge.

Hidrómetro Instrumento de prueba utilizado para verificar la gravedad específica del electrolito y así determinar el estado de carga de la batería.

Hydrostatic lock Liquid entering the cylinder and preventing the piston from moving upward.

Cierre hidrostático La entrada de líquido en el cilindro que impide el movimiento ascendente del pistón.

Impedance The combined opposition to current created by the resistance, capacitance, and inductance of a test meter or circuit.

Impedancia Opocisión combinada a la corriente generada por la resistencia, la capacitancia y la inductancia de un instrumento de prueba o de un circuito.

Instrument voltage regulator (IVR) Provides a constant voltage to the gauge regardless of the voltage output of the charging system.

Instrumento regulador de tensión Le provee tensión constante al calibrador, sin importar cual sea la salida de tensión del sistema de carga.

Insulated circuit resistance test A voltage drop test that is used to locate high resistance in the starter circuit.

Prueba de la resistencia de un circuito aislado Prueba de la caída de tensión utilizada para localizar alta resistencia en el circuito de arranque.

Insulated side The portion of the circuit from the positive side of the source to the load component.

Lado aislado Parte del circuito que va del lado positivo de la fuente al componente de carga.

Insulator A substance that is not capable of supporting the flow of electricity.

Aislador Sustancia que no es capaz de soportar el flujo de electricidad.

Integrated circuit (IC chip) A complex circuit of thousands of transistors, diodes, resistors, capacitors, and other electronic devices that are formed onto a small silicon chip. As many as 30,000 transistors can be placed on a chip that is 1/4 inch (6.35 mm) square.

Circuito integrado (Fragmento CI) Circuito complejo de miles de transistores, diodos, resistores, condensadores, y otros dispositivos electrónicos formados en un fragmento pequeño de silicio. En un fragmento de 1/4 de pulgada (6,35 mm) cuadrada, pueden colocarse hasta 30.000 transistores.

Jack stands Support devices used to hold the vehicle off the floor after it has been raised by the floor jack.

Soportes de gato Dispositivos de soporte utilizados para sostener el vehículo sobre el suelo después de haber sido levantado con el gato de pie.

Lamp A device that produces light as a result of current flow through a filament. The filament is enclosed within a glass envelope and is a type of resistance wire that is generally made from tungsten.

Lámpara Dispositivo que produce luz como resultado del flujo de corriente a través de un filamento. El filamento es un tipo de alambre de resistencia hecho por lo general de tungsteno, que es encerrado dentro de una bombilla.

Light-emitting diode (LED) A gallium-arsenide diode that converts the energy developed when holes and electrons collide during normal diode operation into light.

Diodo emisor de luz Diodo semiconductor de galio y arseniuro que convierte en luz la energía producida por la colisión de agujeros y electrones durante el funcionamiento normal del diodo.

Liquid crystal display (LCD) A display that sandwiches electrodes and polarized fluid between layers of glass. When voltage is applied to the electrodes, the light slots of the fluid are rearranged to allow light to pass through.

Visualizador de cristal líquido Visualizador digital que consta de dos láminas de vidrio selladas, entre las cuales se encuentran los electrodos y el fluido polarizado. Cuando se aplica tensión a los electrodos, se rompe la disposición de las moléculas para permitir la formación de carácteres visibles.

Magnetic pulse generator Sensor that uses the principle of magnetic induction to produce a voltage signal. Magnetic pulse generators are commonly used to send data concerning the speed of the monitored component to the computer.

Generador de impulsos magnéticos Sensor que funciona según el principio de inducción magnética para producir una señal de tensión. Los generadores de impulsos magnéticos se utilizan comúnmente para transmitir datos a la computadora relacionados a la velocidad del componente regulado.

Magnetism An energy form resulting from atoms aligning within certain materials, giving the materials the ability to attract other metals.

Magnetismo Forma de energía que resulta de la alineación de átomos dentro de ciertos materiales y que le da a éstos la capacidad de atraer otros metales.

Maxi-fuse A circuit protection device that looks similar to blade-type fuses except they are larger and have a higher amperage capacity. Maxi-fuses are used because they are less likely to cause an underhood fire when there is an overload in the circuit. If the fusible link burns in two, it is possible that the "hot" side of the fuse could come into contact with the vehicle frame and the wire could catch on fire.

Maxifusible Dispositivo de protección del circuito parecido a un fusible de tipo de cuchilla, pero más grande y con mayor capacidad de amperaje. Se utilizan maxifusibles porque existen menos probabilidades de que ocasionen un incendio debajo de la capota cuando ocurra una sobrecarga en el circuito. Si el cartucho de fusible se quemase en dos partes, es posible que el lado "cargado" del fusible entre en contacto con el armazón del vehículo y que el alambre se encienda.

Molded connector An electrical connector that usually has one to four wires that are molded into a one-piece component.

Conectador moldeado Conectador eléctrico que por lo general tiene hasta un máximo cuatro alambres que se moldean en un componente de una sola pieza.

No-crank A term used to mean that when the ignition switch is placed in the START position, the starter does not turn the engine.

Sin arranque Término utilizado para expresar que cuando el botón conmutador de encendido está en la posición START, el arrancador no enciende el motor.

No-crank test Diagnostic test performed to locate any opens in the starter or control circuits.

Prueba sin arranque Prueba diagnóstica llevada a cabo para localizar aberturas en los circuitos de arranque o de mando.

Normally closed (NC) switch A switch designation denoting that the contacts are closed until acted upon by an outside force.

Conmutador normalmente cerrado Nombre aplicado a un conmutador cuyos contactos permanecerán cerrados hasta que sean accionados por una fuerza exterior.

Normally open (NO) switch A switch designation denoting that the contacts are open until acted upon by an outside force.

Conmutador normalmente abierto Nombre aplicado a un conmutador cuyos contactos permanecerán abiertos hasta que sean accionados por una fuerza exterior.

Occupational safety glasses Eye protection that is designed with special high-impact lens and frames, and side protection.

Gafas de protección para el trabajo Gafas diseñadas con cristales y monturas especiales resistentes y provistas de protección lateral.

Odometer A mechanical counter in the speedometer unit that indicates total miles accumulated on the vehicle.

Odómetro Aparato mecánico en la unidad del velocímetro con el que se cuentan las millas totales recorridas por el vehículo.

Ohm Unit of measure for resistance. One ohm is the resistance of a conductor such that a constant current of one ampere in it produces a voltage of one volt between its ends.

Ohmio Unidad de resistencia eléctrica. Un ohmio es la resistencia de un conductor si una corriente constante de 1 amperio en el conductor produce una tensión de 1 voltio entre los dos extremos.

Ohmmeter A test meter used to measure resistance and continuity in a circuit.

Ohmiómetro Instrumento de prueba utilizado para medir la resistencia y la continuidad en un circuito.

Ohm's law Defines the relationship between current, voltage, and resistance.

Ley de Ohm Define la relación entre la corriente, la tensión y la resistencia.

Open circuit A term used to indicate that current flow is stopped. By opening the circuit, the path for electron flow is broken.

Circuito abierto Término utilizado para indicar que el flujo de corriente ha sido detenido. Al abrirse el circuito, se interrumpe la trayectoria para el flujo de electrones.

Open circuit voltage test Used to determine the battery's state of charge. It is used when a hydrometer is not available or cannot be used.

Prueba de la tensión en un circuito abierto Sirve para determinar el estado de carga de la batería. Esta prueba se lleva a cabo cuando no se dispone de un hidrómetro o cuando el mismo no puede utilizarse.

Overload Excess current flow in a circuit.

Sobrecarga Flujo de corriente superior a la que tiene asignada un circuito.

Parallel circuit A circuit that provides two or more paths for electricity to flow.

Circuito en paralelo Circuito que provee dos o más trayectorias para que circule la electricidad.

Parasitic loads Electrical loads that are still present when the ignition switch is in the OFF position.

Cargas parásitas Cargas eléctricas que todavía se encuentran presente cuando el botón conmutador de encendido está en la posición OFF.

Park switch Contact points located inside the wiper motor assembly that supply current to the motor after the wiper control switch has been turned to the PARK position. This allows the motor to continue operating until the wipers have reached their PARK position.

Conmutador PARK Puntos de contacto ubicados dentro del conjunto del motor del frotador que le suministran corriente al motor después de que el conmutador para el control de los frotadores haya sido colocado en la posición PARK. Esto permite que el motor continue su funcionamiento hasta que los frotadores hayan alcanzado la posición original.

Passive seatbelt system Seatbelt operation that automatically puts the shoulder and/or lap belt around the driver or occupant. The automatic seatbelt is moved by dc motors that move the belts by means of carriers on tracks.

Sistema pasivo de cinturones de seguridad Función de los cinturones de seguridad que automáticamente coloca el cinturón superior y/o inferior sobre el conductor o pasajero. Motores de corriente continua accionan los cinturones automáticos mediante el uso de portadores en pistas.

Photocell A variable resistor that uses light to change resistance.

Fotocélula Resistor variable que utiliza luz para cambiar la resistencia.

Phototransistor A transistor that is sensitive to light.

Fototransistor Transistor sensible a la luz.

Photovoltaic diodes Diodes capable of producing a voltage when exposed to radiant energy.

Diodos fotovoltaicos Diodos capaces de generar una tensión cuando se encuentran expuestos a la energía de radiación.

Pick-up coil The stationary component of the magnetic pulse generator consisting of a weak permanent magnet that is wound around by fine wire. As the timing disc rotates in front of it, the changes of magnetic lines of force generate a small voltage signal in the coil.

Bobina captadora Componente fijo del generador de impulsos magnéticos compuesta de un imán permanente débil devanado con alambre fino. Mientras gira el disco sincronizador enfrente de él, los cambios de las líneas de fuerza magnética generan una pequeña señal de tensión en la bobina.

Piezoresistive sensor A sensor that is sensitive to pressure changes.

Sensor piezoresistivo Sensor susceptible a los cambios de presión.

Pneumatic tools Power tools that are powered by compressed air.

Herrimientas neumáticas Herramientas mecánicas accionadas por aire comprimido.

Potentiometer A variable resistor that acts as a circuit divider, providing accurate voltage drop readings proportional to movement.

Potenciómetro Resistor variable que actúa como un divisor de circuito para obtener lecturas de perdidas de tensión precisas en proporción con el movimiento.

Power tools Tools that use forces other than those generated from the body. They can use compressed air, electricity, or hydraulic pressure to generate and multiply force.

Herramientas mecánicas Herramientas que utilizan fuerzas distintas a las generadas por el cuerpo. Dichas fuerzas pueden ser el aire comprimido, la electricidad, o la presión hidráulica para generar y multiplicar la fuerza.

Primary wiring Conductors that carry low voltage and low current. The insulation of primary wires is usually thin.

Hilos primarios Hilos conductores de tensión y corriente bajas. El aislamiento de hilos primarios es normalmente delgado.

Program A set of instructions that the computer must follow to achieve desired results.

Programa Conjunto de instrucciones que la computadora debe seguir para lograr los resultados deseados.

PROM (programmable read only memory) Memory chip that contains specific data that pertains to the exact vehicle that the computer is installed in. This information may be used to inform the CPU of the accessories that are equipped on the vehicle.

PROM (memoria de sólo lectura programable) Fragmento de memoria que contiene datos específicos referentes al vehículo particular en el que se instala la computadora. Esta información puede utilizarse para informar a la UCP sobre los accesorios de los cuales el vehículo está dotado.

Protection device Circuit protector that is designed to "turn off" the system that it protects. This is done by creating an open to prevent a complete circuit.

Dispositivo de protección Protector de circuito diseñado para "desconectar" el sistema al que provee protección. Esto se hace abriendo el circuito para impedir un circuito completo.

Prove-out circuit A function of the ignition switch that completes the warning light circuit to ground through the ignition switch when it is in the START position. The warning light is on during engine cranking to indicate to the driver that the bulb is working properly.

Circuito de prueba Función del boton conmutador de encendido que completa el circuito de la luz de aviso para que se ponga a tierra a través del botón conmutador de encendido cuando éste se encuentra en la posición START. La luz de aviso se encenderá durante el arranque del motor para avisarle al conductor que la bombilla funciona correctamente.

Pulse width The length of time in milliseconds that an actuator is energized.

Duración de impulsos Espacio de tiempo en milisegundos en el que se excita un accionador.

Pulse width modulation On/off cycling of a component. The period of time for each cycle does not change, only the amount of on time in each cycle changes.

Modulación de duración de impulsos Modulación de impulsos de un componente. El espacio de tiempo de cada ciclo no varía; lo que varía es la cantidad de trabajo efectivo de cada ciclo.

Rectification The converting of ac current to dc current.

Rectificación Proceso a través del cual la corriente alterna es transformada en una corriente continua.

Relay A device that uses low current to control a high current circuit. Low current is used to energize the electromagnetic coil, while high current is able to pass over the relay contacts.

Relé Dispositivo que utiliza corriente baja para controlar un circuito de corriente alta. La corriente baja se utiliza para excitar la bobina electromagnética, mientras que la corriente alta puede transmitirse a través de los contactos del relé.

Reserve-capacity rating An indicator, in minutes, of how long the vehicle can be driven with the headlights on, if the charging system should fail. The reserve-capacity rating is determined by the length of time, in minutes, that a fully charged battery can be discharged at 25 amperes before battery cell voltage drops below 1.75 volts per cell.

Clasificación de capacidad en reserva Indicación, en minutos, de cuánto tiempo un vehículo puede continuar siendo conducido, con los faros delanteros encendidos, en caso de que ocurriese una falla en el sistema de carga. La clasificación de capacidad en reserva se determina por el espacio de tiempo, en minutos, en el que una batería completamente cargada puede descargarse a 25 amperios antes de que la tensión del acumulador de la batería disminuya a un nivel inferior de 1,75 amperios por acumulador.

Resistance Opposition to current flow.
Resistencia Oposición que presenta un conductor al paso de la corriente eléctrica.

Resistive shorts Shorts to ground that pass through a form of resistance first.
Cortocircuitos resistivos Cortocircuitos a tierra que primero pasan por una forma de resistencia.

Resistor block A series of resistors with different values.
Bloque resistor Serie de resistores que tienen valores diferentes.

Reversed-bias A positive voltage is applied to the N-type material and negative voltage is applied to the P-type material of a semiconductor.
Polarización inversa Tensión positiva aplicada al material N y tensión negativa aplicada al material P de un semiconductor.

Rheostat A two-terminal variable resistor used to regulate the strength of an electrical current.
Reóstato Resistor variable de dos bornes utilizado para regular la resistencia de una corriente eléctrica.

Rotor The component of the ac generator that is rotated by the drive belt and creates the rotating magnetic field of the ac generator.
Rotor Parte rotativa del generador de corriente alterna accionada por la correa de transmisión y que produce el campo magnético rotativo del generador de corriente alterna.

Safety goggles Eye protection device that fits against the face and forehead to seal off the eyes from outside elements.
Gafas de seguridad Dispositivo protector que se coloca delante de los ojos para preservarlos de elementos extraños.

Safety stands See Jack stands.
Soportes de seguridad Véase soportes de gato.

Scanner A diagnostic test tool that is designed to communicate with the vehicle's on-board computer.
Dispositivo de exploración Herramienta de prueba diagnóstica diseñada para comunicarse con la computadora instalada en el vehículo.

Sealed-beam headlight A self-contained glass unit that consists of a filament, an inner reflector, and an outer glass lens.
Faro delantero sellado Unidad de vidrio que contiene un filamento, un reflector interior y una lente exterior de vidrio.

Secondary wiring Conductors, such as battery cables and ignition spark plug wires, that are used to carry high voltage or high current. Secondary wires have extra thick insulation.
Hilos secundarios Conductores, tales como cables de batería e hilos de bujías del encendido, utilizados para transmitir tensión o corriente alta. Los hilos secundarios poseen un aislamiento sumamente grueso.

Semiconductors An element that is neither a conductor nor an insulator. Semiconductors are materials that conduct electric current under certain conditions, yet will not conduct under other conditions.
Semiconductores Elemento que no es ni conductor ni aislante. Los semiconductores son materiales que transmiten corriente eléctrica bajo ciertas circunstancias, pero no la transmiten bajo otras.

Sender unit The sensor for the gauge. It is a variable resistor that changes resistance values with changing monitored conditions.
Unidad emisora Sensor para el calibrador. Es un resistor variable que cambia los valores de resistencia según cambian las condiciones reguladas.

Sensor Any device that provides an input to the computer.
Sensor Cualquier dispositivo que le transmite información a la computadora.

Series circuit A circuit that provides a single path for current flow from the electrical source through all the circuit's components, and back to the source.
Circuito en serie Circuito que provee una trayectoria única para el flujo de corriente de la fuente eléctrica a través de todos los componentes del circuito, y de nuevo hacia la fuente.

Series-parallel circuit A circuit that has some loads in series and some in parallel.
Circuito en series paralelas Circuito que tiene unas cargas en serie y otras en paralelo.

Servomotor An electrical motor that produces rotation of less than a full turn. A feedback mechanism is used to position itself to the exact degree of rotation required.
Servomotor Motor eléctrico que genera rotación de menos de una revolución completa. Utiliza un mecanismo de realimentación para ubicarse al grado exacto de la rotación requerida.

Short An electrical fault that allows for electrical current to bypass its normal path.
Cortocircuito Falla eléctrica que permite que la corriente eléctrica se desvíe de su trayectoria normal.

Shutter wheel A metal wheel consisting of a series of alternating windows and vanes. It creates a magnetic shunt that changes the strength of the magnetic field from the permanent magnet of the Hall-effect switch or magnetic pulse generator.
Rueda obturadora Rueda metálica compuesta de una serie de ventanas y aspas alternas. Genera una derivación magnética que cambia la potencia del campo magnético, del imán permanente del conmutador de efecto Hall o del generador de impulsos magnéticos.

Slow charging Battery charging rate between 3 and 15 amps for a long period of time.
Carga lenta Indice de carga de la batería de entre 3 y 15 amperios por un largo espacio de tiempo.

Slow cranking A term used to mean that the starter drive engages the ring gear, but the engine turns too slowly to start.
Arranque lento Término utilizado para expresar que el mecanismo de transmisión de arranque engrana la corona, pero que el motor se enciende de forma demasiado lenta para arrancar.

Soft codes Codes are those that have occurred in the past, but were not present during the last BCM test of the circuit.
Códigos suaves Códigos que han ocurrido en el pasado, pero que no estaban presentes durante la última prueba BCM del circuito.

Soldering The process of using heat and solder (a mixture of lead and tin) to make a splice or connection.
Soldadura Proceso a través del cual se utiliza calor y soldadura (una mezcla de plomo y de estaño) para hacer un empalme o una conexión.

Solderless connectors Hollow metal tubes that are covered with insulating plastic. They can be butt connectors or terminal ends.
Conectadores sin soldadura Tubos huecos de metal cubiertos de plástico aislante. Pueden ser extremos de conectadores o de bornes.

Solenoid An electromagnetic device that uses movement of a plunger to exert a pulling or holding force.
Solenoide Dispositivo electromagnético que utiliza el movimiento de un pulsador para ejercer una fuerza de arrastre o de retención.

Solenoid circuit resistance test Diagnostic test used to determine the electrical condition of the solenoid and the control circuit of the starting system.
Prueba de la resistencia de un circuito solenoide Prueba diagnóstica utilizada para determinar la condición eléctrica del solenoide y del circuito de mando del sistema de arranque.

Specific gravity The weight of a given volume of a liquid divided by the weight of an equal volume of water.
Gravedad específica El peso de un volumen dado de líquido dividido por el peso de un volumen igual de agua.

Speedometer An instrument panel gauge that indicates the speed of the vehicle.
Velocímetro Calibrador en el panel de instrumentos que marca la velocidad del vehículo.

Splice The joining of single wire ends or the joining of two or more electrical conductors at a single point.
Empalme La unión de los extremos de un alambre o la unión de dos o más conductores eléctricos en un solo punto.

Splice clip A special connector used along with solder to assure a good connection. The splice clip is different from solderless connectors in that it does not have insulation.
Grapa para empalme Conectador especial utilizado junto con la soldadura para garantizar una conexión perfecta. La grapa para empalme se diferencia de los conectadores sin soldadura porque no está provista de aislamiento.

Starter drive The part of the starter motor that engages the armature to the engine flywheel ring gear.
Transmisión de arranque Parte del motor de arranque que engrana la armadura a la corona del volante de la máquina.

State of charge The condition of a battery's electrolyte and plate materials at any given time.
Estado de carga Condición del electrolito y de los materiales de la placa de una batería en cualquier momento dado.

Stator The stationary coil of the ac generator in which current is produced.
Estátor Bobina fija del generador de corriente alterna donde se genera corriente.

Stepped resistor A resistor that has two or more fixed resistor values.
Resistor de secciones escalonadas Resistor que tiene dos o más valores de resistencia fija.

Stepper motor An electrical motor that contains a permanent magnet armature with two or four field coils. Can be used to move the controlled device to whatever location is desired. By applying voltage pulses to selected coils of the motor, the armature will turn a specific number of degrees. When the same voltage pulses are applied to the opposite coils, the armature will rotate the same number of degrees in the opposite direction.
Motor paso a paso Motor eléctrico que contiene una armadura magnética fija con dos o cuatro bobinas inductoras. Puede utilizarse para mover el dispositivo regulado a cualquier lugar deseado. Al aplicárseles impulsos de tensión a ciertas bobinas del motor, la armadura girará un número específico de grados. Cuando estos mismos impulsos de tensión se aplican a las bobinas opuestas, la armadura girará el mismo número de grados en la dirección opuesta.

Sulfation A chemical action within the battery that interferes with the ability of the cells to deliver current and accept a charge.
Sulfatado Acción química dentro de la batería que interfiere con la capacidad de los acumuladores de transmitir corriente y recibir una carga.

Tachometer An instrument that measures the speed of the engine in revolutions per minute (rpm).
Tacómetro Instrumento que mide la velocidad del motor en revoluciones por minuto (rpm).

Thermistor A solid-state variable resistor made from a semiconductor material that changes resistance in relation to temperature changes.

Termistor Resistor variable de estado sólido hecho de un material semiconductor que cambia su resistencia en relación con los cambios de temperatura.

Three-coil gauge A gauge design that uses the interaction of three electromagnets and the total field effect upon a permanent magnet to cause needle movement.
Calibrador de tres bobinas Calibrador diseñado para utilizar la interacción de tres electroimanes y el efecto inductor total sobre un imán permanente para producir el movimiento de la aguja.

Three-minute charge test A reasonably accurate method for diagnosing a sulfated battery on conventional batteries.
Prueba de carga de tres minutos Método bastante preciso en baterías convencionales para diagnosticar una batería sulfatada.

Trouble codes Output of the self-diagnostics program in the form of a numbered code that indicates faulty circuits or components. Trouble codes are two or three digital characters that are displayed in the diagnostic display if the testing and failure requirements are both met.
Códigos indicadores de fallas Datos del programa autodiagnóstico en forma de código numerado que indica los circuitos o los componentes defectuosos. Dichos códigos se componen de dos o tres carácteres digitales que se muestran en el visualizador diagnóstico si se llenan los requisitos de prueba y de falla.

Troubleshooting The diagnostic procedure of locating and identifying the cause of the fault. It is a step-by-step process of elimination by use of cause-and-effect.
Detección de fallas Procedimiento diagnóstico a través del cual se localiza e identifica la falla. Es un proceso de eliminación que se lleva a cabo paso a paso por medio de causa y efecto.

Two-coil gauge A gauge design that uses the interaction of two electromagnets and the total field effect upon an armature to cause needle movement.
Calibrador de dos bobinas Calibrador diseñado para utilizar la interacción de dos electroimanes y el efecto inductor total sobre una armadura para generar el movimiento de la aguja.

Vacuum distribution valve A valve used in vacuum-controlled concealed headlight systems. It controls the direction of vacuum to various vacuum motors or to vent.
Válvula de distribución al vacío Válvula utilizada en el sistema de faros delanteros ocultos controlado al vacío. Regula la dirección del vacío a varios motores al vacío o sirve para dar salida del sistema.

Vacuum fluorescent display (VFD) A display type that uses anode segments coated with phosphor and bombarded with tungsten electrons to cause the segments to glow.
Visualización de fluorescencia al vacío Tipo de visualización que utiliza segmentos ánodos cubiertos de fósforo y bombardeados de electrones de tungsteno para producir la luminiscencia de los segmentos.

Variable resistor A resistor that provides for an infinite number of resistance values within a range.
Resistor variable Resistor que provee un número infinito de valores de resistencia dentro de un margen.

Vehicle Identification Number (VIN) A number that is assigned to a vehicle for identification purposes. The identification plate is usually located on the cowl, next to the left upper instrument panel.
Número de identificación del vehículo Número asignado a cada vehículo para fines de identificación. Por lo general, la placa de identificación se ubica en la bóveda, al lado del panel de instrumentos superior de la izquierda.

Vehicle lift points The areas that the manufacturer recommends for safe vehicle lifting. They are the areas that are structurally strong enough to sustain the stress of lifting.

Puntos para elevar el vehículo Áreas específicas que el fabricante recomienda para sujetar el vehículo a fin de lograr una elevación segura. Son las áreas del vehículo con una estructura suficientemente fuerte para sostener la presión de la elevación.

Volt The unit used to measure the amount of electrical force.

Voltio Unidad práctica de tensión para medir la cantidad de fuerza eléctrica.

Voltage The difference or potential that indicates an excess of electrons at the end of the circuit the farthest from the electromotive force. It is the electrical pressure that causes electrons to move through a circuit. One volt is the amount of pressure required to move one amp of current through one ohm of resistance.

Tensión Diferencia o potencial que indica un exceso de electrones al punto del circuito que se encuentra más alejado de la fuerza electromotriz. La presión eléctrica genera el movimiento de electrones a través de un circuito. Un voltio equivale a la cantidad de presión requerida para mover un amperio de corriente a través de un ohmio de resistencia.

Voltage drop A resistance in the circuit that reduces the electrical pressure available after the resistance. The resistance can be either the load component, the conductors, any connections, or unwanted resistance.

Caída de tensión Resistencia en el circuito que disminuye la presión eléctrica disponible después de la resistencia. La resistencia puede ser el componente de carga, los conductores, cualquier conexión o resistencia no deseada.

Voltage regulator Used to control the output voltage of the ac generator, based on charging system demands, by controlling field current.

Regulador de tensión Dispositivo cuya función es mantener la tensión de salida del generador de corriente alterna, de acuerdo a las variaciones en la corriente de carga, controlando la corriente inductora.

Voltmeter A test meter used to read the pressure behind the flow of electrons.

Voltímetro Instrumento de prueba utilizado para medir la presión del flujo de electrones.

Warning light A lamp that is illuminated to warn the driver of a possible problem or hazardous condition.

Luz de aviso Lámpara que se enciende para avisarle al conductor sobre posibles problemas o condiciones peligrosas.

Watt The unit of measure of electrical power, which is the equivalent of horsepower. One horsepower is equal to 746 watts.

Watio Unidad de potencia eléctrica, equivalente a un caballo de vapor. 746 watios equivalen a un caballo de vapor (CV).

Wattage A measure of the total electrical work being performed per unit of time.

Vataje Medida del trabajo eléctrico total realizado por unidad de tiempo.

Weather-pack connector An electrical connector that has rubber seals on the terminal ends and on the covers of the connector half to protect the circuit from corrosion.

Conectador resistente a la intemperie Conectador que tiene sellos de caucho en los extremos de los bornes y en las cubiertas de la parte del conectador para proteger el circuito contra la corrosión.

Wiring diagram An electrical schematic that shows a representation of actual electrical or electronic components and the wiring of the vehicle's electrical systems.

Esquema de conexiones Esquema en el que se muestran las conexiones internas de los componentes eléctricos o electronicos reales y las de los sistemas eléctricos del vehículo.

Wiring harness A group of wires enclosed in a conduit and routed to specific areas of the vehicle.

Cableado preformado Conjunto de alambres envueltos en un conducto y dirigidos hacia áreas específicas del vehículo.

7 Heating and Air Conditioning Systems

Pretest

The purpose of this pretest is to determine the amount of review that you may require prior to writing the ASE Heating and Air Conditioning Systems Test. If you answer all the pretest questions correctly, complete the questions and study the information in this chapter to prepare for the ASE Heating and Air Conditioning Systems Test.

If two or more of your answers to the pretest questions are incorrect, complete a study of Chapters 2 through 13 in *Today's Technician Automotive Air Conditioning Classroom and Shop Manuals*, published by Delmar Publishers, plus a study of the questions and information in this chapter.

The pretest answers are located at the end of the pretest; these answers also are in the answer sheets supplied with this book.

1. Technician A says the line from the condenser to the evaporator should feel warm in the A/C mode.
   Technician B says the accumulator and the compressor suction line should feel cool in the A/C mode.
   Who is correct?
   A. A only
   B. B only
   C. Both A and B
   D. Neither A nor B

2. The sight glass appears clear when an R-12 refrigerant system is operating in the A/C mode. This condition may indicate
   A. air and moisture in the system.
   B. a refrigerant overcharge.
   C. low refrigerant charge.
   D. excessive oil in the system.

3. A refrigerant system has low high-side and high low-side pressure. This could indicate
   A. a defective A/C compressor.
   B. a low refrigerant charge.
   C. a restricted receiver/drier.
   D. a restricted TXV valve.

4. Technician A says a flame-type leak detector may be used on an R-134a refrigerant system.
   Technician B says an R-12 refrigerant system may be charged with R-12 containing a red leak detecting dye.
   Who is correct?
   A. A only
   B. B only
   C. Both A and B
   D. Neither A nor B

5. After a refrigerant system is evacuated, the low-side gauge indicates 29 in. Hg. The vacuum gauge reading rises 1 in. Hg. in 10 minutes.
   Technician A says the refrigerant system has a leak that must be repaired.
   Technician B says to install a partial charge and leak test the system.
   Who is correct?
   A. A only
   B. B only
   C. Both A and B
   D. Neither A nor B

6. During a low-side refrigerant charging procedure on an R-12 system, the low-side manifold gauge set valve should be adjusted to maintain the system pressure at
   A. 20 psi (138 kPa).
   B. 40 psi (275 kPa).
   C. 55 psi (379 kPa).
   D. 75 psi (517 kPa).

7. Technician A says the lubrication in an R-12 system is a polyalkalene glycol (PAG) oil.
   Technician B says a polyalkalene glycol (PAG) oil is hydroscopic.
   Who is correct?
   A. A only
   B. B only
   C. Both A and B
   D. Neither A nor B

8. The fuse in a compressor clutch circuit blows repeatedly. The cause of this problem could be
   A. a loose ground connection on the clutch coil.
   B. a shorted winding in the clutch coil.
   C. a intermittent open circuit in the clutch coil.
   D. a grounded circuit in the clutch control relay winding.

9. The clearance is less than specified between the compressor clutch plate and the pulley friction surface.
   Technician A says another shim is required behind the armature.
   Technician B says another shim is required behind the armature retaining nut.
   Who is correct?
   A. A only
   B. B only
   C. Both A and B
   D. Neither A nor B

10. In the A/C mode frost is forming on the tube between the condenser and the receiver/drier. The cause of this problem could be
    A. restricted condenser air passages.
    B. a restricted receiver/drier.
    C. a restricted fixed orifice tube.
    D. a restriction in the condenser to receiver/drier tube.

11. A refrigerant system has high pressures indicated on the low-side and high-side gauges, and frost is forming on the TXV valve. A shop towel soaked in hot water is placed on the TXV and the system pressures drop, but still remain above normal. The most likely cause of this problem is
    A. moisture in the refrigerant system.
    B. a refrigerant overcharge.
    C. a restricted TXV valve.
    D. a restricted evaporator.

12. A vehicle has a very rotten smell in the passenger compartment.
Technician A says the evaporator case drain may be plugged.
Technician B says the heater core may be leaking.
Who is correct?
A. A only
B. B only
C. Both A and B
D. Neither A nor B

13. When using refrigerant recovery/recycling equipment
A. R-12 and R-134a refrigerants may be mixed in a refrigerant system.
B. mineral oil and PAG oil may be mixed in a refrigerant system.
C. the refrigerant container specified by manufacturer must be used.
D. the equipment must have an SAE J1930 approval.

14. All of these statements about R-12 and R-134a refrigerant are true EXCEPT
A. R-12 is stored in white containers.
B. R-134a is stored in blue containers.
C. R-134a and R-12 may released to the atmosphere.
D. R-12 is harmful to the earth's ozone layer.

Answers to Pretest

1. C, 2. B, 3. A, 4. B, 5. D, 6. B, 7. B, 8. B, 9. A, 10. D, 11. A, 12. A, 13. C, 14. C

A/C System Diagnosis and Repair

ASE Tasks, Questions, and Related Information

In this chapter each task in the Heating and Air Conditioning Systems category is provided followed by a question and some information related to the task. If you answer any question incorrectly, study this information very carefully until you understand the correct answer. For additional information on any task refer to *Today's Technician Automotive Air Conditioning Classroom and Shop Manuals*, published by Delmar Publishers.

Question answers and analysis are provided at the end of this chapter and in the answer sheets provided with this book.

Task 1 **Diagnose the cause of unusual operating noises of the A/C system; determine needed repairs.**

1. An air conditioning (A/C) compressor has a growling noise only with the compressor clutch engaged. The cause of this noise could be
A. a defective internal compressor bearing.
B. a defective pulley bearing.
C. a low refrigerant charge.
D. excessive refrigerant system pressure.

Hint *A squealing noise may be caused by a loose, dry, or worn A/C compressor drive belt. This noise may be worse during fast engine acceleration. Worn or dry blower motor bushings may cause a squealing noise when the blower motor is running; this noise may occur when the engine is first started after sitting overnight.*

A rattling noise from the compressor may be caused by a loose or worn clutch hub, or loose compressor mounting bolts.

A thumping, banging noise from the compressor may be caused by liquid refrigerant entering the compressor, refrigerant system blockage, or incorrect system pressures. Internal compressor damage may cause heavy knocking noises from the compressor. A growling noise with the compressor clutch engaged or disengaged may be caused by a worn compressor pulley bearing. If the growling noise is evident only when the clutch is engaged, the compressor internal bearings may be worn.

Task 2   Identify the system type and conduct a performance test on the A/C system; determine needed repairs.

2. An A/C performance test is performed on a fixed orifice tube, cycling clutch A/C system.

    Technician A says during the performance test the blower speed control should be in the low position.

    Technician B says during the performance test the driver's window should be left down.

    Who is correct?

    A. A only
    B. B only
    C. Both A and B
    D. Neither A nor B

3. If an A/C system is operating properly at the end of an A/C performance test the temperature at the dash outlets should be

    A. 20° to 25°F.
    B. 35° to 40°F.
    C. 40° to 50°F.
    D. 60° to 65°F.

Hint   *The technician must know whether the A/C system is an R-12 or R-134a system. Most of the major components in an R-134a refrigerant system have light blue labels indicating these components are designed for operation in an R-134a system. Vehicles with R-134a A/C systems have a light blue underhood A/C decal. R-12 refrigerant systems have Schrader-type service valves, but R-134a systems have metric-thread, quick disconnect service valves.*

The type of compressor should be identified as a variable displacement or constant displacement. When the A/C system is turned on most variable-displacement compressor clutches are engaged continually, whereas constant-volume compressor clutches cycle on and off.

A/C systems may be classified according to the type of expansion device in the evaporator inlet line. Most A/C systems have a thermostatic expansion valve (TXV) (Figure 7-1), or a fixed orifice tube (FOT) (Figure 7-2), in the evaporator inlet line to control refrigerant flow into the evaporator. Some A/C systems have a suction throttling device between the evaporator and the compressor to control compressor inlet pressure. These suction throttling devices may include a suction throttling valve (STV), pilot-operated-absolute valve, evaporator pressure regulator valve (EPR), and evaporator temperature regulator (ETR) valve.

Many A/C systems have combination valves that usually contain two of the pressure control valves. For example, many older General Motors vehicles have a valves in receiver (VIR) assembly that contains TXV and POA valves plus a receiver-drier. Some Ford A/C systems have a combination valve containing a TXV and an STV.

A/C control systems may be identified as mechanical, semiautomatic, and automatic. Mechanical A/C systems have a slide-type lever or rotary switch to manually control the in-car temperature. In a semiautomatic A/C systems, some of the control features, such as in-car temperature, are handled automatically. An automatic air conditioning system usually has the temperature reading displayed on a digital reading, and the system automatically provides the requested temperature. The blower speed may be automatically controlled in an automatic A/C system.

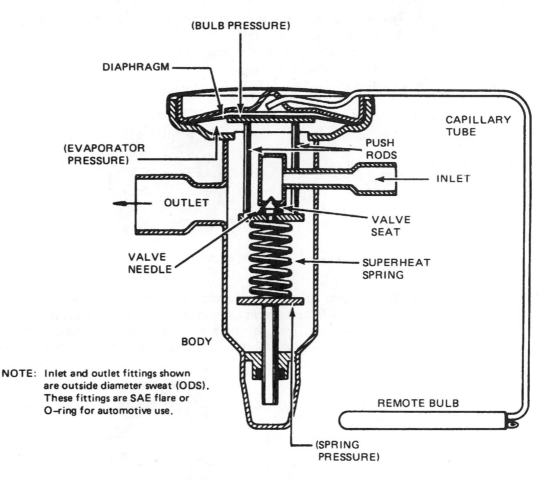

Figure 7-1 Thermostatic expansion valve (TXV)

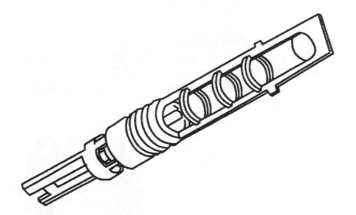

Figure 7-2 Fixed orifice tube (FOT) *(Courtesy of Ford Motor Company)*

Task 3   Diagnose A/C system problems indicated by refrigerant flow past the sight glass (for systems using a sight glass); determine needed repairs.

4. The sight glass in an R-12 A/C system contains bubbles. This problem may be caused by
   A. a low refrigerant charge.
   B. an excessive refrigerant charge.
   C. excessive oil in the refrigerant system.
   D. moisture in the refrigerant system.

5. The sight glass contains oil streaks (Figure 7-3). The cause of this problem could be
   A. excessive oil in the refrigerant system.
   B. a lack of oil in the refrigerant system.
   C. a low refrigerant charge.
   D. moisture in the refrigerant system.

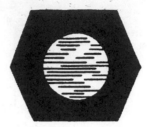

Figure 7-3 Oil streaks in the sight glass *(Courtesy of Chrysler Corporation)*

Hint   *On R-12 systems the sight glass indications are only valid when the ambient temperature is above 70°F (21°C). If the ambient temperature is below 70°F, it is normal for bubbles to appear in the sight glass. A clear sight glass may indicate the proper refrigerant charge (Figure 7-4). However, a clear sight glass may also indicate an excessive refrigerant charge, or no refrigerant charge.*

Figure 7-4 Clear sight glass *(Courtesy of Chrysler Corporation)*

When the sight glass contains bubbles and/or foam, the refrigerant charge is low, and air has likely entered the system (Figure 7-5). If the refrigerant charge is low and the refrigerant system oil is circulating through the system, oil streaks appear on the sight glass. A cloudy sight glass indicates moisture in the refrigerant system. This problem usually is caused by a defective desiccant in the receiver drier or a leak in the system.

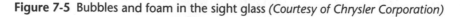

Figure 7-5 Bubbles and foam in the sight glass *(Courtesy of Chrysler Corporation)*

Many R-134a systems do not have a sight glass. Most vehicle manufacturers agree that certain conditions must be present when diagnosing R-134a systems from the sight glass indications. These conditions include
   • *High side pressure below 240 psi (1, 570 kPa)*
   • *Ambient temperature below 95°F (35°C)*
   • *Temperature control set in the lowest position*

- *Humidity below 70 percent*
- *High blower speed*
- *Engine speed set at 1,500 rpm*
- *A/C turned on*
- *Recirculation air activated*

R-134a systems have a normal refrigerant charge if the sight glass indicates a stream of very small bubbles that disappear when the engine rpm is increased. When the sight glass does not indicate any bubbles, the refrigerant system is overcharged. A constant stream of bubbles mixed with foam indicates a low refrigerant charge. If the refrigerant charge is very low, fog may appear in the sight glass. When the sight glass is severely fogged, mineral oil for an R-12 system may have been added to the R-134a system.

Task 4   Diagnose A/C system problems indicated by pressure gauge readings; determine needed repairs.

6. The gauge pressures in Figure 7-6 occur on an R-12 TXV valve, cycling clutch A/C system with an ambient temperature of 80°F. The cause of these readings could be
   A. air and moisture in the refrigerant system.
   B. a defective A/C compressor.
   C. a low refrigerant charge.
   D. the TXV valve restricted.

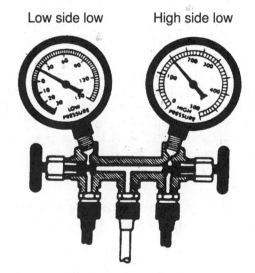

Low side low   High side low

Figure 7-6 Manifold gauge set readings *(Courtesy of Chrysler Corporation)*

7. In an R-12, TXV valve, cycling clutch A/C system with an ambient temperature of 80°F, the low-side pressure is 55 psi (345 kPa)) and the high-side pressure is 260 psi (1,793 kPa). There is no indication of frosting on any of the refrigerant system components, and the discharge air is slightly cool.
   Technician A says the cause of this problem may be air and moisture in the refrigerant system.
   Technician B says the cause of this problem may be a restricted TXV valve.
   Who is correct?
   A. A only
   B. B only
   C. Both A and B
   D. Neither A nor B

8. A refrigerant system has a thermostatic expansion valve (TXV) and an evaporator pressure regulator (EPR) valve. While diagnosing this system the high-side gauge indicates 160 psi (1,103 kPa), the low-side gauge reads 40 psi (275 kPa), and the auxiliary low-side gauge reading is 5 psi (34 kPa). The cause of these readings could be

A. a defective thermostatic expansion valve (TXV).

B. a defective evaporator pressure regulator (EPR) valve.

C a restricted evaporator core.

D. a defective compressor valve.

Hint   *When diagnosing an A/C refrigerant system from the low-side and high-side refrigerant system pressures, pressure specifications must be available in relation to ambient temperature. Some vehicle or A/C equipment manufacturers provide pressure specifications in relation to condenser and evaporator temperature. Diagnosis of refrigerant systems from the system pressures may be summarized as follows:*

Excessive low-side and high-side pressures—refrigerant overcharge, condenser restricted, air and/or moisture in the refrigerant system, TXV valve stuck open, engine overheating, or defective cooling fan.

Low-side and high-side pressures lower than specified—low refrigerant charge, TXV valve stuck closed, restricted line from the condenser to the evaporator. If the TXV valve is stuck closed or a restriction occurs in the line between the condenser and the evaporator, frosting occurs at the restriction.

A defective compressor may cause a low-side gauge reading that is higher than specified and a high-side gauge reading that is lower than specified.

Some R-12 refrigerant systems have a suction throttling valve (STV), or an evaporator pressure regulator (EPR) valve connected between the evaporator and the compressor inlet to control evaporator pressure. On some of these systems an auxiliary low-side gauge is connected to the compressor inlet. The difference in the low-side gauge and auxiliary low-side gauge readings indicates the pressure drop across the EPR or STV valve. On most refrigerant systems this pressure difference should not exceed 6 psi (41 kPa). Excessive pressure drop across the EPR or STV valve indicates the valve is sticking or restricted. If the low-side gauge reading is lower than specified and the auxiliary low-side gauge reading exceeds the low-side gauge reading, the EPR or STV valve is stuck open.

Task 5   Diagnose A/C system problems indicated by visual, smell, and touch procedures; determine needed repairs.

9. When a fixed orifice tube (FOT) cycling clutch A/C system is operating at 82°F ambient temperature, the compressor clutch cycles seven times per minute and the evaporator outlet line is warm. There is no frost on any of the A/C system components. The cause of this problem could be

A. a low refrigerant charge.

B. a flooded evaporator.

C. a restricted TXV valve.

D. a restricted receiver drier.

Hint   *The refrigerant system components should be visually inspected for frosting. Frost on the receiver drier usually indicates an internal restriction in this component. Since the receiver drier is connected between the condenser and the evaporator, it should normally feel warm. Frosting of the TXV valve indicates this valve is restricted or sticking closed. Frost formation on the evaporator outlet usually indicates a flooded evaporator caused by excessive refrigerant charge or the TXV valve stuck open. These problems also may cause frost formation on the compressor suction hose. On a refrigerant system with a pilot-operated absolute (POA) valve, frosting of the compressor suction hose is normal.*

If the refrigerant system has an accumulator, it should feel cold because it is connected between the evaporator and the compressor. Both the evaporator inlet and outlet should feel cold when the refrigerant system is operating normally. If the evaporator outlet is warm, the refrigerant charge may be low.

High-side refrigerant systems should normally feel hot or warm, and low-side components should be cold or cool. Since high-side components may be very hot, use caution when touching these components. If the line from the condenser to the TXV valve or FOT is cold, there may be a restriction in the high side.

A strong odor similar to a rotten-egg smell in the passenger compartment usually is caused by a plugged drain on the evaporator case. When this drain is plugged the water collects and stagnates in the evaporator case resulting a strong odor.

Task 6   Leak test A/C system; determine needed repairs.

10. An oily residue is present on the fittings of the hose connected from the compressor to the condenser.
    Technician A says this residue may be caused by excessive oil in the refrigerant system.
    Technician B says this residue may be caused by a leak at the hose fittings.
    Who is correct?
    A. A only
    B. B only
    C. Both A and B
    D. Neither A nor B

Hint   *Refrigerant systems may be leak tested with dye, a flame-type leak detector, or an electronic leak detector. An R-12 system may be charged with R-12 that contains dye. This dye mixes with the system lubricant. After charging the A/C system must be operated for at least 15 minutes. If the system has a leak, the dye appears on lines, fittings, or components that are leaking. Longer A/C system operation may be necessary to locate small leaks.*

A flame-type leak detector may be used on R-12 systems. If the system has a leak, the chlorine in the R-12 causes the halide torch flame to turn green, bright blue, or purple depending on the size of the leak. Since the burning of R-12 creates a phosgene gas, this type of leak testing must be done in a well-ventilated area. There is no chlorine in R-134a refrigerant, so this method of leak detection does not work on R-134a systems.

Electronic leak detectors provide an audible beeping noise when the leak detector probe is placed near the leak source. Since R-12 and R-134a refrigerant contain different chemicals, each of these systems requires a different electronic leak detector, or one that can be switched to test each of these systems.

Task 7   Identify and recover A/C system refrigerant.

11. While discussing refrigerant identification
    Technician A says that recycled R-12 is sold in white containers marked with a DOT code.
    Technician B says that R-134a refrigerant is sold in blue containers.
    Who is correct?
    A. A only
    B. B only
    C. Both A and B
    D. Neither A nor B

Hint   *Neither R-12 nor R-134a refrigerant may be vented to the atmosphere. Refrigerant recovery and recycling equipment is available to recover, recycle, and recharge R-12 or R-134a systems (Figure 7-7). New or recycled R-12 is sold in white containers, and the recycled R-12 containers have a Department of Transportation (DOT) code. The sale of both refrigerants is restricted to certified A/C service facilities.*

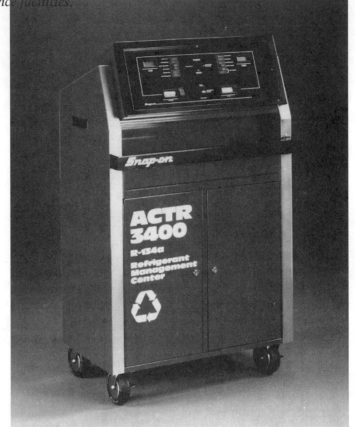

Figure 7-7 R-134a refrigerant recovery, recycle, and recharging equipment *(Courtesy of Snap-on Tools Corporation)*

Task 8   Evacuate A/C system.

12. Evacuating a refrigerant system removes
    A. moisture from the system.
    B. rust particles from the system.
    C. aluminum particles from the system.
    D. desiccant particles from the system.

Hint   *Prior to the refrigerant system evacuation procedure the manifold gauge set must be connected to the service fittings. A vacuum pump is connected to the center hose on the manifold gauge set, and this pump usually is operated for 30 minutes with the service valves open and the low side gauge valve open. After 5 minutes of vacuum pump operation the low-side gauge should indicate 20 in. Hg. (67.6 kPa), and the high-side gauge should read below zero unless it is restricted by a stop pin. If the high-side gauge does not drop below zero, refrigerant system blockage is indicated. When system blockage is indicated, stop the evacuation procedure and repair the blockage.*

After 15 minutes of vacuum pump operation, the low-side gauge should indicate 24 to 26 in. Hg. (81 to 88 kPa) if there are no leaks in the refrigerant system. When the low-side gauge is less than this value, close the low-side valve and observe the gauge. If the low-side gauge needle rises slowly, a refrigerant system leak is indicated. When this condition is present, stop the evacuation, and install a partial refrigerant charge to locate the leak.

When there are no refrigerant system leaks, continue the evacuation for 30 minutes. Close the low-side gauge set valve, shut off the vacuum pump, and disconnect the center gauge set hose from the pump. Replace the protective caps on the pump inlet and outlet fittings.

After the evacuation procedure the low-side gauge should not rise faster than 1 in. Hg. in 5 minutes. If the low-side gauge rises faster than 1 in. Hg. in 5 minutes, the refrigerant system is leaking. Install a partial refrigerant charge and leak test the system.

Task 9 Clean A/C components and hoses.

13. Technician A says that some manufacturers recommend the installation of an in-line filter between the evaporator and the compressor as an alterative to refrigerant system flushing.
    Technician B says an in-line filter containing a fixed orifice tube may be installed and the original orifice tube left in the system.
    Who is correct?
    A. A only
    B. B only
    C. Both A and B
    D. Neither A nor B

Hint   *Refrigerant systems may be flushed to remove debris such as aluminum particles from a failed compressor. Rather than system flushing some vehicle manufacturers recommend the installation of an in-line filter between the condenser and the evaporator to remove debris. These in-line filters are available with or without an internal fixed orifice tube (FOT). If an in-line filter containing an FOT is installed, the original FOT must be removed from the system.*

The refrigerant system must be discharged prior to flushing. The complete system or individual components may be flushed with dry nitrogen. Before flushing individual components disconnect the refrigerant lines from the component and cap the ends of the lines. When flushing the complete system, remove the receiver-drier and bypass it with an appropriate hose. After system flushing, the receiver-drier must be replaced.

Task 10 Charge A/C system with refrigerant (liquid or vapor).

14. Technician A says a high-side charging procedure should be completed with the engine running.
    Technician B says if liquid refrigerant enters the compressor, damage to the compressor may result.
    Who is correct?
    A. A only
    B. B only
    C. Both A and B
    D. Neither A nor B

Hint   *Before charging a refrigerant system the recovery and evacuation procedures must be completed. Always follow the charging procedure recommended by the vehicle manufacturer. High-side (liquid) or low-side (vapor) charging procedures may be recommended. The manifold gauge set hoses or hoses from the refrigerant recovery, recycling, and charging equipment must be connected to service fittings. Be sure both low-side and high-side gauge valves are closed. Connect the center hose on the manifold gauge set to the proper refrigerant container. Open the valve on the refrigerant container to charge the center hose with refrigerant. Open the high-side gauge valve, and observe the low-side gauge. Then close the high-side gauge valve. If the low-side gauge does not move from a vacuum to a pressure, the refrigerant system is restricted. Repair the restriction problem before proceeding with the charging procedure.*

When the system blockage is satisfactory, open the high-side gauge valve to proceed with the high-side charging procedure. Charging is completed when the specified weight of refrigerant has entered the system. Close the high-side gauge valve and the refrigerant container valve. Remove the compressor belt and rotate the compressor by hand for several revolutions to be sure there is no liquid refrigerant in the system. Install and tighten the compressor belt. Start the engine and hold the engine speed at fast idle. Adjust the A/C controls to maximum cooling, and complete an A/C performance test.

When low-side charging is recommended, the system is charged with the engine running at the specified rpm for refrigerant system charging. Set the A/C controls to maximum cooling and high blower speed. Open the low-side gauge valve to allow refrigerant to enter the system. Adjust the low-side gauge valve so the low-side pressure does not exceed 40 psi (275 kPa). Continue charging the system until the specified refrigerant weight has entered the system. Close the low-side gauge valve, and complete an A/C system performance test.

After either charging operation is completed, back seat the service valves and disconnect the manifold gauge set hoses. Replace all the protective caps and covers.

Task 11   **Identify lubricant type; inspect level in A/C system.**

15. The oil required in an R-134a refrigerant system is
    A. a polyalkylene glycol (PAG) oil.
    B. a synthetic engine oil.
    C. a synthetic mineral oil.
    D. a 10W-30 engine oil.

Hint   *An A/C refrigerant system must contain the specified amount of refrigerant oil to lubricate the compressor components and prevent premature compressor wear. An excessive amount of oil in a refrigerant system reduces the cooling efficiency of the system.*

An R-12 A/C system requires a mineral oil with a YN-9 designation, whereas an R134a system with a reciprocating compressor must have a synthetic polyalkylene glycol (PAG) oil designated as YN-12. A different type of PAG oil is used with a rotary A/C compressor. If the oils used in R-12 and R-134a systems are interchanged, compressor damage will result. Both types of refrigerant oils are hydroscopic which means these oils absorb moisture very easily. The PAG oil used in R-134a systems is more hydroscopic than the mineral oil in R-12 systems. Refrigerant oil containers must be kept tightly capped at all times.

A dipstick is required to measure the oil level in some compressors. The refrigerant system must be discharged or the compressor isolated before the compressor oil level is measured. Other compressors must be removed and the oil drained before adding new oil. Always follow the vehicle manufacturer's recommended oil level checking procedure and add the specified amount of oil. When individual refrigerant system components are replaced, install the specified amount of the proper refrigerant oil in the component. Refrigerant oil may be added to the system from pressurized cans. The specified amount of refrigerant oil may also be added to the system between the discharging and evacuation procedures.

Refrigeration System Component Diagnosis and Repair, Compressor and Clutch

ASE Tasks, Questions, and Related Information

Task 1   **Diagnose A/C system problems that cause the protection devices (pressure, thermal, and PCM) to interrupt system operation; determine needed repairs.**

16. The purpose of the refrigerant system component 5 in Figure 7-8 is to
    A. protect the system components from excessive pressure.
    B. shut off the compressor if the refrigerant charge is low.
    C. cycle the compressor on and off in relation to system pressure.
    D. shut off the compressor if the refrigerant temperature is excessive.

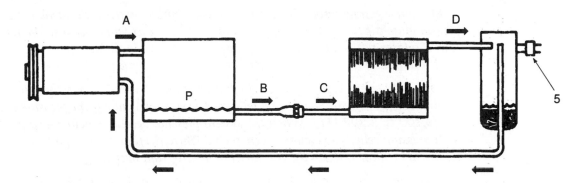

Figure 7-8 Refrigerant system components *(Courtesy of Chevrolet Motor Division, General Motors Corporation)*

17. A refrigerant system is low on refrigerant, and there is oil around the high-pressure relief valve.
    Technician A says the air passages through the condenser may be restricted.
    Technician B says the refrigerant system may be overcharged.
    Who is correct?
    A. A only
    B. B only
    C. Both A and B
    D. Neither A nor B

Hint   *Some refrigerant systems use a pressure cycling switch to cycle the compressor on and off in relation to the low side pressure. In cycling clutch orifice tube (CCOT) systems the pressure cycling switch usually is mounted in the accumulator between the evaporator and the compressor. This switch closes and turns on the compressor when the refrigerant system pressure is at or above 46 psi (315 kPa). The A/C switch in the instrument panel must be turned on to supply voltage to the pressure cycling switch. The pressure cycling switch opens and turns off the compressor when the system pressure decreases to 25 psi (175 kPa). This cycling action maintains the evaporator temperature at 33°F (1°C).*

Some refrigerant systems have a clutch cycling switch that cycles the compressor on and off in relation to evaporator outlet temperature (Figure 7-9). This control device may be called a thermostatic switch. A capillary tube is connected from the clutch cycling switch to the evaporator outlet pipe.

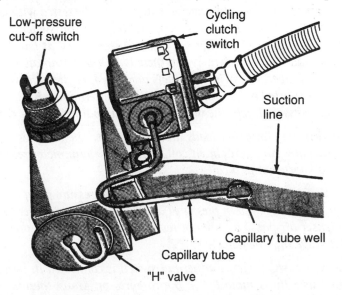

Figure 7-9 Thermostat clutch cycling switch and low pressure cut-off switch *(Courtesy of Chrysler Corporation)*

Many refrigerant system have a low pressure cutoff switch that opens the circuit to the compressor clutch if the system pressure drops below a preset value because the refrigerant charge is low. If a large refrigerant leak occurs, the oil may be lost with the refrigerant. When the compressor is allowed to run under this condition, compressor damage may occur.

Some refrigerant systems have a high-pressure relief valve that is sometimes mounted in the receiver drier. This valve opens and relieves the system pressure if the pressure exceeds 450 to 550 psi (3,100 to 3,792 kPa). When the system pressure decreases below 450 psi (3,100 kPa), the high-pressure relief valve closes. Extremely high refrigerant system pressures may be caused by restricted air flow through the condenser or a refrigerant overcharge.

Some older model refrigerant systems contained a superheat switch mounted in the compressor and a thermal fuse. If a refrigerant leak occurs, high system temperature closes the superheat switch contacts. Under this condition excessive current flow blows the thermal fuse and opens the compressor clutch circuit. Since the refrigerant system oil may have been lost with the refrigerant, the blown thermal fuse protects the compressor from running without lubrication.

In many computer-controlled A/C systems the powertrain control module (PCM) operates a relay that supplies voltage to the compressor clutch. All the input sensor signals are sent to the PCM. In some systems these inputs include a refrigerant pressure signal. If the input signals indicate an abnormal condition, the PCM does not energize the compressor clutch.

Task 2   Inspect, test, and replace A/C system pressure and thermal protection devices.

18. If A/C system in Figure 7-10 the ignition switch is on, and the A/C switch is in the AUTO position. The ambient temperature is 75°F (24°C), and the compressor clutch is inoperative. There is 12V at terminals B, C, and S on the thermal fuse. The cause of the inoperative compressor clutch could be

   A. an open thermal fuse.

   B. an open compressor clutch coil.

   C. a defective superheat switch.

   D. a defective thermal fuse heater.

Hint   *The thermal fuse and superheat switch may be tested with an ohmmeter or voltmeter. With the ignition switch and A/C switch on, and the ambient switch closed, there should be 12V at terminals B, C, and S on the thermal fuse. If there is 12V at terminal B and 0V at terminals C and S, the thermal fuse is blown. When there is 12V at terminals B and C, but a lower voltage at terminal S, the superheat switch contacts are closed, or the wire from the thermal fuse to the superheat switch is shorted to ground. If there is 12V at terminals B, C, and S on the thermal fuse, but 0V at the compressor clutch, the wire from the thermal fuse to the compressor clutch is open.*

When the ohmmeter leads are connected to the compressor clutch terminals an infinite reading indicates an open clutch coil, whereas an ohmmeter reading below the specified value indicates a shorted clutch coil.

If there is evidence of oil around the high-pressure relief valve and the system is low on refrigerant, check the system pressures and inspect the condenser for restricted air passages. When the condenser air passages are not restricted and the system pressures are normal, the high pressure relief valve may be defective.

In some A/C systems a thermal switch is connected in series with the compressor clutch. This switch usually is mounted in the compressor. Many thermal switches open at 257°F (125° C), and close at 230°F (110°C).

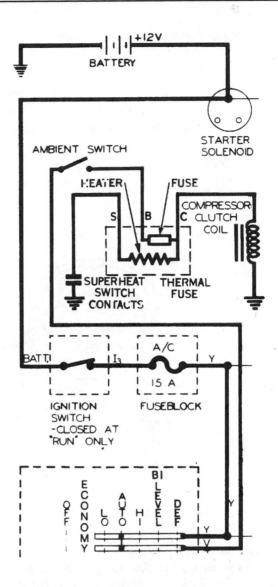

Figure 7-10 Compressor clutch circuit with thermal fuse and superheat switch *(Courtesy of Cadillac Motor Division, General Motors Corporation)*

Task 3   Inspect, adjust, and replace A/C compressor drive belts and pulleys.

19. A intermittent squealing noise is heard on acceleration with the A/C control switch in the on or off position. The most likely cause of this problem is
A. a loose power steering belt.
B. a loose A/C compressor belt.
C. a loose air pump belt.
D. a worn A/C compressor pulley bearing.

Hint   *Since the friction surfaces are on the sides of a V-belt, this type of belt must not be bottomed in the A/C compressor pulley. If the compressor belt is loose or bottomed in the pulley, the belt may slip. Erratic compressor operation and inadequate passenger compartment cooling may be caused by a loose compressor belt. A slipping compressor belt may cause belt squealing especially on acceleration with the A/C on and the compressor clutch engaged.*

The compressor belt should be inspected for cracked, oil-soaked, glazed, and torn or split conditions. The belt tension should be measured with a belt tension gauge positioned at the center of the longest belt span. V-belt tension may be adjusted by loosening the adjustment bolt and moving the compressor until the proper tension is obtained. Tighten the adjustment bolt to the specified torque. Most ribbed V-belts have a spring-loaded tensioner pulley.

Task 4 Inspect, test, service, and replace A/C compressor clutch components or assembly.

20. The measurement in Figure 7-11 is more than specified.

    Technician A says this condition may cause an intermittent scrapping noise with the engine running and the compressor clutch energized.

    Technician B says to correct this condition another shim should be added behind the pulley armature plate.

    Who is correct?

    A. A only
    B. B only
    C. Both A and B
    D. Neither A nor B

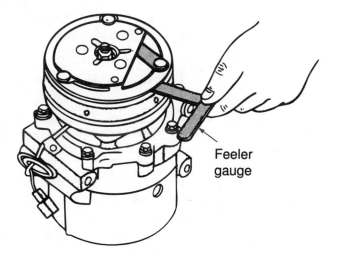

Feeler
gauge

Figure 7-11   A/C compressor clutch measurement *(Courtesy of Chrysler Corporation)*

Hint   *A special tool or two box-end wrenches may be used to hold the pulley while the compressor shaft nut is removed. After this nut is removed, the armature plate and shims may be removed. Remove the pulley snap ring, pulley, and field coil core (Figure 7-12).*

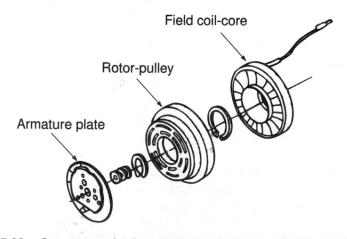

Field coil-core

Rotor-pulley

Armature plate

Figure 7-12   Compressor clutch components *(Courtesy of Chrysler Corporation)*

The pulley and armature plate frictional surfaces should be inspected for wear and oil contamination. Replace contaminated or worn components. Check the hub bearing for roughness, grease leakage, and looseness. Replace the bearing if any of these conditions are present.

When the field coil is installed, the nipple in the back of the coil must be aligned with the locating indentation in the front compressor cover. The pulley snap ring must be installed with the bevelled side facing outward.

After the shims and armature are installed, the clearance between the armature and pulley frictional surface must be measured with a feeler gauge. If the clearance is excessive, remove a shim from behind the armature. Recheck the armature to pulley clearance after the shaft nut is tightened to the specified torque.

Some vehicle manufacturers recommend checking the compressor clutch circuit with a voltmeter and an ohmmeter. With the compressor clutch engaged, the voltage supplied to the clutch coil must be within 2V of battery voltage. If the voltage supplied to the coil is zero, check the clutch fuse link or fuse. If the voltage supplied to the clutch coil is less than specified, check the resistance in the circuit from the battery to the coil.

When the clutch is engaged an ammeter connected in series with the coil should indicate 2.0 to 4.15 amperes. A low ammeter reading indicates excessive resistance in the coil or the ground circuit. If the ammeter reading is higher than specified, the coil is shorted.

Task 5   Identify lubricant type; inspect and correct level in A/C compressor.

21. When replacing an A/C compressor the refrigerant system is discharged and 2 oz. of oil is recovered from the system. When the old compressor is removed, 2 oz. of oil is drained from the compressor. The amount of oil drained from the new compressor is 6 oz. When the new compressor is installed, the amount of oil added to the compressor should be
    A. 1 oz.
    B. 2 oz.
    C. 4 oz.
    D. 6 oz.

Hint

An R-12 A/C system requires a mineral oil with a YN-9 designation, whereas an R134a system with a reciprocating compressor must have a synthetic polyalkylene glycol (PAG) oil designated as YN-12. A different type of PAG oil is used with a rotary A/C compressor. If the oils used in R-12 and R-134a systems are interchanged, compressor damage will result. Both types of refrigerant oils are hydroscopic, which means these oils absorb moisture very easily. The PAG oil used in R-134a systems is more hydroscopic than the mineral oil in R-12 systems. Refrigerant oil containers must be kept tightly capped at all times.

The oil level in some A/C compressors may be checked with a dipstick. When the refrigerant is recovered from a system, the amount of oil recovered should be measured and an equal amount of new oil added to the system. Vehicle manufacturers usually specify the total amount of refrigerant oil required in the system, and the amount of oil required in each component. When any refrigerant system component is replaced, the required amount of refrigerant oil is the total system oil capacity minus the oil capacity of the components that have not been replaced plus the amount of oil recovered during the discharge procedure.

Task 6   Inspect, test, service, or replace A/C compressor.

Task 7   Inspect, repair, or replace A/C compressor mountings.

22. Technician A says component 6 in Figure 7-13 (next page) is a seal protector that is placed over the compressor shaft.
    Technician B says the seal seat O-ring must be installed before the compressor shaft seal.
    Who is correct?
    A. A only
    B. B only
    C. Both A and B
    D. Neither A nor B

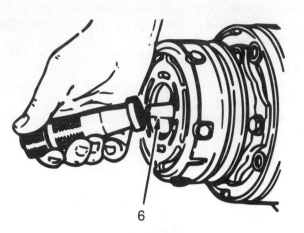

Figure 7-13   Compressor service tool *(Courtesy of Chevrolet Motor Division, General Motors Division)*

Hint   *Inspect the compressor for oil deposits in the pulley area, around the line connections and at the pressure relief valve. Oil deposits in the pulley area indicate a refrigerant system leak at the shaft seal. Check the compressor belt for proper tension, wear, cracks, or an oil-soaked condition. Inspect all the compressor mounts and mounting bolts for wear and proper torque. A rattling noise may be caused by loose compressor mounts. This noise will likely be worse when the compressor clutch is engaged.*

A growling noise that occurs only when the compressor is operating is likely caused by a worn bearing in the compressor. A defective pulley bearing also causes a growling noise with the compressor clutch disengaged.

A defective compressor may be indicated by low high-side pressure, and high low-side pressure on the manifold gauge set.

Discharge the refrigerant system or isolate the compressor before removing the compressor. If the refrigerant system has stem-type service valves these valves may be front seated to isolate the compressor from the refrigerant system. Always follow the recommended isolation procedure in the vehicle manufacturer's service manual. When any refrigerant system component is disconnected, cap all lines and fittings to prevent moisture entry.

Refrigerant System Component Diagnosis and Repair, Evaporator, Condenser, and Related Components

ASE Tasks, Questions, and Related Information

Task 1   Inspect, repair, or replace A/C system mufflers, hoses, lines, filters, fittings, and seals.

23. When servicing refrigerant lines the tool in Figure 7-14 is used to
    A. disconnect spring lock couplings.
    B. connect spring lock couplings.
    C. connect and disconnect spring lock couplings.
    D. crimp male bare-type fittings.

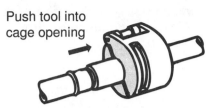

Push tool into
cage opening
→

Figure 7-14   Refrigerant line service tool *(Courtesy of Ford Motor Company)*

Hint   *A defective refrigerant line may be replaced with a complete new line. Male and female barbed fittings are available to repair damaged refrigerant lines or hoses (Figure 7-15). Some refrigerant line connections are sealed with O-rings and retained with spring lock couplings (Figure 7-16). A special tool is required to release these spring lock couplings. Other refrigerant line fittings have a ferrule and an O-ring (Figure 7-17).*

Clamp

Hose

Insert fitting

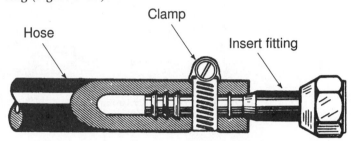

Figure 7-15   Female barbed fitting for refrigerant hose repair

B—Install new o-ring seals-
use only specified
o-ring seals

A—Clean fittings

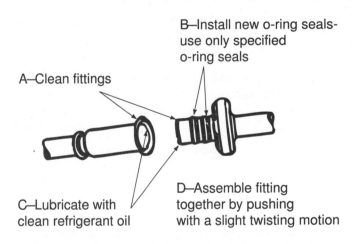

D—Assemble fitting
together by pushing
with a slight twisting motion

C—Lubricate with
clean refrigerant oil

Figure 7-16   Refrigerant line spring lock coupling *(Courtesy of Ford Motor Company)*

Liquid line   Nut   Ferrule

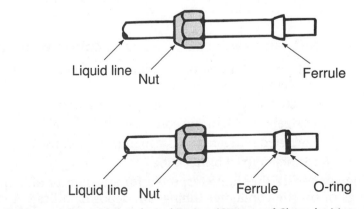

Liquid line   Nut   Ferrule   O-ring

Figure 7-17   Refrigerant line with ferrule and O-ring *(Courtesy of Chevrolet Motor Division, General Motors Corporation)*

Some refrigeration systems have a filter in the line between the condenser and the evaporator. Some of these filters contain an orifice tube. This type of filter must be installed in the proper direction (Figure 7-18).

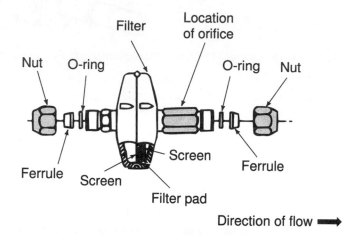

Figure 7-18 Refrigerant filter with orifice tube *(Courtesy of Chevrolet Motor Division, General Motors Corporation)*

Task 2   Inspect A/C condenser for proper air flow.

24. The air passages through an A/C condenser are severely restricted.

    Technician A says this may cause refrigerant discharge from the high-pressure relief valve.

    Technician B says this may cause excessive high-side pressure and low-side pressure.

    Who is correct?

    A. A only

    B. B only

    C. Both A and B

    D. Neither A nor B

Hint   *Debris in the condenser air passages causes excessive high-side and low-side pressures and reduced cooling from the A/C system. This problem may also cause the high-pressure relief valve to discharge refrigerant. The condenser air passages may be blown out with compressed air or washed with a water hose. A plastic rod and a soft-bristled brush may be used to remove debris that is stuck tightly in the condenser air passages. Bent condenser fins must be straightened with a plastic rod or needlenosed pliers.*

Task 3   Inspect, test, and replace A/C system condenser and mountings.

25. Frost is forming on one of the condenser tubes near the bottom of the condenser. This problem could be caused by

    A. restricted air flow passages in the condenser.

    B. a refrigerant leak in the condenser.

    C. a restricted refrigerant passage in the condenser.

    D. a restricted fixed orifice tube.

    The condenser should be inspected for any sign of oil deposits on the inlet or outlet fittings or on the condenser tubing. Oil deposits indicate a refrigerant leak. Frosting on any of the condenser tubing indicates a refrigerant passage restriction. This condition results in excessive high-side and low-side pressures, and inadequate cooling.

Task 4   Inspect and replace receiver/drier or accumulator/drier.

26. A receiver/drier is located between the condenser and the evaporator.

    Technician A says the receiver/drier should be changed if the outlet is colder than the inlet.

    Technician B says the receiver/drier should be changed if the refrigerant in the sight glass appears red.

    Who is correct?

    A. A only

    B. B only

    C. Both A and B

    D. Neither A nor B

Hint   *If the inlet and outlet pipes have a significant temperature difference, the receiver/drier is restricted and must be replaced. Frost forming on the receiver/drier indicates an internal restriction which requires receiver/drier replacement. Receiver/drier replacement also is necessary if there is moisture in the refrigeration system indicated by bubbles and foam in the sight glass or rust contamination in the system.*

Blue or gray particles in the sight glass indicate the desiccant in the receiver/drier is disintegrating and circulating through the refrigeration system. This condition also requires receiver/drier replacement. If the refrigerant in the sight glass is red or yellow, leak-detecting dye has been added to the refrigeration system. This condition does not require corrective action. The refrigeration system must be discharged before receiver/drier removal.

Task 5   Inspect, test, and replace expansion valve (TXV).

27. An A/C system blows cool air when the vehicle is started with an ambient temperature of 80°F. After the vehicle is driven for about 10 miles the system stops blowing cool air, and the TXV is frosted. When the A/C system is shut off for 5 minutes and turned on again it blows cool air for another 10 miles. The most likely cause of this problem is

    A. a defective TXV capillary tube.

    B. a refrigerant overcharge.

    C. restricted condenser refrigerant passage.

    D. moisture in the refrigeration system.

Hint   *Moisture in the refrigeration system may freeze in the TXV resulting in intermittent A/C system operation. If the TXV is stuck closed or the inlet screen is contaminated with debris, frosting may occur on the TXV. This condition causes the air discharged from the evaporator to be warm or slightly cool, and the low-side pressure to be low.*

The TXV may stick in the open position, or a defective capillary tube may cause this valve to remain open. Under this condition the evaporator floods, causing inadequate evaporator cooling, frost on the evaporator suction line, and high low-side pressure.

If there is very little cool air from the evaporator and the low-side pressure is low with TXV frosting and normal high-side pressure, place a shop towel soaked in hot water around the TXV valve. When this action causes the low-side pressure to increase to normal, there is moisture in the refrigeration system and ice formation in the TXV.

When the hot shop towel around the TXV does not change the low-side pressure, remove the remote bulb and warm it in the hand. If the low-side pressure rises, the remote bulb is improperly positioned. When the low-side pressure does not increase, replace the TXV.

If the low-side gauge reading is higher than normal with normal high-side pressure and frosting of the evaporator suction tube, remove the remote bulb and place it in a pan of ice water. When this action causes the low-side pressure to decrease to normal, the remote bulb may be improperly positioned or insulated. Reposition and reinsulate the remote bulb, and check the low-side pressure. If the low-side pressure is still higher than normal, replace the TXV.

Task 6   Inspect and replace the orifice tube.

28. All of these statements are true about using the tool shown in Figure 7-19 to remove a complete orifice tube assembly EXCEPT
    A. pour a small amount of refrigerant oil on top of the orifice tube to lubricate the O-rings.
    B. rotate the T-handle to engage the notch in the tool in the orifice tube tangs.
    C. rotate the T-handle to remove the orifice tube from the evaporator inlet pipe.
    D. hold the T-handle and rotate the outer sleeve on the tool to remove the orifice tube.

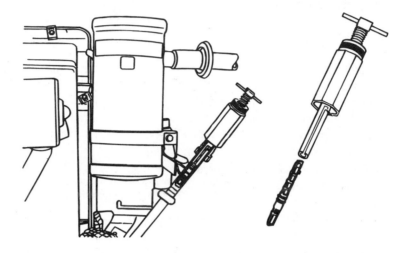

Figure 7-19 Refrigeration system service tool *(Courtesy of Ford Motor Company)*

Hint   *A restricted orifice tube may cause lower-than-specified low-side pressure, frosting of the orifice tube, and inadequate cooling from the evaporator. When these conditions are present, place a shop towel soaked in hot water around the orifice tube. If the low-side pressure increases, there is moisture freezing in the orifice tube. Under this condition the refrigeration system must be discharged, evacuated, and recharged.*

When the orifice tube is warmed with a shop towel and the low-side pressure does not increase, inspect and clean or replace the orifice tube.

Task 7   Inspect, test, or replace the evaporator.

29. Technician A says restricted refrigerant passages in the evaporator may cause frosting of the evaporator outlet pipe.
    Technician B says restricted refrigerant passages in the evaporator may cause much higher than specified low-side pressures.
    Who is correct?
    A. A only
    B. B only
    C. Both A and B
    D. Neither A nor B

Hint   *Evaporator replacement is necessary if it is leaking or restricted. Evaporator leaks may be detected with an electronic tester with an audible beep. Remove the resistor assembly in the blower circuit from the A/C heater case, and insert the leak tester probe through this opening in the case to check for evaporator leaks. Refrigerant leaks in the evaporator also are indicated by evidence of oil in the area of the leak. Evaporator restriction is indicated by a low-side pressure that is considerably lower than specified and inadequate cooling with a normal TXV or orifice tube.*

Always discharge the refrigeration system and disconnect the negative battery cable before removing the evaporator. Since the evaporator core usually is mounted in the A/C-heater case, this assembly must be removed from the vehicle to remove the evaporator.

Task 8   Inspect, clean, and repair evaporator, housing, and water drain.

30. The inside of the windshield has an oily film, A/C cooling is inadequate.
Technician A says this oil film may be caused by a plugged A/C heater case drain.
Technician B says this oil film may be caused by a leak in the evaporator core.
Who is correct?

    A. A only
    B. B only
    C. Both A and B
    D. Neither A nor B

Hint   *If the water drain is plugged on the A/C-heater case, water collects in the bottom of this case. After a period of time this water becomes very stale and produces a pungent odor in the passenger compartment. In some cases it may be possible to push a plastic rod up through the A/C-heater case drain pipe from the lower end of the pipe and clean out the obstruction. In other cases it may be necessary to remove the A/C-heater case and clean the water drain opening and pipe. The refrigeration system must be discharged and the coolant drained before the A/C-heater case is removed.*

If the evaporator core has a leak, an oily film may appear on the inside of the windshield and the discharge air temperature from the evaporator becomes warmer than specified. With the A/C system operating this leak may be located with an electronic tester.

Task 9   Inspect, test, and replace evaporator pressure/temperature control systems and devices.

31. While discussing refrigerant systems with a suction throttling valve (STV), pilot operated absolute (POA) valve, or evaporator pressure regulator (EPR) valve
Technician A says on some of these refrigerant systems the compressor runs continually in the A/C mode.
Technician B says an excessive pressure drop across the EPR valve indicates this valve is sticking open.
Who is correct?

    A. A only
    B. B only
    C. Both A and B
    D. Neither A nor B

Hint   *Some refrigeration systems have an evaporator pressure control device connected at the evaporator outlet. This type of pressure control valve may be called a suction throttling valve (STV) or a pilot operated absolute (POA) valve. Some General Motors vehicles have a valves in receiver (VIR) assembly that contains the TXV valve POA valve and receiver/drier (Figure 7-20, next page). The flows from the condenser into the receiver/drier in the VIR. After leaving the receiver/drier, the refrigerant flows through the TXV to the evaporator. When the refrigerant returns from the evaporator, it flows through the POA valve to the compressor (Figure 7-21, next page). The POA valve contains a pulsating piston that controls evaporator outlet pressure. In some A/C systems with a POA or STV valve the compressor runs continually when the driver selects the A/C mode and the ambient temperature is above a specific value.*

Other A/C systems have an evaporator pressure regulator (EPR) valve mounted in the compressor inlet. The EPR valve performs basically the same function as the STV or POA valve. Some systems with an EPR valve have a compressor inlet service valve and the usual low-side and high-side service valves. An auxiliary gauge in the manifold gauge set may be connected to the compressor inlet service valve. If the refrigerant system and the EPR valve are operating normally, the low-side and high-side gauges should indicate the specified pressures and the pressure difference between the low-side gauge and the compressor inlet gauge should not exceed 6 psi (41 kPa). The pressure difference in these two gauge readings is the pressure drop across the EPR valve.

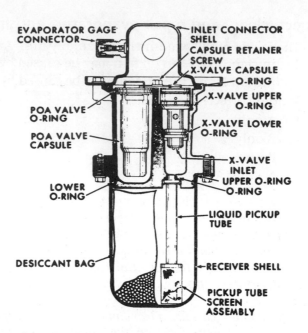

Figure 7-20 Valves in receiver assembly containing the TXV valve, POA valve, and receiver/drier *(Courtesy of Cadillac Motor Car Division, General Motors Corporation)*

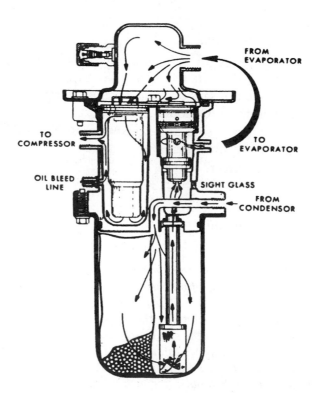

Figure 7-21 Refrigerant flow through the valves in receiver (VIR) assembly *(Courtesy of Cadillac Motor Car Division, General Motors Corporation)*

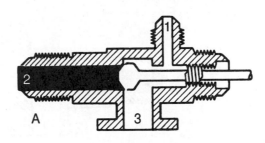

Figure 7-22 Stem-type service valve

Task 10 Identify, test, and replace A/C system service valves (gauge connections).

32. With a stem-type service valve in the position indicated in Figure 7-22
   A. the refrigerant system may be diagnosed with a manifold gauge set.
   B. the refrigerant system operates normally with no pressure at the gauge ports.
   C. the refrigerant system is isolated from the compressor for compressor removal.
   D. the refrigerant system may be discharged, evacuated, and recharged.

Hint   *R-12 refrigerant systems may have Schrader-type or stem-type service valves. Both types of service valves contain dust caps to keep dirt out of the valves. The Schrader-type service valve contains a valve that is similar to a tire valve. When the manifold gauge set hoses are connected to the service valves, a pin in the hose depresses the Schrader valve (Figure 7-23).*

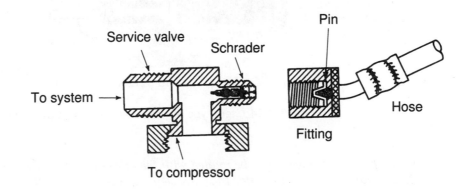

Figure 7-23 Schrader-type service valve

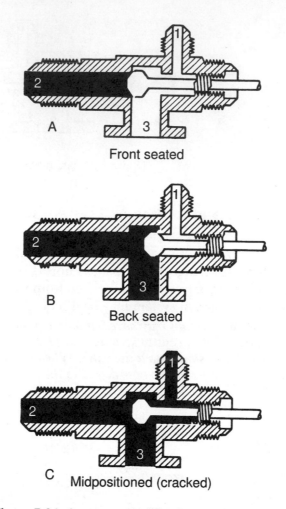

Figure 7-24 Stem-type service valve in different positions

Stem-type service valves must be rotated with the proper tool to move the internal valve stem. When the valve stem is front-seated the valve stem completely blocks the refrigerant flow through the valve (view A Figure 7-24). This valve position may be used to isolate the refrigerant system from the compressor for compressor removal. Severe compressor damage may result with the service valves front-seated and the gauge port capped.

When the stem-type service valve is back seated, the refrigerant flows through the service valve for normal operation (view B Figure 7-24). In this valve position refrigerant system pressure is not available at the gauge port.

If the stem-type service valve is in the midposition, or cracked position, refrigerant flows normally through the valve, and system pressure is available at the gauge port for diagnostic purposes (view C Figure 7-24).

R-134a refrigerant systems have quick-disconnect service valves. An R-12 service valve, (component A Figure 7-25) is compared to an R-134a service valve, (component B Figure 7-25). Quick disconnect hose fittings are located on R-134a manifold gauge sets.

Task 11   Inspect and replace A/C system high-pressure relief device.

33. Component 4 in Figure 7-26 discharges refrigerant at approximately
    A. 200 psi (1379 kPa).
    B. 325 psi (2240 kPa).
    C. 375 psi (2585 kPa).
    D. 475 psi (3275 kPa).

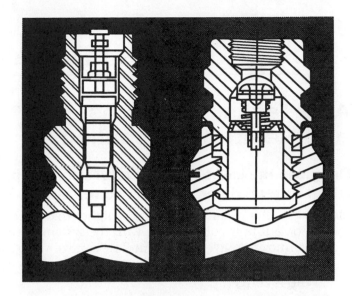

Figure 7-25 R-12 service valve item A, compared to an R-134a service valve item B *(Courtesy of BET, Inc.)*

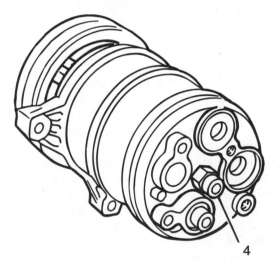

4

Figure 7-26 External compressor components *(Courtesy of Pontiac Motor Division, General Motors Corporation)*

Hint   *If the area around the high-pressure relief valve shows evidence of refrigeration oil, check the refrigeration system pressures. When these pressures are higher than normal, correct the causes of the high pressures such as a refrigerant overcharge, air and moisture in the system, or restricted condenser air passages. If the refrigeration pressures are normal, replace the high-pressure relief valve. The refrigeration system must be discharged before the high-pressure relief valve is removed.*

Heating and Engine Cooling Systems Diagnosis and Repair

ASE Tasks, Questions, and Related Information

Task 1   **Diagnose the cause of temperature control problems in the heater/ventilation system; determine needed repairs.**

    34. The discharge air temperature is higher than specified in an A/C-heater system with a coolant flow control valve with the system operating in the maximum A/C mode.

      Technician A says the coolant flow control valve may be stuck open.

      Technician B says the coolant flow control valve vacuum hose may have a leak.

      Who is correct?

      A. A only

      B. B only

      C. Both A and B

      D. Neither A nor B

Hint   *If the heater supplies cold air continually, the engine thermostat may be stuck open, or the temperature door may be stuck in the cold position. A coolant flow control valve stuck in the closed position or restricted also causes cold air discharge from the heater.*

    Tape a thermometer to the upper radiator hose and allow the engine to idle for 15 minutes. The thermometer reading should be within a few degrees or the specified thermostat opening temperature. If the thermometer reading is considerably lower than the thermostat rating, the thermostat is defective.

    Move the A/C-heater controls from full-cold to the maximum heat position and observe the temperature door linkage. If this linkage does not move, the door is sticking or the actuator system is inoperative.

    When the A/C-heater controls are in the heat position, check the temperature on both sides of the coolant flow control valve. The hose temperature should be the same on both sides of the valve. If this valve is restricted or closed, the temperature at the valve inlet is much higher than the temperature at the outlet.

    If the A/C-heater system supplies warm or hot air continually, check the refrigeration system pressures. When these pressures are not within specifications, repair the refrigeration system. If the refrigeration pressures are within specification and the air discharge is warm, the temperature blend door may be stuck.

Task 2   **Diagnose window fogging problems; determine needed repairs.**

    35. The inside of the windshield has a sticky film.

      Technician A says the engine coolant level should be checked.

      Technician B says the heater core may be leaking.

      Who is correct?

      A. A only

      B. B only

      C. Both A and B

      D. Neither A nor B

Hint   *Windshield fogging may be caused by a leaking heater core, or a plugged A/C-heater case drain which allows water to collect in this case. If the A/C-heater case drain is plugged and water has collected in the case, the water becomes stagnant and provides a very pungent odor in the passenger compartment. If the heater core is leaking, there is a loss of coolant from the cooling system.*

A leaking evaporator core may cause an oil film on the inside of the windshield. If the evaporator core is leaking, the refrigeration system pressures are lower than specified once the system has lost some refrigerant. Check evaporator leaks with probe of an electronic leak detector inserted through the resistor assembly opening in the A/C-heater case.

Task 3   Perform cooling system tests; determine needed repairs.

36. A customer complains about engine coolant loss. The cooling system is pressurized at 15 psi (103 kPa) for 15 minutes. There is no visible sign of coolant leaks in the engine or passenger compartments, but the pressure on the tester gauge decreases to 5 psi (34 kPa). This problem could be caused by the following defects EXCEPT
    A. a leaking heater core.
    B. a leaking transmission cooler.
    C. a leaking head gasket.
    D. a cracked cylinder head.

Hint   *A pressure tester may be connected to the radiator filler neck to check for cooling system leaks. Operate the tester pump and apply 15 psi to the cooling system. Inspect the cooling system for external leaks with the system pressurized. If the gauge pressure drops more than specified by the vehicle manufacturer, the cooling system has a leak. If there are no visible external leaks, check the front floor mat for coolant dripping out of the heater core. When there are no external leaks, check the engine for combustion chamber leaks.*

The radiator pressure cap may be tested with the pressure tester. When the tester pump is operated, the cap should hold the rated pressure. Always relieve the pressure before removing the tester.

Task 4   Inspect and replace engine cooling and heater system hoses and belts.

37. The vacuum valve in the radiator cap is stuck closed. The result of this problem could be
    A. collapsed upper radiator hose after the engine is shut off.
    B. excessive cooling system pressure at normal engine temperature.
    C. engine overheating when operating under heavy load.
    D. engine overheating during extended idle periods.

Hint   *All cooling system hoses should be inspected for soft spots, swelling, hardening, chafing, leaks, and collapsing. If any of these conditions are present, hose replacement is necessary. Hose clamps should be inspected to make sure they are tight. Some radiator hoses contain a wire coil inside them to prevent hose collapse as the coolant temperature decreases. Remember to include heater hoses, and the bypass hose, in the hose inspection. Prior to hose removal the coolant must be drained from the radiator.*

Since the friction surfaces are the sides of a V-belt, the belt must be replaced if the sides are worn and the belt is contacting the bottom of the pulley.

The belt tension may be checked with the engine shut off, and a belt tension gauge placed over the belt at the center of the longest belt span. A loose, or worn, belt may cause a squealing noise when the engine is accelerated.

The belt tension also may be checked by measuring the amount of belt deflection with the engine shut off. Use your thumb to depress the belt at the center of the belt span. If the belt tension is correct, the belt should have 1/2 in. deflection per foot of belt span.

Ribbed V-belts usually have a spring-loaded belt tensioner, with a belt wear indicator scale on the tensioner housing. If a power steering pump belt requires tightening always pry on the pump ear not on the housing.

Task 5   Inspect, test, and replace radiator, pressure cap, coolant recovery system, and water pump.

38. The coolant level in the coolant recovery container is normal when the engine is cold. This level becomes much higher than normal after the vehicle has been driven for 45 minutes.

Technician A says some of the radiator tubes may be restricted.

Technician B says the radiator cap may be defective.

Who is correct?

A. A only

B. B only

C. Both A and B

D. Neither A nor B

Hint   *The radiator cap should be inspected for a damaged sealing gasket, or vacuum valve. If the pressure cap sealing gasket, or seat, are damaged, the engine will overheat, and coolant is lost to the coolant recovery system. Under this condition the coolant recovery container becomes over-filled with coolant.*

If the cap vacuum valve is sticking, a vacuum may occur in the cooling system after the engine is shut off and the coolant temperature decreases. This vacuum may cause collapsed cooling system hoses. A pressure tester may be used to test the pressure cap, and pressure test the entire cooling system.

The coolant level should be at the appropriate mark on the recovery container, depending on engine temperature.

With the engine shut off grasp the fan blades, or the water pump hub, and try to move the blades from side-to-side. This action checks for looseness in the water pump bearing. If there is any side-to-side movement in the bearing, water pump replacement is required.

Check for coolant leaks, rust, or residue at the water pump drain hole in the bottom of the pump, and at the inlet hose connected to the pump. When coolant is dripping from the pump drain hole, replace the pump. The water pump may be tested with the pressure tester connected to the radiator filler neck.

Task 6   Inspect, test, and replace thermostat, bypass, and housing.

39. A port-fuel-injected engine has an excessively rich air-fuel ratio. This problem could be caused by

A. engine overheating.

B. a defective radiator cap.

C. the engine thermostat stuck open.

D. the coolant control valve stuck open.

Hint   *The thermostat may be submerged with a thermometer in a container filled with water. Heat the water while observing the thermostat valve and the thermometer. The thermostat valve should begin to open when the temperature on the thermometer is equal to the rated temperature stamped on the thermostat. Replace the thermostat if it does not open at the rated temperature. Many thermostats are marked for installation in the proper direction. Inspect the bypass hose for cracks, deterioration, and restrictions, and replace the hose if these conditions are present.*

Task 7   Inspect, recover coolant; flush, and refill system with proper coolant.

40. All these statements about cooling system service are true EXCEPT

A. when the cooling system pressure is increased the boiling point is decreased.

B. if more antifreeze is added to the coolant, the boiling point is increased.

C. a good quality ethylene glycol antifreeze contains antirust and corrosion inhibitors.

D. coolant solutions must be recovered, recycled, or handled as hazardous waste.

Hint   *If the radiator tubes, and coolant passages in the block and cylinder head, are restricted with rust and other contaminants, these components may be flushed. Cooling system flushing equipment is available for this purpose. Always operate the flushing equipment according to the equipment manufacturer's directions, and be sure that your service procedure conforms to pollution laws in your state. Engine coolant must be recycled or handled as a hazardous waste material. Coolant recovery and reconditioning machines are available to remove harmful particles and restore corrosion additives so the coolant can be returned to the cooling system.*

Task 8   **Clean, inspect, test, and replace fan (both electrical and mechanical), fan clutch, fan belts, fan shroud, and air dams.**

41. In the electric cooling fan circuit in Figure 7-27 the low-speed and high-speed fans do not operate unless the air conditioning is turned on. The cause of this problem could be
    A. a blown 10 A, 5C fuse in the instrument panel fuse block.
    B. a defective engine coolant temperature sensor.
    C. a defective high speed coolant fan relay.
    D. a defective A/C pressure fan switch.

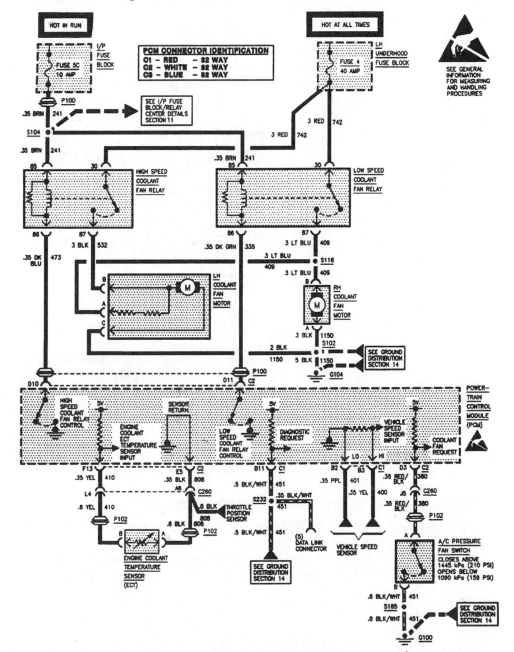

Figure 7-27 Electric cooling fan circuit *(Courtesy of Pontiac Motor Division, General Motors Corporation)*

Hint   *If the radiator shroud is loose, improperly positioned, or broken, air flow through the radiator is reduced, and engine overheating may result.*
The viscous-drive fan clutch should be visually inspected for leaks. If there are oily streaks radiating outward from the hub shaft, the fluid has leaked out of the clutch.

With the engine shut off rotate the cooling fan by hand. If the viscous clutch allows the fan blades to rotate easily both hot and cold, the clutch should be replaced. A slipping viscous clutch results in engine overheating. If there is any looseness between the viscous clutch and the shaft, replace the viscous clutch.

If the electric-drive cooling fan does not operate at the coolant temperature specified by the vehicle manufacturer, engine overheating will result especially at idle and lower speeds when air flow through the radiator is reduced.

Task 9   **Inspect, test, and replace heater coolant control valve (manual, vacuum, and electric types).**

42. In Figure 7-28 the adjustment being performed is
    A. the air mix door adjustment.
    B. the manual coolant control valve adjustment.
    C. the ventilation door control rod adjustment.
    D. the defroster door control rod adjustment.

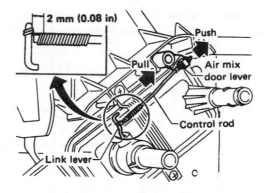

Figure 7-28 A/C-heater system adjustment *(Courtesy of Nissan Motor Co., Ltd.)*

Hint   *The heater coolant control valve may be operated manually, or by an electric solenoid. Some coolant control valves are operated by vacuum supplied from the intake manifold through a vacuum solenoid. In many A/C systems the heater coolant control valve is closed in the maximum A/C mode. This valve remains open in other modes. If the heater coolant control valve is stuck closed or restricted, there is reduced or no heat from the heater core. The hoses on each side of the heater coolant control valve should be the same temperature. When the inlet hose is considerably hotter than the outlet hose, the coolant control valve is restricted. If the coolant control valve is sticking open, cooling may be inadequate in the maximum A/C mode.*

Task 10   **Inspect, flush, and replace heater core.**

43. A gurgling noise is heard inside the passenger compartment. The noise is coming from the A/C-heater case.
    Technician A says the heater core coolant passages may be restricted.
    Technician B says the coolant level in the cooling system may be low.
    Who is correct?
    A. A only
    B. B only
    C. Both A and B
    D. Neither A nor B

Hint   *If the heater core is restricted, passenger compartment heating is reduced. Restricted air passages through the core also reduce passenger compartment heating. After the heater hoses are removed, the heater core may be flushed with a water hose. If the heater core is severely restricted, it may be removed and sent to a radiator shop for flushing with a special cleaning solution.*

Operating Systems and Related Controls Diagnosis and Repair, Electrical

ASE Tasks, Questions, and Related Information

Task 1 **Diagnose the cause of failures in the electrical control system of heating, ventilating, and A/C systems; determine needed repairs.**

44. An A/C-heater system does not discharge cool air in the A/C mode. The refrigerant system pressures are normal. The temperature door does not move when the A/C-heater controls are moved from maximum A/C to maximum heat (Figure 7-29), but the blower motor operation is normal. All of these defects could be the cause of the problem EXCEPT
   A. an open circuit at terminal 6 on the electronic door actuator.
   B. a blown 15A, number 8 fuse in the fuse junction panel.
   C. an open circuit at terminal C2-15 in the automatic temperature control module.
   D. an open circuit at terminal 8 on the electronic door actuator motor.

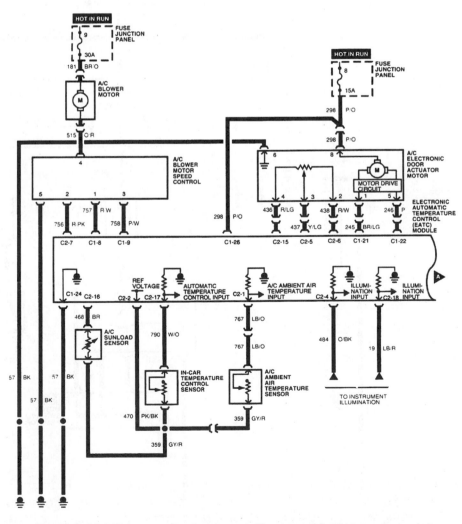

Figure 7-29 A/C-heater electrical system *(Courtesy of Ford Motor Company)*

Hint   *Many computer-controlled A/C systems have self-diagnostic capabilities that provide diagnostic trouble codes (DTCs) representing faults in a specific area. The technician usually has to perform some voltmeter or ohmmeter tests to locate the exact cause of the problem. For example, a DTC representing the in-car temperature sensor may be obtained. The technician has to perform ohmmeter tests on this sensor and the connecting wires to find the exact cause of the problem. If the A/C-heater system does not have self diagnostic capabilities, the technician has perform voltmeter and ohmmeter tests to locate the source of the problem.*

Task 2   Inspect, test, repair, and replace A/C-heater blower motors, resistors, switches, relay/modules, wiring, and protection devices.

45. A blower motor operates at high speed but does not operate at any other speed (Figure 7-30). The cause of this problem could be
    A. an open circuit at terminal 2 in the resistor assembly.
    B. an open circuit in the blower motor switch ground connection.
    C. an open circuit at terminal 4 in the blower switch.
    D. an open circuit at terminal 4 in the resistor assembly.

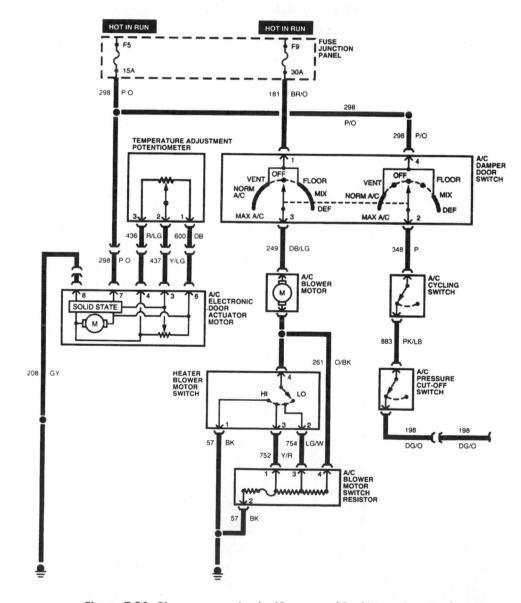

Figure 7-30 Blower motor circuit *(Courtesy of Ford Motor Company)*

Hint   *In many blower motor circuits the motor switch and a resistor assembly are connected on the ground side of the motor. When the heater or A/C system is turned on, voltage is supplied to one brush in the blower motor. Current flows from the other motor brush through the blower motor switch and resistor assembly to ground. When the blower switch is in the high position, the blower motor brush is ground directly through the switch contacts. If one of the lower speeds is selected, the motor brush is grounded through the proper resistor in the resistor assembly to provide the blower speed selected. The resistor assembly usually is mounted in the A/C-heater case so air flow past the resistors provides a cooling action.*

In some computer-controlled A/C systems the blower motor speed is controlled by the A/C computer and the blower speed control module. The A/C computer commands the blower speed control module to provide the proper blower speed. The blower speed module usually controls the blower speed with a pulse width modulated (PWM) voltage signal.

Task 3   **Inspect, test, repair, and replace A/C compressor clutch, relay/modules, wiring, sensors, switches, diodes, and protection devices.**

46. The A/C compressor clutch in Figure 7-31 is deenergized after the vehicle is driven for about 30 minutes, and it does not engage again until the engine is shut off for a period of time. The engine coolant temperature and refrigerant system pressures are normal. The most likely cause of this problem could be

    A. a defective engine coolant temperature sensor.
    B. a defective A/C pressure cutoff switch.
    C. a defective clutch control relay.
    D. a defective compressor clutch coil.

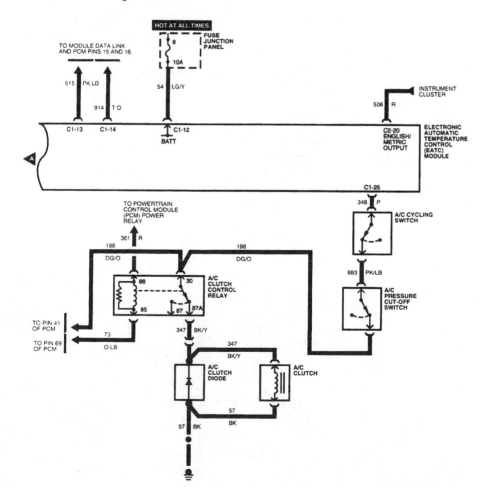

Figure 7-31 Compressor clutch circuit *(Courtesy of Ford Motor Company)*

Hint   *When the A/C mode is selected in many computer-controlled A/C systems, the A/C computer supplies 12V to the pressure cycling switch. Voltage is supplied through this switch to the A/C pressure cutoff switch and the clutch control relay contacts. A signal is transmitted to the powertrain control module (PCM) from the wire between the A/C pressure switch and the relay contacts. This signal informs the PCM the A/C mode is selected.*

The PCM scans the input signals to determine if compressor clutch operation is appropriate. If the engine coolant temperature sensor indicates very high temperature, or the throttle position sensor indicates wide open throttle, the PCM does not ground the compressor clutch winding. Under this condition the compressor clutch remains inoperative. The PCM does not energize the clutch control winding for a brief time period after the engine is started. If the PCM detects a low idle speed condition, it does not energize the clutch control relay winding.

If the PCM inputs indicate the proper conditions are present for compressor clutch operation, the PCM grounds the clutch control relay winding. This action closes the relay contacts that supply voltage to the compressor clutch coil to engage the clutch.

The A/C clutch diode reduces voltage spikes when the compressor clutch is shut off. Some PCMs have a power steering switch input. When this switch input indicates high-power steering pump pressure the PCM opens the ground circuit from the clutch control relay winding. This action deenergizes the compressor clutch.

The A/C pressure cutoff switch opens the compressor clutch circuit if the refrigerant system pressure exceeds 399 to 445 psi (2,751 to 3,068 kPa). This action prevents damage to refrigeration system components.

Task 4   Inspect, test, repair, replace, and adjust A/C-related engine control systems.

47. An engine with the carburetor and throttle kicker shown in Figure 7-32 experiences a stalling problem when the engine is idling and the A/C compressor is engaged.
    Technician A says the throttle kicker diaphragm may be leaking.
    Technician B says the PCM may not be energizing the kicker solenoid.
    Who is correct?

    A. A only
    B. B only
    C. Both A and B
    D. Neither A nor B

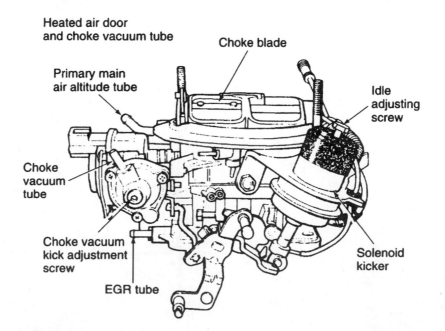

Figure 7-32 Throttle kicker assembly *(Courtesy of Chrysler Corporation)*

Hint   *Some computer-controlled carburetors have a vacuum-operated throttle kicker, and an idle stop solenoid. The throttle kicker maintains engine idle speed when the engine accessory load is increased, such as during A/C compressor clutch operation. This kicker also maintains idle speed during warmup, after the fast idle cam had dropped away from the fast idle screw. Vacuum to the throttle kicker is controlled by an electric solenoid, which in turn is controlled by the power-train control module (PCM). This type of throttle kicker is not adjustable, but the kicker should increase the idle rpm slightly when the A/C compressor clutch is engaged.*

The electric solenoid in the throttle kicker is energized while the ignition switch is turned on. This solenoid maintains the throttle in the specified idle position, and allows the throttle to drop closed against an idle stop screw to prevent engine after-running when the ignition switch is turned off.

Some carbureted engines have a thermal vacuum switch (TVS) connected in the distributor vacuum advance hoses (Figure 7-33). Under normal engine operating temperatures the TVS supplies ported vacuum from above the throttle to the vacuum advance. If the compressor clutch is engaged with the engine idling at high ambient temperatures the engine temperature may increase above the normal range. Under this condition the TVS supplies full intake manifold vacuum to the vacuum advance to increase idle rpm and decrease engine temperature.

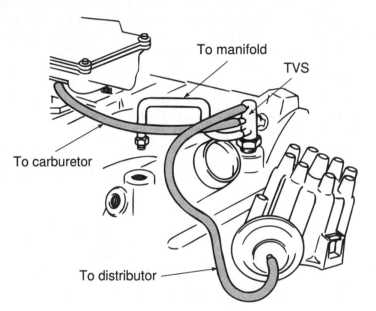

Figure 7-33 Thermal vacuum switch

Task 5   Inspect, test, repair, replace, and adjust load sensitive A/C compressor cutoff systems.

48. While discussing compressor clutch control in an A/C system with a pressure transducer (Figure 7-34, next page)

    Technician A says the pressure transducer is connected in the compressor clutch circuit.

    Technician B says the pressure transducer contains a set of contacts that are normally closed.

    Who is correct?

    A. A only

    B. B only

    C. Both A and B

    D. Neither A nor B

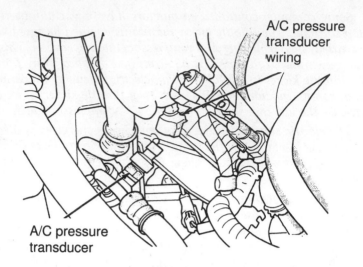

Figure 7-34 A/C system with pressure transducer *(Courtesy of Chrysler Corporation)*

Hint   *Some A/C compressor clutch circuits have a high-pressure cutoff switch. This switch opens the compressor clutch circuit and prevents compressor clutch operation if the refrigeration system pressure exceeds 430 psi (2,960 kPa). This high-pressure cutoff switch prevents damage to the compressor or other refrigeration system components.*

Other A/C systems have an A/C pressure transducer mounted in the compressor discharge line near the compressor. The PCM sends a 5V signal to the pressure transducer, and the potentiometer in the pressure transducer sends a voltage signal to the PCM in relation to refrigeration system pressure (Figure 7-35). A ground wire is connected from the pressure transducer to the PCM. If the refrigeration system pressure becomes excessively high, the PCM opens the circuit from the compressor clutch relay winding to ground in response to the pressure transducer signal. Under this condition the compressor clutch relay contacts open, and the compressor clutch is deenergized. The PCM also uses the pressure transducer for cooling fan control.

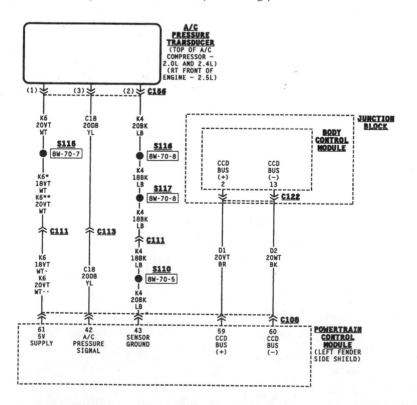

Figure 7-35 Pressure transducer to PCM wiring connections *(Courtesy of Chrysler Corporation)*

Some compressor clutch circuits contain a thermal limiter switch that senses compressor surface temperature. If the compressor surface temperature becomes excessive, the thermal limiter switch opens the compressor clutch circuit to deenergize the clutch.

Task 6 Inspect, test, repair, and replace engine cooling/condenser fan motors, relays/modules, switches, sensors, wiring, and protection devices.

49. The cooling fan system in Figure 7-36 operates normally on high speed, but there is no low speed fan operation under any operating condition. The cause of this problem could be

A. an open circuit in the wire between the number 30 terminals in the two fan relays.

B. a blown 40 amp maxifuse in the LH maxifuse center.

C. a blown 10 amp cooling fan/TCC fuse in the fuseblock.

D. an open circuit at terminal B on the A/C head pressure switch.

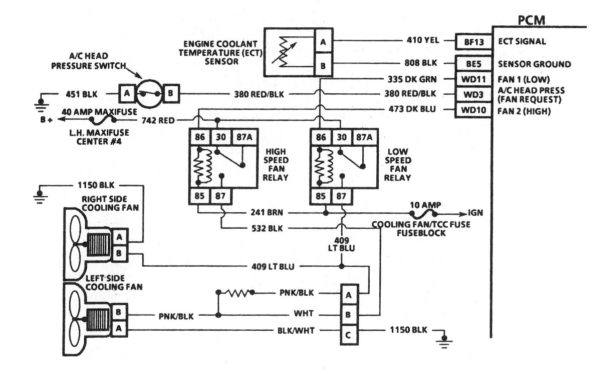

Figure 7-36 Cooling fan circuit *(Courtesy of Pontiac Motor Division, General Motors Corporation)*

Hint *Cooling fan circuits vary depending on the vehicle make and model year. Some cooling fan systems have a low-speed and a high-speed cooling fan. The PCM grounds the low-speed or high-speed relay windings in response to the engine coolant temperature and compressor head pressure switch. When the engine coolant reaches 212° F (100°C), the PCM grounds the low-speed fan relay winding. This action closes these relay contacts, and voltage is supplied through the contacts in the low-speed fan motor.*

When the engine coolant reaches 226° F (108°C), or the compressor discharge pressure reaches 210 psi (1,448 kPa), the PCM grounds the high-speed relay winding. Under this condition the relay contacts close and supply voltage to the high-speed fan motor. The compressor head pressure switch normally is closed, and this switch opens at 210 psi (1,448 kPa). When the compressor head pressure switch opens, the PCM grounds the high-speed fan relay winding.

Task 7
Inspect, test, adjust, repair, and replace electric actuator motors, relays/ modules, switches, sensors, wiring, and protection devices.

50. All of these statements about computer-controlled A/C system actuator motors are true EXCEPT
    A. some actuator motors are calibrated automatically in the self-diagnostic mode.
    B. A/C system diagnostic trouble codes represent a fault in a specific component.
    C. the actuator motor control rods must be calibrated manually on some systems.
    D. the actuator motor control rods should only require adjustment after motor replacement or misadjustment.

Hint   *Many computer-controlled A/C systems have self-diagnostic capabilities. On Chrysler LH and LHS cars the self-diagnostic A/C mode is entered by pressing the floor, mix, and defrost buttons simultaneously for a few seconds. The engine must be running with the vehicle stopped and the A/C temperature control set at 75°F (24°C). When the diagnostic mode is entered, the A/C computer calibrates the actuator motors and performs specific system tests. During this time the control head display continues blinking.*

On some cars, such as a Nissan Maxima, the self-diagnostic mode in the automatic A/C system is entered by starting the engine and pressing the off button in the A/C control head for 5 seconds. The off button must be pressed within 10 seconds after the engine is started, and the fresh vent lever must be in the off position. The self diagnostic mode may be cancelled by pressing the auto button or turning off the ignition switch.

The self-diagnostic tests are completed in five steps, and the up arrow for temperature setting on the A/C control head is pressed to move to the next step. When the down arrow for temperature setting is pressed, the diagnostic system returns to the previous step. The five steps in the diagnostic tests follow.

- *Step 1 checks the LEDs and segments in the A/C control head.*
- *Step 2 checks the input sensor signals.*
- *Step 3 checks the mode door position switch.*
- *Step 4 checks the mode door electric actuators.*
- *Step 5 checks the temperature detected by each input sensor.*

On some systems the actuator door control rods may be adjusted. These adjustments are necessary after components, such as door actuator motors, are replaced. The procedure for the mode door control rod adjustment follows.

- *With the mode door installed on the A/C-heater case and the wiring connector attached to the actuator, disconnect the door motor rod from the slide link.*
- *Enter step four in the self diagnostic mode so 41 is displayed in the A/C control head.*
- *Move the slide link by hand until the mode door is in the vent mode (Figure 7-37).*
- *Connect the motor rod to the slide link.*
- *Continue pressing the DEF button until all six modes have been obtained in step 4, and be sure the door moves to the proper position in each mode.*

Task 8
Inspect, test, service, or replace heating, ventilating, and A/C control panel assemblies.

51. All of these statements about A/C control panel service are true EXCEPT
    A. the negative battery cable must be removed before A/C control panel removal.
    B. the refrigeration system must be discharged before the A/C control panel is removed.
    C. if the vehicle is air-bag-equipped, wait the specified time period after negative battery cable removal.
    D. self-diagnostic tests may indicate a defective A/C control panel in a computer-controlled A/C system.

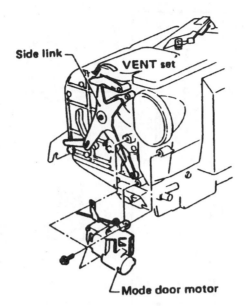

Side link
VENT set
Mode door motor

Figure 7-37 Mode door control rod adjustment *(Courtesy of Nissan Motor Co., Ltd.)*

Hint   *Self-diagnostic tests should indicate a defective A/C control panel. In many systems the A/C computer is contained in the control panel. Always disconnect the negative battery cable before removing the A/C control panel assembly. If the vehicle is equipped with an air bag, wait for the time specified by the vehicle manufacturer after the negative battery cable is disconnected before starting the A/C control panel removal procedure. Remove the instrument panel molding around the A/C control panel, and then remove the A/C control panel retaining screws. Pull the A/C control panel out of the instrument panel as far as possible. Disconnect the electrical connectors from the A/C control panel and remove the control panel.*

Operating Systems and Related Controls Diagnosis and Repair, Vacuum/Mechanical

ASE Tasks, Questions, and Related Information

Task 1   **Diagnose the cause of failures in the vacuum and mechanical switches and controls of the heating, ventilating, and A/C systems; determine needed repairs.**

52. When testing the A/C vacuum system in Figure 7-38 (next page), 18 in. Hg. is supplied to the vacuum distribution hose connected to the A/C control vacuum valve. Under this condition there should be 18 in. Hg. supplied to
   A. the panel door actuator with the switch in the off position.
   B. the outside recirculate door actuator with the switch in the normal A/C position.
   C. the outside recirculate door actuator with the switch in the floor position.
   D. the panel door actuator with the switch in the maximum A/C position.

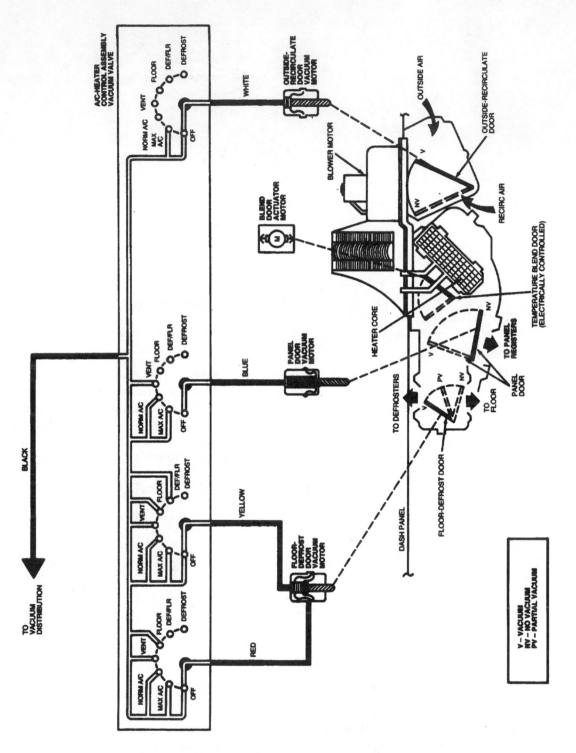

Figure 7-38 A/C vacuum system *(Courtesy of Ford Motor Company)*

Hint *Vacuum hoses and switches may be tested for leaks with a vacuum pump and gauge. When vacuum is supplied to one end of a vacuum hose and the other end of the hose is plugged, the hose should hold 15 to 20 in. Hg. without leaking. The same method may be use to test vacuum switches.*

Task 2 Inspect, test, service, or replace heating, ventilating, and A/C control panel assemblies.

Task 3   Inspect, test, adjust, and replace heating, ventilating, and A/C control cables and linkages.

53. While adjusting the temperature control cable in Figure 7-39
    A. the black cable attaching flag must be removed from the flag receiver.
    B. the self-adjusting clip must be removed from the blend-air door crank.
    C. the blend-air door crank is rotated fully counterclockwise.
    D. the temperature control lever must be in the maximum heat position.

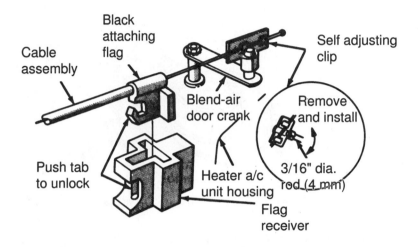

Figure 7-39   Temperature door cable *(Courtesy of Chrysler Corporation)*

Hint   *Many A/C control cables require an adjustment if they have been disconnected or replaced. With the self-adjusting clip installed on the temperature control cable and the blend-air door crank arm, install the black temperature control cable attaching flag. Hold the temperature lever in the maximum cold position and rotate the temperature blend-air door crank arm fully counterclockwise by hand. During this procedure the self-adjusting clip reaches the proper position on the temperature control cable.*

Task 4   Inspect, test, and replace heating, ventilating, and A/C vacuum actuators (diaphragms/motors) and hoses.

54. An A/C system has vacuum-operated mode doors except the temperature door which is operated by an electric actuator (refer to Figure 7-38). In the A/C mode air is discharged from the panel ducts for a short while and then the air discharge switches to the floor ducts. The other modes operate properly. The panel door actuator motor is tested and proven to be satisfactory.
    Technician A says the blue vacuum hose from the A/C control switch to the panel door actuator may be leaking.
    Technician B says the black vacuum hose connected from the vacuum source to the A/C control switch may be leaking.
    Who is correct?
    A. A only
    B. B only
    C. Both A and B
    D. Neither A nor B

Hint   *Connect the vacuum pump to each vacuum actuator and supply 15 to 20 in. Hg. to the actuator (Figure 7-40). Check the vacuum actuator rod to be sure it moves freely. Close the vacuum pump valve and observe the vacuum gauge. The gauge reading should remain steady for at least 1 minute. If the gauge reading drops slowly, the actuator is leaking.*

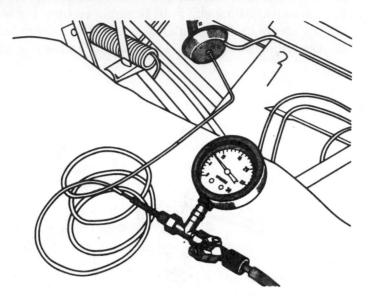

Figure 7-40 Testing mode door actuator *(Courtesy of Chrysler Corporation)*

Task 5   Identify, inspect, test, and replace heating, ventilating, and A/C vacuum reservoir, check valve, and restrictors.

55. An A/C system has a vacuum reservoir, check valve, and vacuum-operated mode door actuators including the blend-air door (Figure 7-41). While operating in the A/C mode and climbing a long hill with the throttle nearly wide open, the air discharge temperature gradually becomes warm and the air discharge switches from the panel to the floor ducts. The A/C system operates normally under all other conditions. The cause of this problem could be

A. a leaking panel door vacuum actuator.

B. a defective vacuum reservoir check valve.

C. a leaking blend-air door vacuum actuator.

D. a leaking intake manifold gasket.

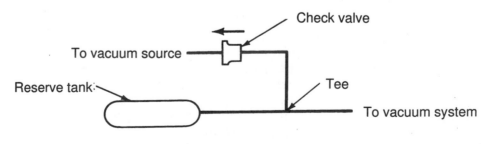

Figure 7-41 Vacuum reservoir and check valve

Hint   *Vacuum reservoirs may be connected in A/C systems to maintain the vacuum supply during periods of wide open throttle operation when the intake manifold vacuum is very low. A check valve usually is connected between the reservoir and the vacuum source to trap the vacuum in the reservoir when the intake manifold vacuum decreases. Some A/C vacuum systems have a restrictor that delays the vacuum supply to certain components.*

When vacuum is supplied from a vacuum pump to the reservoir it should hold 15 to 20 in. Hg. without leaking. A check valve should hold vacuum one way, but leak air through it in the opposite direction.

Task 6   Inspect, test, adjust, repair, or replace heating, ventilating, and A/C ducts, doors, and outlets.

56. The outside air-recirculation door is stuck in position A (Figure 7-42).

    Technician A says under this condition outside air is drawn into the A/C-heater case.

    Technician B says under this condition some in-vehicle air leaks past the door.

    Who is correct?

    A. A only

    B. B only

    C. Both A and B

    D. Neither A nor B

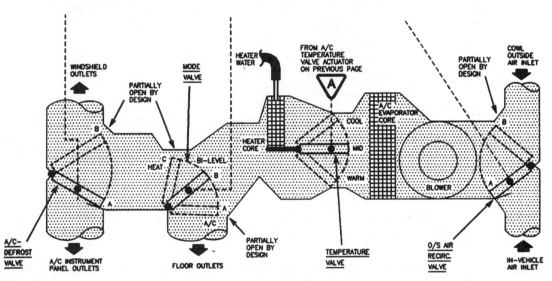

Figure 7-42 A/C-heater case and mode doors *(Courtesy of Pontiac Motor Division, General Motors Corporation)*

Hint   *All the mode doors in the A/C-heater case should move freely. Some mode doors are designed to provide a certain amount of air leakage past the door. For example, when the defrost door is directing air to the panel ducts, some air leaks past the door to the defrost ducts. The A/C-heater case and outlet ducts must not have any air leaks.*

Automatic and Semiautomatic Heating, Ventilating, and A/C Systems

ASE Tasks, Questions, and Related Information

Task 1   **Diagnose temperature control problems; determine needed repairs.**

57. In a semiautomatic A/C system the temperature control is set at 70°F (21°C), and the in-car temperature is 80°F (27°C) after driving the car for 1 hour. The refrigerant system pressures are normal.
Technician A says the in-car sensor may be defective.
Technician B says the temperature door may be sticking.
Who is correct?
    A. A only
    B. B only
    C. Both A and B
    D. Neither A nor B

Hint   *When diagnosing improper temperature control always visually inspect the A/C system for loose or damaged electrical connections and leaking, loose, or deteriorated vacuum hoses. Check the coolant level and engine coolant temperature. Test the refrigerant system pressures. Check the temperature door and related control system.*

Many automatic and semiautomatic A/C systems have self-diagnostic capabilities that provide diagnostic trouble codes (DTCs). These DTCs represent a fault in a specific area such as a sensor or electric actuator. The technician must perform voltmeter or ohmmeter tests to locate the exact cause of the problem.

Task 2   **Diagnose blower system problems; determine needed repairs.**

58. The blower motor in Figure 7-43 operates only at low speed. When the fan speed control in the A/C controls is changed from low speed to high speed the voltage at blower motor terminal A changes from 4V to 13.5V. The cause of this problem could be
    A. a defective blower motor.
    B. an open blower motor ground.
    C. a defective HVAC power module.
    D. an open circuit at programmer terminal D2.

Hint   *Blower motor circuits vary depending on the vehicle make and model year. The technician must use the wiring diagram and diagnostic procedure for the system being diagnosed. Check all the wiring connections and fuse or fuse link in the blower motor circuit. Be sure the blower motor ground is satisfactory.*

Check the voltage supplied to the blower motor. If this voltage is less than specified, repair the blower circuit. Supply 12V and a ground connection to the blower motor terminals with the motor wiring connector disconnected. If the motor operates properly with 12V supplied, the motor is satisfactory. When the blower speed is slow replace the motor.

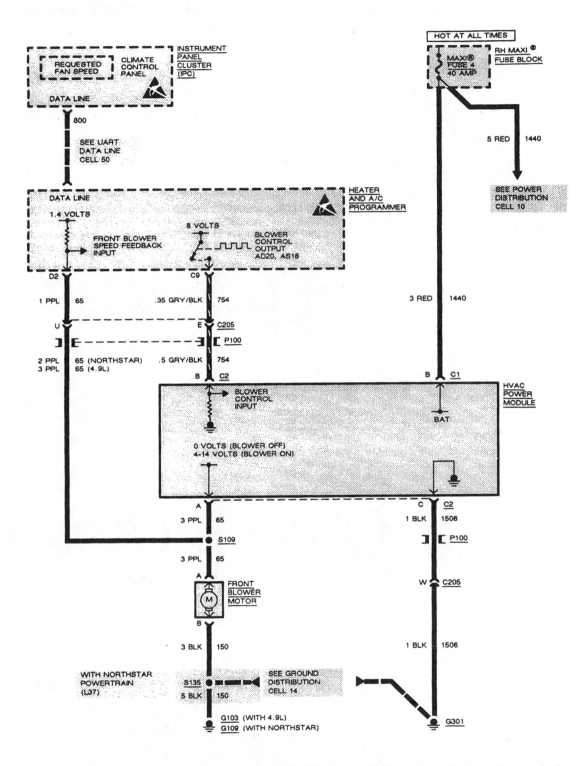

Figure 7-43 Blower motor circuit *(Courtesy of Cadillac Motor Division, General Motors Corporation)*

Task 3   Diagnose air distribution system problems; determine needed repairs.

59. The computer-controlled A/C system in Figure 7-44 will not go into the recirculation air mode. All the other modes operate normally. The vacuum supplied to the outside air recirculation door vacuum actuator is 1 in. Hg. with the engine idling. Technician A says to check the check valves, vacuum tank, and vacuum supply to the HVAC programmer.

Technician B says to check the programmer vacuum switch and hose to the inoperative door actuator.

Who is correct?

A. A only
B. B only
C. Both A and B
D. Neither A nor B

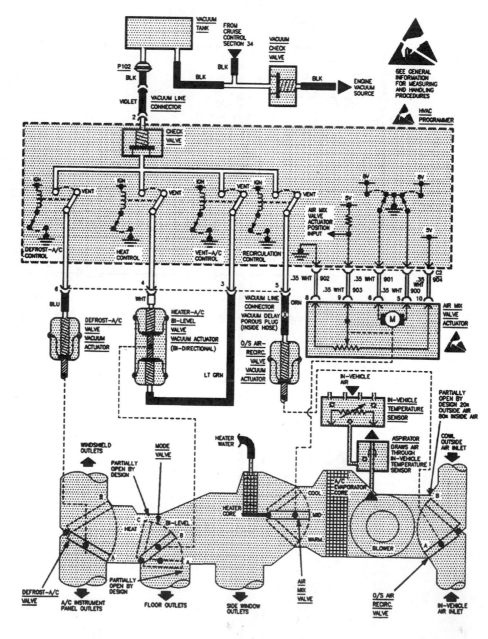

Figure 7-44 Air distribution system, computer-controlled A/C system (*Courtesy of Pontiac Motor Division, General Motors Corporation*)

Hint   *In some computer-controlled A/C systems, the intake manifold vacuum is supplied through a check valve and vacuum tank to the HVAC programmer. Another check valve may be located in the programmer. A vacuum switch supplies vacuum to the appropriate mode door actuator. If none of the mode doors operate properly, check the vacuum source to the programmer. When only one mode door does not operate properly, check that mode door, actuator, vacuum switch and connecting wires.*

Task 4   Diagnose compressor clutch control system; determine needed repairs.

60. The compressor clutch is inoperative in the system shown in Figure 7-45. The switches, relay, compressor clutch coil, and connecting wires are tested and proven satisfactory. When the A/C button is pressed in the A/C control panel, voltage is supplied to the clutch control relay contacts.

   Technician A says there may be an open circuit at terminal 69 on the PCM.

   Technician B says the number 18 15A fuse may be open in the fuse junction panel.

   Who is correct?

   A. A only
   B. B only
   C. Both A and B
   D. Neither A nor B

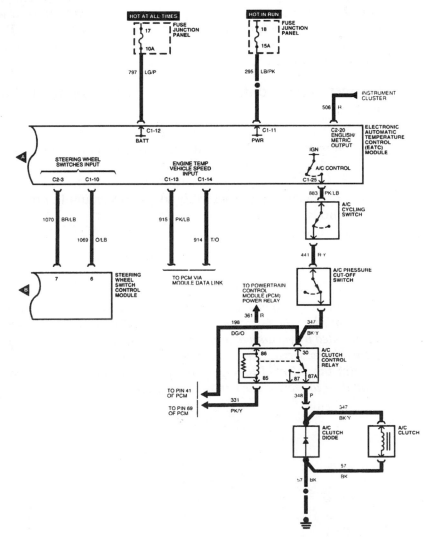

Figure 7-45 Compressor clutch circuit, computer-controlled A/C system *(Courtesy of Ford Motor Company)*

Hint   *In some computer-controlled compressor clutch systems, voltage is supplied through the cycling switch, the pressure cutoff switch and the clutch control relay to the compressor clutch coil. The cycling switch turns the compressor on and off in relation to evaporator suction pressure. The pressure cutoff switch opens the compressor clutch circuit to protect the system if refrigerant pressures become excessive. The clutch control relay is operated by the PCM.*

Task 5   Inspect, test, adjust, or replace climate control temperature and sunload sensors.

61. When discussing in-vehicle temperature sensor testing with an ohmmeter
    Technician A says as the sensor temperature increases the sensor resistance should decrease.
    Technician B says at 50°F (10°C) the sensor should have minimum resistance.
    Who is correct?
    A. A only
    B. B only
    C. Both A and B
    D. Neither A nor B

Hint   *Many A/C system sensors, such as the in-vehicle sensor and ambient sensor, contain thermistors. The resistance of these sensors increases as sensor temperature decreases. Each sensor must have the resistance specified by the vehicle manufacturer at various temperatures. The sunload sensor usually is mounted on top of the instrument panel. This sensor contains a photovoltaic diode that sends a varying current signal to the A/C computer in relation to the amount of sunlight applied to the sensor.*

Task 6   Inspect, test, adjust, and replace temperature blend door/power servo system.

62. When diagnosing the power servo system in Figure 7-46, the temperature control is set in the maximum cold position. Solenoid 4 and solenoid 5 are open, and there is vacuum supplied to power servo 2.
    Technician A says this is a normal condition.
    Technician B says there should be no vacuum to power servo 2.
    Who is correct?
    A. A only
    B. B only
    C. Both A and B
    D. Neither A nor B

Hint   *In some A/C systems the temperature blend door is controlled by an electric door actuator that is controlled by the A/C computer (Figure 7-47). In other A/C systems the temperature blend door is controlled by vacuum actuator that may be called a power servo. A varying vacuum is supplied to this power servo from a solenoid or solenoids controlled by the A/C computer. The computer operates the solenoid to supply the proper vacuum to the servo and position the temperature blend door to provide the temperature selected by the driver.*

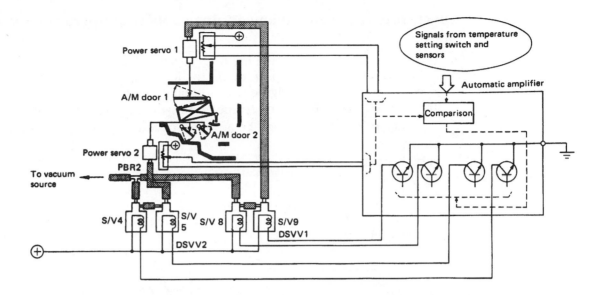

CL: Solenoid valve is closed
OP: Solenoid valve is open

AIR MIX DOOR CONTROL

			HOT side	HOLD	COLD side
Air mix door 1	Operation of solenoid valve	S/V8	CL	CL	OP
		S/V9	OP	CL	OP
Air mix door 2		S/V4	CL	CL	OP
		S/V5	OP	CL	OP

Figure 7-46 Computer-controlled A/C system with vacuum solenoids and power servos *(Courtesy of Nissan Motor Co., Ltd.)*

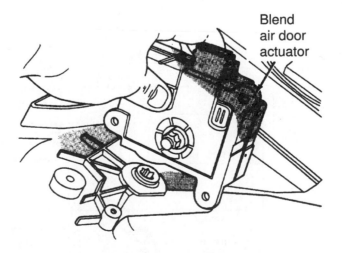

Figure 7-47 Temperature blend door electric actuator *(Courtesy of Chrysler Corporation)*

Task 7   Inspect, test, and replace low engine coolant temperature blower control system.

63. When a cold engine is started the A/C blower in Figure 7-48 starts immediately with the control in the defrost position. In the lo, auto, hi, or bilevel positions the blower does not start until the engine is at normal operating temperature.

Technician A says the engine temperature switch may be defective.

Technician B says there may be an open circuit in the engine temperature switch ground.

Who is correct?

A. A only

B. B only

C. Both A and B

D. Neither A nor B

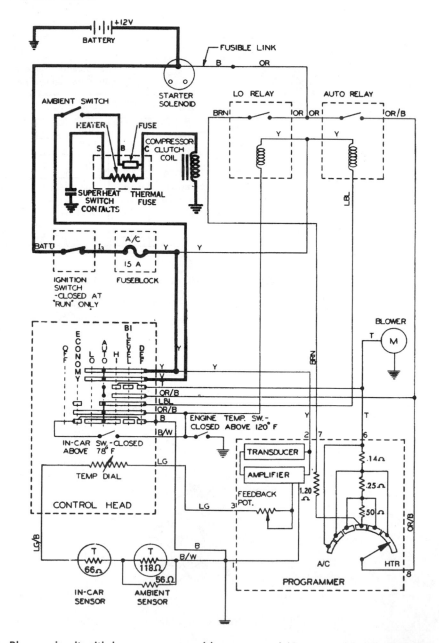

Figure 7-48 Blower circuit with low temperature blower control (*Courtesy of Cadillac Motor Car Division, General Motors Corporation*)

Hint   *Some blower motor circuits have a low temperature cutoff switch that opens the blower motor circuit below a specific temperature. This temperature switch may sense engine metal or coolant temperature. If the A/C controls are placed in the defrost mode, the low temperature cutoff switch is bypassed and the blower operates normally.*

Task 8   Inspect, test, and replace heater water valve and controls.

64. In Figure 7-49 vacuum is supplied from a hand pump to the water valve solenoid and battery voltage is supplied to the solenoid terminals resulting in an audible click from the solenoid. The system holds 16 in. Hg. The water valve does not move. All of these defects may be the cause of the problem EXCEPT
    A. a seized water control valve in the heater hose.
    B. a plugged vacuum hose between the solenoid and the valve.
    C. a jammed linkage from the actuator to the water valve.
    D. a seized plunger in the water valve control solenoid.

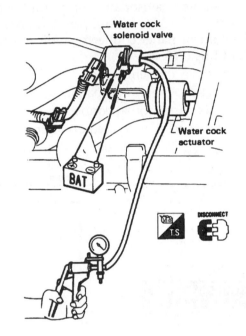

Figure 7-49   Heater coolant control valve *(Courtesy of Nissan Motor Co., Ltd.)*

Hint   *Some A/C systems have a coolant control valve that shuts off the coolant flow through the heater core under certain conditions. This valve may be operated mechanically, electrically, or by engine vacuum. In the vacuum-operated systems a computer-controlled solenoid keeps the vacuum shut off to the coolant control valve actuator. When the A/C system is placed in the maximum A/C mode the A/C computer energizes the solenoid and vacuum is supplied through the solenoid to the coolant valve actuator. Under this condition the actuator closes the coolant control valve and shuts off the coolant flow through the heater core to maximize the cold air flow through the evaporator.*

Task 9   Inspect, test, and replace electric and vacuum motors, solenoids, and switches.

65. When diagnosing a computer-controlled A/C system, a diagnostic trouble code (DTC) is obtained indicating a fault in the temperature blend door actuator motor. Technician A says the first step in the repair procedure is to replace the temperature blend door actuator.
Technician B says the first step in the repair procedure is to check the temperature blend door for a sticking condition.

Who is correct?

A. A only

B. B only

C. Both A and B

D. Neither A nor B

Hint   *Most computer-controlled A/C systems have self-diagnostic capabilities. DTCs may be obtained representing various system components. These DTCs represent a fault in a specific area. For example, a DTC representing an electric temperature blend door actuator indicates a fault in this area. The technician must check the door for sticking and test the motor and connecting wires to determine the exact cause of the problem.*

Task 10   Inspect, test, and replace ATC control panel.

66. A DTC representing the ambient sensor is obtained in the circuit shown in Figure 7-50. The ambient sensor and connector 2 are connected, and connector 7 is disconnected from the ATC control panel. An ohmmeter connected to terminals 9 and 18 in the control panel connector indicates the specified resistance. The most likely cause of this DTC is

A. a defective ambient sensor.

B. a defective A/C control panel.

C. a loose connection at connector 2 terminal 7.

D. an open circuit at connector 7 terminal 18.

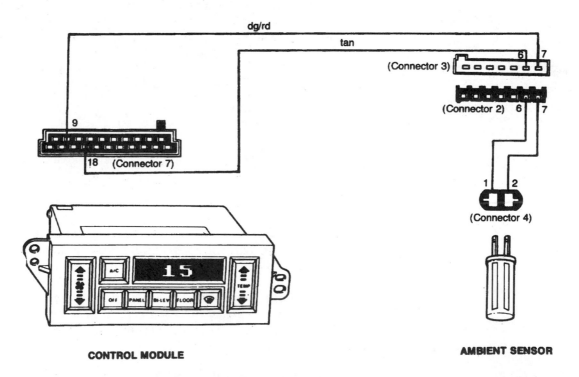

Figure 7-50   ATC control panel and ambient sensor wiring *(Courtesy of Chrysler Corporation)*

Hint   *Some A/C systems have DTCs representing the control panel. In other systems the technician has to test the circuit represented by the DTC and prove that all circuit components are satisfactory. When all the components are satisfactory in the circuit represented by the DTC, the A/C control panel is likely the cause of the DTC.*

Task 11   Inspect, test, adjust, or replace ATC microprocessor (climate control computer/programmer).

Task 12   Check and adjust calibration of ATC system.

67. When measuring the resistance in the A/C computer ground (Figure 7-51) a voltmeter is connected from computer terminal C1-24 to ground. With the ignition switch on, the maximum voltage reading should be
    A.   .1V
    B.   .2V
    C.   .5V
    D.   .8V

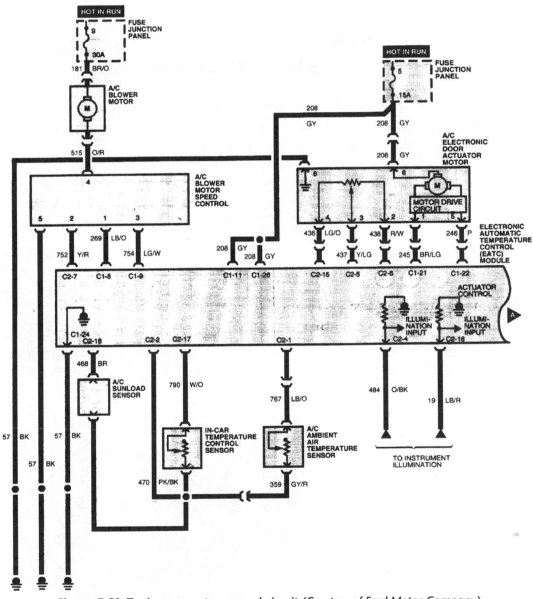

Figure 7-51 Testing computer ground circuit *(Courtesy of Ford Motor Company)*

Hint   *Before replacing any A/C computer always be sure the voltage supply wires and ground wire(s) connected to the computer are satisfactory. With the ignition switch on, the maximum voltage drop across computer ground wires should be .1V. The voltage supply wires should supply 12V to the computer.*

When some computer-controlled A/C systems are placed in the diagnostic mode the A/C computer automatically preforms calibration procedures on the electric door actuators. Some other systems require a calibration procedure on the electric door actuators if the actuators or other major components have been replaced. This calibration procedure involves placing the A/C controls in a specific mode with the electric actuator disconnected. The mode door is then moved by hand to a specific position, and the linkage is then connected between the actuator and the door.

Refrigerant Recovery, Recycling, and Handling

ASE Tasks, Questions, and Related Information

Task 1   **Maintain and verify correct operation of certified equipment.**

68. All of these statements about A/C recovery/recycling equipment are true EXCEPT
    A. the equipment label must indicate UL approval.
    B. the equipment label must indicate SAE J1991 approval.
    C. any size and type of refrigerant storage container over 10 lbs. may be used in this equipment.
    D. R-12 and R-134a refrigerants or refrigerant oils must not be mixed in the recovery/recycling process.

Hint   *Refrigerant recovery/recycling equipment must be used according to the equipment and vehicle manufacturer's recommended procedures (Figure 7-52). Never mix R-12 and R-134a refrigerants or refrigerant oils. This action causes system damage. The recovery/recycling equipment must have a label indicating it meets SAE J1991 standards for this type of equipment. The recovery/recycling equipment must also have an underwriter's laboratory (UL) label. The refrigerant storage tank specified by the A/C recovery/recycling equipment manufacturer must be installed in the recovery/recycling equipment. The tank valve is designed for use with the recovery/recycling equipment, and the unit's overfill limitation system is designed for a specific storage tank.*

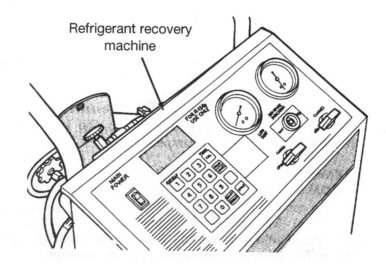

Figure 7-52 Refrigerant recovery/recycling equipment *(Courtesy of Chrysler Corporation)*

Task 2   Identify and recover A/C system refrigerant.

69. After the recovery process the low-side pressure increases above zero after 5 minutes. This condition indicates
    A. there is still some refrigerant in the system.
    B. there is excessive oil in the refrigerant system.
    C. the refrigerant system is leaking.
    D. there is excessive moisture in the refrigerant system.

Hint   *The type of refrigerant in an A/C system may be identified by the underhood refrigerant charge tag. In some R-134a refrigerant systems the clutch cycling switch is a different color compared to R-12 systems. For example, on some Ford products the R-134a cycling switch is yellow. Quick-disconnect service fittings are used in R-134a refrigerant systems, whereas R-12 systems have Schrader-type service valves.*

The first step in the recovery procedure is to connect the low-side and high-side hoses from the equipment to the service valves. Check the low-side and high-side pressures to be sure there is refrigerant in the system. Do not continue with the recovery process if there is no refrigerant in the system. Open the low-side and high-side valves on the control panel, and open the vapor and liquid valves on the storage tank. Follow the equipment manufacturer's recommended procedure to drain any oil from the separator.

Plug in the unit and turn on the main power switch. Press recover on the key pad. The unit clears itself of refrigerant and displays CL-L on the display. After the refrigerant is cleared the unit beings the recovery process; this process continues until CPL is displayed. The initial recovery is completed when the refrigerant system is evacuated to 17 in. Hg. When CPL is displayed the unit flashes the amount of refrigerant and oil recovered. If the low-side gauge pressure increases above zero after 5 minutes, there is still refrigerant in the system and further recovery is required. If the unit flashes full during the recovery process, the unit tank is full, and it must be replaced with the proper empty tank.

Task 3   Recycle refrigerant.

70. The moisture warning light on the recovery/recycling equipment indicates yellow during the recycling process. This indicates
    A. there is excessive oil in the refrigerant.
    B. the filter/drier cartridge on the tester must be changed.
    C. there is excessive noncondensable gases in the refrigerant.
    D. the vacuum pump is not producing enough vacuum.

Hint   *The controls and key pad vary on different makes of recovery/recycling equipment. Always follow the equipment manufacturer's recommended procedure. Some recovery/recycling equipment automatically recycles the refrigerant during a 20-minute evacuation process. The technician can manually select a longer process. When the recovery procedure is completed, press vacuum to start the vacuum pump. The display shows recycle after 5 seconds.*

When the display indicates 17 minutes remaining in the procedure, press hold/continue to stop the vacuum pump. A zero low-side reading indicates a refrigerant system leak. Repair the leak and start the recycling process over again. If the low-side gauge indicates 27 to 30 in. Hg. (91 to 101 kPa) close the low-side and high-side valves and wait a few minutes. If this vacuum is not maintained repair the refrigerant system leak. When the vacuum is maintained, open the low-side and high-side valves and press hold/continue to start the vacuum pump. When the procedure has continued for 20 minutes the display shows CPL. Some recovery/recycling equipment automatically vents noncondensable gases during the recovery process. This venting provides an audible hissing sound.

Task 4   Label and store refrigerant.

71. Technician A says that refrigerant storage containers must be evacuated to 17 in. Hg. before the refrigerant is placed in the container.
    Technician B says refrigerant containers may be filled to 90 percent or their gross weight rating.

Who is correct?
A. A only
B. B only
C. Both A and B
D. Neither A nor B

Hint   *Refrigerant must never be stored in disposable containers. Storage containers for R-12 must be stamped DOT 4B4 or DOT 4BW. Evacuate the storage container to 17 in. Hg. before putting refrigerant in the container. The refrigerant container must only be filled to 60 percent of its gross weight rating. Refrigerant containers should be stored in a cool location at approximately room temperature.*

Task 5   Test recycled refrigerant for noncondensable gases.

72. When checking a refrigerant container for noncondensable gases
    A. the container may be stored at 80°F (27°C) for 6 hours before the test.
    B. the container may be stored near a shop window.
    C. a thermometer should be placed against the container surface.
    D. if the pressure is lower than specified the refrigerant is ready for use.

Hint   *Before a refrigerant container is checked for noncondensable gases, store the container away from the presence of sunlight at 65°F (18°C) for 12 hours. Place a thermometer within 4 inches of the container surface and use a pressure gauge to measure the container pressure. Compare the temperature and pressure readings to the temperature/pressure chart (Figure 7-53) If the pressure of the stored refrigerant is less than specified on the chart, the refrigerant is ready for use. When the pressure of the stored refrigerant is more than indicated on the chart, slowly vent the vapor from the top of the container into the recovery/recycling unit until the pressure is less than specified on the chart.*

Standard Temperature/pressure chart

°F	PSI	°F	PSI	°F	PSI	°F	PSI	°F	PSI
65	74	75	87	85	102	95	118	105	136
66	75	76	88	86	103	96	120	106	138
67	76	77	90	87	105	97	122	107	140
68	78	78	92	88	107	98	124	108	142
69	79	79	94	89	108	99	125	109	144
70	80	80	96	90	110	100	127	110	146
71	82	81	98	91	111	101	129	111	148
72	83	82	99	92	113	102	130	112	150
73	84	83	100	93	115	103	132	113	152
74	86	84	101	94	116	104	134	114	154

Figure 7-53   Refrigerant temperature/pressure chart *(Courtesy of National Institute for Automotive Service Excellence (ASE))*

Answers and Analysis

1. A Since the pulley is turning with the compressor clutch engaged or disengaged, a defective pulley bearing provides a growling noise with the clutch disengaged. Therefore, B is wrong.

A low refrigerant charge or an excessive refrigerant charge do not cause a compressor growling noise, so C and D are wrong.

A defective internal compressor bearing causes a growling noise only with the compressor clutch engaged, since this is the only time the internal compressor components are rotating. A is right.

2. D During an A/C performance test the blower speed control should be in the high position and all the vehicle windows should be up. Therefore, both A and B are wrong and D is the correct response.

3. C During an A/C performance test the temperature at the dash outlets should be 40°F to 50°F. Therefore, C is right and A, B, and D are wrong.

4. A Bubbles in the sight glass indicate there is a low refrigerant charge in the refrigerant system. Therefore, B, C, and D are wrong and A is right.

5. C Oil streaks in the sight glass indicated a low refrigerant charge, so A, B, and D are wrong and C is correct.

6. C Air and moisture in the refrigerant system or a restricted TXV valve cause higher than specified system pressures. Therefore, A and D are wrong.

A defective compressor may cause low high-side pressure and high low-side pressure, and so B is wrong.

A low refrigerant charge causes reduced low-side and high-side pressures, so C is correct.

7. A A restricted TXV valve may cause high refrigerant pressures, but this condition also results in TXV frosting. Therefore, B is wrong.

Air and moisture in the refrigerant system may cause high system pressures, so A is correct.

8. B The maximum pressure difference across the EPR valve usually is 6 psi (41 kPa). Since the gauge readings indicate 35 psi (241 kPa) pressure difference across the EPR valve, a defective EPR valve is indicated. Therefore, A, C, and D are wrong and B is right.

9. A A restricted receiver/drier usually causes frosting of this component. A flooded evaporator causes frosting of the evaporator outlet and compressor suction pipes. A restricted TXV usually causes frosting of this component. Therefore, B, C, and D are wrong.

A low refrigerant charge may cause faster than normal clutch cycling without frosting of any components, so A is correct.

10. B An oily residue on refrigerant system components indicates a refrigerant leak in the area of the oily residue. Therefore, A is wrong and B is right.

11. C R-12 refrigerant is sold in white containers, and R-134a is marketed in light blue containers. Both A and B are right and C is the correct answer.

12. A Evacuating a refrigerant system removes moisture from the system. Therefore, B, C, and D are wrong and A is right.

13. A Some manufacturers recommend installing an in-line filter as an alternate to refrigerant system flushing. If this filter contains an orifice tube, the original orifice tube in the system must be removed. Therefore, A is right and B is wrong.

14. B The high-side charging procedure must be completed with the engine not running. If liquid refrigerant enters the compressor, this component may be damaged. Therefore, A is wrong and B is correct.

15. A The oil required with R-134a refrigerant is a polyalkylene glycol (PAG) oil, so B, C, and D are wrong and A is right.

16. C   The purpose of the cycling switch (Item 5 in Figure 7-8) is to cycle the compressor clutch on and off in relation to refrigerant system pressure. Therefore, A, B, and D are wrong and C is right.

17. C   Refrigerant discharge from the high-pressure relief valve may be caused by restricted condenser air passages, or a refrigerant system overcharge, since both these problems cause high system pressures. Therefore, both A and B are correct and C is the right answer.

18. B   If there is 12V at terminals B, C, and S, the 12V at terminal B indicates there is voltage supplied to the compressor clutch and connecting wire. If the clutch is inoperative, the compressor clutch coil or the wire from the coil to the fuse is open. Therefore, B is right.

19. A   A loose A/C compressor belt will likely provide a squealing noise with the compressor clutch engaged. Therefore, B is wrong.

Since the air pump does not require much power to turn it, this belt is not likely to provide a squealing noise on acceleration. C is wrong.

A worn compressor bearing provides a growling noise under all engine operating conditions with the clutch engaged, so D is wrong.

A loose power steering belt may cause a squealing noise on acceleration, so A is right.

20. A   If the compressor clutch clearance is more than specified, the clutch may slip while the compressor is engaged. This action may cause a scrapping noise, so A is correct.

When the compressor clutch clearance is more than specified, a shim must be removed from behind the armature plate, so B is wrong.

21. C   If there is 2 oz. of oil recovered from the system and 2 oz. of oil drained from the old compressor, add 4 oz. of oil to the replacement compressor before installation. C is right.

22. C   The seal seat O-ring must be installed before the shaft seal and component 6 is a seal protector. Therefore, both A and B are correct and C is the right answer.

23. A   The tool in Figure 7-14 is used to disconnect spring lock couplings, so B, C, and D are wrong and A is correct.

24. C   If the condenser air passages are severely restricted, it may result in high refrigerant system pressures and refrigerant discharge from the high-pressure relief valve. Both A and B are correct and C is the right answer.

25. C   When frost is forming on one of the condenser tubes near the bottom of the condenser, the refrigerant passage is restricted at that location in the condenser. Therefore, A, B, and D are wrong and C is right.

26. A   If the refrigerant in the sight glass appears red, leak-detecting dye has been added to the refrigerant. This condition does not require any corrective action, so B is wrong.

If the receiver/drier outlet is colder than the inlet this unit is restricted and should be changed. A is correct.

27. D   The symptoms described in the question indicate moisture freezing in the TXV valve. Therefore, D is correct.

28. C   This question asks us to select the response that is not true. When using the tool to remove and orifice tube, pour a small amount of refrigerant oil on top of the orifice tube, and engage the notch in the tool in the orifice tube. Hold the T-handle and turn the outer sleeve and rotate the outer sleeve to remove the orifice tube. Therefore, statements A, B, and D are correct, but none of these are the requested answer.

Since statement C is not a proper orifice tube removal procedure, this is the requested answer.

29. A   Restricted evaporator refrigerant passages may cause frosting of the evaporator outlet pipe. Therefore, A is right.

This problem may cause lower-than-specified low-side pressures, so B is wrong. A is the correct answer.

30. B   A plugged evaporator case drain may cause windshield fogging, but this problem would not result in an oily film on the windshield. Therefore, A is wrong.

An oil film on the windshield may be caused by a refrigerant leak in the evaporator core that may allow some refrigerant oil to escape in the evaporator case. B is correct.

31. A   In some refrigerant systems with an EPR, POA, or STV valve, the compressor runs continually in the A/C mode, so A is right.

Excessive pressure drop across the EPR valve indicates this valve is sticking closed, so B is wrong.

32. C   With the service valve in the position shown in Figure 7-22, the refrigerant system is isolated from the compressor for compressor removal. Therefore, A, B, and D are wrong and C is right.

33. D   Component 4 in Figure 7-26 is a high-pressure relief valve that discharges refrigerant at approximately 475 psi (3,275 kPa) and thus A, B, and C are wrong and D is right.

34. C   The coolant control valve should be closed in the maximum A/C mode. If this valve is stuck open, heat from the heater core may reduce evaporator cooling to some extent. A leaking vacuum hose connected to the coolant control valve also allows this valve to remain open in the maximum A/C mode. Therefore, both A and B are right and C is the correct response.

35. C   A sticky film on the inside of the windshield may be caused by a coolant leak in the heater core, so A is right. Under this condition the coolant level in the radiator should be checked. Both A and B are correct and C is the right answer.

36. A   In this question we are asked for the defect that would not cause the problem of coolant loss with no visible leaks under the hood or in the passenger compartment. A leaking transmission cooler, head gasket, or cracked cylinder head could result in this problem. Therefore, B, C, and D are not the requested answer.

A leaking heater core usually causes coolant to leak onto the front floor mat, and thus A is not the cause of the problem. A is the right answer.

37. A   A sticking vacuum valve in the radiator cap may cause a collapsed upper radiator hose after the engine is shut off. Therefore, B, C, and D are wrong and A is the right answer.

38. C   Restricted radiator tubes may cause engine overheating and excessive coolant in the coolant recovery container. Therefore, A is correct.

A defective radiator cap may allow excessive coolant flow into the recovery container, so B also is correct and C is the right answer.

39. C   If the thermostat is stuck open in a fuel-injected engine, the coolant never reaches normal operating temperature. Under this condition the engine coolant temperature sensor sends a low coolant temperature signal to the PCM. This sensor input results in a rich air-fuel ratio. Therefore, A, B, and D are wrong and C is correct.

40. A   This question asks for the statement that is not true. When more antifreeze is added to the coolant, the boiling point is increased and coolant solutions must be recovered and recycled. Most ethylene glycol antifreeze contains antirust and corrosion inhibitors. Therefore, B, C, and D are correct, but these are not the requested answer.

When the cooling system pressure is increased, the boiling point also increases. Therefore, statement A is not true and this is the requested answer.

41. B   A blown 10 A, 5 C fuse in the fuse block would prevent both fan motors from operating even with the A/C on. Therefore, A is wrong.

A defective high-speed coolant fan relay only prevents high speed fan operation, so C is wrong.

Since the cooling fans operate with the A/C on, the A/C pressure fan switch must be operating.

A defective engine coolant temperature sensor that indicates low coolant temperature may cause the PCM not to ground the fan relay windings and thus prevent the operation of the cooling fans except when the A/C pressure switch closes. B is correct.

42. B   The manual coolant control valve adjustment is illustrated in Figure 7-28. Therefore, A, C, and D are wrong and B is right.

43. C   A gurgling noise in the heater core may be caused by a low coolant level in the cooling system or a restricted heater core. Therefore, both A and B are correct, and C is the right answer.

44. C   This question asks for the response that is not the cause of the problem. An open circuit at terminal 6 or 8 on the electronic door actuator, or a blown 15 A, number 8 fuse, would cause the temperature blend door actuator to be inoperative. Therefore, A, B, and D are correct, but none of these are the requested response.

Terminal C2-15 is connected to the feedback circuit from the electronic temperature door actuator to the automatic temperature control module. An open circuit at this terminal may cause inaccurate temperature blend door position, but the door does move. Therefore, C is correct.

45. A   An open circuit at the blower switch ground or at the terminal causes the blower to be completely inoperative. Therefore, B is wrong.

An open circuit at resistor terminal 4 causes some other blower speeds other than high speed, so D is wrong.

An open circuit at switch terminal 4 prevents any blower speed operated through the switch, but voltage still is available through the circuit to resistor terminal 4 to provide very slow blower speed with the ignition switch on the A/C switch in any position but off. Therefore, C is wrong.

An open circuit at resistor terminal 2 prevents current flow through any of the resistors, but high blower speed is available through the high-speed switch contacts directly to ground. A is correct.

46. A   A defective A/C pressure cutoff switch opens the compressor clutch circuit if the refrigerant system pressures are extremely high. However, the question says these pressures are normal, so B is wrong.

A defective clutch control relay, or clutch coil are not likely to cause the clutch to become deenergized in relation to temperature. Therefore, C and D are wrong.

A defective engine coolant temperature sensor may indicate engine overheating to the PCM when the engine is at normal operating temperature. This signal causes the PCM to open the ground circuit on the clutch control relay winding and deenergize the clutch. A is correct.

47. C   A leaking throttle kicker diaphragm or an inoperative kicker solenoid may cause lower than specified idle rpm with the A/C on and engine stalling. Both A and B are right, and C is the right answer.

48. D   The A/C pressure transducer contains a potentiometer that is connected to the PCM. Therefore, both A and B are wrong and D is the correct answer.

49. A   A blown maxifuse in the LH maxifuse holder, or a blown 10 A cooling fan/TCC fuse in the fuseblock cause both cooling fans to be inoperative, so B and C are wrong.

An open circuit at terminal B on the A/C head pressure switch would only prevent cooling fan operation when the A/C pressure is high, thus D is wrong.

An open circuit between the number 30 terminals on the cooling fan relays prevents voltage supply to the low-speed fan relay contacts, but voltage is still available at the high-speed fan relay contacts. Therefore, A is correct.

50. B   This question asks for the statement that is not true. Some actuator motors are calibrated automatically in the self-diagnostic mode, whereas other actuator motors must be calibrated manually. These actuators should only require calibration after motor replacement or misadjustment. Therefore, statements A, C, and D are correct, but they are not the requested answer.

Diagnostic trouble codes indicate a fault in a certain area not in a specific component. Statement B is wrong and this is the requested answer.

51. B   This question asks for the statement that is not true. The negative battery cable must be removed and the technician must wait a specified length of time before A/C control panel removal. Self-diagnostic tests may indicate a defective A/C control panel. Statements A, C, and D are right, but none of these is the requested answer.

The refrigeration system does not require discharging before A/C panel removal. Therefore, B is wrong and this is the requested answer.

52. D   With 18 in. IIg. supplied to the control valve switch assembly, the same vacuum should be applied to the panel door actuator with the switch in the maximum A/C position. Therefore, D is right, and A, B, and C are wrong.

53. C   While adjusting the temperature control cable the attaching flag and the adjusting clip must be installed, and the temperature control lever must be in the maximum cold position. Therefore, A, B, and D are wrong.

During this adjustment the temperature blend door crank must be rotated fully counterclockwise. C is right.

54. A   If the black vacuum hose that supplies vacuum from the intake manifold to the control switch is leaking, none of the mode doors operate properly. Therefore, B is wrong.

A leak in the blue vacuum hose from the control switch to the panel door actuator may cause this actuator to gradually switch from the panel ducts to the floor ducts. A is correct.

55. B   A leaking panel door actuator would cause the air discharge to switch from the panel to the floor ducts any time not just when climbing a hill, so A is wrong.

A leaking temperature blend door actuator causes the system to change temperature any time not just when climbing a long hill, thus C is wrong.

A leaking intake manifold gasket may affect all the vacuum-operated mode doors because this condition causes low-source vacuum. D is wrong.

A leaking check valve does not trap the vacuum in the reserve tank when climbing a long hill. This action may cause the temperature blend door to move to the warm air position and the air discharge to switch to the floor ducts. B is right.

56. A   If the outside recirculation door is stuck in position A, outside air is drawn into the A/C heater case and there is no leakage of in-car air past this door. Therefore, A is right and B is wrong.

57. C   A sticking temperature blend door or a defective in-car sensor may cause the in-car temperature to be above the driver-selected temperature. Both A and B are correct and C is the right answer.

58. A   An open blower motor ground would cause this motor to be completely inoperative, so B is wrong. Since the voltage at the blower motor is normal the HVAC programmer must be satisfactory, thus C is wrong.

Since programmer terminal D2 is only connected to a feedback wire from the blower circuit, an open circuit at this terminal would not cause continual low blower speed. D is wrong.

A defective blower motor may cause continual low blower speed, so A is correct.

59. B   A defect in the check valves, vacuum tank, and vacuum supply affects all the vacuum-operated doors, so A is wrong. The programmer vacuum switch and the hose to the outside/recirculation actuator should be checked first to correct this problem. B is right.

60. A   If the number 18, 15 A fuse is open there is no voltage supplied to the compressor clutch, so B is wrong. Since the PCM must ground the clutch control relay winding through PCM terminal 69, an open circuit at this terminal makes it impossible for the PCM to ground this relay winding. A is correct.

61. A   The in-vehicle sensor resistance is at a minimum when the temperature in the vehicle is hot. Therefore, B is wrong. As the temperature of the in-vehicle sensor increases the sensor resistance should decrease. Therefore, A is correct.

62. A   As indicated in the chart in Figure 7-46 both solenoids 4 and 5 should be open with the temperature control in the cold position and vacuum is supplied to power servo 2. This is a normal condition, so A is right and B is wrong.

63. A   If the temperature switch ground is open, the blower motor would not operate except in the defrost mode. Therefore, B is wrong.

The engine temperature switch should close and turn on the blower at 120° F (49° C). A defective engine temperature switch may delay the blower operation to a higher engine temperature. A is correct.

64. D   This question asks for the defect that is not the cause of the problem. A seized water control valve, a plugged vacuum hose between the solenoid and the valve, or a jammed actuator linkage may cause the water control not to move with 16 in. Hg. supplied to the solenoid. Therefore, A, B, and C may cause the problem, but none of these are the requested answer.

Since the solenoid provides an audible click, the solenoid plunger is not sticking and D is not a cause of the problem. D is the requested answer.

65. B   When a fault code is obtained representing the temperature blend door, the first step in the diagnostic procedure should be to check the temperature blend door for a sticking condition. A is wrong and B is right.

66. B   Since the resistance reading in the ambient sensor and connecting wires is normal, this sensor and connecting wires are satisfactory. Therefore, A, C, and D are wrong. Since the only other component in this circuit is the A/C control panel, this unit must be the problem. B is correct.

67. A   The specified voltage drop across computer ground wires usually is .1V. Therefore, B, C, and D are wrong and A is right.

68. C   This question asks for the statement that is not true. A/C recover/recycling equipment must have a UL approval, SAE J1991 approval, and refrigerant oils for R-12 and R-134a must not be mixed. Therefore, statements A, B, and D are true, but none of these are the requested answer.

The refrigerant container specified by the recovery/recycling equipment manufacturer must be used in this type of equipment to be sure the container has proper capacity and valving. C is right.

69. A   After the recovery process, if the low-side gauge rises above 0 psi, there is some refrigerant remaining in the system. Therefore, B, C, and D are wrong, and A is correct.

70. B   If the moisture warning light indicates yellow during the recycling process the refrigerant contains excessive moisture and the filter/drier cartridge in the recovery/recycling equipment must be changed. A, C, and D are wrong, and B is right.

71. A   Refrigerant storage containers must be filled to 60 percent of their gross weight rating, so B is wrong. Refrigerant storage containers must be evacuated to 17 in. Hg. before refrigerant is placed in the container. A is correct.

72. D   When checking a refrigerant container for noncondensable gases, the container should be stored out of the presence of sunlight at 65°F (18°C) for 12 hours, and a thermometer should be placed 4 inches from the container surface. Therefore, A, B, and C are wrong.

If the container pressure is less than specified, the refrigerant is ready for use. D is correct.

Glossary

Access valve See Service port and Service valve.
Válvula de acceso Ver Service port [Orificio de servicio] y Service value [Válvula de servicio].

Accumulator A tank located in the tailpipe to receive the refrigerant that leaves the evaporator. This device is constructed to ensure that no liquid refrigerant enters the compressor.
Acumulador Tanque ubicado en el tubo de escape para recibir el refrigerante que sale del evaporador. Dicho dispositivo está diseñado de modo que asegure que el regfrigerante líquido no entre en el compresor.

Actuator A device that transfers a vacuum or electric signal to a mechanical motion. An actuator typically performs an on/off or open/close function.
Accionador Dispositivo que transfiere una señal de vacío o una señal eléctrica a un movimiento mecánico. Típicamente un accionador lleva a cabo la función de modulación de impulsos o la de abrir y cerrar.

Adapter A device or fitting that permits different size parts or components to be fastened or connected to each other.
Adaptador Dispositivo o ajuste que permite la sujección o conexión entre sí de piezas de tamaños diferentes.

Aftermarket A term generally given to a device or accessory that is added to a vehicle by the dealer after original manufacture, such as an air conditioning system.
Postmercado Término dado generalmente a un dispositivo o accesorio que el distribuidor de automóviles agrega al automóvil después de la fabricación original, como por ejemplo un sistema de acondicionamiento de aire.

Air gap The space between two components such as the rotor and armature of a clutch.
Espacio de aire El espacio entre dos componentes, como por ejemplo el rotor y la armadura de un embrague.

Ambient sensor A thermistor used in automatic temperature control units to sense ambient temperature. Also see Thermistor.
Sensor ambiente Termistor utilizado en unidades de regulación automática de temperatura para sentir la temperatura ambiente. Ver también Thermistor [Termistor].

Approved power source A power source that is consistent with the requirements of the equipment so far as voltage, frequency, and ampacity are concerned.
Fuente aprobada de potenica Fuente de potencia que cumple con los requisitos del equipo referente a la tensión, frecuencia, y ampacidad.

Armature The part of the clutch that mounts onto the crankshaft and engages with the rotor when energized.
Armadura La parte del embrague que se fija al cigüeñal y se engrana al exitarse el rotor.

Asbestos A silicate of calcium (Ca) and magnesium (Mg) mineral that does not burn or conduct heat. It has been determined that asbestos exposure is hazardous to health and must be avoided.
Asbesto Mineral de silicato de calcio (Ca) y magnesio (Mg) que no se quema ni conduce el calor. Se ha establecido que la exposición al asbesto es nociva y debe evitarse.

Atmospheric Pressure Air pressure at a given altitude. At sea level, atmospheric pressure is 14.696 psia (101.329 kPa absolute).

Presión atmosférica La presión del aire a una dada altitud. Al nivel del mar, la presión atmosférica es de 14,696 psia (101.329 kPa absoluto).

AUTO Abbreviation for automatic.
AUTO Abreviatura del automático.

Back seat (service valve) Turning the valve stem to the left (ccw) as far as possible back seats the valve. The valve outlet to the system is open and the service port is closed.
Asentar a la izquierda (válvula de servicio) El girar el vástago de la válvula al punto más a la izquierda posible asienta a la izquierda la válvula. La salida de la válvula al sistema está abierta y el orificio de servicio está cerrado.

Barb fitting A fitting that slips inside a hose and is held in place with a gear-type clamp. Ridges (barbs) on the fitting prevent the hose from slipping off.
Accesorio arponado Ajuste que se inserta dentro de una manguera y que se sujeta en su lugar con una abrazadera de tipo engranaje. Proyecciones (púas) en el ajuste impiden que se deslice la manguera.

BCM An abbreviation for Blower Control Module
BCM Abreviatura de Módulo regulador del soplador.

Belt See V-belt, V-groove belt, and Serpentine belt.
Correa Ver V-belt [Correa en V], V-groove belt [Correa ranurada en V], y Serpentine belt [Correa serpentina].

Belt tension Tightness of a belt or belts, usually measured in foot-pounds (ft-lb) or Newton-meters (N•m).
Tensión de la correa Tensión de una correa o correas, medida normalmente en libras-pies (ft-lb) o metros-Newton (N·m).

Blower See Squirrel-cage blower.
Soplador Ver Squirrel-cage blower [Soplador con jaula de ardilla].

Blower motor See Motor.
Motor de soplador Ver motor.

Blower relay An electrical device used to control the function or speed of a blower motor.
Relé del soplador Dispositivo eléctrico utilizado para regular la función o velocidad de un motor de soplador.

Boiling point The temperature at which a liquid changes to a vapor.
Punto de ebullición Temperatura a la que un líquido se convierte en vapor.

Break a vacuum The next step after evacuating a system. The vacuum should be broken with refrigerant or other suitable dry gas, not ambient air or oxygen.
Romper un vacío El paso que inmediatamente sigue la evacuación de un sistema. El vacío debe de romperse con refrigerante u otro gas seco apropiado, y no con aire ambiente u oxígeno.

Bypass An alternate passage that may be used instead of the main passage.
Desviación Pasaje alternativo que puede utilizarse en vez del pasaje principal.

Bypass hose A hose that is generally small and is used as an alternate passage to bypass a component or device.
Manguera desviadora Manguera que generalmente es pequeña y se utiliza como pasaje alternativo para desviar un componente o dispositivo.

CAA Clean Air Act.
CAA Ley para Aire Limpio.

Can tap A device used to pierce, dispense, and seal small cans of refrigerant.
Macho de roscar para latas Dispositivo utilizado para perforar, distribuir, y sellar pequeñas latas de refrigerante.

Can tap valve A valve found on a can tap that is used to control the flow of refrigerant.
Válvula de macho de roscar para latas Válvula que se encuentra en un macho de roscar para latas utilizada para regular el flujo de refrigerante.

Cap A protective cover. Also used as an abbreviation for capillary (tube) or capacitor.
Tapadera Cubierta protectiva. Utilizada también como abreviatura del tubo capilar o capacitador.

Cap tube A tube with a calibrated inside diameter and length used to control the flow of refrigerant. In automotive air conditioning systems, the tube connecting the remote bulb to the expansion valve or to the thermostat is called the capillary tube.
Tubo capilar Tubo de diámetro interior y longitud calibrados; se utiliza para regular el flujo de refrigerante. En sistemas automotrices para el acondicionamiento de aire el tubo que conecta la bombilla a distancia con la válvula de expansión o con el termóstato se llama el tubo capilar.

Carbon seal face A seal face made of a carbon composition rather than from another material such as steel or ceramic.
Frente de carbono de la junta hermética Frente de la junta hermética fabricada de un compuesto de carbono en vez de otro material, como por ejemplo el acero o material cerámico.

Caution A notice to warn of potential personal injury situations and conditions.
Precaución Aviso para advertir situaciones y condiciones que podrían causar heridas personales.

CCW Counterclockwise.
CCW Sentido inverso al de las agujas del reloj.

Celsius A metric temperature scale using zero as the freezing point of water. The boiling point of water is 100°C (212°F).
Celsio Escala de temperatura métrica en la que el cero se utlza como el punto de congelación de agua. El punto de ebullición de agua es 100°C (212°F).

Ceramic seal face A seal face made of a ceramic material instead of steel or carbon.
Frente cerámica de la junta hermética Frente de la junta hermética fabricada de un material cerámico en vez del acero o carbono.

Certified Having a certificate. A certificate is awarded or issued to those that have demonstrated appropriate competence through testing and/or practical experience.
Certificado El poseer un certificado. Se les otorga o emite un certificado a los que han demostrado una cierta capacidad por medio de exámenes y/o experiencia práctica.

CFC-12 See Refrigerant-12.
CFC-12 Ver Refrigerante-12.

Charge A specific amount of refrigerant or oil by volume or weight.
Carga Cantidad especícfa de refrigerante o de aceite por volumen o peso.

Check valve A device located in the liquid line or inlet to the drier. The valve prevents liquid refrigerant from flowing the opposite way when the unit is shut off.
Válvula de retención Dispositivo ubicado en la línea de líquido o en la entrada al secador. Al cerrarse la unidad, la válvula impide que el refrigerante líquido fluya en el sentido contrario.

Clean Air Act A Title IV amendment signed into law in 1990 which established national policy relative to the reduction and elimination of ozone-depleting substances.
Ley para Aire Limpio Enmienda Título IV firmado y aprobado en 1990 que estableció la política nacional relacionada con la redución y eliminación de sustancias que agotan el ozono.

Clockwise A term referring to a clockwise (cw), or left to right rotation or motion.
Sentido de las agujas del reloj Término que se refiere a un movimiento en el sentido correcto de las agujas del reloj (cw por sus siglas en inglés), es decir, rotación o movimiento desde la izquierda hacia la derecha.

Clutch An electro-mechanical device mounted on the air conditioning compressor used to start and stop compressor action, thereby controlling refrigerant circulating through the system.
Embrague Dispositivo electromecánico montado en el compresor del acondicionador de aire y utilizado para arrancar y detener la acción del compresor, regulando así la circulación del refrigerante a través del sistema.

Clutch Coil The electrical part of a clutch assembly. When electrical power is applied to the clutch coil, the clutch is engage to start and stop compressor action.
Bobina del embrague La parte eléctrica del conjunto del embrague. Cuando se aplica una potencia eléctrica a la bobina del embrague, éste se engrana para arrancar y detener la acción del compresor.

Compound gauge A gauge that registers both pressure and vacuum (above and below atmospheric pressure); used on the low side of the systems.
Manómetro compuesto Calibrador que registra tanto la presión como el vacío (a un nivel superior e inferior a la presión atmosférica); utilizado en el lado de baja presión de los sistemas.

Compression fitting A type of fitting used to connect two or more tubes of the same or different diameter together to form a leak-proof joint.
Ajuste de compresión Tipo de ajuste utilizado para sujetar dos o más tubos del mismo tamaño o de un tamaño diferente para formar una junta hermética contra fugas.

Compression ring A ring-like part of a compression fitting used for a seal between the tube and fitting.
Aro de compresión Pieza parecida a un anillo del ajuste de compresión utilizada como una junta hermética entre el tubo y el ajuste.

Compression nut A nut-like device used to seat the compression ring into the compression fitting to ensure a leak-proof joint.
Tuerca de compresión Dispositivo parecido a una tuerca utilizado para asentar el anillo de compresión dentro del ajuste de compresión para asegurar una junta hermética contra fugas.

Compressor shaft seal An assembly consisting of springs, snap rings, O-rings, shaft seal, seal sets, and gasket. The shaft seal is mounted on the compressor crankshaft and permits the shaft to be turned without a loss of refrigerant or oil.
Junta hermética del árbol del compresor Conjunto que consiste de muelles, anillos de muelles, juntas tóricas, una junta hermética del árbol, conjuntos de juntas herméticas, y una guarnición. La junta hermética del árbol está montada en el cigüeñal del compresor y permite que el árbol se gire sin una pérdida de refrigerante o aceite.

Contaminated A term generally used when referring to a refrigerant cylinder or a system that is known to contain foreign substances such as other incompatible or hazardous refrigerants.
Contaminado Témino generalmente utilizado al referirse a un cilindro para refrigerante o a un sistema que es reconocido contener sus-

tancias extrañas, como por ejemplo otros refrigerantes incompatibles o peligrosos.

Counterclockwise (ccw) A direction, right to left, opposite that which a clock turns.
Sentido contrario al de las agujas del reloj Dirección de la derecha hacia la izquierda contraria a la correcta de las agujas del reloj.

Cracked position A mid-seated or open position.
Posición parcialmente asentada Posición abierta o media asentada.

CW Abbreviation for clockwise. Also cw.
CW Abreviatura del sentido de las agujas del reloj. También cw.

Cycle clutch time (total) Time from the moment the clutch engages until it disengages, then reengages. Total time is equal to on time plus off time for one cycle.
Duración del ciclo del embrague (total) Espacio de tiempo medido desde el momento en que se engrana el embrague hasta que se desengrane y se engrane de nuevo. El tiempo total es equivalente al trabajo efectivo más el trabajo no efectivo por un ciclo.

Cycling clutch pressure switch A pressure-actuated electrical switch used to cycle the compressor at a predetermined pressure.
Autómata manométrico del embrague con funcionamiento cíclico Interruptor eléctrico accionado a presión utilizado para ciclar el compresor a una presión predeterminada.

Cycling clutch system An airconditioning system in which the air temperature is controlled by starting and stopping the compressor with a thermostat or pressure control.
Sistema de embrague con funcionamiento cíclico Sistema de acondicionamiento de aire en el cual la temperatura del aire se regula al arrancarse y detenerse el compresor con un termóstato o regulador de presión.

Decal A label that is designed to stick fast when transferred. A decal affixed under the hood of a vehicle is used to identify the type of refrigerant used in a system.
Calcomanía Etiqueta diseñada para pegarse fuertemente al ser transferido. Una calcomanía pegada debajo de la capota se utiliza para identificar el tipo de refrigerante utilizado en un sistema.

Department of Transportation The United States Department of Transportation is a federal agency charged with regulation and control of the shipment of all hazardous materials.
Departamento de Transportes El Departamento de Transportes de los Estados Unidos de América es una agencia federal que tiene a su cargo la regulación y control del transporte de todos los materiales peligrosos.

Dependability Reliability; trustworthiness.
Carácter responsable Digno de confianza; integridad.

Depressing pin A pin located in the end of a service hose to press (open) a Schrader-type valve.
Pasador depresor Pasador ubicado en el extremo de una manguera de servicio para forzar que se abra una válvula de tipo Schrader.

Diagnosis The procedure followed to locate the cause of a malfunction.
Diagnosis Procedimiento que se sigue para localizar la causa de una disfunción.

Disarm To turn off; to disable a device or circuit.
Desarmar Apagar; incapacitar un dispositivo o circuito.

Dry nitrogen The element nitrogen (N) which has been processed to ensure that it is free of moisture.
Nitrógeno seco El elemento nitrógeno (N) que ha sido procesado para asegurar que esté libre de humedad.

Dual Two.
Doble Dos.

Dual System Two systems; usually refers to two evaporators in an air conditioning system; one in the front and one in the rear of the vehicle, driven off a single compressor and condenser system.
Sistema doble Dos sistemas; se refiere normalmente a dos evaporadores en un sistema de acondicionamiento de aire; uno en la parte delantera y el otro en la parte trasera del vehículo; los dos son accionados por un solo sistema compresor condensador.

Duct A tube or passage used to provide a means to transfer air or liquid from one point or place to another.
Conducto Tubo o pasaje utilizado para proveer un medio para transferir aire o líquido desde un punto o lugar a otro.

EATC Electronic Automatic Temperature Control.
EATC Regulador Automático y Electrónico de Temperatura.

ECC Electronic Climate Control.
ECC Regulador Electrónico de Clima.

English fastener Any type fastener with English size designations, numbers, decimals, or fractions of an inch.
Asegurador inglés Cualquier tipo de asegurador provisto de indicaciones, números, decimales, o fracciones de una pulgada del sistema inglés.

Environmental Protection Agency (EPA) An agency of the U.S. government charged that is charged with the responsibility of protecting the environment and enforcing the Clean Air Act (CAA) of 1990.
Agencia para la Protección del Medio Ambiente (EPA) Agencia del gobierno estadounidense que tiene a su cargo la responsabilidad de proteger el medio ambiente y ejecutar la Ley para Aire Limpio (CAA por sus siglas en inglés) de 1990.

EPA Environmental Protection Agency.
EPA Agencia para la Protección del Medio Ambiente.

Etch An intentional or unintentional erosion of a metal surface generally caused by an acid.
Atacar con ácido Desgaste previsto o imprevisto de una superficie metálico, ocacionado generalmente por un ácido.

Etching See Etch.
Ataque con ácido Ver Etch [Atacar con ácido].

Evacuate To create a vacuum within a system to remove all traces of air and moisture.
Evacuar El dejar un vacío dentro de un sistema para remover completamente todo aire y humedad.

Evacuation See Evacuate.
Evacuación Ver Evacuate [Evacuar].

Evaporator core The tube and fin assembly located inside the evaporator housing. The refrigerant fluid picks up heat in the evaporator core when it changes into a vapor.
Núcleo del evaporador El conjunto de tubo y aletas ubicado dentro del alojamiento de evaporador. El refrigerante acumula calor en el núcleo del evaporador cuando se convierte en vapor.

Expansion tank An auxiliary tank that is usually connected to the inlet tank or a radiator and which provides additional storage space for heated coolant. Often called a coolant recovery tank.
Tanque de expansión Tanque auxiliar que normalmente se conecta al tanque de entrada o a un radiador y que provee almacenaje adicional del enfriante calentado. Llamado con frecuencia tanque para la recuperación del enfriante.

External On the outside.
Externo Al exterior.

External snap ring A snap ring found on the outside of a part such as a shaft.
Anillo de muelle exterior Anillo de muelle que se encuentra en el exterior de una pieza, como por ejemplo un árbol.

Fan relay A relay for the cooling and/or auxiliary fan motors.
Relé del ventilador Relé para los motores de enfriamiento y/o los auxiliares.

Federal Clean Air Act See Clean Air Act.
Ley Federal para Aire Limpio Ver Clean Air Act [Ley para Aire Limpio].

Fill neck The part of the radiator on which the pressure cap is attached. Most radiators, however, are filled via the recovery tank.
Cuello de relleno La parte del radiador a la que se fija la tapadera de presión. Sin embargo, la mayoría de radiadores se llena por medio del tanque de recuperación.

Filter A device used with the drier or as a separate unit to remove foreign material from the refrigerant.
Filtro Dispositivo utilizado con el secador o como unidad separada para extraer material extraño del refrigerante.

Filter drier A device that has a filter to remove foreign material from the refrigerant and a desiccant to remove moisture from the refrigerant.
Secador del filtro Dispositivo provisto de un filtro para remover el material extraño del refrigerante y un desecante para remover la humedad del refrigerante.

Flange A projecting rim, collar, or edge on an object used to keep the object in place or to secure it to another object.
Brida Cerco, collar o extremo proyectante ubicado sobre un objeto utilizado para mantener un objeto en su lugar o para fijarlo a otro objeto.

Flare A flange or cone-shaped end applied to a piece of tubing to provide a means of fastening to a fitting.
Abocinado Brida o extremo en forma cónica aplicado a una pieza de tubería para proveer un medio de asegurarse a un ajuste.

Forced air Air that is moved mechanically such as by a fan or blower.
Aire forzado Aire que se mueve mecánicamente, como por ejemplo por un ventilador o soplador.

Front seat Closing off the line leaving the compressor open to the service port fitting. This allows service to the compressor without purging the entire system. Never operate the system with the valves front seated.
Asentar a la derecha El cerrar la línea dejando abierto el compresor al ajuste del orificio de servicio, lo cual permite prestar servicio al compresor sin purgar todo el sistema. Nunca haga funcionar el sistema con las válvulas asentadas a la derecha.

Functional test See Performance Test.
Prueba funcional Ver Performance Test [Prueba de rendimiento].

Fusible link A type of fuse made of a special wire that melts to open a circuit when current draw is excessive.
Cartucho de fusible Tipo de fusible fabricado de un alambre especial que se funde para abrir un circuito cuando ocurre una sobrecarga del circuito.

Gasket A thin layer of material or composition that is placed between two machined surfaces to provide a leakproof seal between them.
Guarnición Capa delgada de material o compuesto que se coloca entre dos superficies maquinadas para proveer una junta hermética para evitar fugas entre ellas.

Gauge A tool of a known calibration used to measure components. For example, a feeler gauge is used to measure the air gap between a clutch rotor and armature.

Calibrador Herramienta de una calibración conocida utilizada para la medición de componentes. Por ejemplo, un calibrador de espesores se utiliza para medir el espacio de aire entre el rotor del embrague y la armadura.

Graduated container A measure such as a beaker or measuring cup that has a graduated scale for the measure of a liquid.
Recipiente graduado Una medida, como por ejemplo un cubilete o una taza de medir, provista de una escala graduada para la medición de un líquido.

Ground A general term given to the negative (-) side of an electrical system.
Tierra Término general para indicar el lado negativo (-) de un sistema eléctrico.

Grounded An intentional or unintentional connection of a wire, positive (+) or negative (-), to the ground. A shortcircuit is said to be grounded.
Puesto a tierra Una conexión prevista o imprevista de un alambre, positiva (+) o negativa (-), a la tierra. Se dice que un cortocircuito es puesto a tierra.

Gross weight The weight of a substance or matter that includes the weight of its container.
Peso bruto Peso de una sustancia o materia que incluye el peso de su recipiente.

HCFC Hydrochlorofluorocarbon refrigerant.
HCFC Refrigerante de hidroclorofluorocarbono.

Header tanks The top and bottom tanks (downflow) or side tanks (crossflow) of a radiator. The tanks in which coolant is accumulated or received.
Tanques para alimentación por gravedad Los tanques superiores e inferiores (flujo descendente) o los tanques laterales (flujo transversal) de un radiador. Tanques en los cuales el enfriador se acumula o se recibe.

Heater core A radiator-like heat exchanger located in the case/duct system through which coolant flows to provide heat to the vehicle interior.
Núcleo del calentador Intercambiador de calor parecido a un radiador y ubicado en el sistema de caja/conducto a través del cual fluye el enfriador para proveer calor al interior del vehículo.

Heat exchanger An apparatus in which heat is transferred from one medium to another on the principle that heat moves to an object with less heat.
Intercambiador de calor Aparato en el que se transfiere el calor de un medio a otro, lo cual se basa en el principio que el calor se atrae a un objeto que tiene menos calor.

HI The designation for high as in blower speed or system mode.
HI Indicación para indicar marcha rápida, como por ejemplo la velocidad de un soplador o el modo de un sistema.

High-side gauge The right side gauge on the manifold used to read refrigerant pressure in the high side of the system.
Calibrador del lado de alta presión El calibrador del lado derecho del múltiple utilizado para medir la presión del refrigerante en el lado de alta presión del sistema.

High-side hand valve The high side valve on the manifold set used to control flow between the high side and service ports.
Válvula de mano del lado de alta presión Válvula del lado de alta presión que se encuentra en el conjunto del múltiple, utilizada para regular el flujo entre el lado de alta presión y los orificios de servicio.

High-side service valve A device located on the discharge side of the compressor; this valve permits the service technician to check the high-side pressures and perform other necessary operations.

Válvula de servicio del lado de alta presión Dispositivo ubicado en el lado de descarga del compresor; dicha válvula permite que el mecánico verifique las presiones en el lado de alta presión y lleve a cabo otras funciones necesarias.

High-side switch See Pressure switch.
Autómata manométrico del lado de alta presión Ver Pressure switch [Autómata manométrico].

High-torque clutch A heavy-duty clutch assembly used on some vehicles known to operate with higher-than-average head pressure.
Embrague de alto par de torsión Conjunto de embrague para servicio pesado utilizado en algunos vehículos que funcionan con una altura piezométrica más alta que la normal.

Hot A term given the positive (+) side of an energized electrical system. Also refers to an object that is heated.
Cargado/caliente Término utilizado para referirse al lado positivo (+) de un sistema eléctrico excitado. Se refiere también a un objeto que es calentado.

Hot knife A knife-like tool that has a heated blade. Used for separating objects; e.g., evaporator cases.
Cuchillo en caliente Herramienta parecida a un cuchillo provista de una hoja calentada. Utilizada para separar objetos; p.e. las cajas de evaporadores.

Hub The central part of a wheel-like device such as a clutch armature.
Cubo Parte central de un dispositivo parecido a una rueda, como por ejemplo la armadura del embrague.

Hygiene A system of rules and principles intended to promote and preserve health.
Higiene Sistema de normas y principios cuyo propósito es promover y preservar la salud.

Hygroscopic Readily absorbing and retaining moisture.
Higroscópico Lo que absorbe y retiene fácilmente la humedad.

Idler A pulley device that keeps the belt whip out of the drive belt of an automotive air conditioner. The idler is used as a means of tightening the belt.
Polea loca Polea que mantiene la vibración de la correa fuera de la correa de transmisión de un acondicionador de aire automotriz. Se utiliza la polea loca para proveerle tensión a la correa.

Idler pulley A pulley used to tension or torque the belt(s).
Polea tensora Polea utilizada para proveer tensión o par de torsión a la(s) correa(s).

Idle speed The speed (rpm) at which the engine runs while at rest (idle).
Marcha mínima Velocidad (rpm) a la que no hay ninguna carga en el motor (marcha mínima).

In-car temperature sensor A thermistor used in automatic temperature control units for sensing the in-car temperature. Also see Thermistor.
Sensor de temperatura del interior del vehículo Termistor utilizado en unidades de regulación automática de temperatura para sentir la temperatura del interior del vehículo. Ver también Thermistor [Termistor].

Insert fitting A fitting that is designed to fit inside, such as a barb fitting that fits inside a hose.
Ajuste inserto Ajuste diseñado para insertarse dentro de un objeto, como por ejemplo un ajuste arponado que se inserta dentro de una manguera.

Internal Inside; within.
Interno Al interior, dentro de una cosa.

Internal snap ring A snap ring used to hold a component or part inside a cavity or case.
Anillo de muelle interno Anillo de muelle utilizado para sujetar un componente o una pieza dentro de una cavidad o caja.

Jumper A wire used to temporarily bypass a device or component for the purpose of testing.
Barreta Alambre utilizado para desviar un dispositivo o componente de manera temporal para llevar a cabo una prueba.

Kilogram A unit of measure in the metric system. One kilogram is equal to 2.205 pounds in the English system.
Kilogramo Unidad de medida en el sistema métrico. Un kilogramo equivale a 2,205 libras en el sistema inglés.

Kilopascal A unit of measure in the metric system. One kilopascal (kPa) is equal to 0.145 pound per square inch (psi) in the English system.
Kilopascal Unidad de medida en el sistema métrico. Un kilopascal (kPa) equivale a 0,145 libras por pulgada cuadrada en el sistema inglés.

kPa kilopascal.
kPa kilopascal.

Liquid A state of matter; a column of fluid without solids or gas pockets.
Líquido Estado de materia; columna de fluido sin sólidos ni bolsillos de gas.

Low-refrigerant switch A switch that senses low pressure due to a loss of refrigerant and stops compressor action. Some alert the operator and/or set a trouble code.
Interruptor para advertir un nivel bajo de refrigerante Interruptor que siente una presión baja debido a una pérdida de refrigerante y que detiene la acción del compresor. Algunos interruptores advierten al operador y/o fijan un código indicador de fallas.

Low-side gauge The left-side gauge on the manifold used to read refrigerant pressure in the low side of the system.
Calibrador del lado de baja presión El calibrador en el lado izquierdo del múltiple utilizado para medir la presión del refrigerante del lado de baja presión del sistema.

Low-side hand valve The manifold valve used to control flow between the low side and service ports of the manifold.
Válvula de mano del lado de baja presión Válvula de distribución utilizada para regular el flujo entre el lado de baja presión y los orificios de servicio del colector.

Low-side service valve A device located on the suction side of the compressor which allows the service technician to check low-side pressures and perform other necessary service operations.
Válvula de servicio del lado de baja presión Dispositivo ubicado en el lado de succión del compresor; dicha válvula permite que el mecánico verifique las presiones del lado de baja presión y lleve a cabo otras funciones necesarias de servicio.

Manifold A device equipped with a hand shutoff valve. Gauges are connected to the manifold for use in system testing and servicing.
Múltiple Dispositivo provisto de una válvula de cierre accionada a mano. Calibradores se conectan al múltiple para ser utilizados para llevar a cabo pruebas del sistema y para servicio.

Manifold and gauge set A manifold complete with gauges and charging hoses.
Conjunto del múltiple y calibrador Múltiple provisto de calibradores y mangueras de carga.

Manifold hand valve Valves used to open and close passages through the manifold set.
Válvula de distribución accionada a mano Válvulas utilizadas para abrir y cerrar conductos a través del conjunto del múltiple.

Manufacturer A person or company whose business is to produce a product or components for a product.
Fabricante Persona o empresa cuyo propósito es fabricar un producto o componentes para un producto.

Manufacturer's procedures Specific step-by-step instructions provided by the manufacturer for the assembly, disassembly, installation, replacement, and/or repair of a particular product manufactured by them.
Procedimientos del fabricante Instrucciones específicas a seguir paso por paso; dichas instrucciones son suministradas por el fabricante para montar, desmontar, instalar, reemplazar, y/o reparar un producto específico fabricado por él.

MAX A mode, maximum, for heating or cooling. Selecting MAX generally overrides all other conditions that may have been programmed.
MAX (Máximo) Modo máximo para calentamiento o enfriamiento. El seleccionar MAX generalmente anula todas las otras condiciones que pueden haber sido programadas.

Metric fastener Any type fastener with metric size designations, numbers or millimeters.
Asegurador métrico Cualquier asegurador provisto de indicaciones, números o milímetros.

Mid-positioned The position of a stem-type service valve where all fluid passages are interconnected. Also referred to as "cracked."
Ubicación central Posición de una válvula de servicio de tipo vástago donde todos los pasajes que conducen fluidos se interconectan. Llamado también parcialmente asentada.

Motor An electrical device that produces a continuous turning motion. A motor is used to propel a fan blade or a blower wheel.
Motor Dispositivo eléctrico que produce un movimiento giratorio continuo. Se utiliza un motor para impeler las aletas del ventilador o la rueda del soplador.

Mounting boss See Flange.
Protuberancia de montaje Ver Flange [Brida].

Mounting flange See Flange.
Brida de montaje Ver Flange [Brida].

MSDS Material Safety Data Sheet.
MSDS Hojas de información sobre la seguridad de un material.

Mushroomed A condition caused by pounding of a punch or a chisel, producing a mushroom-shaped end which should be ground off to ensure maximum safety.
Hinchado Condición ocasionada por el golpeo de un punzón o cincel, lo cual hace que el extremo vuelva en forma de un hongo y que debe ser afilado para asegurar máxima seguridad.

Net weight The weight of a product only; container and packaging not included.
Peso neto Peso de sólo el producto mismo; no incluye el recipiente y encajonamiento.

Neutral On neither side; the position of gears when force is not being transmitted.
Neutro Que no está en ningún lado; posición de los engranajes cuando no se transmite la potencia.

Noncycling clutch An electro-mechanical compressor clutch that does not cycle on and off as a means of temperature control; it is used to turn the system on when cooling is desired and off when cooling is not desired.
Embrague sin funcionamiento cíclico Embrague electromecánico del compresor que no se enciende y se apaga como medio de regular la temperatura; se utiliza para arrancar el sistema cuando se desea enfriamiento y para detener el sistema cuando se desea enfriamiento.

Observe To see and note; to perceive; notice.
Observar Ver y anotar; percibir; fijarse en algo.

OEM Original Equipment Manufacturer.
OEM Fabricante Original del Equipo.

Off-the-road Generally refers to vehicles that are not licensed for road use, such as harvesters, bulldozers, and so on.
Fuera de carretera Generalmente se refiere a vehículos que no son permitidos operar en la carretera, como por ejemplo cosechadoras, rasadoras, ecétera.

Ohmmeter An electrical instrument used to measure the resistance in ohms of a circuit or component.
Ohmiómetro Instrumento eléctrico utilizado para medir la resistencia en ohmios de un circuito o componente.

Open Not closed. An open switch, for example, breaks an electrical circuit.
Abierto No cerrado. Un interruptor abierto corta un circuito eléctrico, por ejemplo.

Orifice A small hole. A calibrated opening in a tube or pipe to regulate the flow of a fluid or liquid.
Orificio Agujero pequeño. Apertura calibrada en un tubo o cañería para regular el flujo de un fluido o de un líquido.

O-ring A synthetic rubber or plastic gasket with a round- or square-shaped cross-section.
Junta tórica Guarnición sintética de caucho o de plástico provista de una sección transversal en forma redonda o cuadrada.

OSHA Occupational Safety and Health Administration.
OSHA Dirección para la Seguridad y Salud Industrial.

Outside temperature sensor See Ambient sensor.
Sensor de la temperatura ambiente Ver Ambient sensor [Sensor ambiente].

Overcharge Indicates that too much refrigerant or refrigeration oil is added to the system.
Sobrecarga Indica que una cantidad excesiva de refrigerante o aceite de refrigeración ha sido agregada al sistema.

Overload Anything in excess of the design criteria. An overload will generally cause the protective device such as a fuse or pressure relief to open.
Sobrecarga Cualquier cosa en exceso del criterio de diseño. Generalmente una sobrecarga causará que se abra el dispositivo de protección, como por ejemplo un fusible o alivio de presión.

Ozone friendly Any product that does not pose a hazard or danger to the ozone.
Sustancia no dañina al ozono Cualquier producto que no es peligrosa o amenaza al ozono.

Park Generally refers to a component or mechanism that is at rest.
Reposo Generalmente se refiere a un componente o mecanismo que no está funcionando.

PCM Power control module.
PCM Módulo regulador del transmisor de potencia.

Performance test Readings of the temperature and pressure under controlled conditions to determine if an air conditioning system is operating at full efficiency.
Prueba de rendimiento Lecturas de la temperatura y presión bajo condiciones controladas para determinar si un sistema de acondicionamiento de aire funciona a un rendimiento completo.

Piercing pin The part of a saddle valve that is used to pierce a hole in the tubing.

Pasador perforador Parte de la válvula de silleta utilizada para perforar un agujero en la tubería.

Pin-type connector A single or multiple electrical connector that is round- or pin-shaped and fits inside a matching connector.
Conectador de tipo pasador Conectador eléctrico único o múltiple en forma redonda o en forma de pasador que se inserta dentro de un conectador emparejado.

Poly belt See Serpentine belt.
Correa poli Ver Serpentine belt [Correa serpentina].

Polyol ester (ESTER) A synthetic oil-like lubricant that is occasionally recommended for use in an HFC-134a system. This lubricant is compatible with both HFC-134a and CFC-12.
Polioléster Lubrificante sintético parecido a aceite que se recomienda de vez en cuando para usar en un sistema HFC 134a. Dicho lubrificante es compatible tanto con HFC 134a como CFC 12.

Positive pressure Any pressure above atmospheric.
Presión positiva Cualquier presión sobre la de la atmosférica.

Pound A weight measure, 16 ounces. A term often used when referring to a small can of refrigerant, although the can does not necessarily contain 16-ounces.
Libra Medida de peso, 16 onzas. Término utilizada con frecuencia al referirse a una lata pequeña de refrigerante, aunque es posible que la lata contenga menos de 16 onzas.

Pound of refrigerant A term used by some technicians when referring to a small can of refrigerant that actually contains less than 16 ounces.
Libra de refrigerante Término utilizado por algunos mecánicos al referirse a una lata pequeña de refrigerante que en realidad contiene menos de 16 onzas.

Power module Controls the operation of the blower motor in an automatic temperature control system.
Transmisor de potencia Regula el funcionamiento del motor del soplador en un sistema de control automático de temperatura.

Predetermined A set of fixed values or parameters that have been programmed or otherwise fixed into an operating system.
Predeterminado Valores fijos o parámetros que han sido programados o de otra manera fijados en un sistema de funcionamiento.

Pressure gauge A calibrated instrument for measuring pressure.
Manómetro Instrumento calibrado para medir la presión.

Pressure switch An electrical switch that is activated by a predetermined low or high pressure. A high pressure switch is generally used for system protection; a low pressure switch may be used for temperature control or system protection.
Autómata manométrico Interruptor eléctrico accionado por una baja o alta presión predeterminada. Generalmente se utiliza un autómata manométrico de alta presión para la protección del sistema; puede utilizarse uno de baja presión para la regulación de temperatura o protección del sistema.

Propane A flammable gas used as a propellant for the halide leak detector.
Propano Gas inflamable utilizado como propulsor para el detector de fugas de halogenuro.

Psig Pounds per square inch gauge.
Psig Calibrador de libras por pulgada cuadrada.

Purge To remove moisture and/or air from a system or a component by flushing with a dry gas such as nitrogen (N) to remove all refrigerant from the system.

Purgar Remover humedad y/o aire de un sistema o un componente al descargarlo con un gas seco, como por ejemplo el nitrógeno (N), para remover todo el refrigerante del sistema.

Purity test A static test that may be performed to compare the suspect refrigerant pressure to an appropriate temperature chart to determine its purity.
Prueba de pureza Prueba estática que puede llevarse a cabo para comparar la presión del refrigerante con un gráfico de temperatura apropiado para determinar la pureza del mismo.

Radiation The transfer of heat without heating the medium through which it is transmitted.
Radiación La transferencia de calor sin calentar el medio por el cual se transmite.

Ram air Air that is forced through the radiator and condenser coils by the movement of the vehicle or the action of the fan.
Aire admitido en sentido de la marcha Aire forzado a través de las bobinas del radiador y del condensador por medio del movimiento del vehículo o la acción del ventilador.

Rebuilt To build after having been disassembled, inspected, and worn and damaged parts and components are replaced.
Reconstruido Fabricar después de haber sido desmontado y revisado, y luego reemplazar las piezas desgastadas y averiadas.

Receiver/drier A tank-like vessel having a desiccant and used for the storage of refrigerant.
Receptor/secador Recipiente parecido a un tanque provisto de un desecante y utilizado para el almacenaje de refrigerante.

RECIR An abbreviation for the recirculate mode, as with air.
RECIR Abreviatura del modo recirculatorio, como por ejemplo con aire.

Recovery system A term often used to refer to the circuit inside the recovery unit used to recycle and/or transfer refrigerant from the air conditioning system to the recovery cylinder.
Sistema de recuperación Término utilizado con frecuencia para referirse al circuito dentro de la unidad de recuperación interior utilizada para reciclar y/o transferir el refrigerante del sistema de acondicionamiento de aire al cilindro de recuperación.

Recovery tank An auxiliary tank, usually connected to the inlet tank of a radiator, which provides additional storage space for heated coolant.
Tanque de recuperación Tanque auxiliar que normalmente se conecta al tanque de entrada de un radiador, lo cual provee almacenaje adicional para el enfriante calentado.

Refrigerant-12 The refrigerant used in automotive air conditioners, as well as other air conditioning and refrigeration systems. The chemical name of refrigerant-12 is dichlorodifluoromethane. The chemical symbol is $CC_{l2}F_2$.
Refrigerante 12 Refrigerante utilizado tanto en acondicionadores de aire automotrices como en otros sistemas de acondicionamiento de aire y refrigeración. El nombre químico del refrigerante 12 es diclorodiflorometano, y el símbolo químico es $CC_{l2}F_2$.

Relay An electrical switch device that is activated by a low-current source and controls a high-current device.
Relé Interruptor eléctrico que es accionado por una fuente de corriente baja y regula un dispositivo de corriente alta.

Reserve tank A storage vessel for excess fluid. See Recovery tank, Receiver/drier and Accumulator.
Tanque de reserva Recipiente de almacenaje para un exceso de fluido. Ver Recovery tank [Tanque de recuperación], Receiver/drier [Receptor/secador] y Accumulador [Acumulador].

Resistor A voltage-dropping device that is usually wire wound and provides a means of controlling fan speeds.

Resistor Dispositivo de caída de tensión que normalmente es devando con alambre y provee un medio de regular la velocidad del ventilador.

Respirator A mask or face shield worn in a hazardous environment to provide clean fresh air and/or oxygen.

Mascarilla Máscara o protector de cara que se lleva puesto en un ambiente peligroso para proveer aire limpio y puro y/o oxígeno.

Responsibility Being reliable and trustworthy.

Responsibilidad Ser confiable y fidedigno.

Restricted Having limitations. Keeping within limits, confines, or boundaries.

Restringido Que tiene limitaciones. Mantenerse dentro de límites, confines, o fronteras.

Restrictor An insert fitting or device used to control the flow of refrigerant or refrigeration oil.

Limitador Pieza inserta o dispositivo utilizado para regular el flujo de refrigerante o aceite de refrigeración.

Rotor The rotating or freewheeling portion of a clutch; the belt slides on the rotor.

Rotor Parte giratoria o con marcha a rueda libre de un embrague; la correa se desliza sobre el rotor.

RPM Revolutions per minute; also, rpm or r/min.

RPM Revoluciones por minuto; también rpm o r/min.

Running design change A design change made during a current model/year production.

Cambio al diseño corriente Un cambio al diseño hecho durante la fabricación del modelo/año actual.

RV Recreational vehicle.

RV Vehículo para el recreo.

Saddle valve A two-part accessory valve that may be clamped around the metal part of a system hose to provide access to the air conditioning system for service.

Válvula de silleta Válvula accesoria de dos partes que puede fijarse con una abrazadera a la parte metálica de una manguera del sistema para proveer acceso al sistema de acondicionamiento de aire para llevar a cabo servicio.

SAE Society of Automotive Engineers.

SAE Sociedad de Ingenieros Automotrices.

Safety Freedom from danger or injury; the state of being safe.

Seguridad Libre de peligro o daño; calidad o estado de seguro.

Schrader valve A spring-loaded valve similar to a tire valve. The Schrader valve is located inside the service valve fitting and is used on some control devices to hold refrigerant in the system. Special adapters must be used with the gauge hose to allow access to the system.

Válvula Schrader Válvula con cierre automático parecida al vástago del neumático. La válvula Schrader está ubicada dentro del ajuste de la válvula de servicio y se utiliza en algunos dispositivos de regulación para guardar refrigerante dentro del sistema. Deben utilizarse adaptadores especiales con una manguera calibrador para permitir acceso al sistema.

Seal Generally refers to a compressor shaft oil seal; matching shaft-mounted seal face and front head-mounted seal seat to prevent refrigerant and/or oil from escaping. May also refer to any gasket or O-ring used between two mating surfaces for the same purpose.

Junta hermética Generalmente se refiere a la junta hermética del árbol del compresor; la frente de junta hermética montada en el árbol y el asiento de junta hermética montado en el cabezal delantero emparejados para evitar la fuga de refrigerante y/o de aceite. Puede referirse también a cualquier guarnición o junta tórica utilizada entre dos superficies emparejadas para el mismo propósito.

Seal seat The part of a compressor shaft seal assembly that is stationary and matches the rotating part, known as the seal face or shaft seal.

Asiento de la junta hermética Parte del conjunto de la junta hermética del árbol del compresor que es inmóvil y que se empareja a la parte rotativa; conocido como la frente de junta hermética o la junta hermética del árbol.

Serpentine belt A flat or V-groove belt that winds through all of the engine accessories to drive them off the crankshaft pulley.

Correa serpentina Correa plana o con ranuras en V que atraviesa todos los accesorios del motor para forzarlos fuera de la polea del cigüeñal.

Service port A fitting found on the service valves and some control devices; the manifold set hoses are connected to this fitting.

Orificio de servicio Ajuste ubicado en las válvulas de servicio y en algunos dispositivos de regulación; las mangueras del conjunto del colector se conectan a este ajuste.

Service procedure A suggested routine for the step-by-step act of troubleshooting, diagnosing and/or repairs.

Procedimiento de servico Rutina sugerida para la acción a seguir paso a paso para detectar fallas, diagnosticar, y/o reparar.

Service valve See High-side (Low-side) service valve.

Válvula de servicio Ver High-side (Low-side) service valve [Válvula de servicio del lado de alta presión (baja presión)].

Shaft key A soft metal key that secures a member on a shaft to prevent it from slipping.

Chaveta del árbol Chaveta de metal blando que fija una pieza a un árbol para evitar su deslizamiento.

Shaft seal See Compressor shaft seal.

Junta hermética del árbol Ver Compressor shaft seal [Junta hermética del árbol del compresor].

Short Of brief duration; e.g., short cycling. Also refers to an intentional or unintentional grounding of an electrical circuit.

Breve/corto De una duración breve; p.e., funcionamiento cíclico breve. Se refiere también a un puesto a tierra previsto o imprevisto de un circuito eléctrico.

Shut-off valve A valve that provides positive shut-off of a fluid or vapor passage.

Válvula de cierre Válvula que provee el cierre positivo del pasaje de un fluido o un vapor.

Snap ring A metal ring used to secure and retain a component to another component.

Anillo de muelle Anillo metálico utilizado para fijar y sujetar un componente a otro.

Society of Automotive Engineers A professional organization of the automotive industry. Founded in 1905 as the Society of Automobile Engineers, the SAE is dedicated to providing technical information and standards to the automotive industry. Present goals are to assure a skilled engineering and technical work force for the year 2000 and beyond. The goal, known as VISION 2000, encompasses all of SAE's educational programs, including student competitions, scholarships, teacher recognition, and more.

Sociedad de Ingenieros Automotrices Organización profesional de la industria automotriz. Establecido en 1905 como la Sociedad de Ingenieros de Automóviles (SAE por sus siglas en inglés), dicha sociedad se dedica a proveerle información técnica y normas a la industria automotriz. Sus metas actuales son asegurar una fuerza laboral capacitada en la ingeniería y en el campo técnico para el año 2000 y después. La meta, conocida como VISION 2000, abarca todos los programas educativos de la SAE, e incluye concursos entre estudiantes, el otorgar becas, el reconocer al profesor, y más.

Solenoid See Solenoid valve.
Solenoide Ver Solenoid valve [Válvula de solenoide].

Solenoid valve An electromagnetic valve controlled remotely by electrically energizing and deenergizing a coil.
Válvula de solenoide Válvula electromagnética regulada a distancia por una bobina al excitar y deexcitar una bobina electrónicamente.

Solid state Referring to electronics consisting of semiconductor devices and other related nonmechanical components.
Estado sólido Se refiere a componentes electrónicos que consisten en dispositivos semiconductores y otros componentes relacionados no mecánicos.

Spade-type connector A single or multiple electrical connector that has flat spade-like mating provisions.
Conectador de tipo azadón Conectador único o múltiple provisto de dispositivos planos de tipo azadón para emparejarse.

Specifications Design characteristics of a component or assembly noted by the manufacturer. Specifications for a vehicle include fluid capacities, weights, and other pertinent maintenance information.
Especificaciones Características de diseño de un componente o conjunto indicadas por el fabricante. Las especificaciones para un vehículo incluyen capacidades del fluido, pesos, y otra información pertinente para mantenimiento del vehículo.

Spike In our application, an electrical spike. An unwanted momentary high-energy electrical surge.
Impulso afilado En nuestro campo, un impulso afilado eléctrico. Una elevación repentina eléctrica de alta energía no deseada.

Spring lock fitting A special fitting using a spring to lock the mating parts together forming a leak-proof joint.
Ajuste de cierre automático Ajuste especial utilizando un resorte para cerrar piezas emparejadas para formar así una junta hermética contra fugas.

Squirrel-cage blower A blower wheel designed to provide a large volume of air with a minimum of noise. The blower is more compact than the fan and air can be directed more efficiently.
Soplador con jaula de ardilla Rueda de soplador diseñada para proveer un gran caudal de aire con un mínimo de ruido. El soplador es más compacto que el ventilador y el aire puede dirigirse con un mayor rendimiento.

Stabilize To make steady.
Estabilizar Quedarse detenida una cosa.

Stratify Arrange or form into layers. To fully blend.
Estratificar Arreglar o formar en capas. Mezclar completamente.

Subsystem A system within a system.
Subsistema Sistema dentro de un sistema.

Sun load Heat intensity and/or light intensity produced by the sun.
Carga del sol Intensidad calorífica y/o de la luz generada por el sol.

Superheat switch An electrical switch activated by an abnormal temperature-pressure condition (a superheated vapor); used for system protection.
Interruptor de vapor sobrecalentado Interruptor eléctrico accionado por una condición anormal de presión y temperatura (vapor sobrecalentado); utilizado para la protección del sistema.

System All of the components and lines that make up an air conditioning system.
Sistema Todos los componentes y líneas que componen un sistema de acondicionamiento de aire.

Tank See Header tanks and Expansion tank.
Tanque Ver Header tanks [Tanques para alimentación por gravedad] y Expansion tank [Tanque de expansión].

Tare weight The weight of the packaging material. See Net weight and Gross weight.
Taraje Peso del material de encajonamiento. Ver Net weight [Peso neto] y Gross weight [Peso bruto].

Temperature door A door within the case/duct stem to direct air through the heater and/or evaporator core.
Puerta de temperatura Puerta ubicada dentro del vástago de caja/conducto para conducir el aire a través del núcleo del calentador y/o del evaporador.

Temperature switch A switch actuated by a change in temperature at a predetermined point.
Interruptor de temperatura Interruptor accionado por un cambio de temperatura a un punto predeterminado.

Tension gauge A tool for measuring the tension of a belt.
Manómetro para tensión Herramienta para medir la tensión de una correa.

Thermistor A temperature-sensing resistor that has the ability to change values with changing temperature.
Termistor Resistor sensible a temperatura que tiene la capacidad de cambiar valores al ocurrir un cambio de temperatura.

Torque A turning force; for example, the force required to seal a connection; measured in (English) foot-pounds (ft-lb) or inch-pounds (in-lb); (metric) newton-meters (N·m).
Par de torsión Fuerza de torcimiento; por ejemplo, la fuerza requerida para sellar una conexión; medido en libras-pies (ft-lb) (inglesas) o en libras pulgadas (in-lb); metros-newton (N·m) (métricos).

Triple evacuation A process of evacuation that involves three pump-downs and two system purges with an inert gas such as dry nitrogen (N).
Evacuación triple Proceso de evacuación que involucra tres envíos con bomba y dos purgas del sistema con un gas inerte, como por ejemplo el nitrógeno seco (N).

Troubleshoot The act or art of diagnosing the cause of various system malfunctions.
Detección de fallas Procedimiento o arte de diagnosticar la causa de varias fallas del sistema.

TXV Thermostatic expansion valve.
TXV Válvula de expansión termostática.

Ultraviolet (uv) The part of the electromagnetic spectrum emitted by the sun that lies between visible violet light and X-rays.
Ultravioleta Parte del espectro electromagnético generado por el sol que se encuentra entre la luz violeta visible y los rayos X.

Vacuum gauge A gauge used to measure below atmospheric pressure.
Vacuómetro Calibrador utilizado para medir a una presión inferior a la de la atmósfera.

Vacuum motor A device designed to provide mechanical control by the use of a vacuum.

Motor de vacío Dispositivo diseñado para proveer regulación mecánica mediante un vacío.

Vacuum pump A mechanical device used to evacuate the refrigeration system to rid it of excess moisture and air.

Bomba de vacío Dispositivo mecánico utilizado para evacuar el sistema de refrigeración para purgarlo de un exceso de humedad y aire.

Vacuum signal The presence of a vacuum.

Señal de vacío Presencia de un vacío.

V-belt A rubber-like continuous loop placed between the engine crankshaft pulley and accessories to transfer rotary motion of the crankshaft to the accessories.

Correa en V Bucle continuo parecido a caucho ubicado entre la polea del cigüeñal del motor y los accesorios para transferir el movimiento giratorio del aquél a éstos.

Ventilation The act of supplying fresh air to an enclosed space such as the inside of an automobile.

Ventilación Proceso de suministrar el aire fresco a un espacio cerrado, como por ejemplo al interior de un automóvil.

V-groove belt See V-belt.

Correa con ranuras en V Ver V-belt [Correa en V].

Voltmeter A device used to measure volt(s).

Voltímetro Dispositivo utilizado para la medición de voltios.

Wiring harness A group of wires wrapped in a shroud for the distribution of power from one point to another point.

Cableado preformado Grupo de alambres envuelto por una gualdera para distribuir potencia de un punto a otro.

8 Engine Performance

Pretest

The purpose of this pretest is to determine the amount of review that you may require prior to writing the ASE Engine Performance Test. If you answer all the pretest questions correctly, complete the questions and study the information in this chapter to prepare for the ASE Engine Performance Test.

If two or more of your answers to the pretest questions are incorrect, complete a study of Chapters 2 through 17 in *Today's Technician Engine Performance Classroom and Shop Manuals*, published by Delmar Publishers, plus a study of the questions and information in this chapter.

The pretest answers are located at the end of the pretest; and these answers also are in the answer sheets supplied with this book.

1. An engine has a hollow rapping noise during acceleration. This noise is worse when the engine is cold. The cause of this problem could be
   A. worn piston pins.
   B. loose pistons.
   C. worn connecting rod bearings.
   D. loose flywheel bolts.

2. An engine has gray exhaust especially after the engine is restarted after a brief, hot, shutdown.
   Technician A says the head gasket may be leaking.
   Technician B says the piston rings may be worn.
   Who is correct?
   A. A only
   B. B only
   C. Both A and B
   D. Neither A nor B

3. During a cylinder leakage test number 4 cylinder has 45 percent leakage and air is escaping from the PCV valve opening in the rocker arm cover. The cause of this problem could be
   A. a burned exhaust valve.
   B. a bent intake valve.
   C. a broken valve spring.
   D. worn piston rings.

4. A fuel-injected vehicle fails an emission test for high CO and HC emissions.
   Technician A says the fuel pressure may be low.
   Technician B says the O_2 sensor signal may be continually low.
   Who is correct?

   A. A only
   B. B only
   C. Both A and B
   D. Neither A nor B

5. When diagnosing an electronic distributor ignition system, a 12V test light connected from the negative primary coil terminal to ground flutters while cranking the engine. A test spark plug does not fire when it is connected from the coil secondary wire to ground while cranking the engine. The cause of this problem could be
   A. a defective pickup coil.
   B. a defective ignition module.
   C. a defective ignition coil.
   D. a primary circuit.

6. An electronic distributor has a Hall Effect pickup.
   Technician A says this type of pickup may be tested with an ohmmeter.
   Technician B says this type of pickup produces an analog voltage signal.
   Who is correct?
   A. A only
   B. B only
   C. Both A and B
   D. Neither A nor B

7. A vehicle with an electronic distributorless ignition system has a loss of power with no engine misfiring.
   Technician A says the basic ignition timing may be late.
   Technician B says the exhaust system may be restricted.
   Who is correct?
   A. A only
   B. B only
   C. Both A and B
   D. Neither A nor B

8. A carbureted engine has a hesitation on low-speed acceleration. The cause of this problem could be
   A. late basic ignition timing.
   B. excessive distributor advance.
   C. excessive accelerator pump stroke.
   D. excessive fuel pump pressure.

9. A fuel-injected engine with an electronic distributor ignition system backfires during acceleration, but the engine idles smoothly. The most likely cause of this problem is
   A. defective injectors.
   B. low fuel pump pressure.
   C. a cracked distributor cap.
   D. defective spark plugs.

10. A fuel-injected engine has a severe surging problem only at speeds above 55 mph (88 kmh). Engine operation is normal at idle and low speeds.
    Technician A says there may be low voltage at the fuel pump.
    Technician B says the inertia switch may have high resistance.
    Who is correct?
    A. A only
    B. B only
    C. Both A and B
    D. Neither A nor B

11. A port-fuel-injected engine has a rough idle and hard starting problems. The cause of these problems could be
    A. a leaking fuel pump check valve.
    B. dripping injectors.
    C. a leaking pressure regulator.
    D. low fuel pump pressure.

12. A vehicle fails an emissions test for high NOX emissions. All of these problems could be the cause of the high emissions EXCEPT
    A. a plugged EGR exhaust passage.
    B. a defective knock sensor.
    C. an open winding in the EGR solenoid.
    D. a defective air charge temperature sensor.

13. When diagnosing an EVAP system with a scan tool, the PCM never provides an on command to the EVAP solenoid at any engine or vehicle speed.
    Technician A says to check the ECT sensor signal to the PCM.
    Technician B says to check the vacuum hoses from the intake to the EVAP canister.
    Who is correct?
    A. A only
    B. B only
    C. Both A and B
    D. Neither A nor B

14. When performing a battery capacity test
    A. the battery should be discharged at one-third of the cold cranking rating.
    B. the battery voltage should be 12.8V prior to the capacity test.
    C. the load should be maintained on the battery for 20 seconds.
    D. at 70°F (21°C) the voltage should be above 9.6V at the end of the test.

Answers To Pretest

1. B, 2. A, 3. D, 4. B, 5. C, 6. D, 7. B, 8. A, 9. C, 10. C, 11. B, 12. D, 13. A, 14. D

General Engine Diagnosis

ASE Tasks, Questions, and Related Information

In this chapter each task in the Engine Performance category is provided followed by a question and some information related to the task. If you answer any question incorrectly, study this information very carefully until you understand the correct answer. For additional information on any task refer to *Today's Technician Engine Performance Classroom and Shop Manuals*, published by Delmar Publishers.

Question answers and analysis are provided at the end of this chapter and in the answer sheets provided with this book.

Task 1 Verify driver's complaint and/or road test vehicle; determine needed repairs.

1. Technician A says one of the first steps in a diagnostic procedure is to identify the problem.
Technician B says listening to the customer's complaint may help to identify the problem.
Who is correct?
   A. A only
   B. B only
   C. Both A and B
   D. Neither A nor B

Hint *The following could be used as a general diagnostic and repair procedure:*
   - *Listen to the customer's complaint, be sure the problem is identified, road test the vehicle if necessary.*
   - *Think of the possible causes of the problem.*
   - *Perform appropriate diagnostic tests.*
   - *Repair the cause of the problem.*
   - *Be sure the problem is eliminated.*

Task 2 Diagnose the cause of unusual engine noise and/or vibration problems; determine needed action.

2. After sitting overnight an engine has a heavy thumping noise when it is started. The cause of this problem could be
   A. worn camshaft lobes.
   B. worn valve lifters.
   C. worn main bearings.
   D. loose piston pins.

Hint *Worn or sticking valve lifters produce a light clicking noise that is most noticeable at idle and low speeds. A heavy clicking noise at 2,000 rpm may be caused by worn camshaft lobes. Worn piston pins produce a heavy, sharp, rapping noise at idle speed. Loose pistons produce a hollow rapping noise that is loudest during acceleration. Piston ring or ring ridge noise is a sharp metallic rapping noise that is worse during acceleration. Loose connecting rod bearings may cause a lighter rapping noise at at vehicle speeds above 35 mph (56 kmh). Loose main bearings cause a heavy thumping noise when the engine is started after being shut off for several hours. A heavy thumping noise with the engine idling may be caused by a loose flywheel or vibration damper.*

Task 3 Diagnose the cause of unusual exhaust color, odor, and sound; determine needed repairs.

3. At idle speed the engine exhaust has a "puff" noise at regular intervals.
Technician A says this problem may be caused by a burned exhaust valve.
Technician B says this problem may be caused by excessive coolant temperature.
Who is correct?
   A. A only
   B. B only
   C. Both A and B
   D. Neither A nor B

Hint *Blue exhaust may be caused by excessive amounts of oil entering the combustion chamber. If the exhaust is black, the air-fuel ratio is too rich. On catalytic-converter-equipped vehicles excessive sulphur smell indicates a rich air-fuel ratio. A "puff" noise in the exhaust at regular intervals indicates cylinder misfire from a compression, ignition, or fuel system defect.*

Task 4   Perform engine manifold vacuum or pressure tests; determine needed action.

4. At idle speed a vacuum gauge connected to the intake manifold indicates a steady 14 in. Hg. (Figure 8-1).

Technician A says this reading is caused by sticking valves.

Technician B says this problem is caused by a burned valve.

Who is correct?

A. A only

B. B only

C. Both A and B

D. Neither A nor B

Figure 8-1   Vacuum gauge reading (*Courtesy of Sun Electric Corporation*)

Hint   *Normal and abnormal vacuum gauge readings are shown in Figure 8-2.*

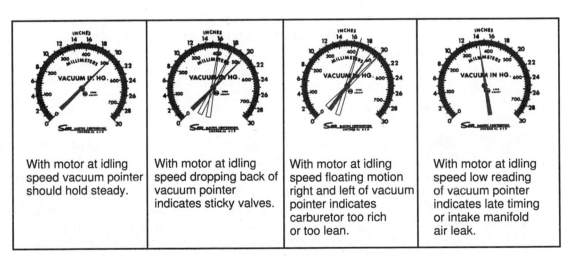

With motor at idling speed vacuum pointer should hold steady.

With motor at idling speed dropping back of vacuum pointer indicates sticky valves.

With motor at idling speed floating motion right and left of vacuum pointer indicates carburetor too rich or too lean.

With motor at idling speed low reading of vacuum pointer indicates late timing or intake manifold air leak.

Figure 8-2   Normal and abnormal vacuum gauge readings (*Courtesy of Sun Electric Corporation*)

Task 5   Perform cylinder power balance test; determine needed repairs.

5. During an engine power balance test at idle speed cylinder number 3 produces 20 rpm decrease, and all the other cylinders produce 75 rpm decrease. When the test is repeated at 2,000 rpm, cylinder number 3 has the same voltage drop as the other cylinders. This problem could be caused by

A. an intake manifold vacuum leak.

B. a defective spark plug.

C. an open spark plug wire.

D. a burned exhaust valve.

Hint   *During the cylinder power balance test, the tester stops each spark plug from firing for a few seconds and the amount of engine rpm decrease is recorded. If all the cylinders have the specified rpm decrease, the cylinders are contributing equally to engine power. When one cylinder has less rpm decrease, that cylinder is not contributing as much to engine power because of a compression, ignition, fuel system defect, or an intake manifold vacuum leak. If a vacuum leak is causing the problem, the cylinder rpm drop may be equal to the other cylinders if the test is repeated at a higher engine rpm.*

Task 6   Perform cylinder compression test; determine needed action.

6. During a compression test cylinder number 6 has 60 psi and all the other cylinders have 135 psi. When a wet test is preformed on cylinder number 6 the compression increases to 120 psi. The cause of this problem could be
   A. a burned exhaust valve.
   B. a bent intake valve.
   C. worn piston rings.
   D. a blown head gasket.

Hint   *During a compression test the ignition and fuel injection system must be disabled. Four compression strokes should be recorded on each cylinder. If some cylinders indicate lower than specified compression, a wet test may be performed by squirting about 3 tablespoons of oil into the cylinder and repeating the test. When the compression reading improves considerably with the oil in the cylinder, worn rings are indicated. If there is very little improvement in the compression reading, the valves or head gasket are probably leaking.*

Task 7   Perform cylinder leakage test; determine needed action.

7. During a leakage test cylinder number 2 indicates 45% leakage and air is escaping from the throttle body.
   Technician A says cylinder number 2 may have a bent intake valve.
   Technician B says cylinder number 2 may have a bottomed intake valve lifter.
   Who is correct?
   A. A only
   B. B only
   C. Both A and B
   D. Neither A nor B

Hint   *The leakage tester supplies a controlled amount of shop air into each cylinder through the spark plug opening. Both valves in the cylinder must be closed during the leakage test. If cylinder leakage exceeds 20 percent, excessive air is leaking past the rings or valves. Air also could be leaking through the cylinder head or head gasket. When air is escaping through from the crankcase through the PCV valve opening in the rocker cover, the rings and cylinders are worn. Air escaping from the throttle body indicates a defective intake valve. If air is escaping from the tailpipe, the exhaust valve is burned. Bubbles in the radiator indicate a leaking head gasket or a cracked cylinder head.*

Task 8   Diagnose engine mechanical, electrical, electronic, fuel, and ignition problems with an ignition oscilloscope and/or engine analyzer; determine needed action.

8. Technician A says some engine analyzers compare test results to specifications and indicate readings that are out of specifications.
   Technician B says some engine analyzers provide the option of manual or automatic test modes.
   Who is correct?
   A. A only
   B. B only
   C. Both A and B
   D. Neither A nor B

Hint   *Many engine analyzers have the capability to test ignition, fuel, battery, charging, starting systems and engine condition. A 4 or 5-gas emissions analyzer may be contained in the engine analyzer. Some engine analyzers provide the option of manual technician-selected test modes or an automatic test mode. Some engine analyzers contain specifications for various vehicles on a diskette or CD. During the test procedures the analyzer compares test readings to specifications and identifies test results that are out of specification.*

Task 9   **Prepare and inspect the vehicle and analyzer for exhaust gas analysis; obtain exhaust gas readings.**

9. While discussing the use of an emissions analyzer to read hydrocarbons (HC), carbon monoxide (CO), carbon dioxide (CO_2) and oxygen (O_2) on a catalytic-converter-equipped vehicle
Technician A says a leak in the exhaust system may cause inaccurate emission analyzer readings.
Technician B says with the tester probe in the tailpipe, the analyzer provides accurate HC and CO readings from the engine.
Who is correct?
A. A only
B. B only
C. Both A and B
D. Neither A nor B

Hint   *Emission analyzers usually have a 15-minute warm-up and calibration period. Many emission analyzers have an automatic calibration function, but some analyzers with analog meters require manual calibration. Since exhaust leaks cause inaccurate emission readings, the exhaust system on the vehicle must be leak-free. The engine should be at normal operating temperature before the emissions test. Since catalytic converters reduce CO, HC, and oxides of nitrogen (NOX) emissions, if the emissions analyzer probe is in the tailpipe, the analyzer does not indicate the actual emissions coming out of the cylinders. The catalytic converter does not affect CO_2 and O_2 levels.*

Ignition System Diagnosis and Repair

ASE Tasks, Questions, and Related Information

Task 1   **Diagnose no-starting, hard starting, engine misfire, poor drivability, spark knock, power loss, poor mileage, and emissions problems on vehicles with distributor and distributorless ignition systems; determine needed repairs.**

10. When diagnosing a no-start condition on an electronic ignition (EI) system there is no spark at any of the spark plugs. A 12V test light does not flutter when connected from the negative primary terminal on each coil to ground with the engine cranking. The cause of this problem could be
A. a defective crankshaft sensor.
B. defective ignition coil assembly.
C. defective spark plug wires.
D. defective spark plugs

11. A vehicle with an EI system has a reduced fuel economy and loss of power complaint (Figure 8-3). The engine runs smoothly, never misfires, and starts easily.

Technician A says the cause of this problem may be an open circuit in the EST wire from the PCM to the coil module.

Technician B says the cause of this problem may be a defective cam sensor.

Who is correct?

A. A only

B. B only

C. Both A and B

D. Neither A nor B

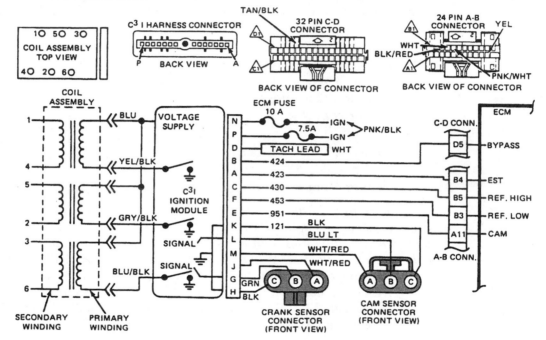

Figure 8-3 Electronic ignition (EI) system *(Courtesy of Buick Motor Division, General Motors Corporation)*

Hint   *For complete ignition system diagnosis refer to* Today's Technician Engine Performance Shop Manual *(pp. 161, 162, 193–198), published by Delmar Publishers.*

12. When diagnosing a no-start condition on an electronic distributor ignition system, a test spark plug fires normally when connected from the coil secondary wire to ground with the engine cranking. The test spark plug fires intermittently when connected from the spark plug wires to ground.

Technician A says the distributor cap or rotor may have a leakage problem.

Technician B says the ignition module may be defective.

Who is correct?

A. A only

B. B only

C. Both A and B

D. Neither A nor B

Hint   *For complete electronic distributor ignition system diagnosis refer to* Today's Technician Engine Performance Shop Manual *(pp. 161, 162, 193–198), published by Delmar Publishers.*

Task 2   Inspect, test repair, or replace ignition primary circuit wiring and components.

13. A pair of ohmmeter leads are connected from one of the distributor pickup leads to ground with the pickup coil connector disconnected from the module. The ohmmeter indicates 14 ohms resistance. This reading indicates

    A. the pickup coil is satisfactory.
    B. the pickup coil winding is shorted.
    C. the pickup coil winding is open.
    D. the pickup coil winding is grounded.

Hint   *The primary ignition coil winding may be tested with an ohmmeter connected to the primary terminals. A reading below the specified value indicates a shorted winding, and a reading higher than specified is caused by a resistance problem or an open circuit.If the ohmmeter leads are connected from one of the primary terminals to ground on the coil container, a low reading indicates a grounded primary winding, whereas an infinite reading proves the primary winding is not grounded. On many coils the secondary winding may be tested with the ohmmeter leads connected from the coil secondary terminal to one of the primary terminals.*

    The pickup coil may be tested for open and short circuits with the ohmmeter leads connected to the pickup leads with these leads disconnected from the module. Connect the ohmmeter leads from one of the pickup coil leads to ground to test the pickup coil for a grounded condition.

Task 3   **Inspect, test, and service distributor.**

    14. When diagnosing the distributor pickup in Figure 8-4, the pickup may be tested with
        A. an ohmmeter connected across the pickup leads with these leads disconnected.
        B. a test light connected across the pickup leads with these leads disconnected.
        C. a voltmeter connected from the pickup signal wire to ground while cranking the engine.
        D. a 12V test light connected from the primary positive terminal to ground while cranking the engine.

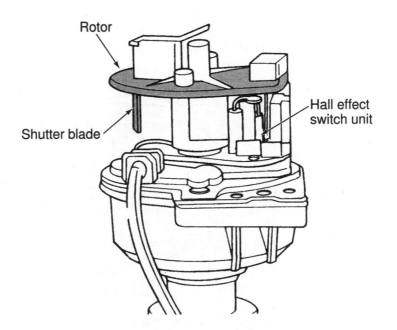

Figure 8-4 Hall Effect distributor pick-up *(Courtesy of Chrysler Corporation)*

Hint   *A Hall-Effect distributor pickup produces a digital voltage signal while cranking the engine. Three wires are connected to a Hall-Effect pickup.These wires include a ground wire, voltage supply wire, and a signal wire. This type of pickup may be tested with a voltmeter connected from the pickup signal wire to ground while cranking the engine. If the pickup is satisfactory the voltmeter should indicate the specified low and high voltage signals.*

Task 4 Inspect, test, service, repair, or replace ignition system secondary circuit wiring and components.

15. When using an oscilloscope to diagnose the secondary ignition system the maximum coil voltage on all spark plug wires is 15 kv. The specified maximum coil voltage is 35 kv.

   Technician A says there may be a high resistance problem in the primary ignition circuit.

   Technician B says the ignition coil may have a secondary insulation leakage defect.
   Who is correct?
   A. A only
   B. B only
   C. Both A and B
   D. Neither A nor B

Hint   *Spark plug wires may be tested with an ohmmeter. If the resistance of any wire exceeds specifications, replace the wire. The normal required secondary coil voltage to fire each spark plug may be checked with an oscilloscope. When the normal required voltage is higher than specified, there is excessive resistance in the secondary circuit, perhaps in spark plug wires or spark plugs. A maximum available coil voltage that is less than specified may be caused by a defective coil, a cracked cap or rotor, or high primary circuit resistance causing low primary current.*

Task 5 Inspect, test, and replace ignition coil(s).

16. In the coil shown in Figure 8-5
   A. 12V are supplied to positive side of each primary winding when the ignition switch is turned on.
   B. the negative side of each primary winding is connected to the secondary winding.
   C. one end of each secondary winding is grounded to the coil case.
   D. the secondary windings are interconnected with each other.

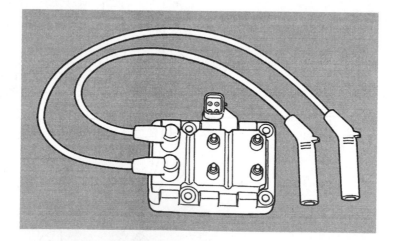

Figure 8-5 EI ignition system coil pack *(Courtesy of Chrysler Corporation)*

Hint   *In a coil used with electronic distributor ignition the primary winding is connected to the primary terminals. The ends of the secondary winding is connected from the coil tower to one of the primary terminals. On some coils one end of the secondary winding is connected to ground on the coil frame or case.*

   In a coil pack used with EI systems 12V are supplied to the positive side of each primary winding when the ignition switch is turned on. The negative side of each primary winding is connected to the ignition module or PCM. The ends of each secondary winding are connected to the two spark plug wire terminals on each coil.

Task 6   Check and adjust ignition system timing and timing advance/retard.

17. Technician A says timing adjustments are required on electronic distributorless ignition (EI) systems.
    Technician B says to set the basic timing on some fuel-injected engines with a distributor, the PCM must be in the limp-in mode.
    Who is correct?
    A. A only
    B. B only
    C. Both A and B
    D. Neither A nor B

Hint   *When checking the basic ignition timing on some fuel-injected engines with electronic distributor ignition, a set timing connector must be disconnected to prevent the PCM from supplying spark advance during this operation. Some manufacturers recommend disconnecting the engine coolant temperature sensor to place the PCM in the limp-in mode while checking basic ignition timing. Basic timing adjustments are not possible on EI systems.*

Task 7   Inspect, test, and replace ignition system pickup sensor or triggering devices.

18. The measurement in Figure 8-6 is more than specified.
    Technician A says this problem may cause cylinder misfiring.
    Technician B says this problem may cause a no-start condition.
    Who is correct?
    A. A only
    B. B only
    C. Both A and B
    D. Neither A nor B

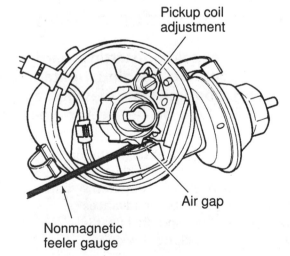

Pickup coil adjustment

Air gap

Nonmagnetic feeler gauge

Figure 8-6 Pick-up coil measurement   (Courtesy of Chrysler Corporation)

Hint   *The pickup coil gap must be measured with a nonmagnetic feeler gauge. If this gap is less than specified, the reluctor high points may strike the pickup coil. This action may cause engine misfiring. Excessive pickup gap may also cause engine misfiring. When this gap is considerably more than specified it may cause a no-start condition.*

Task 8   Inspect, test, and replace ignition control module.

19. When diagnosing a no-start condition, 12V are available at the positive primary coil terminal with the ignition switch on, and the pickup coil tests are satisfactory. When the voltmeter leads are connected from the negative primary coil terminal to ground, the voltmeter always indicates 12V while cranking the engine.
    Technician A says the wire from the negative primary coil terminal to the module may have an open circuit.
    Technician B says the ignition module may have an open circuit.
    Who is correct?
    A. A only
    B. B only
    C. Both A and B
    D. Neither A nor B

Hint   *When a 12V test light connected from the negative primary coil terminal to ground does not flutter while cranking the engine, the ignition module or the pickup coil may be defective. If the ohmmeter tests indicate the pickup is satisfactory, the module is likely defective. A variety of ignition module testers are available to test the module individually.*

Fuel, Air Induction, and Exhaust System Diagnosis and Repair

ASE Tasks, Questions, and Related Information

Task 1   Diagnose hot or cold no-starting, hard starting, poor drivability, incorrect idle speed, poor idle, flooding, hesitation, surging, engine misfire, power loss, stalling, poor mileage, dieseling and emissions problems on vehicles with injection-type or carburetor-type fuel systems; determine needed action.

20. A carbureted engine has a hesitation during low-speed acceleration. The most likely cause of this problem could be
    A. a defective accelerator pump system.
    B. excessive distributor advance.
    C. a higher-than-specified float level.
    D. the choke sticking closed.

Hint   *For complete diagnosis of carburetor problems refer to* Today's Technician Engine Performance Shop Manual *(pp. 248–250), published by Delmar Publishers.*

21. A fuel-injected engine has a higher-than-specified idle speed with the engine at normal operating temperature.
    Technician A says the TPS signal voltage may be higher than specified.
    Technician B says the engine may have an intake manifold vacuum leak.
    Who is correct?
    A. A only
    B. B only
    C. Both A and B
    D. Neither A nor B

22. A customer complains about reduced fuel economy on a fuel-injected engine. The engine runs smoothly, but black smoke is emitted from the tailpipe during engine warm-up. The most likely cause of this problem could be
   A. defective injectors.
   B. defective spark plugs.
   C. higher-than-specified fuel pressure.
   D. defective fuel pump check valve.

Hint   *For complete diagnosis of fuel injection systems refer to* Today's Technician Engine Performance Shop Manual *(pp. 289–292), published by Delmar Publishers.*

Task 2   Perform fuel system pressure and volume tests; determine needed action.

23. When testing the fuel pressure on a port injected engine the fuel pressure is higher than specified.
   Technician A says this problem may be caused by a restricted fuel return line.
   Technician B says this problem may be caused by a fuel pressure regulator that is sticking open.
   Who is correct?
   A. A only
   B. B only
   C. Both A and B
   D. Neither A nor B

Hint   *When performing fuel system pressure and volume tests on a throttle body injection system, the pressure gauge must be connected at the throttle body fuel inlet fitting. On port injected engines connect the fuel pressure gauge to the Schrader valve on the fuel rail. Always relieve the fuel system pressure before attempting to connect the pressure gauge. With the pressure gauge connected, cycle the ignition switch several times or start the engine to read the fuel pressure. Low fuel pressure may be caused by a defective pump, restricted fuel filter or fuel line, or a fuel pressure regulator that is sticking open. High fuel pressure may be caused by a restricted fuel return line or a fuel pressure regulator that is sticking closed.*

Task 3   Inspect fuel tank, tank filter, and gas cap; inspect and replace fuel lines, fittings, and hoses; check fuel for contaminants and quality.

24. A fuel-injected engine is hard to start hot or cold and has an acceleration stumble and a loss of power.
   Technician A says there may be an excessive amount of alcohol mixed with the fuel.
   Technician B says the fuel line from the tank to the filter may be restricted.
   Who is correct?
   A. A only
   B. B only
   C. Both A and B
   D. Neither A nor B

Hint   *At present most fuel-injected engines will operate satisfactorily on 10 percent alcohol mixed with the gasoline. Excessive amounts of alcohol mixed with the gasoline causes a lean air-fuel ratio, hard starting, loss of power, and a hesitation during acceleration. The fuel systems in flex-fuel vehicles are designed to operate on higher percentages of alcohol. Test equipment is available to test the percentage of alcohol mixed with gasoline. A 100 milliliter (mL) cylinder may be filled with 90 mL of gasoline and 10 mL of water. Place a stopper in the cylinder and shake the contents. Remove the stopper to relieve any pressure and wait 5 minutes. If the water content is now 15 percent, there was 5 percent alcohol in the fuel.*

Task 4   Inspect, test, and replace mechanical and electrical fuel pumps and pump control systems; inspect, service, and replace fuel filters.

25. A fuel-injected engine will not start and the fuel pump pressure is zero. When the FP and B+ terminals are connected with a jumper wire the engine starts and runs normally (Figure 8-7). Once the engine starts, the jumper wire may be disconnected and the engine operation is unchanged. The cause of this problem could be
   A. an open circuit in the main relay winding.
   B. a blown 7.5 A IGN fuse.
   C. an open circuit at the B+ fuel pump relay terminal.
   D. a defective circuit opening relay.

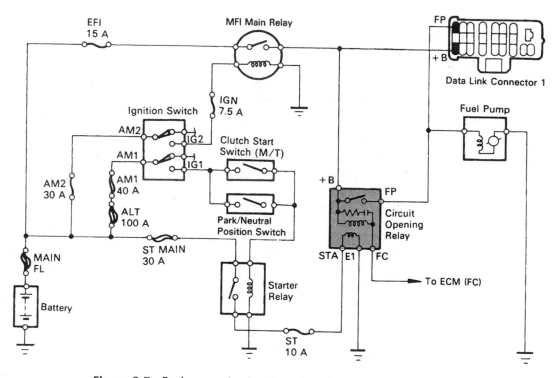

Figure 8-7   Fuel pump circuit   (*Courtesy of Toyota Motor Corporation*)

Hint   *Mechanical fuel pumps should be tested for pressure and volume. A tester is connected at the carburetor inlet fitting to perform these tests. Electric fuel pumps on fuel-injected engines should also be tested for pressure and volume. On a throttle-body-injection system, the fuel pump tester is connected at the throttle body inlet fuel line. In port-fuel-injection systems, a Schrader valve is located on the fuel rail for test equipment connection. On most fuel-injected engines voltage is supplied through a computer-operated relay to the fuel pump. In these systems the computer shuts off the fuel pump if the ignition switch is on for 2 seconds and the engine is not cranked or started.*

Task 5   Inspect, test, and repair or replace fuel pressure regulation system and components of injection-type fuel systems.

26. A fuel-injected vehicle has reduced fuel economy. The engine starts easily, runs smoothly, and has normal power.
   Technician A says the fuel return line may be restricted.
   Technician B says the fuel pump check valve may be leaking.
   Who is correct?
   A. A only
   B. B only
   C. Both A and B
   D. Neither A nor B

Hint   *Higher-than-specified fuel pressure may be caused by a sticking pressure regulator or a restricted return fuel line. A vacuum leak in the hose connected to the fuel pressure regulator on a port-fuel-injected engine causes higher-than-specified fuel pressure at lower engine speeds. High fuel pressure causes a rich air-fuel ratio and reduced fuel economy.*

Low fuel pump pressure causes a loss of power and acceleration stumbles. A defective fuel pump, or a restricted in-tank or in-line filter, causes low fuel pump pressure.

Task 6   **Inspect, test, adjust, and repair or replace cold enrichment system components.**

27. A port-fuel-injected engine with a cold start injector has a rough idle problem and black smoke is emitted from the tail pipe (Figure 8-8). When the electrical connector is disconnected from the cold start injector the engine runs normally. The cause of this problem could be

A. a defective thermo-time switch.
B. a defective PCM.
C. a defective engine coolant temperature sensor.
D. a defective air charge temperature sensor.

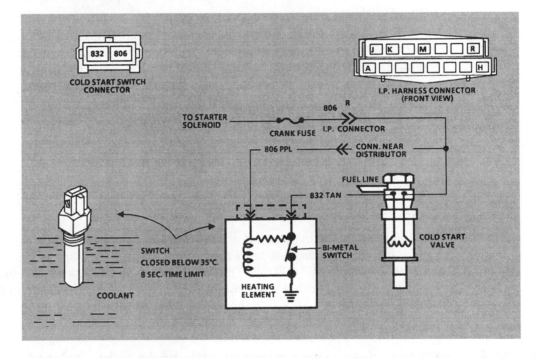

Figure 8-8   Cold start injector circuit *(Courtesy of Chevrolet Motor Division, General Motors Corporation)*

Hint   *Some port-fuel-injected engines have a cold start enrichment valve. This valve is operated by a thermo-time switch that senses engine coolant temperature. The thermo-time switch contains a set of contacts and a bimetal switch. When the coolant temperature is below 95°F (35°C) the closed thermo-time switch contacts supply voltage to the cold start enrichment valve. Under this condition the cold enrichment valve is open and fuel is discharged from this valve into the intake manifold. The bimetal switch action in the thermo-time switch allows the switch contacts to remain closed for a maximum of 8 seconds.*

Task 7   Inspect, test, adjust, and replace acceleration enrichment system components.

28. A throttle body injection system has an acceleration stumble.
    Technician A says the throttle position sensor may be defective.
    Technician B says the fuel pressure regulator may be defective.
    Who is correct?

    A. A only
    B. B only
    C. Both A and B
    D. Neither A nor B

Hint   *The throttle position sensor (TPS) informs the computer regarding the amount and speed of throttle opening. A defective TPS may cause an acceleration stumble. With the ignition switch on the engine not running, connect a voltmeter to the TPS signal wire. The meter should indicate a smooth, steady voltage increase from about 1V to 4.5V as the throttle is opened from the idle position. Low fuel pressure resulting from a defective pressure regulator or fuel pump also causes an acceleration stumble.*

Task 8   Inspect, test, and replace deceleration fuel reduction or shut off system components.

29. A port-injected engine with a deceleration fuel reduction feature in the PCM has a reduced fuel economy complaint. When the emissions are tested this engine also has high hydrocarbon (HC) and (CO) emissions only during deceleration.
    Technician A says to listen to the injectors with a stethoscope as the engine is decelerated.
    Technician B says to test the fuel pressure at idle and 2,500 rpm.
    Who is correct?

    A. A only
    B. B only
    C. Both A and B
    D. Neither A nor B

Hint   *On many engines with deceleration fuel reduction systems the PCM stops grounding the injectors in a specific rpm range during deceleration. Place a stethoscope pickup on any injector and accelerate the engine to 3,000 rpm. Decelerate the engine and listen to the injector clicking. The injector should stop clicking momentarily during deceleration if the deceleration fuel reduction system is operating normally. If the deceleration fuel reduction system is not operating, fuel economy will be reduced while HC and CO emissions are increased.*

Task 9   Remove, clean, and replace throttle body; adjust related linkages.

30. A throttle body injection system has a rough idle and stalling problem when the engine is at normal operating temperature.
    Technician A says the throttle body may require cleaning in the throttle bore area.
    Technician B says the PCV valve may be stuck in the open position.
    Who is correct?

    A. A only
    B. B only
    C. Both A and B
    D. Neither A nor B

Hint   *The throttle body may be cleaned with an approved cleaner. Excessive carbon buildup in the throttle bore area may cause rough idle operation and stalling. A PCV valve that is stuck open causes a similar action to an intake manifold vacuum leak. The PCM senses the extra air entering the engine and supplies more fuel to go with the air, which increases idle speed.*

Task 10   Inspect, test, clean, and replace fuel injectors.

31. During the injector test in Figure 8-9, five of the injectors provide a pressure decrease of 20 psi (138 kPa) and the injector for number 3 cylinder has a 10 psi (69 kPa) decrease. These test results indicate
    A. the plunger in the injector for number 3 cylinder is sticking open.
    B. the orifice in the injector for number 3 cylinder is restricted.
    C. all the injectors are in satisfactory condition.
    D. lower fuel pressure is supplied to the number 3 injector.

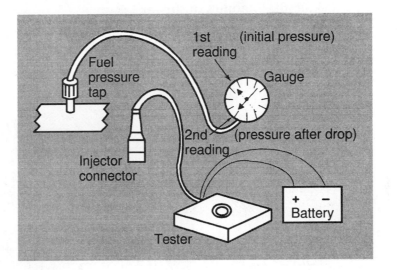

Figure 8-9   Injector test procedure *(Courtesy of Oldsmobile Division, General Motors Corporation)*

Hint   *Defective injectors may cause rough idle operation, stalling, or acceleration stumbles. During an injector balance test a fuel pressure gauge is connected to the Schrader valve in the fuel rail. The injector tester opens each injector for a specific length of time, and the pressure on the fuel gauge should drop the specified amount on each injector. When the pressure drop on an injector is more than specified, the injector plunger is sticking open. If the pressure drop on an injector is less than specified, the injector orifice is restricted.*

Some injectors may be cleaned with a pressurized container of injector cleaner connected to the Schrader valve on the fuel rail. The fuel pump must be disabled and the fuel return line plugged while cleaning injectors.

Task 11   Inspect, service, and repair or replace air filtration system components.

Task 12   Inspect, clean, or replace throttle body mounting plates, air induction system, intake manifold, and gaskets.

32. An engine has a rough idle complaint. A propane cylinder with a precision valve and a hose is used to check for intake vacuum leaks. When propane is charged near the injector in the number 4 cylinder intake port, the engine speed changes and the engine runs smoother.
    Technician A says the intake manifold gasket is leaking.
    Technician B says the lower injector O-ring is leaking.
    Who is correct?
    A. A only
    B. B only
    C. Both A and B
    D. Neither A nor B

Hint   *Intake manifold gaskets and throttle body mounting plates may be checked for leaks with a propane cylinder equipped with a precision metering valve and hose. This equipment is used for the propane enrichment method of adjusting idle mixture screws on carbureted engines. When propane is discharged from the hose on the propane cylinder near an intake manifold vacuum leak, the engine runs smoother and the engine speed changes.*

Task 13   Check/adjust idle speed and fuel mixture.

33. The throttle body adjustment in Figure 8-10 is
    A. a minimum air rate adjustment.
    B. an idle mixture adjustment.
    C. fast idle speed adjustment.
    D. slow idle speed adjustment.

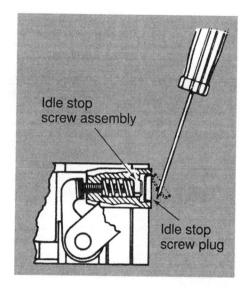

Figure 8-10   Throttle body adjustment *(Courtesy of Chevrolet Motor Division, General Motors Corporation)*

Hint   *Some throttle body assemblies have a minimum air rate adjustment. This adjustment is only required if the throttle body has been replaced. A metal plug must be removed to access the minimum air rate screw. The idle speed control motor must be in the closed position or the air passage through this motor must be plugged before the minimum air rate adjustment. Adjust the minimum air rate screw to the specified rpm.*

Task 14   Remove, clean, inspect/test, and repair or replace vacuum and electrical components and connections of fuel systems.

34. Technician A says the component in Figure 8-11 allows the throttle to close gradually on deceleration.
    Technician B says the component in Figure 8-11 holds the throttle in the specified idle position.
    Who is correct?
    A. A only
    B. B only
    C. Both A and B
    D. Neither A nor B

Hint   *Some fuel-injected engines have a vacuum-operated throttle opener. This opener allows the throttle to close gradually on deceleration to prevent stalling. With the throttle opener vacuum hose removed and plugged, the throttle linkage should strike the opener stem at 1,300 to 1,500 rpm. The opener stem may be adjusted to obtain this rpm.*

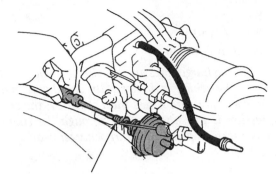

Throttle opener adjusting screw

Figure 8-11   Vacuum actuator *(Courtesy of Toyota Motor Corporation)*

Task 15   **Inspect, service, and replace exhaust manifold, exhaust pipes, mufflers, resonators, tail pipes, and heat shields.**

35. A vehicle has a severe loss of power and a top speed of 55 mph (88 kmh). The engine runs smoothly and does not misfire. The cause of this problem could be
    A. a defective spark plug.
    B. a defective spark plug wire.
    C. defective fuel injectors.
    D. a restricted exhaust pipe.

Hint   *A restricted exhaust system, or a restricted air intake system causes a loss of power and reduced maximum vehicle speed. When this problem is present the engine runs smoothly and does not misfire.*

Task 16   **Perform exhaust back-pressure test; determine needed action.**

36. A port fuel-injected engine runs smoothly at idle speed but misfires during acceleration. The most likely cause of this problem is:
    A. an exhaust gas recirculation (EGR) valve that is stuck open.
    B. a restricted catalytic converter.
    C. dripping fuel injectors.
    D. excessive resistance in a spark plug wire.

Hint   *Excessive exhaust back-pressure may be caused by a restricted exhaust pipe, catalytic converter, or muffler. If the exhaust back-pressure is excessive, engine power and maximum vehicle speed are reduced, but the engine does not misfire. Connect a vacuum gauge to the intake manifold to check for a restricted exhaust system. With the engine idling the manifold vacuum should be 16 to 21 in. hg. (110.32 to 144.79 kPa). When the engine is accelerated to 2,000 rpm, the vacuum should drop momentarily and then recover to 16 to 21 psi (110.32 to 144 kPa). Hold the engine speed at 2,000 rpm. If the vacuum drops below 16 in. hg. (110.32 kPa) after 3 minutes, the exhaust system is restricted.*

Task 17   **Test the operation of turbocharger/supercharger systems; determine needed action.**

37. While testing a turbocharger the maximum boost pressure is 4 psi (27.5 kPa), and the specified boost pressure is 9 psi (62 kPa).
    Technician A says the engine compression may be lower than specified.
    Technician B says the wastegate may be sticking closed.
    Who is correct?

   A. A only
   B. B only
   C. Both A and B
   D. Neither A nor B

Hint   *During a turbocharger pressure boost test a vacuum pressure gauge is connected to the intake manifold, and the vehicle is driven at the speed and engine rpm specified by the vehicle manufacturer. If the boost pressure is less than specified, the engine compression should be tested. When the compression is low, the air flow through the engine and boost pressure are reduced. If the wastegate is sticking open, or the turbocharger bearings are damaged, turbocharger boost pressure is reduced.*

Task 18   Remove, clean, inspect, and repair or replaceturbocharger/supercharger system components.

   38. The vanes on a turbocharger compressor wheel are severely pitted. The cause of this problem could be
   A. a leak in the air intake system.
   B. partially seized turbocharger bearings.
   C. excessive turbocharger shaft end play.
   D. reduced turbocharger coolant circulation.

Hint   *If the turbocharger compressor wheel or turbine wheel housings are scored, the vanes on these wheels have been striking these housings. This problem is caused by excessive turbocharger shaft end play, and the turbocharger or center housing assembly must be replaced. If the compressor wheel vanes are severely pitted, the air intake is leaking.*

Task 19   Identify the causes of turbocharger/supercharger system failure; determine needed action.

   39. A turbocharger has been replaced three times because of bearing failure.
   Technician A says the engine oil may be contaminated from infrequent oil and filter changes.
   Technician B says the coolant passages in the turbocharger housing may be restricted.
   Who is correct?
   A. A only
   B. B only
   C. Both A and B
   D. Neither A nor B

Hint   *The turbocharger bearings are supplied with oil directly from the main oil gallery. If the engine oil is contaminated because of the lack of oil and filter changes, turbocharger bearing life is shortened. In many turbochargers engine coolant is circulated through the turbocharger housing to cool the turbocharger bearings. If the turbocharger housing and bearings become too hot because of inadequate coolant circulation, the oil actually burns in the bearings after a hot engine shut down. This action greatly reduces bearing life.*

Emission Control Systems Diagnosis and Repair, Positive Crankcase Ventilation

ASE Tasks, Questions, and Related Information

Task 1 Diagnose the cause(s) of emissions problems resulting from the failure of the positive crankcase ventilation (PCV) system.

40. With the PCV valve positioned as shown in Figure 8-12, the engine is
    A. operating at wide open throttle.
    B. operating at half open throttle.
    C. backfiring on acceleration.
    D. operating at idle speed.

Idling or deceleration

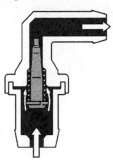

Figure 8-12   PCV valve *(Courtesy of Toyota Motor Corporation)*

Hint   *When the engine is not running, or during an engine backfire, the PCV valve plunger is seated on the housing, and the passage through the valve is closed. If the engine is idling, the high intake manifold vacuum holds the tapered valve plunger nearly closed. When the throttle is opened and the manifold vacuum decreases, the tapered plunger gradually moves downward to provide more PCV valve opening.*

If the PCV valve is stuck closed, the air-fuel ratio is richer and HC and CO emissions are higher. A PCV valve stuck in the open position may cause a leaner air-fuel ratio and increased idle rpm.

Task 2 Test the operation of positive crankcase ventilation (PCV) systems.

41. The inside of the air cleaner is contaminated with engine oil.
    Technician A says the PCV valve clean air filter in the air cleaner may be plugged.
    Technician B says the hose from the PCV valve to the intake manifold may be severely restricted.
    Who is correct?
    A. A only
    B. B only
    C. Both A and B
    D. Neither A nor B

Hint   *If the inside of the air cleaner is contaminated with engine oil, the engine may have excessive blowby, or the PCV valve and connecting hose may be restricted. When the PCV clean air filter in the air cleaner is plugged, higher vacuum is built up in the engine. This action may damage engine gaskets and pull dirt particles past some of the engine gaskets, such as rocker cover or oil pan gaskets, into the engine.*

Task 3   Inspect, service, and replace positive crankcase ventilation (PCV) filter/ breather cap, valve, tubes, orifices, and hoses.

42. An oil filler cap is loose on the rocker arm cover. The engine is in satisfactory condition with the specified compression. This problem could
   A. allow the PCV system to pull dirt particles into the engine.
   B. reduce the amount of flow through the PCV valve.
   C. increase the amount of flow through the PCV valve.
   D. cause oil to be blown out around the oil filler cap.

Hint   *Some manufacturers recommend removing the PCV valve and shaking it to determine if the valve is faulty. While shaking the valve, the plunger in the valve should rattle if the valve is satisfactory. Other manufacturers recommend blowing through it with a length of hose connected to the valve. Air should pass through the valve easily in one direction, but the valve should block air in the opposite direction.*

Emissions Control Systems Diagnosis and Repair, Exhaust Gas Recirculation

ASE Tasks, Questions, and Related Information

Task 1   Diagnose the cause(s) or emissions problems caused by failure of the exhaust gas recirculation (EGR) system.

43. A vehicle fails an emissions test for NOX emissions. The cause of this emission failure could be
   A. an engine thermostat that is stuck open.
   B. a defective engine coolant temperature sensor.
   C. low compression on one cylinder.
   D. restricted exhaust passages under the EGR valve.

Hint   *Oxides or nitrogen emissions are caused by oxygen and nitrogen combining at high combustion chamber temperatures. The EGR valve recirculates some exhaust into the intake manifold. Since there is very little oxygen left in the exhaust, this exhaust does not burn in the combustion chambers and combustion chamber temperatures are lowered. This action reduces NOX emissions.*

Task 2   Test the operation of the exhaust gas recirculation (EGR) system.

44. While diagnosing a positive backpressure EGR valve.
   Technician A says when 18 in. Hg. is supplied to the valve with the engine idling, the valve should open.
   Technician B says when the EGR valve is opened at idle the engine should slow down at least 150 rpm.
   Who is correct?
   A. A only
   B. B only
   C. Both A and B
   D. Neither A nor B

Hint   *When 18 in. Hg. are supplied to a conventional EGR valve with the engine idling, the valve should open and the engine should slow down 150 rpm or stall. If this action does not take place, the EGR valve is sticking or the exhaust passages are plugged with carbon.*

Since a positive backpressure EGR valve contains a vacuum bleed valve, this type of valve does not open when vacuum is supplied to the valve with the engine idling. When the engine speed is increased to 2,000 rpm, exhaust pressure closes the bleed valve. Under this condition the valve should open when vacuum is supplied to the valve vacuum port.

Task 3   Inspect, test, service, and replace valve, EGR tubing, and exhaust passages of exhaust gas recirculation (EGR) systems.

45. When diagnosing the EGR valve in Figure 8-13:
   A. the bleed valve normally is closed with the engine operating at idle speed.
   B. the bleed valve is opened by negative pressure pulses in the exhaust system.
   C. when vacuum is supplied to the valve with the engine stopped, the valve should remain closed.
   D. the bleed valve should be open with the engine operating at 2,000 rpm.

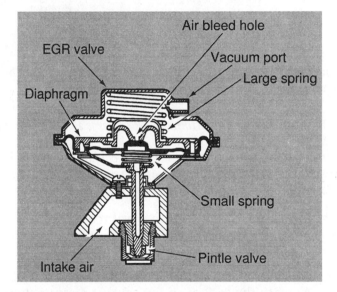

Figure 8-13 Exhaust gas recirculation (EGR) valve *(Courtesy of Chevrolet Motor Division, General Motors Corporation)*

Hint   *Some engines have a negative backpressure EGR valve in which the bleed valve is normally closed. When the engine is idling or operating at low speed, negative pressure pulses in the exhaust system pull the bleed valve open. Under this condition vacuum supplied to the valve is bled off, and the valve remains closed. When the engine speed increases the positive exhaust pressure pulses are closer together and the negative exhaust pressure pulses reduced. This action allows the bleed valve to close and the vacuum supplied to the EGR valve diaphragm pulls the valve open.*

Task 4   Inspect, test, service, and replace vacuum/pressure controls, filters, and hoses of exhaust gas recirculation (EGR) systems.

46. While diagnosing the computer-controlled EGR valve system in Figure 8-14 (next page):
   Technician A says the EGR valve should be open when the engine coolant is cold and the TPS sensor indicates one-half open throttle.
   Technician B says the EGR valve should be open with the engine at normal operating temperature and the throttle wide open.
   Who is correct?
   A. A only
   B. B only
   C. Both A and B
   D. Neither A nor B

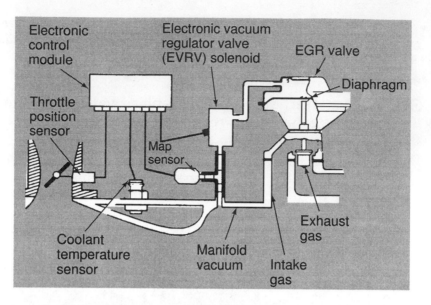

Figure 8-14   Computer-controlled EGR system *(Courtesy of Chevrolet Motor Division, General Motors Corporation)*

Hint   *In most computer-controlled EGR systems the PCM operates a solenoid that supplies vacuum to the EGR valve. Since a cold engine does not have high NOX emissions, the PCM does not energize the EGR solenoid when the engine coolant is cold. When the EGR valve is open, the exhaust recirculated into the intake manifold reduces engine power to some extent. Therefore, when the TPS sensor indicates a wide open throttle, the PCM does not energize the EGR solenoid.*

Task 5   Inspect, test, repair, and replace electrical/electronic sensors, controls, and wiring of exhaust gas recirculation (EGR) systems.

47. While diagnosing the computer-controlled EGR system with dual pressure feedback electronic (DPFE) sensor in Figure 8-15
    A. with the EGR open the exhaust pressure should be the same on each side of the EGR orifice.
    B. the PCM pulses the EGR solenoid on and off to supply the proper vacuum to the EGR valve.
    C. the dual pressure feedback electronic sensor sends a digital signal to the PCM.
    D. the maximum EGR valve opening should occur at 30 to 35 mph with a steady throttle opening.

Hint   *In the computer-controlled EGR system with a DPFE sensor, the PCM pulses the EGR solenoid on and off to supply a precise vacuum to the EGR valve. The DPFE sensor senses the EGR flow from the pressure difference across the EGR orifice. This sensor sends an analog voltage signal to the PCM in relation to the EGR flow. The other input sensors inform the PCM regarding the EGR flow required by the engine. If the actual EGR flow does not match the required flow, the PCM changes the solenoid operation to provide the necessary EGR flow.*

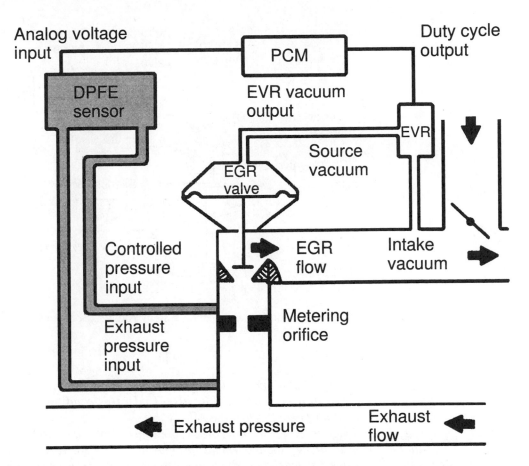

Figure 8-15   Computer-controlled EGR system with DPFE sensor   *(Courtesy of Ford Motor Company)*

Emissions Control Systems Diagnosis and Repair, Exhaust Gas Treatment

ASE Tasks, Questions, and Related Information

Task   1   **Diagnose the cause(s) of emissions problems resulting from failure of the air injection or catalytic converter systems.**

48. A vehicle with a secondary air injection system has high HC and CO emissions and the oxygen sensor voltage is always low (Figure 8-16, next page).
    Technician A says the vacuum hose connected to the air diverter (AIRD) valve may be leaking.
    Technician B says the air pump may be pumping air into the exhaust ports with the engine warmed up.
    Who is correct?
    A. A only
    B. B only
    C. Both A and B
    D. Neither A nor B

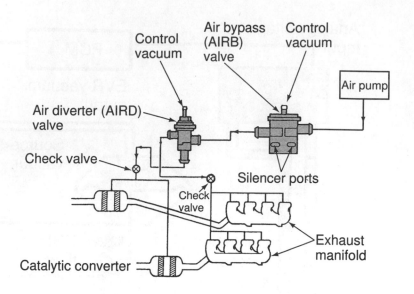

Figure 8-16 Secondary air injection system *(Courtesy of Ford Motor Company)*

Hint   *When the engine is started the PCM does not energize the AIRD or AIRB solenoids and there is no vacuum supplied to the AIRD or AIRB valve. Under this condition air from the air pump is exhausted to the atmosphere through the AIRB valve.*

After the engine runs for a brief time, the PCM energizes both the AIRD and AIRB solenoids, and vacuum is supplied through these solenoids to the AIRD and AIRB valves. Under this condition air from the air pump is directed to the exhaust ports to reduce HC emissions during engine warm-up.

When the engine coolant reaches normal operating temperature, the PCM deenergizes the AIRD solenoid. This action shuts off the vacuum supplied to the AIRD valve, and this valve plunger moves so air is directed to the catalytic converters. If the air pump continues to pump air into the exhaust ports with a warm engine, the extra oxygen in the exhaust stream causes low O_2 sensor voltage. This voltage signal causes the PCM to increase injector pulse width and provide a rich air-fuel ratio. Under this condition HC and CO emissions are high.

Task 2   Test the operation of air injection systems.

49. Refer to Figure 8-16 when answering this question.

    The air from the secondary air injection system is always bypassed to the atmosphere at the AIRB valve. This problem may be caused by all of these defects EXCEPT

    A. a sticking AIRB valve.

    B. a sticking AIRB solenoid plunger.

    C. a sticking AIRD valve.

    D. a leak in the AIRB vacuum hose.

Hint   *If the air flow from the pump is continually exhausted to the atmosphere at the AIRB valve, this valve plunger may be sticking. This problem may also be caused by the plunger in the AIRB solenoid sticking in the closed position. If the vacuum hose connected to the AIRB valve is leaking, the vacuum may not pull the valve plunger upward, and the air flow is exhausted to the atmosphere.*

Task 3   **Inspect, test, service, and replace mechanical components of air injection systems.**

50. The hose connected from the check valve to AIRD valve is severely burned in a secondary air injection system. The cause of this problem could be
    A. a defective check valve.
    B. overheated catalytic converters.
    C. excessive HC emissions.
    D. a defective AIRD valve.

Hint   *The check valves in the secondary air injection system prevent exhaust from entering the system hoses. A burned system hose indicates the check valve is allowing exhaust into the secondary air injection system.*

Task 4   **Inspect, and replace electrical/electronically operated components and circuits of air injection systems.**

51. When the ignition switch is turned on, 12V are supplied to one terminal on the AIRD and AIRB solenoids. The air is bypassed normally to the atmosphere for a few seconds when the engine is started. When this bypass mode is completed, the air flow is always directed downstream. The voltage on the PCM side of the AIRB solenoid is .2V, and the voltage on the PCM side of the AIRD solenoid is 12V.
    Technician A says the wire from the AIRD solenoid to the PCM may have an open circuit.
    Technician B says the PCM may not be grounding the circuit from the AIRD solenoid.
    Who is correct?
    A. A only
    B. B only
    C. Both A and B
    D. Neither A nor B

Hint   *If the air flow is always directed downstream when the bypass mode is completed, the AIRD valve or control circuit is not operating properly. If there is no vacuum supplied to the AIRD valve, check the vacuum hoses, AIRD solenoid, and connecting wires. In the upstream mode both solenoids should be energized, and each solenoid should have 12V at one terminal and a very low voltage at the other terminal. When the voltage is high at the PCM side of the AIRD solenoid, the wire from this solenoid to the PCM is open, or the PCM is not providing a ground for the wire.*

Task 5   **Inspect, test, and replace components of catalytic converter systems.**

52. When diagnosing a catalytic converter with a digital pyrometer, if the converter is operating properly
    A. the converter outlet should be 100°F (55°C) hotter than the inlet.
    B. the converter inlet and outlet should be the same temperature.
    C. the converter inlet should be 50°F (28°C) hotter than the outlet.
    D. the converter outlet should be 50°F (28°C) hotter than the inlet.

Hint   If the catalytic converter rattles when tapped with a soft hammer, the internal components are loose. When this condition is present, the converter should be replaced.
    A digital pyrometer may be used to check the catalytic converter. If the converter is operating properly, the converter outlet temperature should be 100°F (55°C) hotter compared to the inlet temperature.

Emissions Control Systems Diagnosis and Repair, Evaporative Emissions Controls

ASE Tasks, Questions, and Related Information

Task 1 **Diagnose the cause(s) of emissions problems resulting from failure of evaporative emissions control systems.**

53. A vehicle failed an emissions test for HC and CO at idle speed, and the engine has a rough idle condition (Figure 8-17).

    Technician A says the EVAP solenoid plunger may be stuck open.

    Technician B says the wire from the EVAP solenoid to the PCM may be open.

    Who is correct?

    A. A only
    B. B only
    C. Both A and B
    D. Neither A nor B

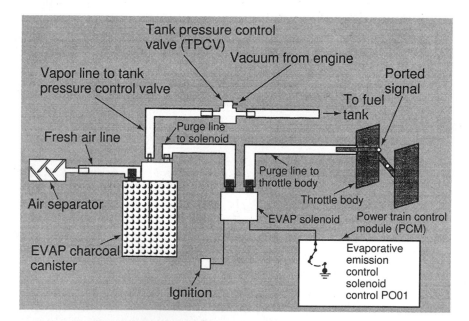

Figure 8-17 Evaporative control system *(Courtesy of Cadillac Motor Car Division, General Motors Corporation)*

Hint *The evaporative (EVAP) emission system prevents fuel tank vapors from escaping to the atmosphere. A hose is connected from the top of the fuel tank to a charcoal canister under the hood. Fuel vapors from the tank flow into the canister where they are absorbed in the charcoal. A tank pressure control or rollover valve usually is connected in this hose. The fuel tank has a filler cap containing pressure and vacuum valves. A purge hose containing a computer-controlled solenoid is connected to the throttle body just above the throttle.*

When the engine is idling below a specific rpm, the purge solenoid remains closed. At a specific engine rpm, the PCM energizes the purge solenoid, and manifold vacuum pulls fuel vapors out of the canister into the intake manifold. Some PCMs energize the purge solenoid at a specific vehicle speed.

Task 2   Test the operation of evaporative emissions control systems.

54. Gasoline is dripping out of the charcoal canister (Figure 8-18).
    Technician A says the canister filter may be plugged.
    Technician B says the hose from the tank to the canister is plugged.
    Who is correct?
    A. A only
    B. B only
    C. Both A and B
    D. Neither A nor B

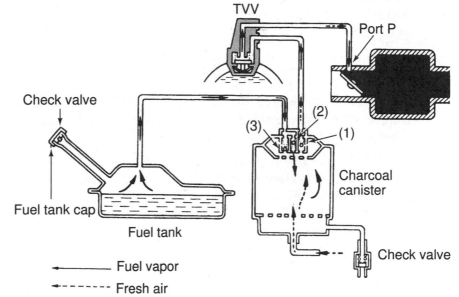

Figure 8-18   EVAP system with thermal vacuum valve (TVV)   *(Courtesy Toyota Motor Corporation)*

Hint   *Some EVAP systems are controlled by a TVV connected in the purge hose from the canister to the intake manifold. When the engine coolant is cold, the TVV remains closed and no canister purging occurs. The TVV opens at a specific coolant temperature, and vacuum is supplied through the TVV to the canister to provide purging action. An air intake and filter are located in the bottom of the canister. If this filter becomes plugged, air does not enter the canister during the purge mode, and higher vacuum is present in the canister. Since air cannot enter the canister, fuel vapors may condense in the canister resulting in gasoline dripping from the canister.*

Task 3   Inspect and replace components and hoses of evaporative emissions control systems.

55. While diagnosing an inoperative EVAP system with a scan tool, the tool indicates the purge solenoid is on at the specified rpm, and there are no diagnostic trouble codes (DTCs) in the PCM memory. There is no vacuum to the canister. This problem may be caused by all of these defects EXCEPT
    A. an open winding in the purge solenoid.
    B. a purge solenoid plunger stuck in the closed position.
    C. a plugged hose from the canister to the intake manifold.
    D. a plugged purge port in the intake manifold.

Hint   *A scan tool may be used to diagnose the EVAP system. If there is a defect in the purge solenoid winding or connecting wires, a DTC usually is set in the PCM memory. At idle speed the scan tester should indicate the purge solenoid is off. With the engine at normal operating temperature and running at the specified rpm, the scan tool should indicate the purge solenoid is on. The scan*

tool only indicates the command from the PCM to the solenoid. The PCM may send an on com-
mand to the purge solenoid, but the solenoid may remain inoperative because of such defects as a
sticking plunger.

Computerized Engine Controls Diagnosis and Repair

ASE Tasks, Questions, and Related Information

Task 1 **Diagnose the causes of emissions or driveability problems resulting from failure of computerized engine controls with no diagnostic trouble codes started.**

56. An vehicle fails an emission test for HC emissions at idle and 2,500 rpm, and the engine has an acceleration stumble. The O_2 sensor voltage is .1V to .3V on the scan tester, and the injector pulse width is more than specified. There are no DTCs in the PCM. When propane is directed into the air intake, the O_2 sensor voltage immediately increases, and the injector pulse width decreases. The most likely cause of this problem is
    A. a defective O_2 sensor.
    B. a defective PCM.
    C. low fuel pump pressure.
    D. an opening in the O_2 sensor signal wire.

Hint   _High HC emissions with normal or low CO emissions may be caused by a lean misfire condi-_
tion. Since the O_2 sensor voltage increases when propane is dispersed into the air intake, this sen-
sor is responding. The PCM is supplying more injector pulse width in response to the low O_2
sensor voltage, so the PCM is satisfactory. If the O_2 sensor signal wire is open, this sensor dis-
plays a 0V signal on the scan tester.

A lean air-fuel ratio and lean misfire condition plus acceleration stumbles may be caused by
low fuel pump pressure.

Task 2 **Retrieve and record stored diagnostic trouble codes.**

57. Technician A says on some vehicles the diagnostic trouble codes (DTCs) may be obtained by cycling the ignition switch on and off three times in a ten second interval. Technician B says on some vehicles the DTCs are read from the flashes of the malfunction indicator (MIL) light.
    Who is correct?
    A. A only
    B. B only
    C. Both A and B
    D. Neither A nor B

Hint   _On some vehicles the diagnostic trouble codes (DTCs) are obtained by connecting a jumper wire_
to the proper terminals in the data link connector (DLC), and then turning on the ignition switch.
On other vehicles the ignition switch is cycled on and off three times in a five second interval to
signal the powertrain control module to enter the diagnostic mode and provide DTCs. On some
vehicles the DTCs may be read by counting the flashes of the malfunction indicator light (MIL) in
the instrument panel. The MIL light flashes each code three times on some vehicles, and codes are
in numerical order. The DTCs may also be read on a scan tester connected to the DLC.

58. While reading diagnostic trouble codes (DTCs) from the flashes of the malfunction indictor (MIL) light
   A. code 42 is provided before code 55.
   B. on some vehicles each code is repeated four times.
   C. the DTCs should be obtained with the engine cold.
   D. a DTC indicates a problem in a specific component.

Hint   *The engine must be at normal operating temperature before attempting to read the diagnostic trouble codes (DTCs). On some vehicles the DTCs are obtained by connecting a jumper wire to the proper terminals in the data link connector (DLC), and then turning on the ignition switch. On other vehicles the ignition switch is cycled on and off three times in a five second interval to signal the powertrain control module to enter the diagnostic mode and provide DTCs. On some vehicles the DTCs may be read by counting the flashes of the malfunction indicator light (MIL) in the instrument panel. The MIL light flashes each code three times on some vehicles, and codes are in numerical order. The DTCs may also be read on a scan tool connected to the DLC. A DTC indicates a defect in a certain area. In many cases further testing with a voltmeter or ohmmeter is required to locate the exact cause of the problem.*

Task 3   Diagnose the causes of emissions or driveability problems resulting from failure of computerized engine controls with stored diagnostic trouble codes.

59. A vehicle fails an emissions test for high NOX. The engine has an EGR valve position sensor. The engine has a rough idle problem, and a diagnostic trouble code (DTC) representing the EGR valve is stored in the powertrain control module (PCM) memory.
   Technician A says the EGR valve may be stuck open.
   Technician B says the EGR passages may be restricted with carbon.
   Who is correct?
   A. A only
   B. B only
   C. Both A and B
   D. Neither A nor B

Hint   *The high HC emissions and normal CO and NOX emissions usually indicate a lean misfire condition. If the high HC emissions only occur at idle and low speed, the misfire must be occurring only at this speed. Defective injectors, low fuel pump pressure, or a defective ECT sensor may cause a lean air-fuel ratio at idle and 2,500 rpm.*

   If the exhaust gas recirculation (EGR) valve is stuck open the exhaust flow into the intake manifold dilutes the mixture at idle and low speed, resulting in a lean misfire condition. Since the EGR valve is normally open at 2,500 rpm it does not cause a lean misfire at this speed.

60. A vehicle has a poor fuel economy complaint. When tested with a scan tool the O_2 sensor voltage is 0.2V. A diagnostic trouble code (DTC) representing the O_2 sensor is stored in the powertrain control module (PCM) memory. The injector pulse width is more than specified. When propane is dispersed into the air intake, the O_2 sensor voltage and the injector pulse width do not change. The cause of this problem is
   A. a defective PCM.
   B. high fuel pressure.
   C. a defective O_2 sensor.
   D. an intake manifold vacuum leak.

Hint   *If the O_2 is satisfactory the sensor voltage changes when propane is dispersed into the air stream. When the O_2 sensor voltage changes, but the computer does not change the injector pulse width the computer may be defective.*

61. An engine fails an emissions test for high CO and HC. These emissions are high at idle and also at higher speed, and the engine has rough idle and hard starting problems. Fuel pressure is normal, but the fuel pressure decreases after the engine is shut off. The scan tool data indicates an O_2 sensor voltage between 0.5V and 0.8V. A diagnostic trouble code (DTC) indicating a rich air-fuel ratio is stored in the powertrain control module (PCM) memory. The cause of these problems could be

A. a leaking fuel pump check valve.

B. a leaking fuel pressure regulator.

C. dripping injectors.

D. a defective engine coolant temperature (ECT) sensor.

Hint   *Dripping injectors cause a rich air-fuel ratio and rough engine idle. Since fuel also drips out of the injectors after the engine is shut off, the fuel rail is empty when a restart is attempted. This condition causes a longer cranking time before the engine starts.*

62. A fuel injected engine has a rough idle condition, and high HC emissions only at idle and low speeds. A diagnostic trouble code (DTC) indicating a lean air-fuel ratio is stored in the powertrain control module (PCM) memory. Emissions of CO and NOX meet emission standards. At 2,500 engine rpm, the HC emissions are normal. The most likely cause of this problem is

A. defective injectors.

B. an EGR valve stuck open.

C. low fuel pump pressure.

D. defective ECT sensor.

Hint   *The high HC emissions and normal CO and NOX emissions usually indicate a lean misfire condition. Since the high HC emissions only occur at idle and low speed, the misfire must be occurring only at this speed. Defective injectors, low fuel pump pressure, or a defective ECT sensor may cause a lean air-fuel ratio at idle and 2,500 rpm. If the EGR valve is stuck open the exhaust flow into the intake manifold dilutes the mixture at idle and low speed, resulting in a lean misfire condition. Since the EGR valve is normally open at 2,500 rpm it does not cause a lean misfire at this speed.*

63. A vehicle has a secondary air injection system that delivers air to the dual-bed catalytic converter. This vehicle fails an emissions test for HC and CO. Engine operation is normal, but the owner complains about excessive fuel consumption. A diagnostic trouble code (DTC) is stored in the powertrain control module (PCM) memory indicating secondary air injection system air flow is always upstream.

Technician A says the secondary air injection diverter (AIRD) solenoid plunger may be sticking open.

Technician B says the secondary air injection diverter (AIRD) valve may

be sticking.

Who is correct?

A. A only

B. B only

C. Both A and B

D. Neither A nor B

Hint   *The secondary air injection system delivers air to the center of many dual-bed catalytic converters. This air flow is necessary to oxidize HC and CO to CO_2 and H_2O in the rear oxidation catalyst bed.*

Task 4   Inspect, test, adjust, and replace computerized engine control system sensors, powertrain control module (PCM), actuators and circuits.

64. An port injected engine is hard to start after the vehicle is not driven for several hours. When tested the fuel pressure drops off when the engine is shut off. There are no visible external fuel leaks. With the fuel pressure line and return fuel line plugged, the fuel pressure still drops off when the engine is shut off. The cause of this problem could be

A. dripping injectors.

B. a leaking fuel pump check valve.

C. a leaking fuel pressure regulator.

D. a leaking upper injector O-ring.

Hint   *When the ECT sensor is tested with an ohmmeter, the sensor resistance should decrease as the temperature increases. If the resistance of the ECT sensor is lower than specified, the computer thinks the engine is warmer than the actual coolant temperature. Under this condition the PCM provides a leaner air-fuel ratio than the engine requires. Conversely if the ECT sensor resistance is higher than specified, the PCM thinks the coolant is colder than the actual coolant temperature. This action causes the PCM to provide a rich air-fuel ratio. The ECT sensor signal affects many computer outputs such as the EGR valve, torque converter clutch, and EVAP purge solenoid.*

Hard starting after the engine is shut off for several hours may be caused by a leaking fuel pump check valve or pressure regulator. Dripping injectors or a defective ECT sensor may cause this problem, but these defects also affect idle operation and emission levels.

65. When testing MAP sensors

Technician A says many MAP sensors act as a barometric pressure sensor when the ignition switch is turned off.

Technician B says on many MAP sensors the signal voltage should increase when the vacuum at the sensor decreases.

Who is correct?

A. A only

B. B only

C. Both A and B

D. Neither A nor B

Hint   *Many manifold absolute pressure (MAP) sensors act as a barometric pressure sensor when the ignition is turned on or when the throttle is wide open. Many MAP sensors send an analog voltage signal to the PCM in relation to intake manifold vacuum. When the ignition switch is turned on the barometric pressure signal from the MAP sensor to the PCM is usually 4V to 5V. If the engine is idling the MAP sensor signal voltage is approximately 1V. When the throttle is opened and manifold vacuum decreases, the MAP sensor signal voltage increases.*

Some MAP sensors provide an AC voltage signal that is measured in hertz (hz). The signal from these MAP sensors varies from approximately 92 hz at idle to 162 hz at wide open throttle. This type of MAP sensor also acts as a barometric pressure sensor with the ignition switch on.

66. When testing a MAP sensor that produces an analog voltage signal, with 18 in. hg. (40.5 kPa absolute) applied to the sensor, the sensor signal is 1.2V. When 13 in. hg. (57.3 kPa absolute) is supplied to the sensor, the sensor signal should be

A. 1.4V.

B. 1.6V.

C. 2.0V.

D. 2.6V.

Task 5   Use and interpret digital multimeter (DMM) readings.

67. An oxygen (O$_2$) sensor voltage signal should be checked with
    A. a digital ohmmeter.
    B. an analog ohmmeter.
    C. a digital voltmeter.
    D. a milliammeter.

Hint   *Some computer inputs, such as the O$_2$ sensor, produce a low voltage signal and a very low current flow. A digital voltmeter must be used to test the O$_2$ sensor voltage. Since an analog voltmeter draws more current compared to a digital voltmeter, O$_2$ sensor damage occurs when this sensor is tested with an analog voltmeter.*

Task 6   Read and Interpret technical literature (service publications and information).

Task 7   Test, remove, inspect, clean, service, and repair or replace power distribution circuits and connections.

68. In the power distribution circuit (Figure 8-19), with the ignition switch in the accessory (ACCY) position, voltage is supplied to the following fuses
    A. O$_2$ heater and HVAC.
    B. instrument and air BG2.
    C. wiper and radio.
    D. instrument and HVAC.

Hint   *The power distribution circuit is the power and ground circuits from the battery through the ignition switch and fuses to the individual circuits on the vehicle.*

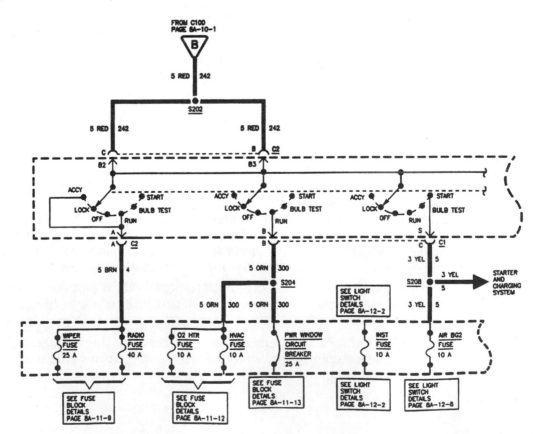

Figure 8-19   Partial power distribution circuit *(Courtesy of Pontiac Motor Division, General Motors Corporation)*

Task 8   **Practice recommended precautions when handling static sensitive devices.**

69. All of these statements are true about installing an EEPROM chip EXCEPT
   A. do not remove the chip from the protective envelope until you are ready to install it.
   B. touch a good ground on the vehicle with your finger prior to installing the chip.
   C. connect a grounding strap from your wrist to a good vehicle ground.
   D. remove any dust from the chip pins with a shop towel prior to installation.

Hint   *Static sensitive components are shipped in an anti-static envelope. This envelope should not be removed before you are ready to install the chip. Touch a good ground on the vehicle, and connect a grounding strap from your wrist or clothing to a vehicle ground, before installing the chip. Do not handle the chip unnecessarily, and do not move around on the vehicle seat when installing the chip.*

Task 9   **Diagnose drivability and emissions problems resulting from failures of interrelated systems (cruise control, security alarms, torque controls, suspension controls, traction control, torque management, A/C, and similar systems).**

70. A torque converter clutch (TCC) is inoperative. The vehicle has a separate PCM and transmission control module (TCM) with interconnected data links. The other transmission functions are normal and the scan tool never indicates a TCC on command from the TCM. Engine drivability is normal.
   Technician A says the data links should be tested with the scan tool.
   Technician B says the ECT sensor signal may not be transmitted from the PCM to the TCM.
   Who is correct?
   A. A only
   B. B only
   C. Both A and B
   D. Neither A nor B

Hint   *Some vehicles have separate computers for engine control, transmission control, ABS and traction control, suspension control, and A/C control. On many vehicles these computers are interconnected by data links. Some input sensor signals are connected to one of these computers and transmitted on the data links to the other computers.*

If the data links are defective, the necessary inputs will not reach the computer or computers that require these inputs. Defective data links may cause malfunctioning of some outputs. For example, the ECT sensor signal is sent from the PCM to the TCM for proper TCC control. If the data links are defective and the ECT signal is not available to the TCM, this computer does not know anything about engine temperature. This condition may cause an inoperative or malfunctioning TCC. On many vehicles the data links may be tested with a scan tester.

Task 10   **Diagnose the causes of emissions or driveability problems resulting from computerized spark controls; determine needed repairs.**

71. An engine has a knock sensor in the right-hand coolant drain plug. With the engine running at 2,000 rpm, the exhaust manifold is tapped above the knock sensor with a small hammer. While observing the timing marks with a timing light, the timing mark does not move.
   Technician A says the knock sensor may be defective.
   Technician B says the engine may have a detonation problem.
   Who is correct?
   A. A only
   B. B only

C. Both A and B
D. Neither A nor B

Hint   *Many spark control systems on fuel injected engines are designed to reduce the spark advance when the engine detonates. If a defective spark control system allows engine detonation, high NOX emissions may occur.*

Engine-Related Service

ASE Tasks, Questions, and Related Information

Task 1   Adjust valves on engines with mechanical or hydraulic valve lifters.

72. All of these statements are true about adjusting mechanical valve lifters EXCEPT
   A. if the exhaust valve clearance is less than specified, premature valve burning may occur.
   B. the piston should be at TDC on the exhaust stroke in the cylinder on which the valves are being adjusted.
   C. excessive intake or exhaust valve clearance causes a clicking noise with the engine idling.
   D. the piston should be at TDC on the compression stroke in the cylinder on which the valves are being adjusted.

Hint   *The engine should be at the temperature specified by the vehicle manufacturer before a valve adjustment. When adjusting either type of valve lifter, the piston should be at TDC on the compression stroke in the cylinder on which the valves are being adjusted. When adjusting hydraulic valve lifters, the adjusting nut should be backed off until there is clearance between the rocker arm and valve stem. Rotate the push rod while tightening the adjusting nut. Continue tightening the adjusting nut until the push rod turning effort increases slightly. Tighten the adjusting nut the specified amount from this position.*

The adjusting nut on mechanical lifters is rotated until the specified clearance is available between the rocker arm and valve stem. This clearance is measured with a feeler gauge.

Task 2   Verify correct camshaft timing; determine needed action.

73. When diagnosing valve timing number 1 piston is positioned at TDC on the exhaust stroke with the timing mark aligned with the 0° position on the timing indicator. The valves in number 1 cylinder should be positioned so
   A. the intake valve is beginning to open and the exhaust valve is beginning to close.
   B. the intake valve is completely open and the exhaust valve is closed.
   C. the exhaust valve is completely open and the intake valve is closed.
   D. the intake valve is beginning to close and the exhaust valve is beginning to open.

Hint   *The valve timing may be verified with number 1 piston positioned at TDC on the exhaust stroke and the timing marks aligned at the zero mark. Since the exhaust stroke is ending and the intake stroke is beginning, the exhaust valve should be closing and the intake valve should be opening in this crankshaft position. This valve action is called valve overlap. If the valves are in any other position, the valve timing is incorrect.*

Task 3   Check engine operating temperature; determine needed action.

74. All of these problems may be caused by an engine thermostat that is stuck open
EXCEPT
   A. a rich air-fuel ratio and reduced fuel economy.
   B. an inoperative EGR system.
   C. improper cooling fan operation.
   D. an improper air charge temperature sensor signal.

Hint   *A thermometer may be taped to the upper radiator hose to check the thermostat operation. When the engine has been idling for 20 minutes, the temperature reading on the thermometer should be close to the specified temperature rating of the thermostat. If the thermostat is sticking closed, the engine overheats. A thermostat that sticks open may cause a rich air-fuel ratio, inoperative emission systems such as EVAP and EGR, improper cooling fan operation, and improper TCC lockup operation.*

Task 4   Perform cooling system pressure tests; check coolant; inspect, test, radiator, thermostat, pressure cap, coolant recovery tank, and hoses; determine needed repairs.

75. The pressure rating on a radiator cap is 16 to 18 psi (110 to 124 kPa). When tested
with a pressure tester (Figure 8-20), the cap should release pressure at
   A. 8 psi (55 kPa).
   B. 10 psi (69 kPa).
   C. 12 psi (83 kPa).
   D. 15 psi (103 kPa).

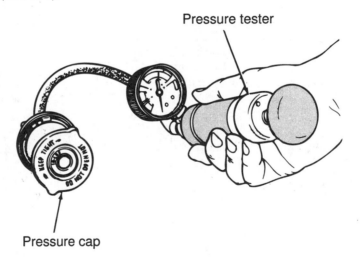

Figure 8-20   Pressure cap testing *(Courtesy of Chrysler Corporation)*

Hint   *A pressure tester may be installed on the radiator filler neck to check cooling system leaks. Pressurize the cooling system for 15 minutes at the pressure stamped on the radiator cap. If the pressure decreases after 15 minutes, the cooling system is leaking. Inspect the cooling system for external leaks and look for coolant dripping out of the heater case onto the front floor mat. If there are no external leaks check for combustion chamber leaks or transmission cooler leaks.*

The radiator cap may be installed on the pressure tester. When the tester pump is operated, the cap should release pressure at or slightly below the cap rating.

76. A new thermostat has been installed. When a thermometer is taped to the upper radiator hose, the thermometer reading is 160°F (71°C) after 20 minutes of engine idle operation. The specified thermostat opening temperature is 195°F (90°C). The cause of this problem could be
   A. the thermostat fits loosely in the housing groove.
   B. a defective ECT sensor.
   C. a restricted bypass hose.
   D. a defective water pump impeller.

Hint   *A thermostat must be installed in the proper direction.Many thermostats are marked with an arrow in the direction of coolant flow. If the thermostat is installed backward, the engine overheats. The thermostat must fit snugly in the housing groove. When the thermostat fits loosely in this groove, coolant bypasses the thermostat, making it ineffective to some extent.*

Task 5   Inspect, test, and replace mechanical/electrical fans, fan clutch, fan shroud/ducting, and fan control devices.

77. An electric cooling fan is inoperative. With the ignition switch on there is 12V supplied to the cooling fan relay contacts and this relay winding (Figure 8-21). When the engine is hot enough to close the temperature switch, the voltage at the switch contacts is .2V. The cause of this problem could be
   A. a defective coolant temperature switch.
   B. an open circuit in the main relay winding.
   C. an open circuit between the fan relay and the motor.
   D. a blown 30A fan fuse.

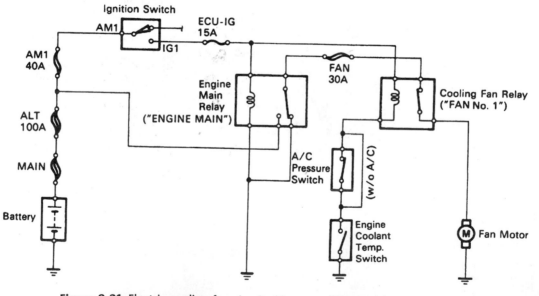

Figure 8-21 Electric cooling fan circuit *(Courtesy of Toyota Motor Corporation)*

Hint   *Some electric cooling fan circuits contain a relay and a temperature switch. When the temperature switch closes, the relay winding is grounded through the switch and this action closes the relay contacts. Under this condition voltage is supplied through the contacts to the fan motor.*

Engine Electrical Systems Diagnosis and Repair, Battery

ASE Tasks, Questions, and Related Information

Task 1   **Inspect, service, and replace battery, battery cables, clamps, and hold-down devices.**

78. A battery is disconnected from the vehicle electrical system on a fuel-injected vehicle. All of the following problems may occur EXCEPT
    A. the adaptive memory in the PCM is erased.
    B. the radio programming is erased.
    C. the interval wiper program is erased.
    D. the electric seat and mirror memory is erased.

Hint   *Before a battery is disconnected, a 12V source may be connected to the cigarette lighter socket to maintain the power supply to the on-board computers. This prevents erasing of the adaptive strategies and computer memories. Always disconnect the negative battery cable first. If the vehicle is equipped with an air bag or bags, wait for the time specified by the manufacturer after the negative cable is disconnected. Always reconnect the negative battery cable last when installing the cables.*

Task 2   **Measure and diagnose the cause(s) of abnormal key-off battery drain; determine needed repairs.**

79. While discussing key-off battery drain testing on a vehicle with port fuel injection and an antilock brake system (ABS)
    Technician A says when the ignition switch is turned off, the current drain increases after the tester switch is opened.
    Technician B says when the ignition switch is turned off, key-off drain should be read immediately on the milliameter.
    Who is correct?
    A. A only
    B. B only
    C. Both A and B
    D. Neither A nor B

Hint   *Computer memories have a very low key-off drain. Other light circuits such as glove compartment lights and trunk lights cause battery drain if they remain on with the ignition key off. Always be sure all electrical systems and components are shut off before testing key-off battery drain. The doors and hood must be closed to be sure the interior lights and underhood light are off.*

A special tester switch is connected between the negative battery cable and the negative battery terminal. An ammeter with a milliampere scale is connected to the tester switch terminals. The switch is closed to start the engine. When the ignition switch is turned off, the tester switch is opened. Always wait the specified length of time before reading the milliameter. Since computers require a brief time interval to enter the sleep mode, the current drain may be higher than specified when the tester switch is opened initially. If the current drain is excessive, individual circuits and components may be disconnected to locate the exact cause of the drain. Removing fuses and circuit breakers is often the most convenient method of disconnecting individual circuits and components.

Task 3

Perform battery capacity (load, high-rate discharge) test; determine needed service.

80. In Figure 8-22 the tester is connected to test
    A. battery open circuit voltage.
    B. starter draw.
    C. battery capacity.
    D. starter circuit voltage drop.

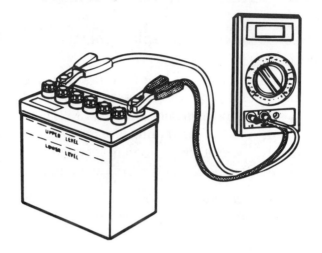

Figure 8-22   Battery test   *(Courtesy of Toyota Motor Corporation)*

Hint   *Prior to the battery capacity test the battery open circuit voltage should be at least 12.2V. The battery is discharged at one-half the cold cranking rating for 15 seconds. With the load still applied after 15 seconds, the battery voltage should be 9.6V with the battery temperature at 70°F (21°C). Vehicle manufacturers provide a battery voltage specification chart in relation to temperature.*

Task 4   **Slow and fast charge batteries.**

81. Technician A says the battery gives off hydrogen gas while charging.
    Technician B says a battery may be fully charged on a fast charger at a high charging rate.
    Who is correct?
    A. A only
    B. B only
    C. Both A and B
    D. Neither A nor B

Hint   *During the fast charging procedure the battery temperature must be kept below 125°F (52°C) to avoid battery damage. When the battery reaches approximately 1.225 specific gravity, the fast charging rate should be reduced to prevent excessive battery gassing. The battery is fully charged when the specific gravity is 1.265. Since explosive hydrogen gas is emitted from the battery during the charging process, do not allow any type of sparks, flames, or sources of ignition near the battery.*

Engine Electrical Systems
Diagnosis and Service, Starting System

ASE Tasks, Questions, and Related Information

Task 1   **Perform starter current draw test; determine needed action.**

82. The starter draw is more than specified and the cranking rpm and voltage are low. The cause of this problem could be
    A. high resistance in the starter field coils.
    B. worn starter bushings.
    C. high resistance in the battery cables.
    D. high resistance in the solenoid windings.

Hint   *High starter current draw with low cranking speed and cranking voltage usually indicate a defect in the starting motor (such as worn bushings), allowing the armature to rub on the field coils. Electrical defects in the armature or field coils also cause this problem.*

Low cranking speed with low current draw and high cranking voltage usually indicate high resistance in the starter circuit (such as corroded battery cables).

Task 2   **Perform starter circuit voltage drop tests; determine needed action.**

83. With the engine cranking the voltmeter connected to the starter circuit reads .2V (Figure 8-23). This reading indicates
    A. the positive battery cable has excessive resistance.
    B. the starter ground circuit has excessive resistance.
    C. the starter relay contacts have excessive resistance.
    D. the positive battery cable has normal resistance.

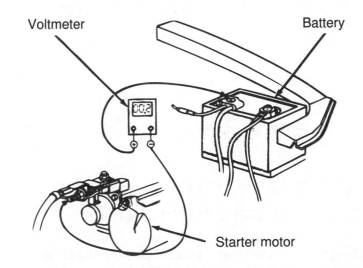

Figure 8-23   Starter circuit voltage test *(Courtesy of Chrysler Corporation)*

Hint   *The voltmeter should be connected from the positive battery terminal to the battery terminal on the starting motor to test voltage drop in the positive battery cable. With the engine cranking if the voltage drop exceeds .2V the battery positive cable has excessive resistance.*

Connect voltmeter leads from the negative battery terminal to the starter case. With the engine cranking, the voltage drop across the battery ground circuit should not exceed .2V.

Task 3 **Inspect, test, and repair or replace components and wires in the starter control circuit.**

84. The ohmmeter is connected from the solenoid terminal to ground on the solenoid case (Figure 8-24). The ohmmeter indicates an infinite reading.
Technician A says the solenoid hold-in winding has an open circuit.
Technician B says the solenoid pull-in winding is shorted to ground.
Who is correct?
A. A only
B. B only
C. Both A and B
D. Neither A nor B

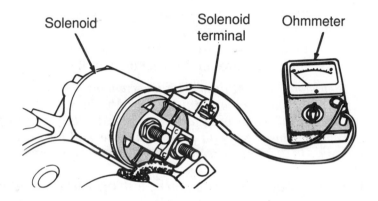

Figure 8-24   Starter solenoid test *(Courtesy of Chrysler Corporation)*

Hint   *The ohmmeter leads are connected from the solenoid terminal to ground when checking the hold-in winding. An infinite reading indicates an open hold-in winding and this winding is shorted if the reading is below specifications.*

Connect the ohmmeter leads from the solenoid terminal to the field coil terminal in the starting motor to test the pull-in winding.

Engine Electrical Systems Diagnosis and Repair, Charging System

ASE Tasks, Questions, and Related Information

Task 1 **Diagnose charging system problems that cause an undercharge, overcharge, or no-charge condition.**

85. The battery gradually becomes undercharged and the vehicle is driven regularly. All of these defects may be the cause of the problem EXCEPT
A. a loose alternator belt.
B. a battery drain.
C. a high regulator voltage.
D. a low regulator voltage.

Hint   *An undercharged battery may be caused by a loose alternator belt, low regulator voltage, defective alternator, or a battery drain. An overcharged battery may be caused by a high regulator voltage. If the alternator has zero output the field circuit may be open, or the fuse link may be burned out between the battery positive cable and the alternator battery terminal.*

Task 2   Inspect, adjust, and replace alternator drive belts, pulleys, and fans.

86. Technician A says an alternator V-belt should be replaced if the belt is contacting the bottom of the pulley.
   Technician B says many alternator ribbed V-belts have an automatic tensioner with a wear indicator.
   Who is correct?
   A. A only
   B. B only
   C. Both A and B
   D. Neither A nor B

Hint   *A V-belt should be replaced if the belt is oil soaked, cracked, frayed, or contacting the bottom of the alternator pulley. A belt tension gauge should indicate the specified tension when the gauge is placed in the center of the longest belt span.*
   Many ribbed V-belts have a spring-loaded automatic belt tensioner with a wear indicator.

Task 3   Inspect and repair or replace charging circuit connections and wires.

87. The fuse link is burned out in the wire between the battery positive terminal and the alternator battery terminal. The most likely cause of this problem is
   A. a shorted alternator stator.
   B. reversed battery polarity.
   C. defective alternator diode.
   D. high regulator voltage.

Hint   *A burned fuse link in the alternator battery wire may be caused by reversed battery polarity when installing the battery or using a booster battery. Replacement fuse links are available, and the old fuse link may be cut out of the wiring harness. Bare the wire in the harness and crimp the replacement fuse link to the wire (Figure 8-25).*

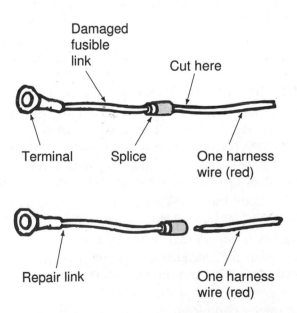

Figure 8-25 Fuse link replacement *(Courtesy of Pontiac Motor Division, General Motors Corporation)*

Answers and Analysis

1. **C** One of the first steps in a diagnostic procedure is to identify the problem; listening to the customer's complaint may help to achieve this objective. Both A and B are correct, and C is the right answer.

2. **C** Worn camshaft lobes may cause a heavy clicking noise at higher engine speeds. Loose piston pins may cause a light rapping noise at idle speed, and sticking or worn valve lifters may cause a light clicking noise at idle and low speed. Therefore, A, B, and D are wrong.

   The heavy thumping noise when the engine is first started may be caused by worn main bearings, so C is right.

3. **A** A "puff" noise in the exhaust at regular intervals may be caused by a burned exhaust valve. Excessive coolant temperature does not result in this symptom. Therefore, A is correct and B is wrong.

4. **D** A low steady vacuum gauge reading may be caused by late ignition timing or an intake manifold vacuum leak. Therefore, both A and B are wrong and D is the correct answer.

5. **A** A lack of rpm decrease during the cylinder power balance test indicates the cylinder is misfiring. A defective spark plug, wire, or a burned exhaust valve would cause misfire at low or high speed, so B, C, and D are wrong.

   Since the cylinder misfire disappeared at 2,000 rpm, an intake manifold vacuum leak is indicated. A is right.

6. **C** There is no increase in the compression reading during the wet test with a burned exhaust valve, bent intake valve, or a blown head gasket. Therefore, A, B, and D are wrong.

   Worn piston rings do cause an increase in compression readings during the wet test. C is correct.

7. **C** During a cylinder leakage test excessive leakage and air escaping from the throttle body may be caused by a bent intake valve or a bottomed intake valve lifter. Both A and B are correct and C is the right answer.

8. **C** Some engine analyzers compare test results to specifications and indicated readings that are out-of-specifications. Manual or automatic test modes may be selected on some engine analyzers. Both A and B are correct and C is the right answer.

9. **A** With the emissions analyzer probe in the tailpipe, the analyzer indicates the emissions coming out of the converter, which could be considerably lower than the emissions coming out of the engine. Therefore, B is wrong.

   A leak in the exhaust system may cause inaccurate emission readings. A is correct.

10. **A** If the 12V test light does not flutter on any of the coils while cranking the engine, the crankshaft sensor, camshaft sensor, or ignition module is defective. Since the crankshaft sensor is the only one of these components in the responses, A is correct.

11. **A** If the cam sensor becomes defective with the engine running, the engine continues to run, but the engine does not restart when it is shut off. Therefore, B is wrong.

   An open circuit in the EST wire results in no spark advance, which results in reduced fuel economy and performance.

12. **A** If the ignition module is defective, the test spark plug would not fire. Therefore, B is wrong.

   A leaking cap and rotor causes the test spark plug to fire normally when connected to the coil secondary wire, but fire intermittently when connected to the spark plug wires. A is correct.

13. **D** An ohmmeter connected from one of the pickup leads to ground should provide an infinite reading. If the ohmmeter provides a low reading, the pickup coil is grounded. Therefore, A, B, and C are wrong and D is right.

14. C   The Hall-Effect pickup may be tested with a voltmeter connected from the pickup signal wire to ground. Therefore, A, B, and D are wrong and C is right.

15. C   A high resistance problem in the primary circuit reduces primary current flow and magnetic strength, and this action causes low maximum secondary coil voltage. Therefore A is right.

A coil with a secondary insulation leakage defect also lowers the maximum available secondary coil voltage. Both A and B are correct and C is the right answer.

16. A   The negative side of each primary winding is connected to the module or PCM. Therefore, B is wrong.

The two ends of each secondary winding are connected to the spark plug wire terminals on each coil, so C is wrong.

The secondary windings are not interconnected, thus D is wrong.

When the ignition switch is turned on, 12V are supplied to the positive side of each primary winding. A is right.

17. B   Timing adjustments are not possible or required on EI-equipped engines. Therefore, A is wrong.

On some fuel-injected engines with a distributor, the PCM must be placed in the limp-in mode before adjusting basic ignition timing. B is correct.

18. C   Excessive pickup gap may cause misfiring or a no-start condition depending on the amount of excessive gap. Both A and B are correct and C is the right answer.

19. C   If the voltmeter connected from the negative primary coil terminal always indicates 12V while cranking the engine, the module is open or the wire from the coil negative terminal to the module is open. Both A and B are correct and C is the right answer.

20. A   Excessive spark advance causes engine detonation, so B is wrong. A high float level, or the choke sticking closed results in a rich air-fuel ratio, but adequate power on acceleration, thus C and D are wrong.

Since the accelerator pump prevents a hesitation on low-speed acceleration, a defective accelerator pump system causes an acceleration stumble. A is correct.

21. B   A higher-than-specified TPS voltage signal may cause a lower than specified idle speed, so A is wrong.

A intake manifold vacuum leak on a fuel injected engine may cause a faster than specified idle speed. B is right.

22. C   Defective injectors may cause a rich air-fuel ratio, but this defect also causes rough idle. A is wrong.

Defective spark plugs cause cylinder misfiring and rough idle, so B is wrong.

A defective fuel pump check valve causes fuel drain back into the tank when the engine is shut off and this results in hard starting. D is wrong.

High fuel pump pressure causes a rich air-fuel ratio, thus C is right.

23. A   A fuel pressure regulator that is sticking open causes low fuel pressure. B is wrong. A restricted fuel return line causes high fuel pressure. A is correct.

24. A   A restricted fuel line may cause an acceleration stumble and power loss, but this defect should not result in hard starting. B is wrong.

Excessive alcohol in the fuel causes a lean air-fuel ratio, hard starting, power loss, and acceleration stumbles, so A is correct.

25. D   If there is an open in the main relay winding, there is no action from the fuel pump when the FP and B+ terminals are connected. A is wrong.

When the 7.5 A, IGN fuse is blown there is no action from the main relay or the fuel pump if the FP and B+ terminals are connected, so B is wrong.

An open circuit at the B+ terminal on the circuit opening relay does not allow the engine to keep running after the jumper wire at the FP and B+ terminals is removed.

An open circuit in the winding connected to the STA and E1 terminals in the circuit opening relay causes no action from the fuel pump while cranking the engine. D is correct.

26. A   A leaking fuel pump check valve causes fuel drain back into the tank after the engine is shut off. This results in hard starting, so B is wrong.

A partially plugged return fuel line causes high fuel pressure, a rich air-fuel ratio, and reduced fuel economy. A is correct.

27. A   The PCM, ECT sensor, or ACT sensor do not affect the cold start injector operation. Therefore, B, C, and D are wrong.

Since the cold start injector is controlled by a thermo-time switch, if this component is defective the cold start injector may stay open longer than specified resulting in a rich air-fuel ratio, rough idle, and black smoke from the tailpipe. A is correct.

28. C   A defective TPS, or a defective pressure regulator allowing lower than specified fuel pressure, may cause an acceleration stumble. Both A and B are correct, and C is the right answer.

29. A   Since the emissions are high only during deceleration and the fuel pressure has no effect on deceleration operation, B is wrong.

An injector deceleration fuel reduction system that does not cut off the injector operation during deceleration may cause high CO and HC emissions during deceleration. A is correct.

30. A   A PCV valve stuck open on a TBI engine acts much like a vacuum leak and causes faster than specified idle speed. B is wrong.

Carbon in the throttle bore area may cause stalling and rough idle operation. A is correct.

31. B   An injector plunger sticking in the open position causes excessive pressure drop during the balance test, so A is wrong.

Since the number 3 injector had more than the specified pressure drop, it has a definite problem, thus C is wrong.

It is impossible to have less pressure at the number 3 injector compared to the other injectors. D is wrong.

A restricted orifice in the number 3 injector causes less pressure drop on this injector, so B is correct.

32. B   Since engine rpm and idle condition both change when the propane hose is placed near the injector, the lower O-ring on the injector must be leaking. A is wrong and B is correct.

33. A   The minimum air rate adjustment is shown in the figure. Therefore, B, C, and D are wrong and A is correct.

34. A   The throttle opener allows the throttle to close gradually on deceleration, so B is wrong and A is right.

35. D   A defective spark plug wire, spark plug, or injectors cause cylinder misfiring, thus A, B, and C are wrong.

A restricted exhaust system causes a severe loss of power and limited maximum vehicle speed. D is right.

36. D   An EGR valve that is stuck open causes rough idle operation. A is wrong.

A restricted catalytic converter limits the top vehicle speed. B is wrong.

Dripping injectors cause rough idle operation. C is wrong.

Excessive resistance in a spark plug wire causes misfiring during acceleration. D is correct.

37. A   A wastegate sticking closed causes high boost pressure, so B is wrong.

Low engine compression causes reduced air and exhaust flow through the engine, resulting in reduced turbocharger shaft speed. Since this action results in reduced boost pressure, A is correct.

38. A Partially seized bearings, or excessive shaft endplay cause reduced turbocharger boost pressure. B and C are wrong.

Reduced coolant flow through the turbocharger causes damaged turbocharger bearings, so D is wrong.

A leak in the air intake system causes compressor vane pitting, so A is right.

39. C Contaminated oil or restricted coolant passages cause damage to the turbocharger bearings. Both A and B are correct, and C is the right answer.

40. D Wide open throttle operation causes a wide PCV valve opening, so A is wrong.

Operation at half-open throttle results in a medium PCV valve opening, thus B is wrong.

A backfire causes the PCV to close, so C is wrong.

The high vacuum during idle operation causes a very small PCV valve opening. D is correct.

41. B A plugged PCV clean air filter causes excessive vacuum in the engine. This may result in damaged gaskets and dirt particles entering the engine past these gaskets. A is wrong.

A plugged hose between the PCV valve and the intake manifold causes excessive pressure in the engine. This action causes oil vapors to be forced through the PCV clean air hose into the air cleaner, so B is correct.

42. A A loose oil filler cap would not change the amount of flow through the PCV valve. B and C are wrong.

Since the engine compression is satisfactory, oil should not be blown out around the loose oil filler cap, thus D is wrong.

A loose oil filler cap allows the PCV system to pull dirt particles into the engine. A is right.

43. D An engine thermostat stuck open, a defective engine coolant temperature sensor, or low compression on one cylinder would not increase NOX emissions. Therefore, A, B, and C are wrong.

Plugged exhaust passages under the EGR valve prevent EGR flow and increase NOX emissions. D is right.

44. B Since the bleed valve normally is open in a positive backpressure EGR valve, supplying vacuum to this valve with the engine idling does not open the valve. A is wrong.

When the EGR valve is opened at idle speed, the engine should slow down 150 rpm, so B is correct.

45. B The bleed valve in a negative backpressure EGR valve is pulled open by negative pressure pulses in the exhaust at low engine speeds, so A is wrong.

When vacuum is supplied to this type of valve with the engine stopped, the valve should open because the bleed valve is closed, thus C is wrong.

If the engine is running at 2,000 rpm the negative pressure pulses in the exhaust are reduced, and the bleed valve closes. D is wrong.

The bleed valve is opened by negative pressure pulses in the exhaust at low engine speeds. B is correct.

46. D The PCM does not open the EGR valve if the engine coolant is cold or if the throttle is wide open. Therefore, both A and B are wrong, and D is the correct answer.

47. B With the EGR valve open, the flow through the orifice creates a pressure drop on the valve side of the orifice, so A is wrong.

The DPFE sensor sends an analog signal to the PCM, thus C is wrong.

The maximum EGR valve opening would occur at medium vehicle speeds and throttle opening. D is wrong.

The PCM pulses the EGR solenoid on and off to supply the proper vacuum to the EGR valve. B is correct.

48. B   If the vacuum hose to the AIRD valve is leaking, the air flow from the pump is directed through this valve to the catalytic converter; this action does not cause a low O_2 sensor signal. Therefore, A is wrong.

When the air flow from the pump is directed to the exhaust ports with the engine warmed up, the O_2 sensor detects the additional oxygen in the exhaust stream. This causes a low O_2 sensor voltage signal and the PCM responds to this signal by providing a rich air-fuel ratio. B is correct.

49. C   This question asks for the defect that is not the cause of the problem. A sticking AIRB valve, a sticking AIRB solenoid plunger, a leak in the AIRB vacuum hose would cause the air flow from the pump to be diverted to the atmosphere at the AIRB valve. Therefore, A, B, and D could be the cause of the problem, but none of these is the requested answer.

A sticking AIRD valve would not cause the air to be diverted to the atmosphere at the AIRB valve. Therefore, C is not the cause of the problem and this is the requested answer.

50. A   If the AIR system hoses are severely burned, one of the check valves is allowing exhaust to enter the system. Therefore, B, C, and D are wrong and A is right.

51. C   In the upstream mode the PCM energizes both the AIRB and AIRD solenoids, and vacuum is supplied to both the AIRB and AIRD valves. If the voltage is 12V on both terminals of the AIRD valve, there is no voltage drop across the winding because there is no current flow through the winding. This condition may be caused by an open wire from this solenoid to the PCM, or the PCM may not be grounding the wire. Both A and B are correct and C is the right answer.

52. A   If the catalytic converter is operating properly, the outlet is about 100°F (55°C) hotter than the inlet. Therefore, B, C, and D are wrong, and A is right.

53. A   If the wire from the EVAP solenoid to the PCM is open, the EVAP solenoid never opens and the canister is never purged. This may allow HC emissions to escape from the canister to the atmosphere, but it does not affect engine idle. Therefore, B is wrong.

When the EVAP solenoid is stuck open, the canister is purged at idle speed, and this causes rough idle operation. A is correct.

54. A   If the hose from the fuel tank to the canister is plugged, there is no vapor flow from the tank to the canister. Since there is nothing entering the canister, gasoline cannot drip from this component. B is wrong.

When the canister filter is plugged gasoline vapors may condense in the canister, and higher vacuum is present in the canister. This condition may cause gasoline dripping from the canister. A is correct.

55. A   This question asks for the response that is not the cause of the problem. A purge solenoid plunger stuck in the closed position, a plugged hose from the canister to the intake manifold, or a plugged purge port in the intake may cause an inoperative EVAP system without any DTCs in the PCM memory. Therefore, B, C, and D may be the cause of the problem, but none of these is the requested answer.

An open winding in the purge solenoid causes an inoperative EVAP system, but this defect also causes a DTC in the PCM memory. Therefore, A is not the cause of the problem and this is the requested answer.

56. C   Since the O_2 sensor responds to the propane flow into the air intake, this sensor is not defective and A is wrong.

When the propane flow into the air intake causes a rich signal from the O_2 sensor, the PCM responds by reducing the injector pulse width, so the PCM is not likely the cause of the problem. B is wrong.

If the O_2 sensor signal wire is open, there is no signal from this sensor to the PCM, thus D is wrong.

Low fuel pump pressure causes a lean air-fuel ratio and lean misfiring. This condition causes high HC emissions at low and high speeds. C is correct.

57. B On some vehicles the diagnostic trouble codes (DTCs) may be obtained by cycling the ignition switch on and off three times in a five second interval. A is wrong.

On some vehicles the DTCs are read from the flashes of the malfunction indicator light (MIL). B is correct.

58. A While reading the DTCs from the flashes of the MIL, each code is repeated three times. B is wrong.

The engine must be at normal operating temperature before obtaining the DTCs. C is wrong.

A DTC indicates a defect in a certain area, but not in a specific component. D is wrong.

Since the codes are supplied in numerical order, code 42 is provided before code 55. A is correct.

59. A An EGR valve that is stuck open causes a rough idle problem, high NOX emissions, and a DTC representing the EGR valve when the EGR valve has a valve position sensor. A is correct.

High NOX emissions occur if the EGR passages are restricted with carbon, but this condition does not cause rough idle or a DTC representing the EGR valve. B is wrong.

60. C High fuel pressure causes a rich air-fuel ratio and high O_2 sensor voltage. B is wrong.

An intake manifold vacuum leak causes a low O_2 sensor voltage, but this voltage signal increases when propane is dispersed into the air intake. D is wrong.

Since the O_2 sensor voltage does not change when propane is dispersed into the air intake, this sensor is defective. C is correct.

The PCM does not change the injector pulse width because the O_2 sensor signal did not change. A is wrong.

61. C A leaking fuel pump check valve may cause hard starting, but this problem does not cause rough idle operation. A is wrong.

A leaking fuel pressure regulator may cause hard starting, but this problem does not cause rough idle operation. B is wrong.

A defective ECT sensor may cause high CO and HC emissions, but this defect does not cause a decrease in fuel pressure after the engine is shut off. D is wrong.

Dripping injectors may cause high CO and HC emissions, rough idle, and hard starting after the engine is shut off, and a decrease in fuel pressure after the engine is shut off. C is correct.

62. B Defective injectors may cause rough idle, but this defect causes high HC and CO emissions at idle and higher speeds. A is wrong.

Low fuel pump pressure or a defective ECT sensor cause a lean air-fuel ratio that may result in lean misfiring and high HC emissions at idle and higher speeds. C and D are wrong.

An EGR valve that is stuck open dilutes the air fuel ratio at idle and slow speeds and this causes a lean misfire and high HC emissions. Since the EGR valve is normally open at 2,500 rpm, the stuck EGR valve does not affect emissions at this speed. B is correct.

63. C If the AIRD solenoid plunger is sticking open, or the AIRD valve is sticking, the air flow from the pump may be continually directed upstream. Since A and B are both correct, C is the correct answer.

64. A If the fuel line is plugged, a leaking fuel pump check valve does not cause the fuel pressure to drop off after the engine is shut off. B is wrong.

If the return fuel line is plugged, a leaking pressure regulator does not cause the fuel pressure to drop off after the engine is shut off. C is wrong.

A leaking upper injector O-ring causes a visible leak between the injector and the fuel rail. D is wrong.

Dripping injectors cause hard starting and fuel pressure drop off after the engine is shut off with the fuel line and fuel return line plugged. A is correct.

65. C Many MAP sensors act as a barometric pressure sensor. On many MAP sensors the voltage signal increases as the vacuum at the sensor decreases. Both A and B are correct, and C is the correct answer.

66. C On many MAP sensors for every 5 in. hg. vacuum change at the sensor, the sensor voltage signal should change .7V to 1 V. If the sensor voltage signal is 1.2V at 18 in. hg., this voltage signal should be 2.0V at 13 in. hg. C is correct.

67. C An O_2 sensor signal must be tested with a digital voltmeter. Therefore, A, B, and D are wrong and C is right.

68. C In Figure 8-31 with the ignition switch in the ACCY position, voltage is supplied only to the wiper and radio fuses, thus A, B, and D are wrong and C is right.

69. D This question asks for the statement that is not true. When handling an EEPROM chip do not remove the chip from the protective envelope until you are ready to install it. Before touching the chip touch a good ground, and connect a ground strap from your wrist to a good vehicle ground. Therefore, A, B, and C are true, but none of these are the requested answer.

The chip pins should not be touched with your fingers or anything else, thus D is not true and this is the requested answer.

70. C If the TCM is not issuing a TCC on command, the ECT sensor signal may not be received by the TCM. This problem could be caused by the ECT signal not being transmitted from the PCM to the TCM. Under this condition the data links should be tested with a scan tester. Both A and B are correct and C is the right answer.

71. C If the timing mark does not move when the exhaust manifold is tapped with a hammer, the knock sensor may be defective. Under this condition, the engine may detonate. Both A and B are correct, so C is the right answer.

72. B This question asks for the statement that is not true. If the exhaust valve clearance is less than specified, the valve may burn prematurely. Excessive valve clearance causes a clicking noise, especially at idle. The piston should be at TDC on the compression stroke in the cylinder on which the valve is being adjusted. Therefore, A, C, and D are true, but none of these is the requested answer.

Both valves are open slightly with the piston at TDC on the exhaust stroke, and the valve clearance must not be adjusted in this piston position. Therefore, B is not true, and this is the requested answer.

73. A With the piston at TDC on the exhaust stroke the intake valve should be starting to open and the exhaust valve should be starting to close. Therefore, B, C, and D are wrong, and A is right.

74. D This question asks for the statement that is not true. If the engine thermostat is stuck open, the air-fuel ratio is rich and fuel economy is reduced because the ECT sensor informs the PCM regarding the lower coolant temperature. This defect also affects the EGR and cooling fan operation. Therefore, A, B, and C are true, but none of these are the requested answer.

Since the air charge sensor sends a signal to the PCM in relation to air intake temperature, this signal is not affected by a thermostat that is stuck open. Therefore, D is not true, and this is the requested answer.

75. D A radiator cap with a 16 to 18 psi (110 to 124 kPa) rating should not release the pressure from the tester until 15 psi (103 kPa). Therefore, A, B, and C are wrong and D is correct.

76. A A defective ECT sensor does not affect the thermostat operation, thus B is wrong.

A restricted bypass hose may cause higher coolant pressure in the block during engine warm-up, and so C is wrong.

A defective water pump impeller causes engine overheating, thus D is wrong.

If the thermostat fits loosely in the housing groove, some of the coolant bypasses the thermostat making it ineffective. A is correct.

77.   C   Since the voltage at the coolant switch contacts is .2V, these contacts must be closed. This indicates the coolant switch is not the cause of the problem, so A is wrong.

If the main relay winding is open, there is no voltage supplied to the cooling fan relay contacts, thus B is wrong.

A blown 30 A fan fuse also causes 0V at the cooling fan relay contacts, so D is wrong.

An open circuit from the cooling fan relay to the fan motor causes no cooling fan operation with the voltages mentioned in the question. C is correct.

78.   C   This question asks for the condition that does not occur when the battery is disconnected. If the battery is disconnected the PCM adaptive strategy, radio programming, and electric seat and mirror memory is erased. Therefore, statements A, B, and D represent conditions that do occur, but none of these is the requested answer.

When the battery is disconnected the interval wiper memory is not erased. Since statement C does not occur, this is the requested answer.

79.   D   The current drain decreases after the tester switch is opened. The key-off drain should not be read until the computers enter the sleep mode, which takes several minutes. Both A and B are wrong, so D is the correct answer.

80.   A   The tester in Figure 8-34 is connected to test battery open circuit voltage, so B, C, and D are wrong, and A is correct.

81.   A   If the battery is fully charged on a fast charger at a high rate, excessive battery gassing occurs, so B is wrong.

Hydrogen gas is emitted from a battery during the charging process, thus A is correct.

82.   B   High resistance in the field coils, or battery cables causes low cranking speed, low current draw, and high cranking voltage. A and C are wrong.

High resistance in the solenoid windings affects the solenoid operation, so D is wrong.

Worn starter bushings allow the armature to rub on the field coils resulting in high current draw with low cranking speed and low cranking voltage. B is correct.

83.   D   The voltmeter leads are connected to the ends of the positive battery cable to test the voltage drop across this cable. A voltage drop of .2V across the positive battery cable indicates a normal resistance. Therefore, A, B, and C are wrong, and D is right.

84.   A   The ohmmeter is connected to test the solenoid hold-in winding. An infinite reading indicates an open winding. Therefore, B is wrong and A is correct.

85.   C   This question asks for the defect that is not the cause of the problem. Battery undercharging may be caused by a loose alternator belt, a battery drain, or a low regulator voltage. Therefore, A, B, and D may be the cause of the problem, but none of these is the requested answer.

High regulator voltage causes battery overcharging, so C is not the cause of the problem and this is the requested answer.

86.   C   An alternator V-belt should be replaced if it is touching the bottom of the pulley. Many ribbed V-belts have a spring-loaded automatic tensioner with a belt wear scale. Both A and B are correct and C is the right answer.

87.   B   A shorted alternator stator or a defective diode reduce alternator output, but these defects would not blow the fuse link. High regulator voltage causes an overcharged battery. Therefore, A, C, and D are wrong.

Reversed battery polarity causes extremely high current flow through the diodes in the forward direction and also through the alternator battery wire. This high current flow burns out the fuse link in the battery wire. B is right.

Glossary

Accelerator pump adjustment An adjustment that sets the accelerator pump height and stroke.
Ajuste de la bomba del acelerador Ajuste que fija la altura y la carrera de la bomba del acelerador.

Actuation test mode A scan tester mode used to cycle the relays and actuators in a computer system.
Modo de prueba de activación Instrumento de pruebas de exploración utilizado para ciclar los relés y los accionadores en una computadora.

Adjusting pads, mechanical lifters Metal discs that are available in various thicknesses, and positioned in the end of the mechanical lifter to adjust valve clearance.
Cojines de ajuste, elevadores mecánicos Discos metálicos disponibles en diferentes espesores que se colocan en el extremo del elevador mecánico para ajustar el espacio libre de la válvula.

Advance-type timing light A timing light that is capable of checking the degrees of spark advance.
Luz de ensayo de regulación del encendido tipo avance Luz de ensayo de regulación del encendido capaz de verificar la cantidad del avance de la chispa.

AIR by-pass (AIRB) solenoid A computer-controlled solenoid that directs air to the atmosphere or to the AIR diverter solenoid.
Solenoide de paso AIR Solenoide controlado por computadora que conduce el aire hacia la atmósfera o hacia el solenoide derivador AIR.

Air charge temperature (ACT) sensor A sensor that sends a signal to the computer in relation to intake air temperature.
Sensor de la temperatura de la carga de aire Sensor que le envía una señal a la computadora referente a la temperatura del aire aspirado.

AIR diverter (AIRD) solenoid A computer-controlled solenoid in the secondary air injection system that directs air upstream or downstream.
Solenoide derivador AIRD Solenoide controlado por computadora en el sistema de inyección secundaria de aire que conduce el aire hacia arriba o hacia abajo.

Air-operated vacuum pump A vacuum pump operated by air pressure that may be used to pump liquids such as gasoline.
Bomba de vacío accionada hidráulicamente Bomba de vacío accionada por la presión del aire que puede utilizarse para bombear líquidos, como por ejemplo la gasolina.

Analog meter A meter with a movable pointer and a meter scale.
Medidor analógico Medidor provisto de un indicador móvil y una escala métrica.

Antidieseling adjustment An adjustment that prevents dieseling when the ignition switch is turned off.
Ajuste antiautoencendido Ajuste que evita el autoencendido cuando el botón conmutador de encendido está apagado.

ASE blue seal of excellence A seal displayed by an automotive repair facility that employs ASE certified technicians.
Sello azul de excelencia de la ASE Logotipo exhibido en talleres de reparación de automóviles donde se emplean mecánicos certificados por la ASE.

ASE technician certification Certification of automotive technicians in various classifications by the National Institute for Automotive Service Excellence (ASE).

Certificación de mecánico de la ASE Certificación de mecánico de automóviles en áreas diferentes de especialización otorgada por el Instituto Nacional para la Excelencia en la Reparación de Automóviles (ASE).

Automatic shutdown (ASD) relay A computer-operated relay that supplies voltage to the fuel pump, coil primary, and other components on Chrysler fuel injected engines.
Relé de parada automática Relé accionado por computadora que les suministra tensión a la bomba del combustible, al bobinado primario, y a otros componentes en motores de inyección de combustible fabricados por la Chrysler.

Barometric (Baro) pressure sensor A sensor that sends a signal to the computer in relation to barometric pressure.
Sensor de la presión barométrica Sensor que le envía una señal a la computadora referente a la presión barométrica.

Belt tension gauge A gauge designed to measure belt tension.
Calibrador de tensión de la correa de transmisión Calibrador diseñado para medir la tensión de una correa de transmisión.

Bimetal sensor A temperature-operated sensor used to control vacuum.
Sensor bimetal Sensor accionado por temperatura utilizado para controlar el vacío.

Block learn A chip responsible for fuel control in a General Motors PCM.
Control de combustible en bloque Pastilla responsable de controlar el combustible en un módulo del control del tren transmisor de potencia de la General Motors.

Blowgun A device attached to the end of an air hose to control and direct air flow while cleaning components.
Soplete Dispositivo fijado en el extremo de una manguera de aire para controlar y conducir el flujo de aire mientras se lleva a cabo la limpieza de los componentes.

Boost pressure The amount of intake manifold pressure created by a turbocharger or supercharger.
Presión de sobrealimentación Cantidad de presión en el colector de aspiración producida por un turbocompresor o un compresor.

Breakout box A terminal box that is designed to be connected in series at Ford PCM terminals to provide access to these terminals for test purposes.
Caja de desenroscadura Caja de borne diseñada para conectarse en serie a los bornes del módulo del control del tren transmisor de potencia de la Ford, con el objetivo de facilitar el acceso a dichos bornes para propósitos de prueba.

Burn time The length of the spark line while the spark plug is firing measured in milliseconds.
Duración del encendido Espacio de tiempo que la línea de chispas de la bujía permanece encendida, medido en milisegundos.

Calibrator package (CAL-PAK) A removable chip in some computers that usually contains a fuel backup program.
Paquete del calibrador Pastilla desmontable en algunas computadoras; normalmente contiene un programa de reserva para el combustible.

Canister purge solenoid A computer-operated solenoid connected in the evaporative emission control system.
Solenoide de purga de bote Solenoide accionado por computadora conectado en el sistema de control de emisiones de evaporación.

Canister-type pressurized injector cleaning container A container filled with unleaded gasoline and injector cleaner and pressurized during the manufacturing process or by the shop air supply.
Recipiente de limpieza del inyector presionizado tipo bote Recipiente lleno de gasolina sin plomo y limpiador de inyectores, presionizado durante el proceso de fabricación o mediante el suministro de aire en el taller mecánico.

Carbon dioxide (CO2) A gas formed as a by-product of the combustion process.
Bióxido de carbono (CO2) Gas que es un producto derivado del proceso de combustión.

Carbon monoxide A gas formed as a by-product of the combustion process in the engine cylinders. This gas is very dangerous or deadly to the human body in high concentrations.
Monóxido de carbono Gas que es un producto derivado del proceso de combustión en los cilindros del motor. Este gas es muy peligroso y en altas concentraciones podría ocasionar la muerte.

Catalytic converter vibrator tool A tool used to remove the pellets from catalytic converters.
Herramienta vibradora del convertidor catalítico Herramienta utilizada para remover los granos gordos de los convertidores catalíticos.

Choke control lever adjustment An adjustment that provides proper choke spring tension.
Ajuste de la palanca de control del estrangulador Ajuste que proporciona la tensión adecuada del resorte del estrangulador.

Choke diaphragm connector rod adjustment An adjustment that assures proper cushioning of the secondary air valves by the vacuum break diaphragm.
Ajuste de la biela con el diafragma del estrangulador Ajuste que asegura el acojinamiento adecuado de las válvulas de aire secundarias mediante el diafragma del interruptor de vacío.

Choke unloader adjustment An adjustment that assures the proper choke position when the throttles are held wide open and the choke spring is cold.
Ajuste del abridor del estrangulador Ajuste que asegura la posición adecuada del estrangulador cuando se mantienen las mariposas abiertas de par en par y el resorte del estrangulador está frío.

Choke vacuum kick adjustment An adjustment that assures the proper choke opening when the vacuum kick diaphragm is bottomed after a cold engine is started.
Ajuste del retardador de vacío del estrangulador Ajuste que asegura la apertura adecuada del estrangulador cuando el diafragma del retardador de vacío se sumerje luego del arranque de un motor frío.

Choke valve angle gauge A gauge with a degree scale and a level bubble for performing carburetor adjustments.
Calibrador del ángulo de la válvula de estrangulación Calibrador provisto de una escala medida en grados y un tubo de burbuja para nivelación con el que se realizan ajustes al carburador.

Closed loop A computer operating mode in which the computer uses the oxygen sensor signal to help control the air-fuel ratio.
Bucle cerrado Modo de funcionamiento de una computadora en el que se utiliza la señal del sensor de oxígeno para ayudar a controlar la relación de aire y combustible.

Comprehensive tests A complete series of battery, starting, charging, ignition, and fuel system tests performed by an engine analyzer.
Pruebas comprensivas Serie completa de pruebas realizadas en los sistemas de la batería, del arranque, de la carga, del encendido, y del combustible con un analizador de motores.

Compression gauge A gauge used to test engine compression.
Manómetro de compresión Calibrador utilizado para revisar la compresión de un motor.

Computed timing test A computer system test mode on Ford products that checks spark advance supplied by the computer.
Prueba de regulación del avance calculado Modo de prueba en una computadora de productos fabricados por la Ford que verifica el avance de la chispa suministrado por la computadora.

Computer-controlled carburetor performance test A test that determines the general condition of a computer-controlled carburetor system.
Prueba de rendimiento del carburador controlado por computadora Prueba que determina la condición general de un sistema de carburador controlado por computadora.

Concealment plug A plug installed over the idle mixture screw to prevent carburetor tampering by inexperienced service personnel.
Tapón obturador Tapón instalado sobre el tornillo de mezcla de la marcha lenta para evitar que mecánicos inexpertos alteren el carburador.

Continuous self-test A computer system test mode on Ford products that provides a method of checking defective wiring connections.
Prueba automática continua Modo de prueba en una computadora de productos fabricados por la Ford que proporciona un método de verificar conexiones defectuosas del alambrado.

Coolant hydrometer A tester designed to measure coolant specific gravity and determine the amount of antifreeze in the coolant.
Hidrómetro de refrigerante Instrumento de prueba diseñado para medir la gravedad específica del refrigerante y determinar la cantidad de anticongelante en el refrigerante.

Cooling system pressure tester A tester used to test cooling system leaks and radiator pressure caps.
Instrumento de prueba de la presión del sistema de enfriamiento Instrumento de prueba utilizado para revisar fugas en el sistema de enfriamiento y en las tapas de presión del radiador.

Custom tests A series of tests programmed by the technician and performed by an engine analyzer.
Pruebas de diseño específico Serie de pruebas programadas por el mecánico y realizadas por un analizador de motores.

Cylinder leakage tester A tester designed to measure the amount of air leaking from the combustion chamber past the piston rings or valves.
Instrumento de prueba de la fuga del cilindro Instrumento de prueba diseñado para medir la cantidad de aire que se escapa desde la cámara de combustión y que sobrepasa los anillos de pistón o las válvulas.

Data link connector (DLC) A computer system connector to which the computer supplies data for diagnostic purposes.
Conector de enlace de datos Conector de computadora al que ésta suministra datos para propósitos diagnósticos.

Diagnostic trouble code (DTC) A code retained in a computer memory representing a fault in a specific area of the computer system.
Códigos indicadores de fallas para propósitos diagnósticos Código almacenado en la memoria de una computadora que representa una falla en un área específica de la computadora.

Diesel particulates Small carbon particles in diesel exhaust.
Partículas de diesel Pequeñas partículas de carbón presentes en el escape de un motor diesel.

Digital EGR valve An EGR valve that contains a computer-operated solenoid or solenoids.

Válvula EGR digital   Una válvula EGR que contiene un solenoide o solenoides accionados por computadora.

Digital meter   A meter with a digital display.
Medidor digital   Medidor con lectura digital.

Distributor ignition (DI) system   SAE J1930 terminology for any ignition system with a distributor.
Sistema de encendido con distribuidor   Término utilizado por la SAE J1930 para referirse a cualquier sistema de encendido que tenga un distribuidor.

Downstream air   Air injected into the catalytic converter.
Aire conducido hacia abajo   Aire inyectado dentro del convertidor catalítico.

EGR pressure transducer (EPT)   A vacuum switching device operated by exhaust pressure that opens and closes the vacuum passage to the EGR valve.
Transconductor de presión EGR   Dispositivo de conmutación de vacío accionado por la presión del escape que abre y cierra el paso del vacío a la válvula EGR.

EGR vacuum regulator (EVR) solenoid   A solenoid that is cycled by the computer to provide a specific vacuum to the EGR valve.
Solenoide regulador de vacío EGR   Solenoide ciclado por la computadora para proporcionarle un vacío específico a la válvula EGR.

Electric throttle kicker   An electric solenoid that controls throttle opening at idle speed and prevents dieseling.
Nivelador eléctrico de la mariposa   Solenoide eléctrico que controla la apertura de la mariposa a una velocidad de marcha lenta y evita el autoencendido.

Electronic fuel injection (EFI)   A generic term applied to various types of fuel injection systems.
Inyección electrónica de combustible   Término general aplicado a varios sistemas de inyección de combustible.

Electronic ignition (EI) system   SAE J1930 terminology for any ignition system without a distributor.
Sistema de encendido electrónico   Término utilizado por la SAE J1930 para referirse a cualquier sistema de encendido que no tenga distribuidor.

Engine analyzer   A tester designed to test engine systems such as battery, starter, charging, ignition, and fuel, plus engine condition.
Analizador de motores   Instrumento de prueba diseñado para revisar sistemas de motores, como por ejemplo los de la batería, del arranque, de la carga, del encendido, y del combustible, además de la condición del motor.

Engine coolant temperature (ECT) sensor   A sensor that sends a voltage signal to the computer in relation to coolant temperature.
Sensor de la temperatura del refrigerante del motor   Sensor que le envía una señal de tensión a la computadora referente a la temperatura del refrigerante.

Engine lift   A hydraulically operated piece of equipment used to lift the engine from the chassis.
Elevador de motores   Equipo accionado hidráulicamente que se utiliza para levantar el motor del chasis.

Evaporative (EVAP) system   A system that collects fuel vapors from the fuel tank and directs them into the intake manifold rather than allowing them to escape to the atmosphere.
Sistema de evaporación   Sistema que acumula los vapores del combustible que escapan del tanque del combustible y los conduce hacia el colector de aspiración en vez de permitir que los mismos se escapen hacia la atmósfera.

Exhaust gas analyzer   A tester that measures carbon monoxide, carbon dioxide, hydrocarbons, and oxygen in the engine exhaust.
Analizador del gas del escape   Instrumento de prueba que mide el monóxido de carbono, el bióxido de carbono, los hidrocarburos, y el oxígeno en el escape del motor.

Exhaust gas recirculation (EGR) valve   A valve that circulates a specific amount of exhaust gas into the intake manifold to reduce NOx emissions.
Válvula de recirculación del gas del escape   Válvula que hace circular una cantidad específica del gas del escape hacia el colector de aspiración para disminuir emisiones de óxidos de nitrógeno.

Exhaust gas recirculation valve position (EVP) sensor   A sensor that sends a voltage signal to the computer in relation to EGR valve position.
Sensor de la posición de la válvula de recirculación del gas del escape   Sensor que le envía una señal de tensión a la computadora referente a la posición de la válvula EGR.

Exhaust gas temperature sensor   A sensor that sends a voltage signal to the computer in relation to exhaust temperature.
Sensor de la temperatura del gas del escape   Sensor que le envía una señal de tensión a la computadora referente a la temperatura del escape.

Fast idle cam position adjustment   An adjustment that positions the fast idle cam properly in relation to the choke valve position.
Ajuste de la posición de árbol de levas de marcha lenta rápida   Ajuste que coloca el árbol de levas de marcha lenta rápida adecuadamente según la posición de la válvula de estrangulación.

Fast idle thermo valve   A valve operated by a thermo-wax element that allows more air into the intake manifold to increase idle speed when the engine is cold.
Termoválvula de la marcha lenta rápida   Válvula accionada por un elemento de termocera que permite que una mayor cantidad de aire entre en el colector de aspiración para aumentar la velocidad de la marcha lenta cuando el motor está frío.

Feeler gauge   Metal strips with a specific thickness for measuring clearances between components.
Calibrador de espesores   Láminas metálicas de un espesor específico para medir espacios libres entre componentes.

Field service mode   A computer diagnostic mode that indicates whether the computer is in open or closed loop on General Motors PCMs.
Modo de servicio de campo   Modo diagnóstico de una computadora que indica si la misma se encuentra en bucle abierto o cerrado en módulos del control del tren transmisor de potencia de la General Motors.

Flash code diagnosis   Reading computer system diagnostic trouble codes (DTCs) from the flashes of the malfunction indicator light (MIL).
Diagnosis con código de destello   La lectura de códigos indicativos de fallas para propósitos diagnósticos de una computadora mediante los destellos de la luz indicadora de funcionamiento defectuoso.

Float drop adjustment   An adjustment that provides the proper maximum downward float position.
Ajuste del descenso del flotador   Ajuste que proporciona la posición descendente máxima adecuada del flotador.

Float level adjustment   A float adjustment that assures the proper level of fuel in the float bowl.
Ajuste del nivel del flotador   Ajuste del flotador que asegura el nivel adecuado de combustible en el depósito del flotador.

Floor jack A hydraulically operated device mounted on casters and used to raise one end or corner of the chassis.

Gato de pie Dispositivo accionado hidráulicamente montado en rolletes y utilizado para levantar un extremo o una esquina del chasis.

Four-gas emissions analyzer An analyzer designed to test carbon monoxide, carbon dioxide, hydrocarbons, and oxygen in the exhaust.

Analizador de cuatro tipos de emisiones Analizador diseñado para revisar el monóxido de carbono, el bióxido de carbono, los hidrocarburos, y el oxígeno en el escape.

Fuel cut rpm The rpm range in which the computer stops operating the injectors during deceleration.

Detención de combustible según las rpm Margen de revoluciones al que la computadora detiene el funcionamiento de los inyectores durante la desaceleración.

Fuel pressure test port A threaded port on the fuel rail to which a pressure gauge may be connected to test fuel pressure.

Lumbrera de prueba de la presión del combustible Lumbrera fileteada que se encuentra en el carril del combustible a la que puede conectársele un calibrador de presión para revisar la presión del combustible.

Fuel pump volume The amount of fuel the pump delivers in a specific time period.

Volumen de la bomba del combustible Cantidad de combustible que la bomba envía dentro de un espacio de tiempo específico.

Fuel tank purging Removing fuel vapors and foreign material from the fuel tank.

Purga del tanque del combustible La remoción de vapores de combustible y de material extraño del tanque del combustible.

Graphite oil An oil with a graphite base that may be used for special lubricating requirements such as door locks.

Aceite de grafito Aceite con una base de grafito que puede utilizarse para necesidades de lubricación especiales, como por ejemplo en cerraduras de puertas.

Hall effect pickup A pickup containing a Hall element and a permanent magnet with a rotating blade between these components.

Captación de efecto Hall Captación que contiene un elemento Hall y un imán permanente entre los cuales está colocada una aleta giratoria.

Hand-held digital pyrometer A tester for measuring component temperature.

Pirómetro digital de mano Instrumento de prueba para medir la temperatura de un componente.

Hand press A hand-operated device for pressing precision-fit components.

Prensa de mano Dispositivo accionado manualmente para prensar componentes con un fuerte ajuste de precisión.

Heated resistor-type MAF sensor A MAF sensor that uses a heated resistor to sense air intake volume and temperature, and sends a voltage signal to the computer in relation to the total volume of intake air.

Sensor MAF tipo resistor térmico Sensor MAF que utiliza un resistor térmico para advertir el volumen y la temperatura del aire aspirado, y que le envía una señal de tensión a la computadora referente al volumen total de aire aspirado.

Hot wire-type MAF sensor A MAF sensor that uses a heated wire to sense air intake volume and temperature, and sends a voltage signal to the computer in relation to the total volume of intake air.

Sensor MAF tipo térmico Sensor MAF que utiliza un alambre térmico para advertir el volumen y la temperatura del aire aspirado, y que le envía una señal de tensión a la computadora referente al volumen total de aire aspirado.

Hydraulic press A hydraulically operated device for pressing precision-fit components.

Prensa hidráulica Dispositivo accionado hidráulicamente que se utiliza para prensar componentes con un fuerte ajuste de precisión.

Hydraulic valve lifters Round, cylindrical, metal components used to open the valves. These components are operated by oil pressure to maintain zero clearance between the valve stem and rocker arm.

Desmontaválvulas hidráulicas Componentes metálicos, cilíndricos y redondos utilizados para abrir las válvulas. Estos componentes se accionan mediante la presión del aceite para mantener cero espacio libre entre el vástago de válvula y el balancín.

Hydrocarbons (HC) Left over fuel from the combustion process.

Hidrocarburos El combustible restante después del proceso de combustión.

Hydrometer A tester designed to measure the specific gravity of a liquid.

Hidrómetro Instrumento de prueba diseñado para medir la gravedad específica de un líquido.

Idle air control by-pass air (IAC BPA) motor An IAC motor that controls idle speed by regulating the amount of air by-passing the throttle.

Motor para el control de la marcha lenta con el paso de aire Un motor IAC que controla la velocidad de la marcha lenta regulando la cantidad de aire que se desvía de la mariposa.

Idle air control by-pass air (IAC BPA) valve A valve operated by the IAC BPA motor that regulates the air by-passing the throttle to control idle speed.

Válvula para el control de la marcha lenta con el paso de aire Válvula accionada por el motor IAC BPA que regula el aire que se desvía de la mariposa para controlar la velocidad de la marcha lenta.

Idle air control (IAC) motor A computer-controlled motor that controls idle speed under all conditions.

Motor para el control de la marcha lenta con aire Motor controlado por computadora que controla la velocidad de la marcha lenta bajo cualquier condición del funcionamiento del motor.

Idle contact switch A switch in the IAC motor stem that informs the computer when the throttle is in the idle position.

Conmutador de contacto de la marcha lenta Conmutador en el vástago del motor IAC que le advierte a la computadora cuándo la mariposa está en la posición de marcha lenta.

Idle stop solenoid An electric solenoid that maintains the throttle in the proper idle speed position and prevents dieseling when the ignition switch is turned off.

Solenoide de detención de la marcha lenta Solenoide eléctrico que mantiene la mariposa en la posición de velocidad de marcha lenta adecuada y evita el autoencendido al desconectarse el botón conmutador de encendido.

Ignition crossfiring Ignition firing between distributor cap terminals or spark plug wires.

Encendido por inducción Encendido entre los bornes de la tapa del distribuidor o los alambres de las bujías.

Ignition module tester An electronic tester designed to test ignition modules.

Instrumento de prueba del módulo del encendido Instrumento de prueba electrónico diseñado para revisar módulos del encendido.

Injector balance tester A tester designed to test port injectors.
Instrumento de prueba del equilibro del inyector Instrumento de prueba diseñado para revisar inyectores de lumbreras.

Inspection, maintenance (I/M) testing Emission inspection and maintenance programs that are usually administered by various states.
Pruebas de inspección y mantenimiento Programas de inspección y mantenimiento de emisiones que normalmente administran diferentes estados.

Integrator A chip responsible for fuel control in a General Motors PCM.
Integrador Pastilla responsable de controlar el combustible en un módulo del control del tren transmisor de potencia de la General Motors.

International System (SI) A system of weights and measures in which each unit may be divided by 10.
Sistema internacional Sistema de pesos y medidas en el que cada unidad puede dividirse entre 10.

Jack stand A metal stand used to support one corner of the chassis.
Soporte de gato Soporte de metal utilizado para apoyar una esquina del chasis.

Key on engine off (KOEO) test A computer system test mode on Ford products that displays diagnostic trouble codes (DTCs) with the key on and the engine stopped.
Prueba con la llave en la posición de encendido y el motor apagado Modo de prueba en una computadora de productos fabricados por la Ford que muestra códigos indicadores de fallas para propósitos diagnósticos cuando la llave está en la posición de encendido y el motor está apagado.

Key on engine running (KOER) test A computer system test mode on Ford products that displays diagnostic trouble codes (DTCs) with the engine running.
Prueba con la llave en la posición de encendido y el motor encendido Modo de prueba en una computadora de productos fabricados por la Ford que muestra códigos indicadores de fallas para propósitos diagnósticos cuando el motor está encendido.

Kilovolts (kV) Thousands of volts.
Kilovoltios (kV) Miles de voltios.

Knock sensor A sensor that sends a voltage signal to the computer in relation to engine detonation.
Sensor de golpeteo Sensor que le envía una señal de tensión a la computadora referente a la detonación del motor.

Knock sensor module An electronic module that changes the analog knock sensor signal to a digital signal and sends it to the PCM.
Módulo del sensor de golpeteo Módulo electrónico que convierte la señal analógica del sensor de golpeteo en una señal digital y se la envía al módulo del control del tren transmisor de potencia.

Lift A device used to raise a vehicle.
Elevador Dispositivo utilizado para levantar un vehículo.

Linear EGR valve An EGR valve containing an electric solenoid that is pulsed on and off by the computer to provide a precise EGR flow.
Válvula EGR lineal Válvula EGR con un solenoide eléctrico que la computadora enciende y apaga para proporcionar un flujo exacto de EGR.

Magnetic-base thermometer A thermometer that may be retained to metal components with a magnetic base.
Termómetro con base magnética Termómetro que puede sujetarse a componentes metálicos por medio de una base magnética.

Magnetic probe-type digital tachometer A digital tachometer that reads engine rpm and uses a magnetic probe pickup.
Tacómetro digital tipo sonda magnética Tacómetro digital que lee las rpm del motor y utiliza una captación de sonda magnética.

Magnetic probe-type digital timing meter A digital reading that displays crankshaft degrees and uses a magnetic-type pickup probe mounted in the magnetic timing probe receptacle.
Medidor de regulación digital tipo sonda magnética Lectura digital que muestra los grados del cigüeñal y utiliza una sonda de captación tipo magnético montada en el receptáculo de la sonda de regulación magnética.

Magnetic sensor A sensor that produces a voltage signal from a rotating element near a winding and a permanent magnet. This voltage signal is often used for ignition triggering.
Sensor magnético Sensor que produce una señal de tensión desde un elemento giratorio cerca de un devanado y de un imán permanente. Esta señal de tensión se utiliza con frecuencia para arrancar el motor.

Magnetic timing offset An adjustment to compensate for the position of the magnetic receptacle opening in relation to the TDC mark on the crankshaft pulley.
Desviación de regulación magnética Ajuste para compensar la posición de la abertura del receptáculo magnético de acuerdo a la marca TDC en la roldana del cigüeñal.

Magnetic timing probe receptacle An opening in which the magnetic timing probe is installed to check basic timing.
Receptáculo de la sonda de regulación magnética Abertura en la que está montada la sonda de regulación magnética para verificar la regulación básica.

Malfunction indicator light (MIL) A light in the instrument panel that is illuminated by the PCM if certain defects occur in the computer system.
Luz indicadora de funcionamiento defectuoso Luz en el panel de instrumentos que el módulo del control del tren transmisor de potencia ilumina si ocurren ciertas fallas en la computadora.

Manifold absolute pressure (MAP) sensor An input sensor that sends a signal to the computer in relation to intake manifold vacuum.
Sensor de la presión absoluta del colector Sensor de entrada que le envía una señal a la computadora referente al vacío del colector de aspiración.

Mechanical valve lifters, or solid tappets Round, cylindrical, metal components mounted between the camshaft lobes and the pushrods to open the valves.
Desmontaválvulas mecánicas, o alzaválvulas sólidas Componentes metálicos, cilíndricos y redondos montados entre los lóbulos del árbol de levas y las varillas de empuje para abrir las válvulas.

Memory calibrator (MEM-CAL) A removable chip in some computers that replaces the PROM and CAL-PAK chips.
Calibrador de memoria Pastilla desmontable en algunas computadoras que reemplaza las pastillas PROM y CAL-PAK.

Meter impedance The total internal electrical resistance in a meter.
Impedancia de un medidor La resistencia eléctrica interna total en un medidor.

Mixture heater An electric heater mounted between the carburetor and the intake manifold that heats the air-fuel mixture when the engine is cold.
Calentador de la mezcla Calentador eléctrico que se monta entre el carburador y el colector de aspiración para calentar la mezcla de aire y combustible cuando el motor está frío.

Muffler chisel A chisel that is designed for cutting muffler inlet and outlet pipes.
Cincel para silenciadores Cincel diseñado para cortar los tubos de entrada y salida del silenciador.

Multiport fuel injection (MFI) A fuel injection system in which the injectors are grounded in the computer in pairs or groups of three or four.
Inyección de combustible de paso múltiple Sistema de inyección de combustible en el que los inyectores se ponen a tierra en la computadora en pares o en grupos de tres o cuatro.

National Institute for Automotive Service Excellence (ASE) An organization responsible for certification of automotive technicians in the US.
Instituto Nacional para la Excelencia en la Reparación de Automóviles Organización que tiene a su cargo la certificación de mecánicos de automóviles en los Estados Unidos.

Negative backpressure EGR valve An EGR valve containing a vacuum bleed valve that is operated by negative pulses in the exhaust.
Válvula EGR de contrapresión negativa Válvula EGR que contiene una válvula de descarga de vacío accionada por impulsos negativos en el escape.

Neutral/drive switch (NDS) A switch that sends a signal to the computer in relation to gear selector position.
Conmutador de mando neutral Conmutador que le envía una señal a la computadora referente a la posición del selector de velocidades.

Nose switch A switch in the IAC motor stem that informs the computer when the throttle is in the idle position.
Conmutador de contacto de la marcha lenta Conmutador en el vástago del motor IAC que le advierte a la computadora cuándo la mariposa está en la posición de marcha lenta.

Oil pressure gauge A gauge used to test engine oil pressure.
Manómetro de la presión del aceite Calibrador utilizado para revisar la presión del aceite del motor.

Open loop A computer operating mode in which the computer controls the air-fuel ratio and ignores the oxygen sensor signal.
Bucle abierto Modo de funcionamiento de una computadora en el que se controla la relación de aire y combustible y se pasa por alto la señal del sensor de oxígeno.

Optical-type pickup A pickup that contains a photo diode and a light emitting diode with a slotted plate between these components.
Captación tipo óptico Captación que contiene un fotodiodo y un diodo emisor de luz entre los cuales está colocada una placa ranurada.

Oscilloscope A cathode ray tube (CRT) that displays voltage waveforms from the ignition system.
Osciloscopio Tubo de rayos catódicos que muestra formas de onda de tensión provenientes del sistema de encendido.

Output state test A computer system test mode on Ford products that turns the relays and actuators on and off.
Prueba del estado de producción Modo de prueba en una computadora de productos fabricados por la Ford que enciende y apaga los relés y los accionadores.

Oxygen (O2) A gaseous element that is present in air.
Oxígeno (O2) Elemento gaseoso presente en el aire.

Oxygen (O2) sensor A sensor mounted in the exhaust system that sends a voltage signal to the computer in relation to the amount of oxygen in the exhaust stream.
Sensor de oxígeno (O2) Sensor montado en el sistema de escape que le envía una señal de tensión a la computadora referente a la cantidad de oxígeno en el caudal del escape.

Park/neutral switch A switch connected in the starter solenoid circuit that prevents starter operation except in park or neutral.
Conmutador PARK/neutral Conmutador conectado en el circuito del solenoide del arranque que evita el funcionamiento del arranque si el selector de velocidades no se encuentra en las posiciones PARK o NEUTRAL.

Parts per million (ppm) The volume of a gas such as hydrocarbons in ppm in relation to one million parts of the total volume of exhaust gas.
Partes por millón (ppm) Volumen de un gas, como por ejemplo los hidrocarburos, en partes por millón de acuerdo a un millón de partes del volumen total del gas del escape.

Photoelectric tachometer A tachometer that contains an internal light source and a photoelectric cell. This meter senses rpm from reflective tape attached to a rotating component.
Tacómetro fotoeléctrico Tacómetro que contiene una fuente interna de luz y una célula fotoeléctrica. Este medidor advierte las rpm mediante una cinta reflectora adherida a un componente giratorio.

Pinging noise A shop term for engine detonation that sounds like a rattling noise in the engine cylinders.
Sonido agudo Término utilizado en el taller mecánico para referirse a la detonación del motor cuyo ruido se asemeja a un estrépito en los cilindros de un motor.

Pipe expander A tool designed to expand exhaust system pipes.
Expansor de tubo Herramienta diseñada para expandir los tubos del sistema de escape.

Polyurethane air cleaner cover A circular polyurethane ring mounted over the air cleaner element to improve cleaning capabilities.
Cubierta de poliuretano del filtro de aire Anillo circular de poliuretano montado sobre el elemento del filtro de aire para facilitar la limpieza.

Port EGR valve An EGR valve operated by ported vacuum from above the throttle.
Válvula EGR lumbrera Válvula EGR accionada por un vacío con lumbreras desde la parte superior de la mariposa.

Port fuel injection (PFI) A fuel injection system with an injector positioned in each intake port.
Inyección de combustible de lumbrera Sistema de inyección de combustible que tiene un inyector colocado en cada una de las lumbreras de aspiración.

Positive backpressure EGR valve An EGR valve with a vacuum bleed valve that is operated by positive pressure pulses in the exhaust.
Válvula EGR de contrapresión positiva Válvula EGR con una válvula de descarga de vacío accionada por impulsos de presión positiva en el escape.

Positive crankcase ventilation (PCV) valve A valve that delivers crankcase vapors into the intake manifold rather than allowing them to escape to the atmosphere.
Válvula de ventilación positiva del cárter Válvula que conduce los vapores del cárter hacia el colector de aspiración en vez de permitir que los mismos se escapen hacia la atmósfera.

Power balance tester A tester designed to stop each cylinder from firing for a brief time and record the rpm decrease.
Instrumento de prueba del equilibro de la potencia Instrumento de prueba diseñado para detener el encendido de cada cilindro por un breve espacio de tiempo y registrar el descenso de las rpm.

Power train control module (PCM) SAE J1930 terminology for an engine control computer.

Módulo del control del tren transmisor de potencia   Término utilizado por la SAE J1930 para referirse a una computadora para el control del motor.

Prelubrication   Lubrication of components such as turbocharger bearings prior to starting the engine.

Prelubrificación   Lubrificación de componentes, como por ejemplo los cojinetes del turbocompresor, antes del arranque del motor.

Pressurized injector cleaning container   A small, pressurized container filled with unleaded gasoline and injector cleaner for cleaning injectors with the engine running.

Recipiente presionizado para la limpieza del inyector   Pequeño recipiente presionizado lleno de gasolina sin plomo y limpiador de inyectores para limpiar los inyectores cuando el motor está encendido.

Programmable read only memory (PROM)   A computer chip containing some of the computer program. This chip is removable in some computers.

Memoria de solo lectura programable (PROM)   Pastilla de memoria que contiene una parte del programa de la computadora. Esta pastilla es desmontable en algunas computadoras.

Propane-assisted idle mixture adjustment   A method of adjusting idle mixture with fuel supplied from a small propane cylinder.

Ajuste de la mezcla de la marcha lenta asistido por propano   Método de ajustar la mezcla de la marcha lenta utilizando el combustible suministrado desde un cilindro pequeño de propano.

Pulsed secondary air injection system   A system that uses negative pressure pulses in the exhaust to move air into the exhaust system.

Sistema de inyección secundaria de aire por impulsos   Sistema que utiliza impulsos de la presión negativa en el escape para conducir el aire hacia el sistema de escape.

Quad driver   A group of transistors in a computer that controls specific outputs.

Excitador cuádruple   Grupo de transistores en una computadora que controla salidas específicas.

Quick-disconnect fuel line fittings   Fuel line fittings that may be disconnected without using a wrench.

Conexiones de la línea del combustible de desmontaje rápido   Conexiones de la línea del combustible que se pueden desmontar sin la utilización de una llave de tuerca.

Radiator shroud   A circular component positioned around the cooling fan to concentrate the air flow through the radiator.

Bóveda del radiador   Componente circular que rodea el ventilador de enfriamiento para concentrar el flujo de aire a través del radiador.

Reference pickup   A pickup assembly that is often used for ignition triggering.

Captación de referencia   Conjunto de captación que se utiliza con frecuencia para el arranque del encendido.

Reference voltage   A constant voltage supplied from the computer to some of the input sensors.

Tensión de referencia   Tensión constante que le suministra la computadora a algunos de los sensores de entrada.

Revolutions per minute (rpm) drop   The amount of rpm decrease when a cylinder stops firing for a brief time.

Descenso de las revoluciones por minuto (rpm)   Cantidad que descienden las rpm cuando un cilindro detiene el encendido por un breve espacio de tiempo.

Ring ridge   A ridge near the top of the cylinder created by wear in the ring travel area of the cylinder.

Reborde del anillo   Reborde cerca de la parte superior del cilindro ocasionado por un desgaste en el área de carrera del anillo del cilindro.

Room temperature vulcanizing (RTV) sealant   A type of sealant that may be used to replace gaskets, or to help to seal gaskets, in some applications.

Compuesto obturador vulcanizador a temperatura ambiente   Tipo de compuesto obturador que puede utilizarse para reemplazar guarniciones, o para ayudar a sellarlas, en algunas aplicaciones.

Scan tester   A tester designed to test automotive computer systems.

Instrumento de pruebas de exploración   Instrumento de prueba diseñado para revisar computadoras de automóviles.

Schrader valve   A threaded valve on the fuel rail to which a pressure gauge may be connected to test fuel pressure.

Válvula Schrader   Válvula fileteada que se encuentra en el carril del combustible a la que puede conectársele un calibrador de presión para revisar la presión del combustible.

Secondary air injection (AIR) system   A system that injects air into the exhaust system from a belt driven pump.

Sistema de inyección secundaria de aire   Sistema que inyecta aire dentro del sistema de escape desde una bomba accionada por correa.

Secondary air valve alignment   An adjustment that provides correct air valve alignment in the secondary bores.

Alineación de la válvula de aire secundaria   Ajuste que proporciona una alineación adecuada de la válvula de aire en los calibres secundarios.

Secondary air valve opening adjustment   An adjustment that provides the proper wide-open air valve position.

Ajuste de la apertura de la válvula de aire secundaria   Ajuste que proporciona una posición abierta de par en par adecuada de la válvula de aire.

Secondary air valve spring adjustment   An adjustment that provides the proper secondary air valve spring tension.

Ajuste del resorte de la válvula de aire secundaria   Ajuste que proporciona la tensión adecuada del resorte de la válvula de aire secundaria.

Secondary lockout adjustment   An adjustment that assures the secondary throttles are locked when the choke is partly or fully closed on a four-barrel carburetor.

Ajuste de fijación secundario   Ajuste que asegura la fijación de las mariposas secundarias cuando el estrangulador está parcial o completamente cerrado en un carburador de cuatro cilindros.

Secondary throttle linkage adjustment   An adjustment that assures the correct primary throttle opening when the secondary throttles begin to open.

Ajuste de la conexión de la mariposa secundaria   Ajuste que asegura la apertura correcta de la mariposa primaria cuando las mariposas secundarias comienzan a abrirse.

Self-powered test light   A test light powered by an internal battery.

Luz de prueba propulsada automáticamente   Luz de prueba propulsada por una batería interna.

Self-test input wire   A diagnostic wire located near the diagnostic link connector (DLC) on Ford vehicles.

Alambre de entrada de prueba automática   Alambre diagnóstico ubicado cerca del conector de enlace diagnóstico en vehículos fabricados por la Ford.

Sequential fuel injection (SFI) A fuel injection system in which the injectors are individually grounded into the computer.
Inyección de combustible en ordenamiento Sistema de inyección de combustible en el que los inyectores se ponen individualmente a tierra en la computadora.

Shop layout The design of an automotive repair shop.
Arreglo del taller de reparación Diseño de un taller de reparación de automóviles.

Silicone grease A heat-dissipating grease placed on components such as ignition modules.
Grasa de silicón Grasa para disipar el calor utilizada en componentes, como por ejemplo módulos del encendido.

Slitting tool A special chisel designed for slitting exhaust system pipes.
Herramienta de hender Cincel especial diseñado para hendir los tubos del sistema de escape.

Snap shot testing The process of freezing computer data into the scan tester memory during a road test and reading this data later.
Prueba instantánea Proceso de capturar datos de la computadora en la memoria del instrumento de pruebas de exploración durante una prueba en carretera y leer dichos datos más tarde.

Specific gravity The weight of a liquid in relation to the weight of an equal volume of water.
Gravedad específica El peso de un líquido de acuerdo al peso de un volumen igual de agua.

Starting air valve A vacuum-operated valve that supplies more air into the intake manifold when starting the engine.
Válvula de aire para el arranque Válvula accionada por vacío que le suministra mayor cantidad de aire al colector de aspiración durante el arranque del motor.

Stethoscope A tool used to amplify sound and locate abnormal noises.
Estetoscopio Herramienta utilizada para amplificar el sonido y localizar ruidos anormales.

Sulfuric acid A corrosive acid mixed with water and used in automotive batteries.
Ácido sulfúrico Ácido sumamente corrosivo mezclado con agua y utilizado en las baterías de automóviles.

Switch test A computer system test mode that tests the switch input signals to the computer.
Prueba de conmutación Modo de prueba de una computadora que revisa las señales de entrada de conmutación hechas a la computadora.

Synchronizer (SYNC) pickup A pickup assembly that produces a voltage signal for ignition triggering or injector sequencing.
Captación sincronizadora Conjunto de captación que produce una señal de tensión para el arranque del encendido o para el ordenamiento del inyector.

Tach-dwellmeter A meter that reads engine rpm and ignition dwell.
Tacómetro y medidor de retraso Medidor que lee las rpm del motor y el retraso del encendido.

Tachometer (TACH) terminal The negative primary coil terminal.
Borne del tacómetro Borne negativo de la bobina primaria.

Temperature switch A mechanical switch operated by coolant or metal temperature.
Conmutador de temperatura Conmutador mecánico accionado por la temperatura del refrigerante o del metal.

Test spark plug A spark plug with the electrodes removed so it requires a much higher firing voltage for testing such components as the ignition coil.
Bujía de prueba Bujía que a consecuencia de habérsele removido los electrodos necesitará mayor tensión de encendido para revisar componentes, como por ejemplo la bobina del encendido.

Thermal vacuum switch (TVS) A vacuum switching device operated by heat applied to a thermo-wax element.
Conmutador de vacío térmico Dispositivo de conmutación de vacío accionado por el calor aplicado a un elemento de termocera.

Thermal vacuum valve (TVV) A valve that is opened and closed by a thermo-wax element mounted in the cooling system.
Válvula térmica de vacío Válvula de vacío que un elemento de termocera montado en el sistema de enfriamiento abre y cierra.

Thermostat tester A tester designed to measure thermostat opening temperature.
Instrumento de prueba del termostato Instrumento de prueba diseñado para medir la temperatura inicial del termostato.

Throttle body injection (TBI) A fuel injection system with the injector or injectors mounted above the throttle.
Inyección del cuerpo de la mariposa Sistema de inyección de combustible en el que el inyector, o los inyectores, están montados sobre la mariposa.

Throttle position sensor (TPS) A sensor mounted on the throttle shaft that sends a voltage signal to the computer in relation to throttle opening.
Sensor de la posición de la mariposa Sensor montado sobre el árbol de la mariposa que le envía una señal de tensión a la computadora referente a la apertura de la mariposa.

Throttle position switch A switch that informs the computer whether the throttle is in the idle position. This switch is usually part of the TPS.
Conmutador de la posición de la mariposa Conmutador que le advierte a la computadora si la mariposa se encuentra en la posición de marcha lenta. Este conmutador normalmente forma parte del sensor de la posición de la mariposa.

Timing connector A wiring connector that must be disconnected while checking basic ignition timing on fuel injected engines.
Conector de regulación Conector del alambrado que debe desconectarse mientras se verifica la regulación básica del encendido en motores de inyección de combustible.

Two-gas emissions analyzer An analyzer designed to measure hydrocarbons and carbon monoxide in the exhaust.
Analizador de dos tipos de emisiones Analizador diseñado para medir los hidrocarburos y el monóxido de carbono en el escape.

United States customary (USC) A system of weights and measures.
Sistema usual estadounidense (USC) Sistema de pesos y medidas.

Upstream air Air injected into the exhaust ports.
Aire conducido hacia arriba Aire inyectado dentro de las lumbreras del escape.

Vacuum delay valve A vacuum valve with a restrictive port that delays a vacuum increase through the valve.
Válvula de retardo de vacío Válvula de vacío con una lumbrera restrictiva que retarda el aumento de vacío a través de la válvula.

Vacuum-operated decel valve A valve that allows more air into the intake manifold during deceleration to improve emission levels.
Válvula de desaceleración accionada por vacío Válvula accionada por vacío que admite una mayor cantidad de aire en el colector de aspiración durante una desaceleración a fin de reducir los niveles de emisiones.

Vacuum pressure gauge A gauge designed to measure vacuum and pressure.
Manómetro de la presión del vacío Calibrador diseñado para medir el vacío y la presión.

Vacuum throttle kicker A vacuum diaphragm with a stem that provides more throttle opening under certain conditions such as deceleration or A/C on.
Nivelador de la mariposa de vacío Diafragma de vacío con un vástago que proporciona una apertura más extensa de la mariposa bajo ciertas condiciones, como por ejemplo una desaceleración o cuando el aire acondicionado está encendido.

Valve overlap The few degrees of crankshaft rotation when both valves are open and the piston is near TDC on the exhaust stroke.
Solape de la válvula Los pocos grados que gira el cigüeñal cuando ambas válvulas están abiertas y el pistón se encuentra cerca del punto muerto superior durante la carrera de escape.

Valve stem installed height The distance between the top of the valve retainer and the valve spring seat on the cylinder head.
Altura instalada del vástago de la válvula Distancia entre la parte superior del retenedor de la válvula y el asiento del resorte de la válvula en la culata del cilindro.

Vane-type MAF sensor A MAF sensor containing a pivoted vane that moves a pointer on a variable resistor. This resistor sends a voltage signal to the computer in relation to the total volume of intake air.
Sensor MAF tipo paleta Sensor MAF con una paleta articulada que mueve un indicador en un resistor variable. Este resistor le envía una señal de tensión a la computadora referente al volumen total de aire aspirado.

Vehicle speed sensor (VSS) A sensor that is usually mounted in the transmission and sends a voltage signal to the computer in relation to engine speed.
Sensor de la velocidad del vehículo Sensor que normalemente se monta en la transmisión y que le envía una señal de tensión a la computadora referente a la velocidad del motor.

Viscous-drive fan clutch A cooling fan drive clutch that drives the fan at higher speed when the temperature increases.
Embrague de mando viscoso del ventilador Embrague de mando del ventilador de enfriamiento que acciona el ventilador para que gire más rápido a temperaturas más altas.

Volt-amp tester A tester designed to test volts and amps in such circuits as battery, starter, and charging.
Instrumento de prueba de voltios y amperios Instrumento de prueba diseñado para revisar los voltios y los amperios en circuitos como por ejemplo, de la batería, del arranque y de la carga.

Wastegate stroke The amount of turbocharger wastegate diaphragm and rod movement.
Carrera de la compuerta de desagüe Cantidad de movimiento del diafragma y de la varilla de la compuerta de desagüe del turbocompresor.

Wet compression test A cylinder compression test completed with a small amount of oil in the cylinder.
Prueba húmeda de compresión Prueba de la compresión de un cilindro llevada a cabo con una pequeña cantidad de aceite en el cilindro.